Student Book, Higher 1

NEW GCSE MATHS
AQA Linear

Matches the 2010 GCSE Specification

Brian Speed • Keith Gordon • Kevin Evans • Trevor Senior • Chris Pearce

CONTENTS

INTRODUCTION

Welcome to Collins New GCSE Maths for AQA Linear Higher Book 1.

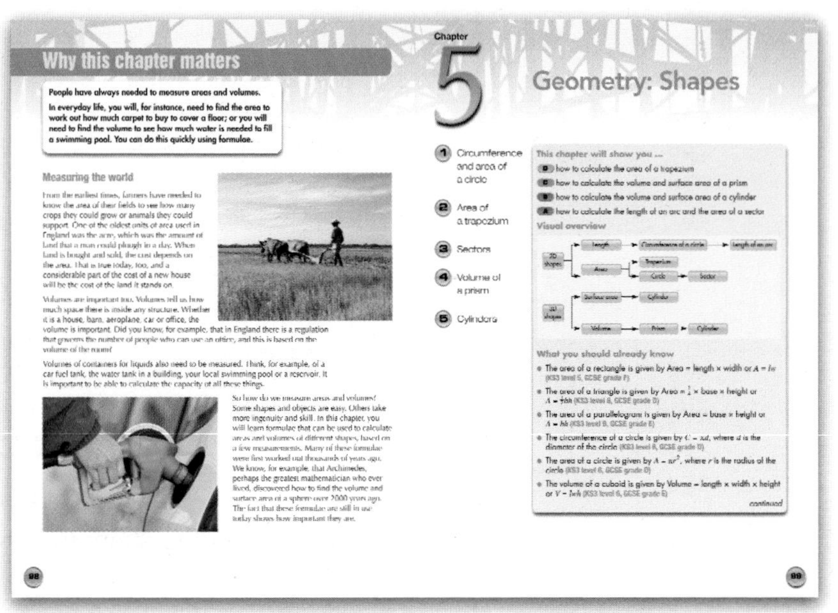

Why this chapter matters

People have always needed to measure areas and volumes. In everyday life, you will, for instance, need to find the area to work out how much carpet to buy to cover a floor; or you will need to find the volume to see how much water is needed to fill a swimming pool. You can do this quickly using formulae.

Measuring the world

Chapter 5 Geometry: Shapes

1 Circumference and area of a circle
2 Area of a trapezium
3 Sectors
4 Volume of a prism
5 Cylinders

This chapter will show you ...

Visual overview

What you should already know

Why this chapter matters

Find out why each chapter is important through the history of maths, seeing how maths links to other subjects and cultures, and how maths is related to real life.

Chapter overviews

Look ahead to see what maths you will be doing and how you can build on what you already know.

Colour-coded grades

Know what target grade you are working at and track your progress with the colour-coded grade panels at the side of the page.

Use of calculators

Questions where you must or could use your calculator are marked with icon.

Explanations involving calculators are based on *CASIO fx–83ES*.

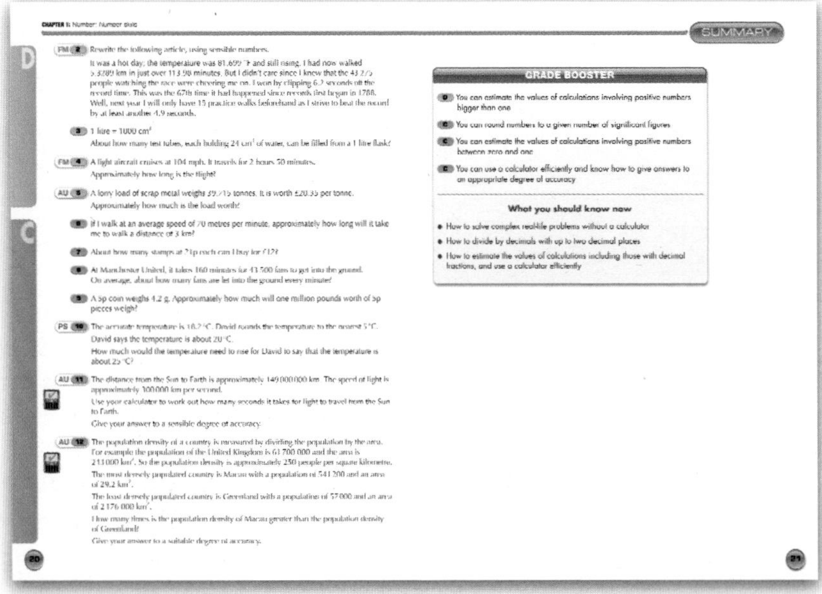

SUMMARY

GRADE BOOSTER

What you should know now

Grade booster

Review what you have learnt and how to get to the next grade with the Grade booster at the end of each chapter.

Worked examples

Understand the topic before you start the exercise by reading the examples in blue boxes. These take you through questions step by step.

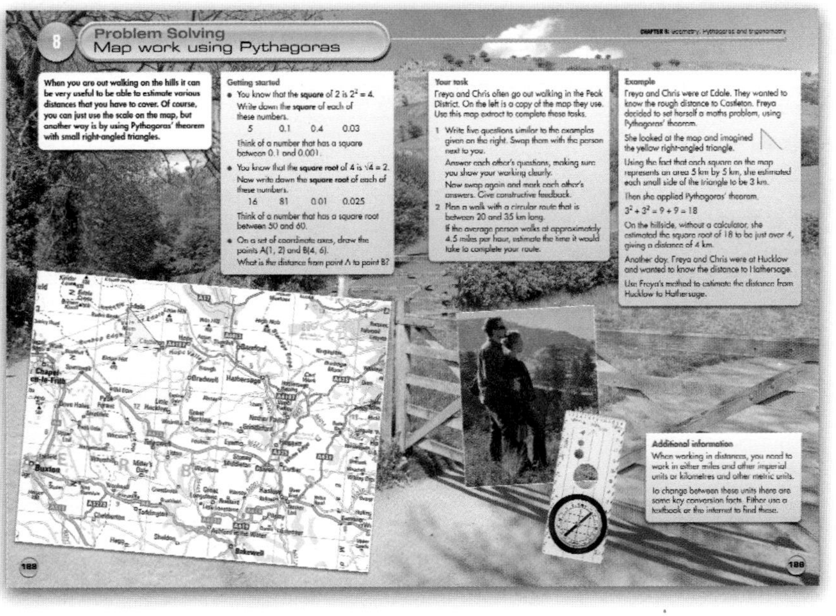

Functional maths

Practise functional maths skills to see how people use maths in everyday life. Look out for practice questions marked FM. There are also extra functional-maths and problem-solving activities at the end of every chapter to build and apply your skills.

New Assessment Objectives

Practise new parts of the curriculum (Assessment Objectives AO2 and AO3) with questions that assess your understanding marked AU and questions that test if you can solve problems marked PS. You will also practise some questions that involve several steps and where you have to choose which method to use; these also test AO2. There are also plenty of straightforward questions (AO1) that test if you can do the maths.

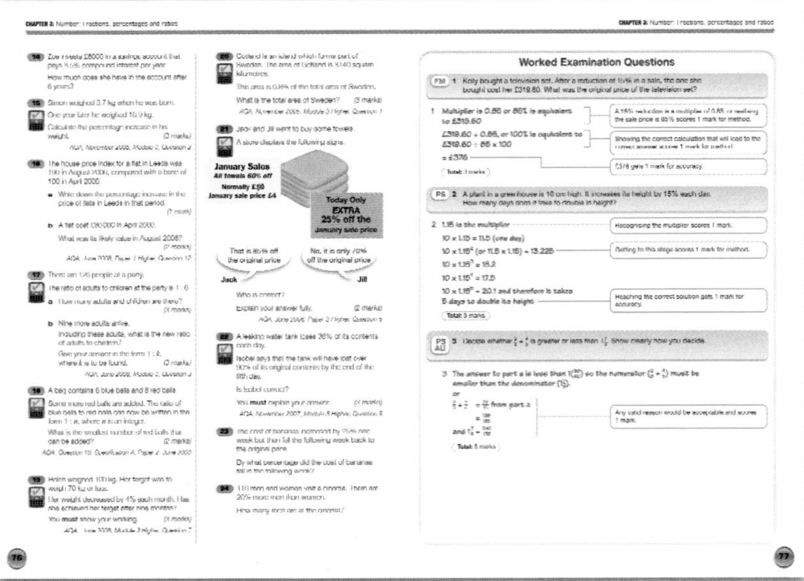

Exam practice

Prepare for your exams with past exam questions and detailed worked exam questions with examiner comments to help you score maximum marks.

Quality of Written Communication (QWC)

Practise using accurate mathematical vocabulary and writing logical answers to questions to ensure you get your QWC (Quality of Written Communication) marks in the exams. The Glossary and worked exam questions will help you with this.

Why this chapter matters

Most jobs require some knowledge of mathematics. You should be competent in the basic number skills of addition, subtraction, multiplication and division using whole numbers, fractions and decimals. You should know when it is sensible to use estimation or approximation and when it is important to have exact answers.

You need to be able to identify the skills required when questions are set in real-life situations in preparation for the world of work.

Jobs or careers using mathematics

How many jobs can you think of that require some mathematics?

Here are a few job ideas.

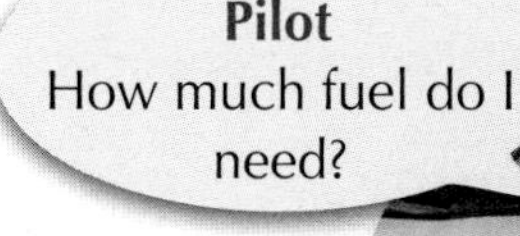

Doctor
How much medicine should I prescribe?

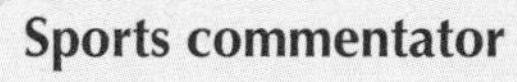

Accountant
How much profit has she made?

If you have chosen a job you would like to do, think of the questions you will need to ask and the mathematics you will require.

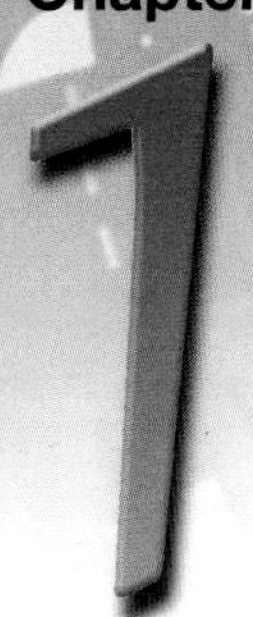

Chapter 1

Number: Number skills

1 Solving real-life problems

2 Multiplication and division with decimals

3 Approximation of calculations

This chapter will show you ...

D how to calculate with integers and decimals

C how to round numbers to a given number of significant figures

Visual overview

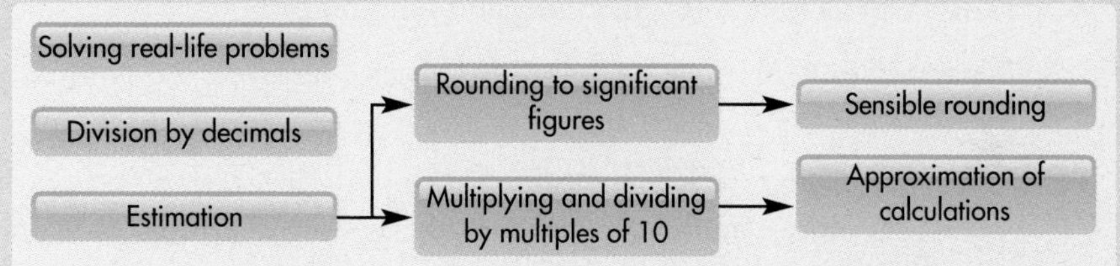

What you should already know

- How to add, subtract, multiply and divide with integers **(KS3 level 4, GCSE grade G)**
- The BIDMAS/BODMAS rule and how to substitute values into simple algebraic expressions **(KS3 level 5, GCSE grade F)**

Quick check

1 Work out the following.

a 6×78 **b** 3×122 **c** $432 \div 8$

d $3 \times 5 \times 20$ **e** $6 \times 34 \div 2$ **f** $5 \times (12 + 7)$

2 Work out the following.

a 23×167 **b** $984 \div 24$ **c** $(16 + 9)^2$

3 Work out the following.

a $2 + 3 \times 5$ **b** $(2 + 3) \times 5$ **c** $2 + 3^2 - 6$

1.1 Solving real-life problems

This section will show you how to:

- solve problems set in a real-life context

Key words

column method (or traditional method)
grid method (or box method)
long division
long multiplication
problem
strategy

In your GCSE examination, you will be given *real-life* **problems** that you have to *read carefully*, *think about* and then plan a **strategy** without using a calculator. These may involve arithmetical skills such as **long multiplication** and **long division**.

There are several ways to do these, so make sure you are familiar with and confident with at least one of them.

The **grid method** (or **box method**) for long multiplication is shown in the first example and the standard **column method** (or **traditional method**) for long division is shown in the second example. In this type of problem it is important to show your working as you will get marks for correct methods.

EXAMPLE 1

A supermarket receives a delivery of 235 cases of tins of beans. Each case contains 24 tins.

a How many tins of beans does the supermarket receive altogether?

b 5% of the tins were damaged. These were thrown away. The supermarket knows that it sells, on average, 250 tins of beans a day. How many days will the delivery of beans last before a new delivery is needed?

a The number of tins is worked out by the multiplication 235 × 24.

Using the grid method:

×	200	30	5
20	4000	600	100
4	800	120	20

$$\begin{array}{r} 4000 \\ 600 \\ 100 \\ +\ 800 \\ 120 \\ 20 \\ \hline 5640 \end{array}$$

Using the column method:

$$\begin{array}{r} 235 \\ \times\ \ 24 \\ \hline 940 \\ 4700 \\ \hline 5640 \end{array}$$

So the answer is 5640 tins.

b 10% of 5640 is 564, so 5% is $564 \div 2 = 282$

This leaves $5640 - 282 = 5358$ tins to be sold.

There are 21 lots of 250 in 5358 (you should know that $4 \times 250 = 1000$), so the beans will last for 21 days before another delivery is needed.

EXAMPLE 2

A party of 613 children and 59 adults are going on a day out to a theme park.

a How many coaches, each holding 53 people, will be needed?

b One adult gets into the theme park free for every 15 children. How many adults will have to pay to get in?

a Altogether there are $613 + 59 = 672$ people.

So the number of coaches needed is $672 \div 53$ (number of seats on each coach).

$$\begin{array}{r} 12 \\ 53\overline{)672} \\ \underline{530} \\ 142 \\ \underline{106} \\ 36 \end{array}$$

The answer is 12 remainder 36. So, there will be 12 full coaches and one coach with 36 people on it. So, they would have to book 13 coaches.

b This is also a division, $613 \div 15$. This can be done quite easily if you know the 15 times table as $4 \times 15 = 60$, so $40 \times 15 = 600$. This leaves a remainder of 13. So 40 adults get in free and $59 - 40 = 19$ adults will have to pay.

EXERCISE 1A

D

1 There are 48 cans of soup in a crate. A supermarket had a delivery of 125 crates of soup.

a How many cans of soup were in this delivery?

b The supermarket is running a promotion on soup. If you buy five cans you get one free. Each can costs 39p. How much will it cost to get 32 cans of soup?

2 Greystones Primary School has 12 classes, each of which has 24 students.

a How many students are there at Greystones Primary School?

b The student–teacher ratio is 18 to 1. That means there is one teacher for every 18 students. How many teachers are there at the school?

3 Barnsley Football Club is organising travel for an away game. 1300 adults and 500 juniors want to go. Each coach holds 48 people and costs £320 to hire. Tickets to the match are £18 for adults and £10 for juniors.

a How many coaches will be needed?

b The club is charging adults £26 and juniors £14 for travel and a ticket. How much profit does the club make out of the trip?

D

4 A first-class letter costs 39p to post and a second-class letter costs 30p. How much will it cost to send 20 first-class and 90 second-class letters?

PS 5 Kirsty collects small models of animals. Each one costs 45p. She saves enough to buy 23 models but when she goes to the shop she finds that the price has gone up to 55p. How many can she buy now?

PS 6 Eunice wanted to save up for a mountain bike that costs £250. She baby-sits each week for 6 hours for £2.75 an hour, and does a Saturday job that pays £27.50. She saves three-quarters of her weekly earnings. How many weeks will it take her to save enough to buy the bike?

PS 7 The magazine *Teen Dance* comes out every month. In a newsagent the magazine costs £2.45. The annual (yearly) subscription for the magazine is £21. How much cheaper is each magazine when bought on subscription?

AU 8 Paula buys a sofa. She pays a deposit of 10% of the cash price and then 36 monthly payments of £12.50. In total she pays £495. How much was the cash price of the sofa?

FM 9 There are 125 people at a wedding. They need to get to the reception.

52 people are going by coach and the rest are travelling in cars. Each car can take up to five people.

What is the least number of cars needed to take everyone to the reception?

C

PS 10 A fish pond in a shop contains 240 fish.

Each week the manager has a delivery from one supplier of 45 new fish that he adds to the pond.

On average he sells 62 fish each week. When his stock falls below 200 fish, he buys in extra fish from a different supplier. After how many weeks will he need to buy in extra fish?

AU 11 A baker supplies bread rolls to a catering company.

The bread rolls are sold in packs of 24 for £1.99 per pack. The catering company want 500 fresh rolls each day. How much will the bill be for one week, assuming they do not work on Sundays?

FM 12 Gavin's car does 8 miles to each litre of petrol. He does 12 600 miles a year of which 4600 is on company business.

Petrol costs 95p per litre.

Insurance and servicing costs £800 a year.

Gavin's company gives him an allowance of 40p for each mile he drives on company business.

How much does Gavin pay towards running his car each year?

1.2 Multiplication and division with decimals

This section will show you how to:

- multiply a decimal number by another decimal number
- divide by decimals by changing the calculation to division by an integer

Key words
decimal places
decimal point
integer

Multiplying two decimal numbers together

Follow these steps to multiply one decimal number by another decimal number.

- First, complete the whole calculation as if the **decimal points** were not there.
- Then, count the total number of **decimal places** in the two decimal numbers. This gives the number of decimal places in the answer.

EXAMPLE 3

Work out: 3.42×2.7

Ignoring the decimal points gives the following calculation.

Now, 3.42 has two decimal places (.42) and 2.7 has one decimal place (.7). So, the total number of decimal places in the answer is three.

$$\begin{array}{r} 342 \\ \times \quad 27 \\ \hline 2394 \\ 6840 \\ \hline 9234 \\ \hline \end{array}$$

So $3.42 \times 2.7 = 9.234$

Dividing by a decimal

EXAMPLE 4

Work out the following. **a** $42 \div 0.2$ **b** $19.8 \div 0.55$

a The calculation is $42 \div 0.2$ which can be rewritten as $420 \div 2$. In this case both values have been multiplied by 10 to make the divisor into a whole number or **integer**. This is then a straightforward division to which the answer is 210.

Another way to view this is as a fraction problem.

$$\frac{42}{0.2} = \frac{42}{0.2} \times \frac{10}{10} = \frac{420}{2} = \frac{210}{1} = 210$$

b $19.8 \div 0.55 = 198 \div 5.5 = 1980 \div 55$

This then becomes a long division problem.

This has been solved by the method of repeated subtraction.

$$\begin{array}{rr} 1980 & \\ -\ 1100 & 20 \times 55 \\ \hline 880 & \\ -\ 440 & 8 \times 55 \\ \hline 440 & \\ -\ 440 & 8 \times 55 \\ \hline 0 & 36 \times 55 \end{array}$$

So $19.8 \div 0.55 = 36$

EXERCISE 1B

D

1 Work out each of these.

a 0.14×0.2 **b** 0.3×0.3 **c** 0.24×0.8 **d** 5.82×0.52

e 5.8×1.23 **f** 5.6×9.1 **g** 0.875×3.5 **h** 9.12×5.1

2 For each of the following:

i estimate the answer by first rounding each number to the nearest whole number

ii calculate the exact answer and then calculate the difference between this and your answers to part **i**.

a 4.8×7.3 **b** 2.4×7.6 **c** 15.3×3.9 **d** 20.1×8.6

e 4.35×2.8 **f** 8.13×3.2 **g** 7.82×5.2 **h** 19.8×7.1

AU **3** **a** Use any method to work out: 26×22

b Use your answer to part **a** to work out the following.

i 2.6×2.2 **ii** 1.3×1.1 **iii** 2.6×8.8

C

AU **4** Lee is trying to work out the answer to 8.6×4.7. His answer is 40.24.

a Without working it out, can you tell whether his answer is correct?

b Tracy says the answer is 46.42.
Without working out the answer can you tell whether her answer is correct?
In each part, show how you decided.

PS **5** Here are three calculations. $26.66 \div 3.1$ $17.15 \div 3.5$ $55.04 \div 8.6$

Which has the largest answer? Show how you know.

6 Work out each of these.

a $3.6 \div 0.2$ **b** $56 \div 0.4$ **c** $0.42 \div 0.3$ **d** $8.4 \div 0.7$ **e** $4.26 \div 0.2$

f $3.45 \div 0.5$ **g** $83.7 \div 0.03$ **h** $0.968 \div 0.08$ **i** $7.56 \div 0.4$

7 Work out each of these.

a $67.2 \div 0.24$ **b** $6.36 \div 0.53$ **c** $0.936 \div 5.2$ **d** $162 \div 0.36$ **e** $2.17 \div 3.5$

f $98.8 \div 0.26$ **g** $0.468 \div 1.8$ **h** $132 \div 0.55$ **i** $0.984 \div 0.082$

8 A pile of paper is 6 cm high. Each sheet is 0.008 cm thick. How many sheets are in the pile of paper?

9 Doris buys a big bag of safety pins. The bag weighs 180 g. Each safety pin weighs 0.6 g. How many safety pins are in the bag?

AU **10** **a** Use any method to work out: $81 \div 3$

b Use your answer to part **a** to work these out.

i $8.1 \div 0.3$ **ii** $0.81 \div 30$ **iii** $0.081 \div 0.3$

FM 11 A party of 24 scouts and their leader went to a zoo. The cost of a ticket for each scout was £2.15, and the cost of a ticket for the leader was £2.60. What was the total cost of entering the zoo?

PS 12 Mark went shopping.
He went into three stores and bought one item from each store.

Music Store		Clothes Store		Book Store	
CDs	£5.98	Shirt	£12.50	Magazine	£2.25
DVDs	£7.99	Jeans	£32.00	Pen	£3.98

In total he spent £43.97. What did he buy?

1.3 Approximation of calculations

This section will show you how to:
- round to a given number of significant figures
- approximate the result before multiplying two numbers together
- approximate the result before dividing two numbers
- round a calculation, at the end of a problem, to give what is considered to be a sensible answer

Key words
approximate
estimation
round
significant figures

Rounding to significant figures

We often use **significant figures** when we want to **approximate** a number with quite a few digits in it. We use this technique in **estimations**.

Look at this table which shows some numbers rounded to one, two and three significant figures (sf).

One sf	8	50	200	90 000	0.000 07	0.003	0.4
Two sf	67	4.8	0.76	45 000	730	0.0067	0.40
Three sf	312	65.9	40.3	0.0761	7.05	0.003 01	0.400

The steps taken to round a number to a given number of significant figures are very similar to those used for rounding to a given number of decimal places.

- From the left, count the digits. If you are rounding to 2 sf, count two digits, for 3 sf count three digits, and so on. When the original number is less than 1, start counting from the first non-zero digit.
- Look at the next digit to the right. When the value of this next digit is less than 5, leave the digit you counted to the same. However if the value of this next digit is equal to or greater than 5, add 1 to the digit you counted to.
- Ignore all the other digits, but put in enough zeros to keep the number the right size (value).

For example, look at the following table, which shows some numbers rounded to one, two and three significant figures, respectively.

Number	Rounded to 1 sf	Rounded to 2 sf	Rounded to 3 sf
45 281	50 000	45 000	45 300
568.54	600	570	569
7.3782	7	7.4	7.38
8054	8000	8100	8050
99.8721	100	100	99.9
0.7002	0.7	0.70	0.700

EXERCISE 1C

D

1 Round each of the following numbers to 1 significant figure.

a 46 313 **b** 57 123 **c** 30 569 **d** 94 558 **e** 85 299

f 0.5388 **g** 0.2823 **h** 0.005 84 **i** 0.047 85 **j** 0.000 876

k 9.9 **l** 89.5 **m** 90.78 **n** 199 **o** 999.99

C

2 Round each of the following numbers to 2 significant figures.

a 56 147 **b** 26 813 **c** 79 611 **d** 30 578 **e** 14 009

f 1.689 **g** 4.0854 **h** 2.658 **i** 8.0089 **j** 41.564

k 0.8006 **l** 0.458 **m** 0.0658 **n** 0.9996 **o** 0.009 82

3 Round each of the following to the number of significant figures (sf) indicated.

a 57 402 (1 sf) **b** 5288 (2 sf) **c** 89.67 (3 sf) **d** 105.6 (2 sf)

e 8.69 (1 sf) **f** 1.087 (2 sf) **g** 0.261 (1 sf) **h** 0.732 (1 sf)

i 0.42 (1 sf) **j** 0.758 (1 sf) **k** 0.185 (1 sf) **l** 0.682 (1 sf)

4 What are the least and the greatest numbers of sweets that can be found in these jars?

a

b

c

5 $3000 \times 400 = 1\,200\,000$

Write down four multiplication problems for which that answer is 1 200 000.

B

6 What are the least and the greatest numbers of people that live in these towns?

Elsecar population 800 (to 1 significant figure)
Hoyland population 1200 (to 2 significant figures)
Barnsley population 165 000 (to 3 significant figures)

PS 7 A joiner estimates that he has 20 pieces of skirting board in stock. This is correct to 1 sf.

He uses three pieces and now has 10 left, to 1 sf. How many pieces could he have had to start with? Work out all possible answers.

PS 8 There are 500 fish in a pond, to 1 sf. What is the least possible number of fish that could be taken from the pond so that there are 400 fish in the pond to 1 sf?

C

AU 9 Karen says that the population of Preston is 132 000 to the nearest thousand. Donte says that the population of Preston is 130 000. Explain why Donte could also be correct.

Multiplying and dividing by multiples of 10

Questions often involve multiplication of multiples of 10, 100 and so on. This method is used in estimation. You should have the skill to do this mentally so that you can check that your answers to calculations are about right. (Approximation of calculations is covered in the next section.)

Use a calculator to work out the following.

a $300 \times 200 =$ **b** $100 \times 40 =$ **c** $2000 \times 0.3 =$

d $0.2 \times 50 =$ **e** $0.2 \times 0.5 =$ **f** $0.3 \times 0.04 =$

Can you see a way of doing these without using a calculator or pencil and paper? Basically, you multiply the non-zero digits and then work out the number of zeros or the position of the decimal point by combining the zeros or decimal places in the original calculation.

Dividing is almost as simple. Try doing the following on your calculator.

a $400 \div 20 =$ **b** $200 \div 50 =$ **c** $1000 \div 0.2 =$

d $300 \div 0.3 =$ **e** $250 \div 0.05 =$ **f** $30\,000 \div 0.6 =$

Once again, there is an easy way of doing these 'in your head'. Look at these examples.

$300 \times 4000 = 1\,200\,000$ $5000 \div 200 = 25$ $20 \times 0.5 = 10$

$0.6 \times 5000 = 3000$ $400 \div 0.02 = 20\,000$ $800 \div 0.2 = 4000$

Can you see a connection between the digits, the number of zeros and the position of the decimal point, and the way in which these calculations are worked out?

EXERCISE 1D

D

1 Without using a calculator, write down the answers to these.

a 200×300 **b** 30×4000 **c** 50×200 **d** 0.3×50 **e** 200×0.7

f 200×0.5 **g** 0.1×2000 **h** 0.2×0.14 **i** 0.3×0.3 **j** $(20)^2$

k $(20)^3$ **l** $(0.4)^2$ **m** 0.3×150 **n** 0.4×0.2 **o** 0.5×0.5

p $20 \times 40 \times 5000$ **q** $20 \times 20 \times 900$

D

2 Without using a calculator, write down the answers to these.

a 2000 ÷ 400 **b** 3000 ÷ 60 **c** 5000 ÷ 200

d 300 ÷ 0.5 **e** 2100 ÷ 0.7 **f** 2000 ÷ 0.4

g 3000 ÷ 1.5 **h** 400 ÷ 0.2 **i** 2000 × 40 ÷ 200

j 200 × 20 ÷ 0.5 **k** 200 × 6000 ÷ 0.3 **l** 20 × 80 × 60 ÷ 0.03

C

AU 3 You are given that 16 × 34 = 544

Write down the value of:

a 160 × 340 **b** 544 000 ÷ 34

PS 4 Match each calculation to its answer and then write out the calculations in order, starting with the smallest answer.

5000 × 4000	600 × 8000	200 000 × 700	30 × 90 000
140 000 000	4 800 000	2 700 000	20 000 000

5 In 2009 there were £28 000 million worth of £20 notes in circulation.

How many notes is this?

Approximation of calculations

How do you approximate the value of a calculation?

What do you actually do when you try to approximate an answer to a problem?

For example, what is the approximate answer to 35.1 × 6.58?

To approximate the answer in this and many other similar cases, you simply round each number to 1 significant figure, then work out the calculation.

So in this case, the approximation is:

35.1 × 6.58 ≈ 40 × 7 = 280

Note: ≈ symbol means 'approximately equal to'.

For the division 89.1 ÷ 2.98, the approximate answer is 90 ÷ 3 = 30

Sometimes when dividing it can be sensible to round to 2 significant figures instead of 1 significant figure. For example:

24.3 ÷ 3.87 using 24 ÷ 4 gives an approximate answer of 6

whereas

24.3 ÷ 3.87 using 20 ÷ 4 gives an approximate answer of 5

Both of these answers would be acceptable in the GCSE examination as they are both sensible answers, but generally rounding to 1 significant figure is easier.

If you are using a calculator, whenever you see a calculation with a numerator and denominator *always* put brackets around the top and the bottom. This is to remind you that the numerator and denominator must be worked out separately before the division. You can work out the numerator and denominator separately but most calculators will work out the answer straight away if you use brackets. You are expected to use a calculator *efficiently*; doing the calculation in stages is not efficient.

EXAMPLE 5

a Find approximate answers to these calculations.

i $\dfrac{213 \times 69}{42}$ ii $\dfrac{78 \times 397}{0.38}$

b Use a calculator to work out the answer. Round this to 3 significant figures.

a i Round each number to 1 significant figure. $\dfrac{200 \times 70}{40}$

Work out the numerator. $= \dfrac{14\,000}{40}$

Divide by the denominator. $= 350$

ii Round each value to 1 significant figure. $\dfrac{80 \times 400}{0.4}$

Work out the numerator. $= \dfrac{32\,000}{0.4}$

$= \dfrac{320\,000}{4}$

Divide by the denominator. $= 80\,000$

b Use a calculator to check your approximate answers.

i $\dfrac{213 \times 69}{42} = \dfrac{(213 \times 69)}{(42)}$

So key in

(2 1 3 × 6 9) ÷ (4 2) =

The display should say 349.9285714 which rounds to 350. This agrees exactly with the estimate.

Note that you do not have to put brackets around the 42 but it is a good habit to get into.

ii $\dfrac{78 \times 397}{0.38} = \dfrac{(78 \times 397)}{(0.38)}$

So key in

(7 8 × 3 9 7) ÷ (0 · 3 8) =

The display should say 81489.47368 which rounds to 81 500. This agrees with the estimate.

EXERCISE 1E

1 Find approximate answers to the following.

a 5435 × 7.31 **b** 5280 × 3.211 **c** 63.24 × 3.514 × 4.2

d 354 ÷ 79.8 **e** 5974 ÷ 5.29 **f** 208 ÷ 0.378

D

2 Use a calculator to work out the answers to question **1**. Round your answers to 1 sf and compare them with the estimates you made.

3 By rounding, find approximate answers to these.

a $\dfrac{573 \times 783}{107}$ **b** $\dfrac{783 - 572}{24}$ **c** $\dfrac{352 + 657}{999}$

d $\dfrac{78.3 - 22.6}{2.69}$ **e** $\dfrac{3.82 \times 7.95}{9.9}$ **f** $\dfrac{11.78 \times 61.8}{39.4}$

4 Use a calculator to work out the answers to question **3**. Round your answers to 1 sf and compare them with the estimates you made.

5 Find the approximate monthly pay of the following people, whose annual salaries are given.

a Paul £35 200 **b** Michael £25 600 **c** Jennifer £18 125 **d** Ross £8420

6 Find the approximate annual pay of the following people, whose earnings are shown.

a Kevin £270 a week **b** Malcolm £1528 a month **c** David £347 a week

AU 7 A farmer bought 2713 kg seed at a cost of £7.34 per kg. Find the approximate total cost of this seed.

8 A greengrocer sold a box of 450 oranges for £37. Approximately how much did each orange sell for?

9 Gold bars weigh 400 ounces (12.44 kg). On 6 October 2009, one gold bar was worth $413 080.

Approximately how much is one ounce of gold worth, in dollars?

FM 10 It took me 6 hours 40 minutes to drive from Sheffield to Bude, a distance of 295 miles. My car uses petrol at the rate of about 32 miles per gallon. The petrol cost £3.51 per gallon.

a Approximately how many miles did I travel each hour?

b Approximately how many gallons of petrol did I use in going from Sheffield to Bude?

c What was the approximate cost of all the petrol I used in the journey to Bude and back again?

11 By rounding, find an approximate answer to each of the following.

a $\dfrac{462 \times 79}{0.42}$ **b** $\dfrac{583 - 213}{0.21}$ **c** $\dfrac{252 + 551}{0.78}$ **d** $\dfrac{296 \times 32}{0.325}$

e $\dfrac{297 + 712}{0.578 - 0.321}$ **f** $\dfrac{893 \times 87}{0.698 \times 0.47}$ **g** $\dfrac{38.3 + 27.5}{0.776}$ **h** $\dfrac{29.7 + 12.6}{0.26}$

i $\dfrac{4.93 \times 3.81}{0.38 \times 0.51}$ **j** $\dfrac{12.31 \times 16.9}{0.394 \times 0.216}$

12 Use a calculator to work out the answers to question **11**. Round your answers to 3 significant figures and compare them with the estimates you made.

C

13 A sheet of paper is 0.012 cm thick. Approximately how many sheets will there be in a pile of paper that is 6.35 cm deep?

PS **14** Kirsty arranges for magazines to be put into envelopes. She sorts out 178 magazines between 10.00 am and 1.00 pm. Approximately how many magazines will she be able to sort in a week in which she works for 17 hours?

AU **15** A box full of magazines weighs 8 kg. One magazine weighs about 15 g. Approximately how many magazines are there in the box?

16 Use your calculator to work out the following. In each case:

i write down the full calculator display of the answer

ii round your answer to 3 significant figures.

a $\dfrac{12.3 + 64.9}{6.9 - 4.1}$ **b** $\dfrac{13.8 \times 23.9}{3.2 \times 6.1}$ **c** $\dfrac{48.2 + 58.9}{3.62 \times 0.042}$

Sensible rounding

Sensible rounding is simply writing or saying answers to questions that have a real-life context so that the answer makes sense and is the sort of thing someone might say in a normal conversation.

For example:

'The distance from Rotherham to Sheffield is 9 miles,' is a sensible statement.
'The distance from Rotherham to Sheffield is 8.7864 miles,' is not sensible.

'Painting a house takes 6 tins of paint,' is sensible.
'Painting a house takes 5.91 tins of paint,' is not sensible.

As a general rule, if it sounds sensible it will be acceptable.

In a question for which you are asked to give an answer to a sensible or appropriate degree of accuracy, use the following rule. Give the answer to the same accuracy as the numbers in the question. So, for example, if the numbers in the question are given to 2 significant figures give your answer to 2 significant figures but remember, unless working out an approximation, do all the working to at least 4 significant figures or use the calculator display.

EXERCISE 1F

D

1 Round each of the figures in these statements to a suitable degree of accuracy.

a I am 1.7359 metres tall.

b It took me 5 minutes 44.83 seconds to mend the television.

c My kitten weighs 237.97 grams.

d The correct temperature at which to drink Earl Grey tea is 82.739 °C.

e There were 34 827 people at the test match yesterday.

f The distance from Wath to Sheffield is 15.528 miles.

g The area of the floor is 13.673 m^2.

FM 2 Rewrite the following article, using sensible numbers.

It was a hot day; the temperature was 81.699 °F and still rising. I had now walked 5.3289 km in just over 113.98 minutes. But I didn't care since I knew that the 43 275 people watching the race were cheering me on. I won by clipping 6.2 seconds off the record time. This was the 67th time it had happened since records first began in 1788. Well, next year I will only have 15 practice walks beforehand as I strive to beat the record by at least another 4.9 seconds.

3 1 litre = 1000 cm^3

About how many test tubes, each holding 24 cm^3 of water, can be filled from a 1 litre flask?

FM 4 A light aircraft cruises at 104 mph. It travels for 2 hours 50 minutes.

Approximately how long is the flight?

AU 5 A lorry load of scrap metal weighs 39.715 tonnes. It is worth £20.35 per tonne.

Approximately how much is the load worth?

6 If I walk at an average speed of 70 metres per minute, approximately how long will it take me to walk a distance of 3 km?

7 About how many stamps at 21p each can I buy for £12?

8 At Manchester United, it takes 160 minutes for 43 500 fans to get into the ground. On average, about how many fans are let into the ground every minute?

9 A 5p coin weighs 4.2 g. Approximately how much will one million pounds worth of 5p pieces weigh?

PS 10 The accurate temperature is 18.2 °C. David rounds the temperature to the nearest 5 °C.

David says the temperature is about 20 °C.

How much would the temperature need to rise for David to say that the temperature is about 25 °C?

AU 11 The distance from the Sun to Earth is approximately 149 000 000 km. The speed of light is approximately 300 000 km per second.

Use your calculator to work out how many seconds it takes for light to travel from the Sun to Earth.

Give your answer to a sensible degree of accuracy.

AU 12 The population density of a country is measured by dividing the population by the area. For example the population of the United Kingdom is 61 700 000 and the area is 243 000 km^2. So the population density is approximately 250 people per square kilometre.

The most densely populated country is Macau with a population of 541 200 and an area of 29.2 km^2.

The least densely populated country is Greenland with a population of 57 000 and an area of 2 176 000 km^2.

How many times is the population density of Macau greater than the population density of Greenland?

Give your answer to a suitable degree of accuracy.

GRADE BOOSTER

D You can estimate the values of calculations involving positive numbers bigger than one

C You can round numbers to a given number of significant figures

C You can estimate the values of calculations involving positive numbers between zero and one

C You can use a calculator efficiently and know how to give answers to an appropriate degree of accuracy

What you should know now

- How to solve complex real-life problems without a calculator
- How to divide by decimals with up to two decimal places
- How to estimate the values of calculations including those with decimal fractions, and use a calculator efficiently

1 Frank earns £12 per hour. He works for 40 hours per week. He saves $\frac{1}{5}$ of his earnings each week. *(1 mark)*

How many weeks will it take him to save £500? *(1 mark)*

AQA, June 2009, Module 3 Higher, Question 4

2 A floor measures 4.75 metres by 3.5 metres. It is to be covered with square tiles of side 25 centimetres.

Tiles are sold in boxes of 16. How many boxes are needed?

3 Gianni wants to tile the wall behind his bath. The total area measures 3.45 metres by 1.85 metres. He is going to use square tiles of side 15 centimetres. Tiles are sold in boxes of 24. How many boxes of tiles does he need?

4 Work out: $\dfrac{21.6 \times 64}{35.1 + 9.57}$

a Write down your full calculator display. *(1 mark)*

b Write your answer to two decimal places. *(1 mark)*

AQA, June 2009, Module 3 Higher, Question 1

5 Work out as a decimal: $\dfrac{4.6^2}{8.6 - 2.7}$

a Write down your full calculator display. *(1 mark)*

b Write your answer to three significant figures. *(1 mark)*

AQA, March 2008, Module 3 Higher, Question 4

6 Use approximations to estimate the value of:

$$\frac{212 \times 7.88}{0.365}$$

7 Ahmed, Briony and Carl used calculators to find the value of:

$$\frac{56.94}{7.16 \times 0.83}$$

Ahmed's answer was 6.5797, Briony had 9.581 and Carl made it 95.81.
Use approximations to find out which one was correct.

8 A shopkeeper buys boxes of printer cartridges. Each box cost £150. Each box contains ten cartridges. He sells each cartridge for £24. He sells all the cartridges and makes a profit of £180. How many cartridges does he sell?

AQA, November 2011 F1 Question 17

9 Here are two ways of buying music downloads.

79p for each track	£14.95 per month All downloads free Minimum contract two years

Jake estimates that he will download 500 tracks over the next two years.

Which is the cheaper option for Jake?

a Each track separately?

b Two year contract?

You must show all your working.

AQA, November 2011 F1 Question 15

10 Estimate the value of $\dfrac{19.8^2}{4.7 \times 9.93}$

AQA, November 2011 H1 Question 4

11 **a** Use your calculator to work out $\dfrac{5.39}{8.34 - 2.17}$

b **i** Write down your full calculator display. *(1 mark)*

b **ii** Give your answer to 3 significant figures. *(1 mark)*

AQA, June 2011 H2 Question 4

12 Sal and Bill each have a whole number of pounds.

Sal says to Bill.

How much do Sal and Bill have?
Show your working.

AQA, November 2010 F2 Question 21

13 Naismith's rule is used to calculate an approximate time for mountain walks.

Allow 1 hour for every 5 km walked, plus 30 minutes for every 300 metres of ascent.

The walk from Dungeon Ghyll to the top of Scafell Pike in the Lake District is a distance of 18 km.
The total ascent is 1400 metres

Lannie calculates an approximate time of 6 hours.
Show that Lannie is correct.

AQA, June 2010 H2 Question 2

Worked Examination Questions

1 Estimate the result of the calculation

$$\frac{195.71 - 53.62}{\sqrt{0.0375}}$$

Show the estimates you make.

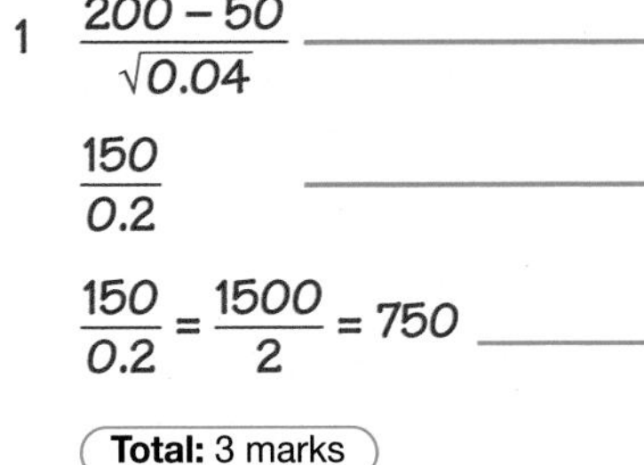

1 $\frac{200 - 50}{\sqrt{0.04}}$ — First round each number to 1 significant figure. Rounding the two numbers in the numerator to 1sf gets 1 mark for method.

$\frac{150}{0.2}$ — Working out the numerator and taking out the square root in the denominator gets 1 mark.

$\frac{150}{0.2} = \frac{1500}{2} = 750$ — Change the problem so it becomes division by an integer 750 gets 1 mark for accuracy.

Total: 3 marks

FM **2** I earn £30 000 in 12 months.

Half of this is spent on tax.

How much do I have left each month?

2 *30 000 ÷ 2 or £15 000* — Work out how much I spend on tax or how much I have left. This is the same calculation as it is half of £30 000 and gets 1 mark for method. Note the answer of £15 000 does not need to be worked out at this stage.

30 000 ÷ 2 ÷ 12 — The next step is to calculate one twelfth of £30 000 ÷ 2. This gets 1 mark for method.

= £1250 — This gets 1 mark for accuracy.

Total: 3 marks

PS **3** Two numbers have been rounded.
The first number is 360 to two significant figures.
The second number is 500 to one significant figure.

What is the smallest sum of the two original numbers?

3 *The smallest that the first number could be is 355.*
The smallest that the second number could be is 450. — Realising that each number needs to be the smallest possible gets 1 mark for method. Either value 355 or 450 gets 1 mark for accuracy.

Smallest sum = 355 + 450 = 805 — Correctly adding the two smallest numbers gets 1 mark for accuracy.

Total: 3 marks

1 Functional Maths
Planning a dinner party

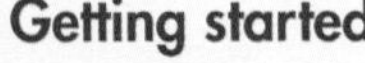

You are planning a dinner party but have not finalised your guest list or menu.

You have decided to make your dessert from scratch because it will be cheaper than buying it from the local shop. Your Aunt Mildred has helped you by giving you four recipes for desserts that all your friends will like: crème caramel, blueberry and lime cheesecake, mango sorbet and chocolate brownies. You have already researched the prices of the ingredients required to make these desserts and are now ready to plan which ones to make!

Getting started

Consider the questions below to get you started.

- How many 150 g portions can I get from 1 kg?
- How many 75-g bars of chocolate would I need to buy if I wanted 0.5 kg of chocolate?
- What is cheaper, three 330-ml cans of lemonade at 59p per can, or a litre bottle of lemonade at £1.80?
- In a meeting one packet of biscuits is bought per three people attending. How many packets would be needed if 20 people attend the meeting?

Your task

You have £50 to spend on your desserts.

Working in pairs, decide on several different numbers of guests that you might have at your dinner party. For each potential number of guests, plan three different combinations of desserts to make, and work out how much these combinations would cost overall and per person.

Ingredient costs

Item	Amount	Cost
Butter	500 g	£2.40
Eggs	6	£1.10
Lime	1	20p
Mango	1	£1.50
Milk	2.272 litres	£1.53
Plain chocolate	200 g	£1.09
Plain flour	1.5 kg	75p
Caster sugar	1 kg	£1.25
Granulated sugar	2 kg	£1.90
Walnuts	200 g	£2.50
Vanilla extract	30 ml	£1.55
Baking powder	50 g	£1.25
Biscuits	275 g	89p
Blueberries	500 g	£3.99
Cream cheese	10 oz	£2.49
Double cream	284 ml	£1.89
Sour cream	284 ml	£1.39
Gelatine sachet	25 g	99p

(Ingredients can only be bought in these amounts or multiples of these amounts.)

Recipes

Crème Caramel – serves 6

500 ml milk
2 eggs
4 egg yolks
225 g caster sugar
1 teaspoon of vanilla extract

Blueberry and Lime Cheesecake – serves 8

10 oz sweet oaty biscuits
4 oz butter
1 lb 2 oz blueberries
$8\frac{1}{2}$ oz caster sugar
Grated zest and juice of two limes
20 oz of cream cheese
$\frac{1}{2}$ pint of double cream
4 tsp powdered gelatine
$\frac{1}{2}$ pint sour cream

Mango Sorbet – serves 8

4 large ripe mangos
Juice of 2 large limes
450 ml sugar syrup
(150 g granulated sugar and 300 ml water)

Chocolate Brownies – serves 15

50 g plain flour
110 g butter
2 eggs
225 g granulated sugar
175 g walnuts
1 level teaspoon of baking powder

Extend your task

1 Find a recipe for your favourite dessert and research the cost of its ingredients. Work out the cost of the dessert per person for each number of guests.

2 You are going to create table decorations for your dinner party. The basic outlines of these decorations can be seen below.

Find the missing lengths required to make these decorations. How much card would you need in order to make enough decorations for each of your guests?

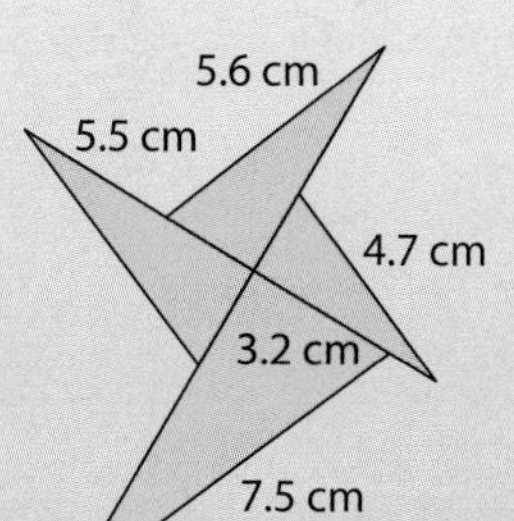

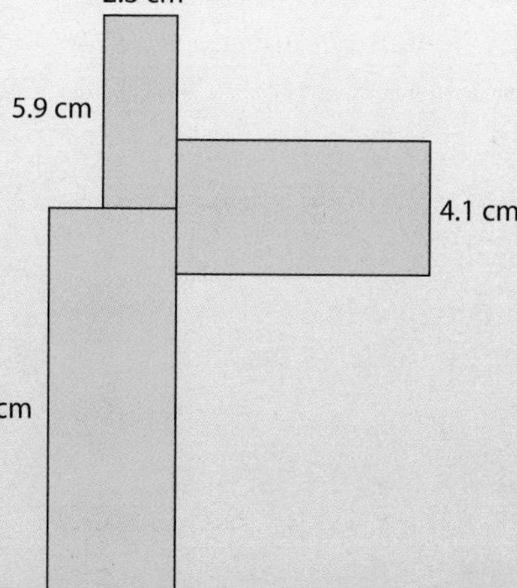

Why this chapter matters

How good is your mental arithmetic? In everyday life you will often need to use mental arithmetic to carry out calculations.

Improving your mental arithmetic skills will make mathematics easier and more enjoyable.

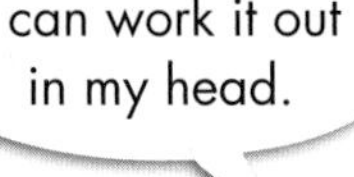

Mental arithmetic

- In 1980 Shakuntala Devi claimed to multiply two 13-digit numbers in 28 seconds.
- In 1999 Alberto Coto of Spain set the record for adding up 100 single-digit numbers in 19.23 seconds.
- In 2009 an 11-year-old Australian correctly answered 129 106 arithmetic questions within 48 hours to win the World Maths Day Challenge; 1.9 million students from over 225 countries entered and altogether answered 452 682 682 mental arithmetic questions correctly.

Quick ways of doing mental arithmetic

1 Multiplying a two-digit number by 11

Work out 43×11.

Write 4 … 3

Now add the two digits.

$4 + 3 = 7$

Put the 7 in the middle, between the 4 and the 3.

Answer: 473

Work out 59×11.

Write 5 … 9

Now add the two digits.

$5 + 9 = 14$

Put the units digit (4) in the middle, between the 5 and the 9, and carry the tens digit (1) to the first number.

$5_1\,4\,9 = 649$

Answer: 649

2 Doubling and halving

To multiply large numbers, when at least one is even, try doubling one number and halving the other.

Work out 24×16.

This is the same as 48×8

This is the same as 96×4

This is the same as 192×2

This is the same as 384×1

Answer: 384

2 Square a number that ends in 5

Work out 45.

Take the number of 10s. (4)

Add 1. (5)

Multiply these two numbers. $4 \times 5 = 20$

Put 25 on the end.

Answer: 2025

Work out 75.

Take the number of 10s. (7)

Add 1. (8)

Multiply these numbers. $7 \times 8 = 56$

Put 25 on the end.

Answer: 5625

Chapter

Number: More number

1 Multiples, factors, prime numbers, powers and roots

2 Prime factors, LCM and HCF

3 Negative numbers

This chapter will show you ...

C how to find prime factors, least common multiples (LCM) and highest common factors (HCF)

Visual overview

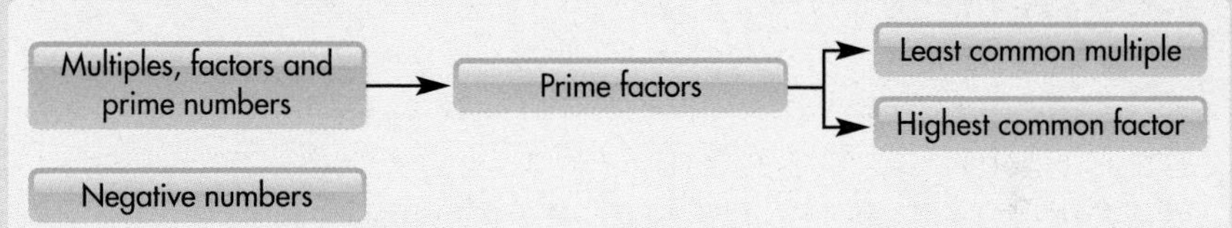

What you should already know

- How to add, subtract, multiply and divide with integers **(KS3 level 4, GCSE grade G)**
- What multiples, factors, square numbers and prime numbers are **(KS3 level 4, GCSE grade G/E)**
- The BIDMAS/BODMAS rule and how to substitute values into simple algebraic expressions **(KS3 level 5, GCSE grade F)**

Quick check

1 Work out the following.

a 9×233 **b** $792 \div 22$ **c** $2^2 \times 3^2$

2 Write down the following.

a A multiple of 7

b A prime number between 10 and 20

c A square number under 80

d The factors of 9

3 Work out the following.

a 8^2 **b** 12^2 **c** $2^2 + 3^2$

2.1 Multiples, factors, prime numbers, powers and roots

This section will show you how to:

- find multiples and factors
- identify prime numbers
- identify square numbers and triangular numbers
- find square roots
- identify cubes and cube roots

Key words

cube roots
cubes
factor
multiple
prime number
square roots
squares
triangle number
triangular number

Multiples. Any number in the multiplication table. For example, the multiples of 7 are 7, 14, 21, 28, 35, … .

Factors. Any whole number that divides exactly into another number. For example, factors of 24 are 1, 2, 3, 4, 6, 8, 12, 24.

Prime numbers. Any number that only has two factors, 1 and itself. For example, 11, 17, 37 are prime numbers.

Squares. A number that comes from multiplying a number by itself. For example, 1, 4, 9, 16, 25, 36 … are square numbers.

Triangular or triangle numbers. Numbers that can make triangular patterns. For example, 1, 3, 6, 10, 15, 21, 28 … are triangular numbers.

Square roots. The square root of a given number is a number which, when multiplied by itself, produces the given number. For example, the square root of 9 is 3, since $3 \times 3 = 9$.

A square root is represented by the symbol $\sqrt{\ }$. For example, $\sqrt{16} = 4$.

Because $-4 \times -4 = 16$, there are always two square roots of every positive number.

So $\sqrt{16} = +4$ or -4. This can be written as $\sqrt{16} = \pm 4$, which is read as plus or minus four.

Cubes. The cube of a number is a number multiplied by itself and then by itself again. For example, the cube of 4 is $4 \times 4 \times 4 = 64$.

Cube roots. The cube root of a number is the number that when multiplied by itself twice gives the cube. For example, the cube root of 27 is 3 because $3 \times 3 \times 3 = 27$ and the cube root of –8 is –2 because $-2 \times -2 \times -2 = -8$.

D

EXERCISE 2A

1 From this box choose the number that fits each of these descriptions. (One number per description.)

12			21
	8		15
13		17	
9			18
	10		6
14		16	

a A multiple of 3 and a multiple of 4

b A square number and an odd number

c A factor of 24 and a factor of 18

d A prime number and a factor of 39

e An odd factor of 30 and a multiple of 3

f A number with four factors and a multiple of 2 and 7

g A number with five factors exactly

h A triangular number and a factor of 20

i An even number and a factor of 36 and a multiple of 9

j A prime number that is one more than a square number

k If you write the factors of this number out in order they make a number pattern in which each number is twice the one before

l An odd triangular number that is a multiple of 7

FM **2** If hot-dog sausages are sold in packs of 10 and hot-dog buns are sold in packs of eight, how many of each do you have to buy to have complete hot dogs with no wasted sausages or buns?

3 Rover barks every 8 seconds and Spot barks every 12 seconds. If they both bark together, how many seconds will it be before they both bark together again?

4 A bell chimes every 6 seconds. Another bell chimes every 5 seconds. If they both chime together, how many seconds will it be before they both chime together again?

PS **5** Fred runs round a running track in 4 minutes. Debbie runs round in 3 minutes. If they both start together on the line at the end of the finishing straight, when will they both be on the same line together again? How many laps will Debbie have run? How many laps will Fred have run?

6 Copy these sums and write out the *next four* lines.

$$1 = 1$$
$$1 + 3 = 4$$
$$1 + 3 + 5 = 9$$
$$1 + 3 + 5 + 7 = 16$$

7 Write down the negative square root of each of these.

a 4 **b** 25 **c** 49 **d** 1 **e** 81

f 121 **g** 144 **h** 400 **i** 900 **j** 169

D

8 Write down the cube root of each of these.

a 1 **b** 27 **c** 64 **d** 8 **e** 1000

f –8 **g** –1 **h** 8000 **i** 64 000 **j** –64

9 The triangular numbers are 1, 3, 6, 10, 15, 21 …

a Continue the sequence until you get the first triangular number that is greater than 100.

b Add consecutive pairs of triangular numbers, starting with 1 + 3 = 4, 3 + 6 = 9. What do you notice?

AU 10 Here are four numbers.

8 28 49 64

Copy and complete the table by putting each of the numbers in the correct box.

	Square number	**Factor of 56**
Cube number		
Multiple of 7		

PS 11 The following numbers are described as triangular numbers.

1, 3, 6, 10, 15

a Investigate why they are called triangular numbers.

b Write down the next five triangular numbers.

12 John is writing out his 4 times table. Mary is writing out her 6 times table. They notice that some answers are the same.

In which other times tables do these common answers also appear?

C

PS 13 **a** $36^3 = 46\,656$. Work out 1^3, 4^3, 9^3, 16^3, 25^3.

b $\sqrt{46\,656} = 216$. Use a calculator to find the square roots of the numbers you worked out in part **a**.

c 216 = 36 × 6. Can you find a similar connection between the answer to part **b** and the numbers cubed in part **a**?

d What type of numbers are 1, 4, 9, 16, 25, 36?

B

14 Write down the values of these numbers.

a $\sqrt{0.04}$ **b** $\sqrt{0.25}$ **c** $\sqrt{0.36}$ **d** $\sqrt{0.81}$

e $\sqrt{1.44}$ **f** $\sqrt{0.64}$ **g** $\sqrt{1.21}$ **h** $\sqrt{2.25}$

15 Use your calculator to work out the answers to these. Give your answers to the nearest whole number.

a $\dfrac{13.7 + 21.9}{\sqrt{0.239}}$ **b** $\dfrac{29.6 \times 11.9}{\sqrt{0.038}}$ **c** $\dfrac{87.5 - 32.6}{\sqrt{0.8} - \sqrt{0.38}}$

2.2 Prime factors, LCM and HCF

This section will show you how to:

- identify prime factors
- identify the least common multiple (LCM) of two numbers
- identify the highest common factor (HCF) of two numbers

Key words
highest common factor (HCF)
index notation
least common multiple (LCM)
prime factor
prime factor tree
product
product of prime factors

Start with a number, such as 110, and find two numbers that, when multiplied together, give that number, for example, 2×55. Are they both prime? No, 55 isn't. So take 55 and repeat the operation, to get 5×11. Are these both prime? Yes. So:

$110 = 2 \times 5 \times 11$

The **prime factors** of 110 are 2, 5 and 11.

This method is not very logical and you need to know your multiplication tables well to use it. There are, however, two methods that you can use to make sure you do not miss any of the prime factors. The next two examples show you how to use the first of these methods.

EXAMPLE 1

Find the prime factors of 24.

Divide 24 by any prime number that goes into it. (2 is an obvious choice.)

Divide the answer (12) by a prime number. Repeat this process until you have a prime number as the answer.

2	24
2	12
2	6
	3

So the prime factors of 24 are 2, 2, 2 and 3.

$24 = 2 \times 2 \times 2 \times 3$

A quicker and neater way to write this answer is to use **index notation**, expressing the answer using powers.

In index notation, the prime factors of 24 are $2^3 \times 3$.

EXAMPLE 2

Find the prime factors of 96.

2	96
2	48
2	24
2	12
2	6
	3

So, the prime factors of 96 are 2, 2, 2, 2, 2 and 3.

$96 = 2 \times 2 \times 2 \times 2 \times 2 \times 3 = 2^5 \times 3$

When 24 is expressed as $2 \times 2 \times 2 \times 3$ it has been written as a **product of its prime factors**.

Another name for the product $2 \times 2 \times 2 \times 3$ or $2^3 \times 3$ is the *prime factorisation* of 24.

The prime factorisation of 96 is $2 \times 2 \times 2 \times 2 \times 2 \times 3$ or $2^5 \times 3$, which is also the product of its prime factors.

The second method uses **prime factor trees**. You start by dividing the number into a pair of factors. Then you divide this, and carry on dividing until you get to prime numbers.

EXAMPLE 3

Find the prime factors of 76.

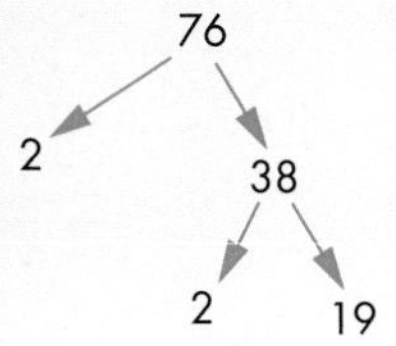

Stop dividing the factors here because 2, 2 and 19 are all prime numbers.

So, the prime factors of 76 are 2, 2 and 19.

$76 = 2 \times 2 \times 19 = 2^2 \times 19$

EXAMPLE 4

Find the prime factors of 420.

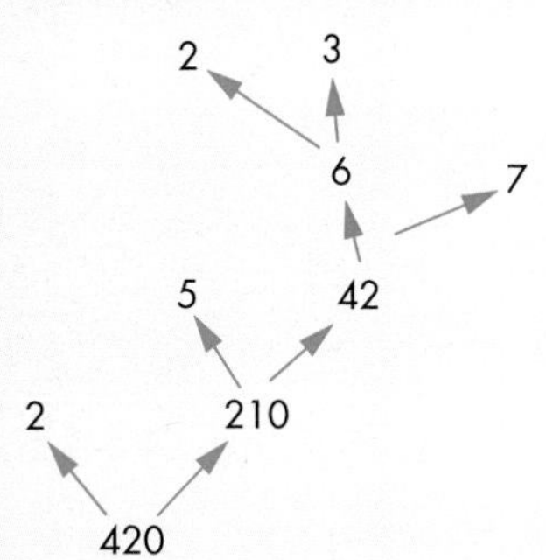

The process can be done upside down to make an upright tree.

So, the prime factors of 420 are 2, 2, 3, 5 and 7.

$420 = 2 \times 5 \times 2 \times 3 \times 7$
$= 2^2 \times 3 \times 5 \times 7$

EXERCISE 2B

1 Copy and complete these prime factor trees.

a

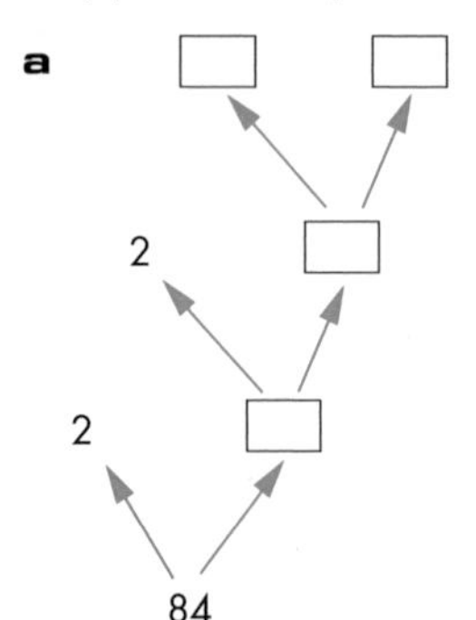

$84 = 2 \times 2$ ... ...

b

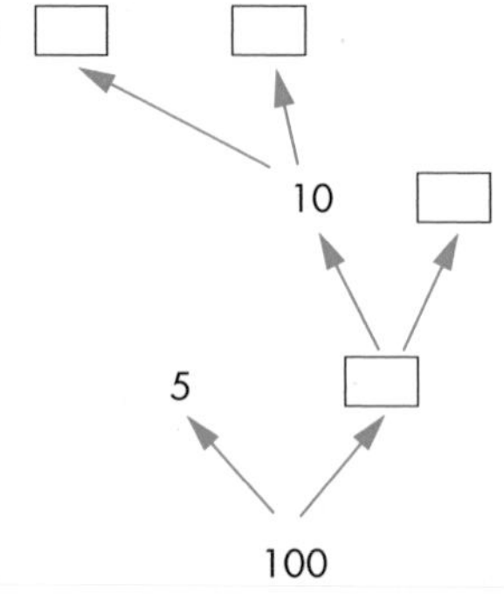

$100 = 5 \times 2$ ... ...

c

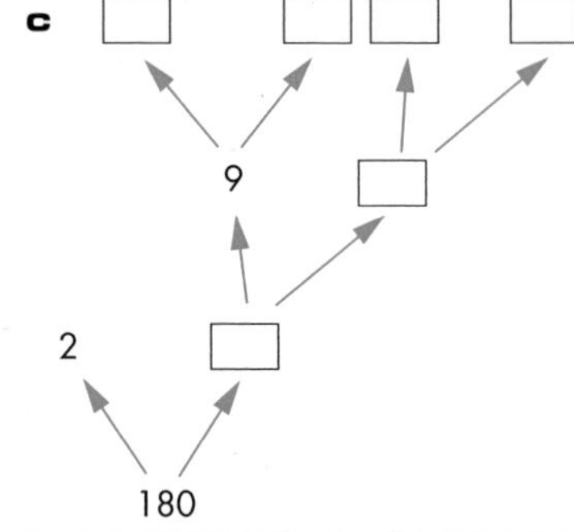

$180 = 2$ ... ...

C

d

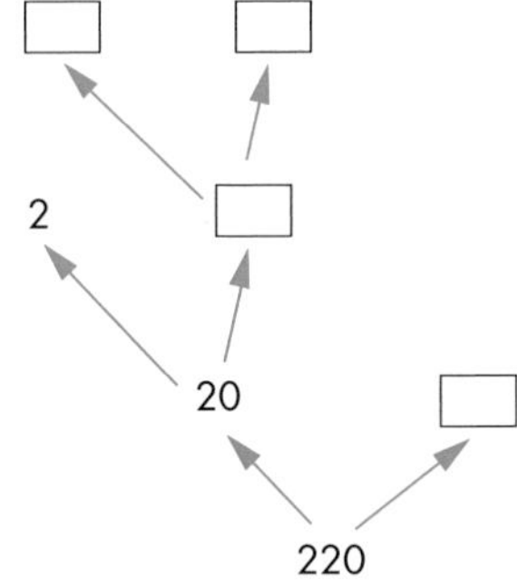

220 = 2 ... ...

e

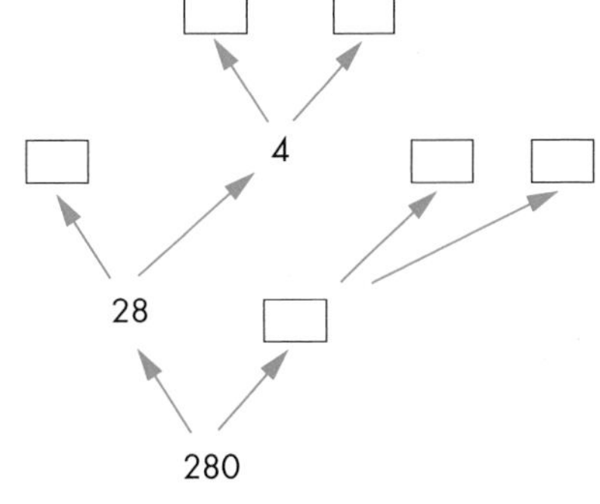

280 = ...

f

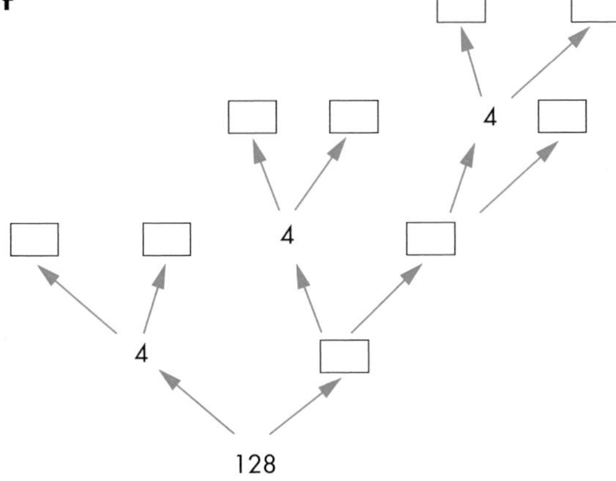

128 = ...

g

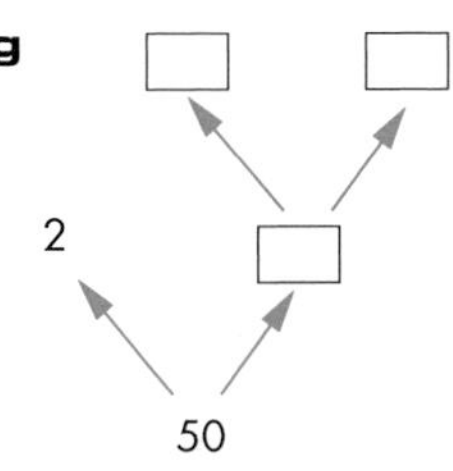

50 = ...

2 Use index notation, for example:

$$100 = 2 \times 2 \times 5 \times 5 = 2^2 \times 5^2$$

and $$540 = 2 \times 2 \times 3 \times 3 \times 3 \times 5 = 2^2 \times 3^3 \times 5$$

to rewrite your answers to question **1**, parts **a** to **g**.

3 Write the numbers from 1 to 50 as products of their prime factors. Use index notation. For example:

$1 = 1$ $\quad 2 = 2$ $\quad 3 = 3$

$4 = 2^2$ $\quad 5 = 5$ $\quad 6 = 2 \times 3$ $\quad \dots$

4 **a** What is special about the numbers 2, 4, 8, 16, 32, …?

b What are the next two terms in this series?

c What are the next three terms in the series 3, 9, 27, …?

d Continue the series 4, 16, 64, …, for three more terms.

e Rewrite all the series in parts **a**, **b**, **c** and **d** in index notation. For example, the first series is:

$2^1, 2^2, 2^3, 2^4, 2^5, 2^6, 2^7, \dots$

C

AU **5** **a** Express 60 as a product of prime factors.

b Write your answer to part **a** in index form.

c Use your answer to part **b** to write 120, 240 and 480 each as a product of prime factors in index form.

PS **6** $1001 = 7 \times 11 \times 13$

$1001^2 = 1002001$

$1001^3 = 1003003001$

a Write 1002001 as a product of prime factors in index form.

b Write 1003003001 as a product of prime factors in index form.

c Write 1001^{10} as a product of prime factors in index form.

7 Harriet wants to share £40 between three of her grandchildren. Explain why it is not possible for them to get equal shares.

Least common multiple

The **least common multiple** or *lowest common multiple* (LCM) of two numbers is the smallest number that appears in the multiplication tables of both numbers.

For example, the LCM of 3 and 5 is 15, the LCM of 2 and 7 is 14 and the LCM of 6 and 9 is 18.

There are two ways of working out the LCM.

EXAMPLE 5

Find the LCM of 18 and 24.

Write out the 18 times table: 18, 36, 54, (72), 90, 108, … .

Write out the 24 times table: 24, 48, (72), 96, 120, …

You can see that 72 is the smallest (least) number in both (common) tables (multiples).

EXAMPLE 6

Find the LCM of 42 and 63.

Write 42 in prime factor form. $42 = 2 \times 3 \times 7$

Write 63 in prime factor form. $63 = 3^2 \times 7$

Write down the smallest number in prime factor form that includes all the prime factors of 42 and of 63.

You need $2 \times 3^2 \times 7$ (this includes $2 \times 3 \times 7$ and $3^2 \times 7$).

Then work it out.

$2 \times 3^2 \times 7 = 2 \times 9 \times 7 = 18 \times 7 = 126$

The LCM of 42 and 63 is 126.

Highest common factor

The **highest common factor** (HCF) of two numbers is the biggest number that divides exactly into both of them.

For example, the HCF of 24 and 18 is 6, the HCF of 45 and 36 is 9 and the HCF of 15 and 22 is 1.

There are two ways of working out the HCF.

EXAMPLE 7

Find the HCF of 28 and 16.

Write out the factors of 28. 1, 2, (4), 7, 14, 28

Write out the factors of 16. 1, 2, (4), 8, 16

You can see that 4 is the biggest (highest) number in both (common) lists (factors).

EXAMPLE 8

Find the HCF of 48 and 120.

Write 48 in prime factor form. $48 = 2^4 \times 3$

Write 120 in prime factor form. $120 = 2^3 \times 3 \times 5$

Write down the biggest number in prime factor form that is in the prime factors of 48 and 120.

You need $2^3 \times 3$ (this is in both $2^4 \times 3$ and $2^3 \times 3 \times 5$).

Then work it out. $2^3 \times 3 = 8 \times 3 = 24$

The HCF of 48 and 120 is 24.

EXERCISE 2C

1 Find the LCM of each pair of numbers.

a 4 and 5 **b** 7 and 8

c 2 and 3 **d** 4 and 7

e 2 and 5 **f** 3 and 5

g 3 and 8 **h** 5 and 6

2 What connection is there between the LCMs and the pairs of numbers in question **1**?

3 Find the LCM of each pair of numbers.

a 4 and 8 **b** 6 and 9

c 4 and 6 **d** 10 and 15

PS 4 Does the connection you found in question **2** still work for the numbers in question **3**? If not, explain why not.

C

C

5 Find the LCM of each pair of numbers.

a 24 and 56 **b** 21 and 35

c 12 and 28 **d** 28 and 42

e 12 and 32 **f** 18 and 27

g 15 and 25 **h** 16 and 36

FM **6** Cheese slices are sold in packs of eight.

Bread rolls are sold in packs of six.

What is the least number of each pack that needs to be bought to have the same number of cheese slices and bread rolls?

7 Find the HCF of each pair of numbers.

a 24 and 56 **b** 21 and 35

c 12 and 28 **d** 28 and 42

e 12 and 32 **f** 18 and 27

g 15 and 25 **h** 16 and 36

i 42 and 27 **j** 48 and 64

k 25 and 35 **l** 36 and 54

B

PS **8** In prime factor form $1250 = 2 \times 5^4$ and $525 = 3 \times 5^2 \times 7$.

a Which of these are common multiples of 1250 and 525?

i $2 \times 3 \times 5^3 \times 7$

ii $2^3 \times 3 \times 5^4 \times 7^2$

iii $2 \times 3 \times 5^4 \times 7$

iv $2 \times 3 \times 5 \times 7$

b Which of these are common factors of 1250 and 525?

i 2×3

ii 2×5

iii 5^2

iv $2 \times 3 \times 5 \times 7$

AU PS **9** The HCF of two numbers is 6.

The LCM of the same two numbers is 72.

What are the numbers?

2.3 Negative numbers

This section will show you how to:
- multiply and divide positive and negative numbers

Key words
negative
order
positive

Multiplying and dividing with negative numbers

The rules for multiplying and dividing with **negative** numbers are very easy.

- When the signs of the numbers are the same, the answer is **positive**.
- When the signs of the numbers are different, the answer is **negative**.

Here are some examples.

$2 \times 4 = 8$ $\quad 12 \div -3 = -4$

$-2 \times -3 = 6$ $\quad -12 \div -3 = 4$

Negative numbers on a calculator

You can enter a negative number into your calculator and check the result.

Enter –5 by pressing the keys **5** and **(–)**. (You may need to press **(–)** or **–** followed by **5**, depending on the type of calculator that you have.) You will see the calculator shows –5.

Now try these two calculations.

$-8 - 7 \rightarrow$ **8** **(–)** **–** **7** **=** –15

$6 - -3 \rightarrow$ **6** **(–)** **–** **3** **=** 9

EXERCISE 2D

1 Write down the answers to the following.

a -3×5 **b** -2×7

c -4×6 **d** -2×-3

e -7×-2 **f** $-12 \div -6$

g $-16 \div 8$ **h** $24 \div -3$

i $16 \div -4$ **j** $-6 \div -2$

k 4×-6 **l** 5×-2

m 6×-3 **n** -2×-8

o -9×-4

D

D

2 Write down the answers to the following.

a $-3 + -6$ **b** -2×-8 **c** $2 + -5$

d 8×-4 **e** $-36 \div -2$ **f** -3×-6

g $-3 - -9$ **h** $48 \div -12$ **i** -5×-4

j $7 - -9$ **k** $-40 \div -5$ **l** $-40 + -8$

m $4 - -9$ **n** $5 - 18$ **o** $72 \div -9$

3 What number do you multiply –3 by to get the following?

a 6 **b** –90 **c** –45

d 81 **e** 21

4 Evaluate the following.

a $-6 + (4 - 7)$ **b** $-3 - (-9 - -3)$ **c** $8 + (2 - 9)$

5 Evaluate the following.

a $4 \times (-8 \div -2)$ **b** $-8 - (3 \times -2)$ **c** $-1 \times (8 - -4)$

PS 6 Write down six different multiplications that give the answer –12.

PS 7 Write down six different division sums that give the answer –4.

AU 8 **a** Work out: 12×-2

b The average temperature drops by two degrees celsius every day for 12 days. By how much has the temperature dropped altogether?

c The temperature drops by six degrees celsius for the next three days. Write down the calculation to work out the total change in temperature over these three days.

PS 9 Put these calculations in order from lowest result to highest.

-15×4 $-72 \div 4$ $-56 \div -8$ 13×-6

Hierarchy of operations

Reminder: The **order** in which you *must* do mathematical operations should follow the BIDMAS or BODMAS rule.

B	Brackets	B	Brackets
I	Indices (Powers)	O	Order (Powers)
D	Division	D	Division
M	Multiplication	M	Multiplication
A	Addition	A	Addition
S	Subtraction	S	Subtraction

Errors are often made because of negative signs or doing calculations in the wrong order. For example:

$2 + 3 \times 4$ is equal to $2 + 12 = 14$ and **not** 5×4.

-6^2 is **not** the same as $(-6)^2$.

$-6^2 = -(6 \times 6) = -36$ but $(-6)^2 = -6 \times -6 = 36$

EXAMPLE 9

Work out each of the following.

a $(8 - 3^2) \times 9 \div (-1 + 4)$ **b** $5 \times [6^2 + (5 - 8)^2]$

a The brackets are calculated first.

$(8 - 3^2) \times 9 \div (-1 + 4) = (8 - 9) \times 9 \div 3 = -1 \times 9 \div 3 = -9 \div 3 = -3$

b This has nested brackets. The inside (round) bracket is calculated first:

$5 \times [6^2 + (5 - 8)^2] = 5 \times [6^2 + (-3)^2] = 5 \times [36 + 9] = 5 \times 45 = 225$

EXERCISE 2E

D

1 Work out each of these. Remember to work out the brackets first.

a $-2 \times (-3 + 5) =$ **b** $6 \div (-2 + 1) =$ **c** $(5 - 7) \times -2 =$

d $-5 \times (-7 - 2) =$ **e** $-3 \times (-4 \div 2) =$ **f** $-3 \times (-4 + 2) =$

2 Work out each of these.

a $-6 \times -6 + 2 =$ **b** $-6 \times (-6 + 2) =$ **c** $-6 \div 6 - 2 =$

d $12 \div (-4 + 2) =$ **e** $12 \div -4 + 2 =$ **f** $2 \times (-3 + 4) =$

g $-(5)^2 =$ **h** $(-5)^2 =$ **i** $(-1 + 3)^2 - 4 =$

j $-(1 + 3)^2 - 4 =$ **k** $-1 + 3^2 - 4 =$ **l** $-1 + (3 - 4)^2 =$

PS 3 Copy each of these and then put in brackets where necessary to make each one true.

a $3 \times -4 + 1 = -11$ **b** $-6 \div -2 + 1 = 6$ **c** $-6 \div -2 + 1 = 4$

d $4 + -4 \div 4 = 3$ **e** $4 + -4 \div 4 = 0$ **f** $16 - -4 \div 2 = 10$

4 $a = -2$, $b = 3$, $c = -5$.

Work out the values of the following.

a $(a + c)^2$ **b** $-(a + b)^2$ **c** $(a + b)c$ **d** $a^2 + b^2 - c^2$

5 Work out each of the following.

a $(6^2 - 4^2) \times 2$ **b** $9 \div (1 - 4)^2$

c $2 \times [8^2 - (2 - 7)^2]$ **d** $[(3 + 2)^2 - (5 - 6)^2] \div 6$

D

AU 6 Use each of the numbers 2, 3 and 4 and each of the symbols –, × and ÷ to make a calculation with an answer –6.

7 Use any four different numbers to make a calculation with answer –8.

C

PS 8 Use the numbers 5, 6, 7, 8 and 9 in order, from smallest to largest, together with one of each of the symbols + , –, × ÷ and two pairs of brackets to make a calculation with an answer of $\frac{25}{8}$.

For example, making a calculation with an answer of $\frac{43}{9}$:

$(5 + 6) - (7 \times 8) \div 9 = \frac{43}{9}$

PS 9 This magic square is made up of negative and positive numbers.

–8	–1	–6
–3	–5	–7
–4	–9	–2

a What is the total of each row, column and diagonal. (Called the magic number).

b You can make your own magic square using negative and/or positive numbers by picking a value for x and a and substituting them into the values in the square on the left below.

$x + 7a$	x	$x + 5a$
$x + 2a$	$x + 4a$	$x + 6a$
$x + 3a$	$x + 8a$	$x + a$

–5	2	–3
0	–2	–4
–1	–6	1

For example if you pick $x = 2$ and $a = -1$, you get the square on the right which has a magic number of –6.

You can look up magic squares on the internet.

A*

10 Try working out $\sqrt{(-100)}$ on your calculator. Depending on the type of calculator you will get an error message or 10i.

The square root of a negative number doesn't exist but mathematicians found that they needed it to do some calculations so they 'invented' it. This was first done in 1572. They called the square root of –1 i which stands for the imaginary number. So $i^2 = -1$. You might think that this is a bit pointless but planes would not fly and rockets would not get to the moon without this idea.

$\sqrt{-4} = \pm 2i$, $\sqrt{-9} = \pm 3i$ etc…

Write down or work out

a $(4i)^2$ **b** $(6i)^2$ **c** $\sqrt{-100}$ **d** $\sqrt{-144}$ **e** i^4

f i^6 **g** i^3 **h** $(-i)^2$ **i** $(-5i)^2$ **j** $(-2i)^4$

GRADE BOOSTER

D You can recognise and work out multiples, factors and primes

D You can multiply and divide with negative numbers

C You can write a number as the product of its prime factors

C You can work out the LCM and HCF of pairs of numbers

What you should know now

- How to write a number in prime factor form and find LCMs and HCFs
- How to find the square roots of some decimal numbers
- How to add, subtract, multiply and divide negative numbers
- Find multiples, factors of numbers
- The square numbers, triangle number series and the prime numbers up to 50

EXAMINATION QUESTIONS

1 **a** Write 42 as the product of its prime factors.

b Find the least common multiple of 28 and 42.

2 As the product of prime factors $60 = 2^2 \times 3 \times 5$

a What number is represented by $2 \times 3^2 \times 5$?

b Find the least common multiple (LCM) of 60 and 48.

c Find the highest common factor (HCF) of 60 and 78.

3 Write 40 as the product of prime factors.

Give your answer in index form. *(3 marks)*

AQA, June 2009, Module 3 Higher, Question 4

4 The Least Common Multiple (LCM) of two numbers is 36.

Find one possible pair for the two numbers. *(1 mark)*

AQA, November 2007, Module 3 Higher, Question 5

5 Mary set up her Christmas tree with two sets of twinkling lights.

Set A would twinkle every 3 seconds.

Set B would twinkle every 4 seconds.

How many times in a minute will both sets be twinkling at the same time?

6 Emily, Jack and Mia are tapping on the desk in time to Tom clapping his hands to a regular beat.

Emily taps on every fifth clap.

Jack taps on every sixth clap.

Mia taps on every eighth clap.

They all start together on the first clap.

How many claps will it be before Emily, Jack and Mia all tap at the same time again?

7 **a** You are given that $8x^3 = 1000$. Find the value of x.

b Write 150 as the product of its prime factors.

8 **a** p and q are prime numbers such that $pq^3 = 250$

Find the values of p and q.

b Find the highest common factor of 250 and 80.

9 **a** a and b are prime numbers such that $(ab)^3 = 1000$. What are the values of a and b?

b p and q are integers such that $(pq)^3 = 216$. Explain why it is possible to find values of p and q.

10 **a** Two numbers a and b are less than 15. The Least Common Multiple (LCM) of a and b is 24.

Give one possible pair for the numbers, a and b. *(2 marks)*

b The numbers x and y are between 10 and 20. The Highest Common Factor (HCF) of x and y is 3.

Give one possible pair for the two numbers, x and y. *(2 marks)*

AQA, November 2011 F2 Question 26

11 The difference between the squares of two whole numbers is sometimes a prime number.

For example $5^2 - 2^2 = 21$ and 21 is not prime

but $4^2 - 3^2 = 7$ and 7 is prime

a Find a different example where the answer is not prime. *(1 mark)*

b Find a different example where the answer is prime. *(1 mark)*

AQA, November 2010 F1 Question 18

12 $A = -9$ and $B = 12$

Work out the value of $\frac{4(A + 3)}{B}$ *(2 marks)*

June 2010 H1 Question 1

13 p is a prime number.

a is the expression $p^2 + 6$ always even, always odd or could it be either odd and even? *(1 mark)*

b n and p are both prime numbers. Work out values of n and p so that $n = p^2 + 6$ *(2 marks)*

AQA, June 2010 F2 Question 26

14 Work out the value of $\frac{a(3b + 1)}{5}$ when $a = -2$ and $b = 3$. *(2 marks)*

AQA, November 2009 H1 Question 2

15 Show that $\sqrt{72}$ is between 8 and 9.

AQA, November 2009 F1 Question 17

16 $A = 6$ and $B = -7$

Work out the value of $\frac{A(B + 2)}{3}$.

AQA, May 2009 H1 Question 2

17 **a** Find the Highest Common Factor (HCF) of 8 and 12. *(2 marks)*

b Find the Least Common Multiple (LCM) of 8 and 12. *(2 marks)*

AQA, November 2008 H1 Question 12

Worked Examination Questions

1 **a** Work out the LCM of 24 and 40

b Work out the HCF of 24 and 40

1 a 24, 48, 72, 96, 120, ….
40, 80, 120, ….
LCM = 120

Write out the times tables for both 24 and 40. This will get 1 method mark.

Pick out the biggest (Highest) number (Multiple) in both lists (Common). This gets 1 accuracy mark

b Factors of 24 = {1, 2, 3, 4, 6, 8, 12, 24}
Factors of 40 = {1, 2, 4, 5, 8, 10, 20, 40}
HCF = 8

Write out the factors for both 24 and 40. This will get 1 method mark.

Pick out the biggest (Highest) number (Factor) in both lists (Common). This gets 1 accuracy mark

Total: 4 marks

2 As the product of prime factors $45 = 3 \times 3 \times 5$ and $70 = 2 \times 5 \times 7$

a Explain why the HCF of 45 and 75 is 5

b Work out the LCM of 45 and 75

2 a The only number in both products of prime factors is 5.

The HCF will be the product of any numbers that are in both sets of products. This is 1 mark

b LCM = 2 x 3 x 3 x 5 x 7

The LCM will be the product of all the **different** numbers that are in both sets of products. This is 1 method mark

= 630

Work out the answer. This is 1 accuracy mark

Total: 3 marks

3 x and y are prime numbers such that $xy^2 = 363$

Work out the values of x and y.

3 363 ÷ 3 = 121

Try to break 363 into a product. 3 is an obvious choice. This is 1 method mark

$\sqrt{121} = 11$

Take the square root of the answer This is 1 method mark

x = 3 and y = 11

Write down both answers. This is 1 accuracy mark.

Total: 3 marks

2 Functional Maths
Creating your own allotment

Today, many people living in towns and cities do not have their own gardens. Allotments give these people the opportunity to enjoy gardening regardless of not having their own garden. Allotments are also increasingly popular as a way of producing home-grown cheap fruit and vegetables. You have just started renting an allotment so that can grow your own fruit and vegetables. The allotment is divided into rows for planting. The whole plot is 3 m wide and each row is 5 m long.

Your task

1 Design the plant layout for the allotment using the information that you will gather from the table. Consider as many different arrangements as possible.

You must explain your choices, stating the assumptions that you have made.

2 You have a shed at your allotment, one wall of which you can use for storage. The wall is 3 m long and 3 m high.

You store most of your gardening tools and equipment in these boxes in your shed.

Type of box	Length (cm)	Width (cm)	Height (cm)	Number of boxes
Flower seeds	10 cm	10 cm	10 cm	10
Vegetable seeds	15 cm	15 cm	10 cm	10
Wire, string, nails	20 cm	15 cm	15 cm	5
Gloves	30 cm	20 cm	15 cm	2
Hand tools (trowels, etc.)	50 cm	40 cm	30 cm	2
Tool attachments	80 cm	70 cm	50 cm	1

You are putting up shelves in your shed. Show how many shelves you would put up and how you would arrange the boxes to fit as many into your shed as possible.

You should consider these points when you design your shelf layout:

- space between shelves
- height from the floor
- where you will place each box
- the convenience of removing items from the boxes
- place the largest box away from the door as it sticks out the most.

Facts

This table shows how much distance is needed between plants and between rows for the key vegetables in your allotment.

Vegetable	Distance between plants (cm)	Distance between rows (cm)
Potatoes	30 cm	50 cm
Carrots	10 cm	20 cm
Broad beans	10 cm	30 cm
Onions	10 cm	20 cm
Cucumber	30 cm	60 cm
Lettuce	20 cm	30 cm

Handy hints

You may find these rules helpful when planning your allotment:

- use one type of vegetable for each complete row
- do not have more than two rows of the same vegetable
- if different plants are next to each other, make sure that the largest necessary distance between rows is left between them
- make sure that vegetables are not planted too near to the edge of the allotment
- use graph paper to represent the allotment
- use a code to represent each plant.

Why this chapter matters

Percentages are used in many places and situations in our everyday lives.

Why use fractions and percentages?

Because:

- basic percentages and simple fractions are quite easy to understand
- they are a good way of comparing quantities
- fractions and percentages are used a lot in everyday life.

Who uses them?

- Shops and businesses
  - Sale → Save $\frac{1}{4}$ off the marked price
  - Special offer → 10% off
- Banks
  - Loans → Interest rate $6\frac{1}{4}$%
  - Savings → Interest rate $2\frac{1}{2}$%
- Salespeople
  - $7\frac{1}{2}$% commission on sales
- Government
  - Half of the workers in this sector are over 55
  - The aim is to cut carbon emissions by one-third by 2020
  - Unemployment has fallen by 1%
  - Inflation is 3.6%
  - Income tax is 20%
  - Value added tax is 20%
- Workers
  - My pay rise is 2.3%
  - My overtime rate is time and a half
- Teachers
  - Test result 67%
  - Three-fifths of our students gain a grade C in GCSE mathematics

Can you think of other examples? You will find several everyday uses in this chapter.

Chapter 3 Number: Fractions, percentages and ratios

This chapter will show you ...

- to D C how to apply the rules of addition and subtraction to fractions
- to D C how to use ratios to solve problems
- D how to increase or decrease a quantity by a percentage
- C how to calculate compound interest
- C how to express one quantity as a percentage of another
- C how to calculate a percentage increase or decrease
- B how to calculate the original value after a percentage increase or decrease

Visual overview

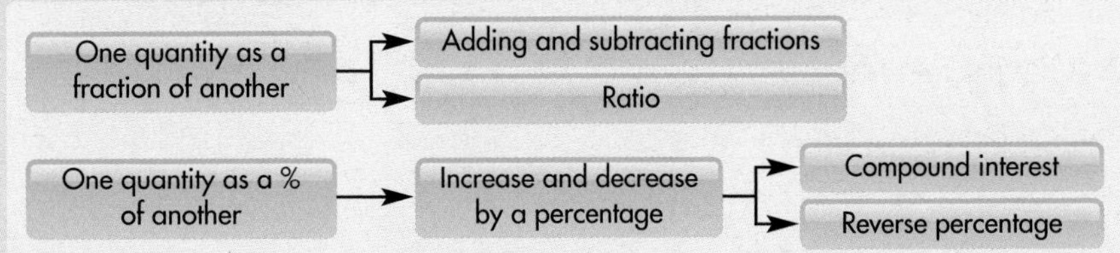

What you should already know

- How to cancel down fractions to their simplest form **(KS3 level 5, GCSE grade F)**
- How to find equivalent fractions, decimals and percentages **(KS3 level 5, GCSE grade F)**
- How to add and subtract fractions with the same denominator **(KS3 level 6, GCSE grade F)**
- How to work out simple percentages, such as 10%, of quantities **(KS3 level 4, GCSE grade F)**
- How to convert a mixed number to a top-heavy fraction and vice versa **(KS3 level 6, GCSE grade F)**

Quick check

1 Cancel the following fractions to their simplest form.

a $\frac{8}{20}$ **b** $\frac{12}{32}$ **c** $\frac{15}{35}$

2 Complete this table of equivalences.

Fraction	Percentage	Decimal
$\frac{3}{4}$		
	40%	
		0.55

3 What is 10% of:

a £230 **b** £46.00 **c** £2.30?

3.1 One quantity as a fraction of another

This section will show you how to:
- find one **quantity** as a fraction of another

Key words
fraction
quantity

An amount often needs to be given as a **fraction** of another amount.

EXAMPLE 1

Write £5 as a fraction of £20.

As a fraction this is written $\frac{5}{20}$. This cancels to $\frac{1}{4}$.

So £5 is one-quarter of £20.

EXERCISE 3A

D

1 Write the first quantity as a fraction of the second.

a 2 cm, 6 cm **b** 4 kg, 20 kg **c** £8, £20 **d** 5 hours, 24 hours

e 12 days, 30 days **f** 50p, £3 **g** 4 days, 2 weeks **h** 40 minutes, 2 hours

2 In a form of 30 students, 18 are boys. What fraction of the form are boys?

3 During March, it rained on 12 days. For what fraction of the month did it rain?

FM **4** Linda earns £120 a week. She saves $\frac{1}{4}$ of her earnings. She is saving for the deposit on a car of £600. How many weeks will it take until she has saved enough?

AU **5** Jon earns £90 and saves £30 of it. Matt earns £100 and saves £35 of it.

Who is saving the greater proportion of their earnings?

AU **6** In two tests Harry gets 13 out of 20 and 16 out of 25. Which is the better mark? Explain your answer.

C

7 Frank gets a pay rise from £120 a week to £135 a week. What fraction of his original pay was his pay rise?

8 When she was born Alice weighed 3 kg. After a month she weighed 4 kg 250 g. By what fraction of what she originally weighed had she increased?

9 After the breeding season a bat colony increased in size from 90 bats to 108 bats. What fraction had the size of the colony increased?

10 After dieting Bart went from 80 kg to 68 kg. What fraction did his weight decrease by?

AU **11** In a class of 30 students, 18 are boys. Half of the boys study French.

What fraction of the class are boys who study French?

Give your answer in its simplest form.

PS **12** The manager of a small company claims that three out of every four of her workers are women.

She employs between 30 and 40 workers altogether.

If her statement is true, how many workers could she have altogether? Write down all possible answers.

3.2 Adding and subtracting fractions

This section will show you how to:

- add and subtract fractions with different denominators

Key words

denominator
equivalent fraction
lowest common denominator

Fractions can only be added or subtracted after they have been changed to **equivalent fractions** with the same **denominator**.

EXAMPLE 2

Work out: $\frac{5}{6} - \frac{3}{4}$

The **lowest common denominator** (LCM of 4 and 6) is 12.

The problem becomes $\frac{5}{6} - \frac{3}{4} = \frac{5}{6} \times \frac{2}{2} - \frac{3}{4} \times \frac{3}{3} = \frac{10}{12} - \frac{9}{12} = \frac{1}{12}$

EXAMPLE 3

Work out: **a** $2\frac{1}{3} + 3\frac{5}{7}$ **b** $3\frac{1}{4} - 1\frac{3}{5}$

The best way to deal with addition and subtraction of mixed numbers is to deal with the whole numbers and the fractions separately.

a $2\frac{1}{3} + 3\frac{5}{7} = 2 + 3 + \frac{1}{3} + \frac{5}{7} = 5 + \frac{7}{21} + \frac{15}{21} = 5 + \frac{22}{21} = 5 + 1\frac{1}{21} = 6\frac{1}{21}$

b $3\frac{1}{4} - 1\frac{3}{5} = 3 - 1 + \frac{1}{4} - \frac{3}{5} = 2 + \frac{5}{20} - \frac{12}{20} = 2 - \frac{7}{20} = 1\frac{13}{20}$

EXERCISE 3B

D

1 Work out the following.

a $\frac{1}{3}+\frac{1}{5}$ **b** $\frac{1}{3}+\frac{1}{4}$ **c** $\frac{1}{5}+\frac{1}{10}$ **d** $\frac{2}{3}+\frac{1}{4}$ **e** $\frac{1}{5}-\frac{1}{10}$ **f** $\frac{7}{8}-\frac{3}{4}$

g $\frac{5}{6}-\frac{3}{4}$ **h** $\frac{5}{6}-\frac{1}{2}$ **i** $\frac{1}{3}+\frac{4}{9}$ **j** $\frac{1}{4}+\frac{3}{8}$ **k** $\frac{7}{8}-\frac{1}{2}$ **l** $\frac{3}{5}-\frac{8}{15}$

C

2 Work out the following.

a $1\frac{7}{18}+2\frac{3}{10}$ **b** $3\frac{1}{3}+1\frac{9}{20}$ **c** $1\frac{1}{8}-\frac{5}{9}$ **d** $1\frac{3}{16}-\frac{7}{12}$

e $\frac{5}{6}+\frac{7}{16}+\frac{5}{8}$ **f** $\frac{7}{10}+\frac{3}{8}+\frac{5}{6}$ **g** $1\frac{1}{3}+\frac{7}{10}-\frac{4}{15}$ **h** $\frac{5}{14}+1\frac{3}{7}-\frac{5}{12}$

3 In a class of children, three-quarters are Chinese, one-fifth are Malay and the rest are Indian. What fraction of the class are Indian?

4 **a** In a class election, half of the students voted for Aminah, one-third voted for Janet and the rest voted for Peter. What fraction of the class voted for Peter?

PS **b** One of the following is the number of students in the class.

25 28 30 32

How many students are in the class?

FM **5** A one-litre bottle of milk is used to fill three glasses with a capacity of an eighth of a litre and one glass with a capacity of a half a litre.

Priya likes milky coffee so has at least 10 cl of milk in each cup. Is there enough milk left for Priya to have two cups of coffee?

6 Because of illness, $\frac{2}{5}$ of a school was absent one day. If the school had 650 students on the register, how many were absent that day?

AU **7** Which is the biggest: half of 96, one-third of 141, two-fifths of 120, or three-quarters of 68?

AU **8** Mick says that $1\frac{1}{3}+2\frac{1}{4}=3\frac{2}{7}$

He is incorrect. What is the mistake that he has made? Work out the correct answer.

AU **9** Here is a calculation.

$\frac{1}{4}+\frac{2}{5}$

Imagine that you are trying to explain to someone over the telephone how to do this calculation.

Write down what you would say.

10 To increase sales, a shop reduced the price of a car stereo set by $\frac{2}{5}$. If the original price was £85, what was the new price?

B

PS **11** At a burger-eating competition, Lionel ate 34 burgers in 20 minutes while Brian ate 26 burgers in 20 minutes. How long after the start of the competition would they have consumed a total of 30 burgers between them?

3.3 Increasing and decreasing quantities by a percentage

This section will show you how to:
- increase and decrease quantities by a percentage

Key words
multiplier
percentage

Increasing by a percentage

There are two methods for increasing a quantity by a **percentage**.

Method 1

Work out the increase and add it on to the original amount.

EXAMPLE 4

Increase £6 by 5%.

Work out 5% of £6: (5 ÷ 100) × 6 = £0.30

Add the £0.30 to the original amount: £6 + £0.30 = £6.30

Method 2

Use a **multiplier**. An increase of 6% is equivalent to the original 100% plus the extra 6%. This is a total of 106% ($\frac{106}{100}$) and is equivalent to the multiplier 1.06.

EXAMPLE 5

Increase £6.80 by 5%.

A 5% increase is a multiplier of 1.05.

So £6.80 increased by 5% is £6.80 × 1.05 = £7.14

EXERCISE 3C 

D

1 What multiplier is used to increase a quantity by:

a 10% **b** 3% **c** 20% **d** 7% **e** 12%?

2 Increase each of the following by the given percentage. (Use any method you like.)

a £60 by 4% **b** 12 kg by 8% **c** 450 g by 5%
d 545 m by 10% **e** £34 by 12% **f** £75 by 20%
g 340 kg by 15% **h** 670 cm by 23% **i** 130 g by 95%
j £82 by 75% **k** 640 m by 15% **l** £28 by 8%

D

FM **3** Kevin is on a salary of £27 500. He is offered a pay rise of 7% or an extra £150 per month. Which should he accept? Show how you decided.

4 In 2005 the population of Melchester was 1 565 000. By 2010 it had increased by 8%. What was the population of Melchester in 2010?

5 A small firm made the same pay increase of 5% for all its employees.

a Calculate the new pay of each employee listed below. Each of their salaries before the increase is given.

Bob, caretaker, £16 500

Anne, tea lady, £17 300

Jean, supervisor, £19 500

Brian, manager, £25 300

AU **b** Explain why the actual pay increases are different for each employee.

6 A bank pays 7% interest on the money that each saver keeps in the bank for a year. Allison keeps £385 in the bank for a year. How much will she have in the bank after the year?

7 In 1980 the number of cars on the roads of Sheffield was about 102 000. Since then it has increased by 90%. Approximately how many cars are there on the roads of Sheffield now?

8 An advertisement for a breakfast cereal states that a special-offer packet contains 15% more cereal for the same price as a normal 500 g packet. How much breakfast cereal is in a special-offer packet?

9 A headteacher was proud to point out that, since he had arrived at the school, the number of students had increased by 35%. How many students are now in the school, if there were 680 when the headteacher started at the school?

10 At a school disco there are always about 20% more girls than boys. If at one disco there were 50 boys, how many girls were there?

FM **11** The Government adds a tax called VAT to the price of most goods in shops. At the moment, it is 20% on all electrical equipment.

Calculate the price of the following electrical equipment after VAT of 20% has been added.

Equipment	*Pre-VAT price*
TV set	£245
Microwave oven	£72
CD player	£115
Personal stereo	£29.50

PS **12** A TV cost £400 before VAT was added.

The VAT rate went from 17.5% to 20% in January 2011.

How much did the cost of the TV increase by?

C

AU 13 Bookshop BookWorms increased its prices by 5%, then increased them by 3%. Bookshop Books Galore increased its prices by 3%, then increased them by 5%.

Which shop's prices increased by the greatest percentage?

a BookWorms **b** Books Galore **c** Both same **d** Cannot tell

Justify your choice.

AU 14 Shop A increased its prices by 4% and then by another 4%. Shop B increased its prices by 8%.

Which shop's prices increased by the greatest percentage?

a Shop A **b** Shop B **c** Both same **d** Cannot tell

Justify your choice.

AU 15 A hi-fi system was priced at £420 at the start of 2008. At the start of 2009, it was 12% more expensive. At the start of 2010, it was 15% more expensive than the price at the start of 2009. What is the price of the hi-fi at the start of 2010?

B

FM 16 The VAT rate in Spain is 18%. A quick way to work out VAT is to divide the pre-VAT price by 6. For example, the VAT on an item costing €120 is approximately €120 ÷ 6 = €20. Show that this approximate method gives the VAT correct to within €10 for pre-VAT prices up to €600.

Decreasing by a percentage

There are two methods for decreasing by a percentage.

Method 1

Work out the decrease and subtract it from the original amount.

EXAMPLE 6

Decrease £8 by 4%.

Work out 4% of £8: (4 ÷ 100) × 8 = £0.32

Subtract the £0.32 from the original amount: £8 – £0.32 = £7.68

Method 2

Use a multiplier. A 7% decrease is equivalent to 7% less than the original 100%, so it represents 100% – 7% = 93% of the original. This is a multiplier of 0.93.

EXAMPLE 7

Decrease £8.60 by 5%.

A decrease of 5% is a multiplier of 0.95.

So £8.60 decreased by 5% is £8.60 × 0.95 = £8.17

EXERCISE 3D 

D

1 What multiplier is used to decrease a quantity by:

a 8% **b** 15% **c** 25% **d** 9% **e** 12%?

2 Decrease each of the following by the given percentage. (Use any method you like.)

a £10 by 6% **b** 25 kg by 8% **c** 236 g by 10%

d 350 m by 3% **e** £5 by 2% **f** 45 m by 12%

g 860 m by 15% **h** 96 g by 13% **i** 480 cm by 25%

j 180 minutes by 35% **k** 86 kg by 5% **l** £65 by 42%

3 A car valued at £6500 last year is now worth 15% less. What is its value now?

4 A new P-plan diet guarantees that you will lose 12% of your weight in the first month. How much should the following people weigh after one month on the diet?

a Gillian, who started at 60 kg **b** Peter, who started at 75 kg

c Margaret, who started at 52 kg

FM **5** A motor insurance firm offers no-claims discounts off the full premium, as follows.

1 year no claims	15% discount off the full premium
2 years no claims	25% discount off the full premium
3 years no claims	45% discount off the full premium
4 years no claims	60% discount off the full premium

Mr Speed and his family are all offered motor insurance from this firm.

Mr Speed has four years' no-claims discount and the full premium would be £440.
Mrs Speed has one year's no-claims discount and the full premium would be £350.
James has three years' no-claims discount and the full premium would be £620.
John has two years' no-claims discount and the full premium would be £750.

Calculate the actual amount each member of the family has to pay for the motor insurance.

6 A large factory employed 640 people. It had to streamline its workforce and lose 30% of the workers. How big is the workforce now?

7 On the last day of the Christmas term, a school expects to have an absence rate of 6%. If the school population is 750 students, how many students will the school expect to see on the last day of the Christmas term?

8 A charity called *Young Ones* said that since the start of the National Lottery they have had a decrease of 45% in the amount of money raised by scratch cards. If before the start of the National Lottery the charity had an annual income of £34 500 from their scratch cards, how much do they collect now?

D

9 Most speedometers in cars have an error of up to 5% below the true reading. When my speedometer says I am driving at 70 mph, what is the lowest speed I could be doing?

FM **10** Kerry wants to buy a sweatshirt (£19), a tracksuit (£26) and some running shoes (£56). If she joins the store's premium club which costs £25 to join she can get 20% off the cost of the goods.

Should she join or not? Give figures to support your answer.

C

FM **11** **a** I read an advertisement in my local newspaper last week which stated: "By lagging your roof and hot water system you will use 18% less fuel." Since I was using an average of 640 units of gas a year, I thought I would lag my roof and my hot water system. How much gas would I expect to use now?

AU **b** I actually used 18% more gas than I expected to use.

Did I use less gas than last year, more gas than last year or the same amount of gas as last year?

Show how you work out your answer.

12 Shops add VAT to the basic price of goods to find the selling price that customers will be asked to pay. In a sale, a shop reduces the selling price by a certain percentage to set the sale price. Calculate the sale price of each of these items.

Item	Basic price	VAT rate	Sale discount	Sale price
TV	£220	20%	14%	
DVD player	£180	20%	20%	

AU PS **13** A shop advertises garden ornaments at £50 but with 10% off in a sale. It then advertises an extra 10% off the sale price.

Show that this is not a decrease in price of 20%.

AU **14** A computer system was priced at £1000 at the start of 2008. At the start of 2009, it was 10% cheaper. At the start of 2010, it was 15% cheaper than the price at the start of 2009. What is the price of the computer system at the start of 2010?

B

PS **15** Show that a 10% decrease followed by a 10% increase is equivalent to a 1% decrease overall.

HINTS AND TIPS

Choose an amount to start with.

PS **16** A cereal packet normally contains 300 g of cereal and costs £1.40.

There are two special offers.

Offer A: 20% more for the same price

Offer B: Same amount for 20% off the normal price

Which is the better offer?

a Offer A **b** Offer B **c** Both same **d** Cannot tell

Justify your choice.

3.4 Expressing one quantity as a percentage of another

This section will show you how to:

- express one quantity as a percentage of another
- work out percentage change

Key words
percentage change
percentage decrease
percentage increase
percentage loss
percentage profit

You express one quantity as a percentage of another by setting up the first quantity as a fraction of the second, making sure that the *units of each are the same*. Then you convert the fraction into a percentage by multiplying by 100%.

EXAMPLE 8

Express £6 as a percentage of £40.

Set up the fraction and multiply by 100%.

$$\frac{6}{40} \times 100\% = 15\%$$

EXAMPLE 9

Express 75 cm as a percentage of 2.5 m.

First, change 2.5 m to 250 cm to get a common unit.

So, the problem now becomes: Express 75 cm as a percentage of 250 cm.

Set up the fraction and multiply by 100%.

$$\frac{75}{250} \times 100\% = 30\%$$

Percentage change

A **percentage change** may be a **percentage increase** or a **percentage decrease**.

$$\text{Percentage change} = \frac{\text{change}}{\text{original amount}} \times 100$$

Use this to calculate **percentage profit** or **percentage loss** in a financial transaction.

EXAMPLE 10

Jake buys a car for £1500 and sells it for £1800. What is Jake's percentage profit?

Jake's profit is £300, so his percentage profit is:

$$\text{percentage profit} = \frac{\text{profit}}{\text{original amount}} \times 100 = \frac{300}{1500} \times 100\% = 20$$

Using a multiplier (or decimal)

To use a multiplier, divide the increase by the original quantity and change the resulting decimal to a percentage.

EXAMPLE 11

Express 5 as a percentage of 40.

Set up the fraction or decimal: $5 \div 40 = 0.125$

Convert the decimal to a percentage: $0.125 = 12.5\%$

EXERCISE 3E

D

1 Express each of the following as a percentage. Give suitably rounded figures where necessary.

- **a** £5 of £20
- **b** £4 of £6.60
- **c** 241 kg of 520 kg
- **d** 3 hours of 1 day
- **e** 25 minutes of 1 hour
- **f** 12 m of 20 m
- **g** 125 g of 600 g
- **h** 12 minutes of 2 hours
- **i** 1 week of a year
- **j** 1 month of 1 year
- **k** 25 cm of 55 cm
- **l** 105 g of 1 kg

2 Liam went to school with his pocket money of £2.50. He spent 80p at the tuck shop. What percentage of his pocket money had he spent?

3 In Greece, there are 3 654 000 acres of agricultural land. Olives are grown on 237 000 acres of this land. What percentage of the agricultural land is used for olives?

4 During the wet year of 1981, it rained in Manchester on 123 days of the year. What percentage of days were wet?

C

5 Find the percentage profit on the following. Give your answers to one decimal place.

	Item	*Retail price* (selling price)	*Wholesale price* (price the shop paid)
a	CD player	£89.50	£60
b	TV set	£345.50	£210
c	Computer	£829.50	£750

6 Before Anton started to diet, he weighed 95 kg. He now weighs 78 kg. What percentage of his original weight has he lost?

7 In 2009 the Melchester County Council raised £14 870 000 in council tax. In 2010 it raised £15 597 000 in council tax. What was the percentage increase?

8 When Blackburn Rovers won the championship in 1995, they lost only four of their 42 league games. What percentage of games did they *not* lose?

C

AU **9** In the year 1900 Britain's imports were as follows.

British Commonwealth	£109 530 000
USA	£138 790 000
France	£53 620 000
Other countries	£221 140 000

a What percentage of the total imports came from each source? Give your answers to 1 decimal place.

b Add up your answers to part **a**. What do you notice? Explain your answer.

AU **10** Calum and Stacey take the same tests. Both tests are out of the same mark.

Here are their results.

	Test A	Test B
Calum	12	17
Stacey	14	20

Whose result has the greater percentage increase from test A to test B? Show your working.

FM **11** A supermarket advertises its cat food as shown.

8 out of 10 cat owners choose our cat food.

Trading standards are checking the claim.

They observe that over one hour, 46 people buy cat food and 38 buy the store's own brand.

Based on these figures is the store's claim correct?

3.5 Compound interest and repeated percentage change

This section will show you how to:
- calculate compound interest
- solve problems involving repeated percentage change

Key words
annual rate
compound interest
multiplier
principal

Banks and building societies usually pay **compound interest** on savings accounts.

When compound interest is used, the interest earned each year is added to the original amount (**principal**) and the new total then earns interest at the **annual rate** in the following year. This pattern is then repeated each year while the money is in the account.

The most efficient way to calculate the total amount in the account after several years is to use a **multiplier**.

EXAMPLE 12

Elizabeth invests £400 in a savings account. The account pays compound interest at 6% each year. How much will she have in the account after three years?

The amount in the account increases by 6% each year, so the multiplier is 1.06.

After 1 year she will have £400 × 1.06 = £424
After 1 year she will have £424 × 1.06 = £449.44
After 1 year she will have £449.44 × 1.06 = £476.41 (rounded)

If you calculate the differences, you can see that the amount of interest increases each year (£24, £25.44 and £26.97).

From this example, you should see that you could have used £400 × $(1.06)^3$ to find the amount after three years. That is, you could have used the following formula for calculating the total amount due at any time:

$$\text{total amount} = P \times \text{multiplier raised to the power } n = P \times \left(1 + \frac{r}{100}\right)^n$$

or $A = P(1 + r)^n$ where P is the original amount invested, r is the percentage interest rate, giving a multiplier of $(1 + r)$, and n is the number of years for which the money is invested.

So, in Example 11, P = £400, $r = 0.06$ and $n = 3$,

and the total amount = £400 × $(1.06)^3$

Using your calculator

You may have noticed that you can do the above calculation on your calculator without having to write down all the intermediate steps.

To add on the 6% each time, just multiply by 1.06 each time. So you can do the calculation as:

4 0 0 × 1 . 0 6 × 1 . 0 6 × 1 . 0 6 =

or

4 0 0 × 1 . 0 6 $x^{\blacksquare}$ 3 =

or

4 0 0 × 1 0 6 % $x^{\blacksquare}$ 3 =

You need to find the method with which you are comfortable and which you understand.

The idea of compound interest does not only concern money. It can be about, for example, the growth in population, increases in salaries, or increases in body weight or height. Also, the idea can involve regular reduction by a fixed percentage: for example, car depreciation, population losses and even water losses. The next exercise shows the extent to which compound interest ideas are used.

EXERCISE 3F

C

1 A baby octopus increases its body weight by 5% each day for the first month of its life. In a safe ocean zoo, a baby octopus was born weighing 10 g.

a What was its weight after:

i 1 day **ii** 2 days **iii** 4 days **iv** 1 week?

b After how many days will the octopus first weigh over 15 g?

2 A certain type of conifer hedging increases in height by 17% each year for the first 20 years. When I bought some of this hedging, it was all about 50 cm tall. How long will it take to grow 3 m tall?

3 The manager of a small family business offered his staff an annual pay increase of 4% for every year they stayed with the firm.

a Gareth started work at the business on a salary of £12 200. What salary will he be on after 4 years?

PS **b** Julie started work at the business on a salary of £9350. How many years will it be until she is earning a salary of over £20 000?

4 Scientists have been studying the shores of Scotland and estimate that due to pollution the seal population of those shores will decline at the rate of 15% each year. In 2006 they counted about 3000 seals on those shores.

a If nothing is done about pollution, how many seals did they expect to be there in

i 2007 **ii** 2008 **iii** 2011?

PS **b** How long will it take for the seal population to be less than 1000?

5 I am told that if I buy a new car its value will depreciate at the rate of 20% each year. If I bought a car in 2009 priced at £8500, what would be the value of the car in:

a 2010 **b** 2011 **c** 2013?

6 At the peak of a drought during the summer, a reservoir in Derbyshire was losing water at the rate of 8% each day. On 1 August this reservoir held 2.1 million litres of water.

a At this rate of losing water, how much would have been in the reservoir on the following days?

i 2 August **ii** 4 August **iii** 8 August

FM **b** The danger point is when the water drops below 1 million litres. When would this have been if things had continued as they were?

7 The population of a small country, Yebon, was only 46 000 in 2001, but it steadily increased by about 13% each year during the 2000s.

a Calculate the population in:

i 2002 **ii** 2006 **iii** 2010.

PS **b** If the population keeps growing at this rate, when will it be half a million?

PS **8** How long will it take to accumulate one million pounds in the following situations?

a An investment of £100 000 at a rate of 12% compound interest

b An investment of £50 000 at a rate of 16% compound interest

PS **9** An oak tree is 60 cm tall. It grows at a rate of 8% per year. A conifer is 50 cm tall. It grows at a rate of 15% per year. How many years does it take before the conifer is taller than the oak?

PS **10** A tree increases in height by 18% per year. When it is 1 year old, it is 8 cm tall. How long will it take the tree to grow to 10 m?

PS **11** Show that a 10% increase followed by a 10% increase is equivalent to a 21% increase overall.

FM AU **12** Here are two advertisements for savings accounts.

Bradley Bank	Monastery Building Society
Invest £1000 for two years and earn 3.2% interest overall.	Invest £1000. Interest rate 1.3% compound per annum. Bonus of 0.5% on balance after 2 years.

Which account is worth more after 2 years?

You **must** show your working.

PS **13** A fish weighs 3 kg and increases in weight by 10% each month. A crab weighs 6 kg but decreases in weight by 10% each month. After how many months will the fish weigh more than the crab?

FM **14** There is a bread shortage.

Each week during the shortage a shop increases its price of bread by 20% of the price the week before.

After how many weeks would the price of the bread have doubled?

PS **15** In a survey exactly 35% of the people surveyed wanted a new supermarket.

What is the least number that could have been surveyed?

3.6 Reverse percentage (working out the original quantity)

This section will show you how to:

- calculate the original amount, given the final amount, after a known percentage increase or decrease

Key words
final amount
multiplier
original amount
unitary method

Reverse percentage questions involve working backwards from the **final amount** to find the **original amount** when you know, or can work out, the final amount as a percentage of the original amount.

Method 1: The unitary method

The **unitary method** has three steps.

Step 1: Equate the final percentage to the final value.

Step 2: Use this to calculate the value of 1%.

Step 3: Multiply by 100 to work out 100% (the original value).

EXAMPLE 13

In a factory, 70 workers were given a pay rise. This was 20% of all the workers. How many workers are there altogether?

20% represents 70 workers.

Divide by 20.
1% represents 70 ÷ 20 workers. (There is no need to work out this calculation yet.)

Multiply by 100.
100% represents all the workers: 70 ÷ 20 × 100 = 350

So there are 350 workers altogether.

EXAMPLE 14

The price of a car increased by 6% to £9116. Work out the price before the increase.

106% represents £9116.

Divide by 106.
1% represents £9116 ÷ 106

Multiply by 100.
100% represents original price: £9116 ÷ 106 × 100 = £8600

So the price before the increase was £8600.

Method 2: The multiplier method

The **multiplier** method involves fewer steps.

Step 1: Write down the multiplier.

Step 2: Divide the final value by the multiplier to give the original value.

EXAMPLE 15

In a sale the price of a freezer is reduced by 12%. The sale price is £220.
What was the price before the sale?

A decrease of 12% gives a multiplier of 0.88.

Dividing the sale price by the multiplier gives £220 ÷ 0.88 = £250

So the price before the sale was £250.

EXERCISE 3G 

1 Find what 100% represents in these situations.

- **a** 40% represents 320 g
- **b** 14% represents 35 m
- **c** 45% represents 27 cm
- **d** 4% represents £123
- **e** 2.5% represents £5
- **f** 8.5% represents £34

2 On a gruelling army training session, only 28 youngsters survived the whole day. This represented 35% of the original group. How large was the original group?

3 VAT is a government tax added to goods and services. With VAT at 20%, what is the pre-VAT price of the following priced goods?

T-shirt	£10.08	Tights	£1.44
Shorts	£6.24	Sweater	£12.90
Trainers	£29.76	Boots	£38.88

4 Howard spends £200 a month on food. This represents 24% of his monthly take-home pay. How much is his monthly take-home pay?

5 Tina's weekly pay is increased by 5% to £315. What was Tina's pay before the increase?

6 The number of workers in a factory fell by 5% to 228. How many workers were there originally?

7 In a sale the price of a TV is reduced to £325.50. This is a 7% reduction on the original price. What was the original price?

C

B

8 If 38% of plastic bottles in a production line are blue and the remaining 7750 plastic bottles are brown, how many plastic bottles are blue?

9 I received £3.85 back from the tax office, which represented the 20% VAT on a piece of equipment. How much did I pay for this equipment in the first place?

10 The diagram shows two cuboids.

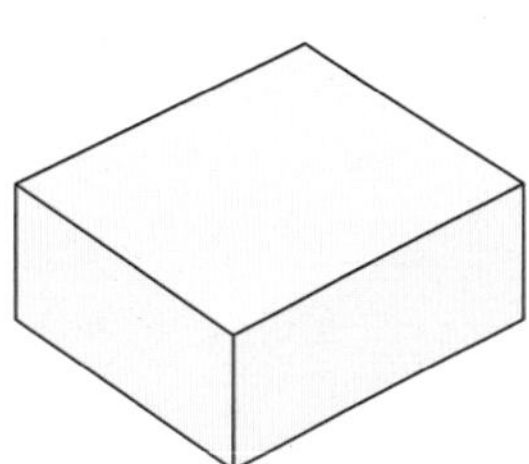

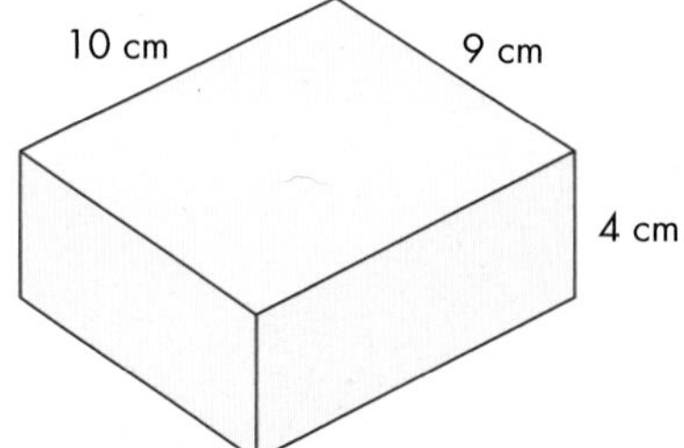

The volume of the larger cuboid is 20% more than the volume of the smaller cuboid.

Work out the volume of the smaller cuboid.

11 A

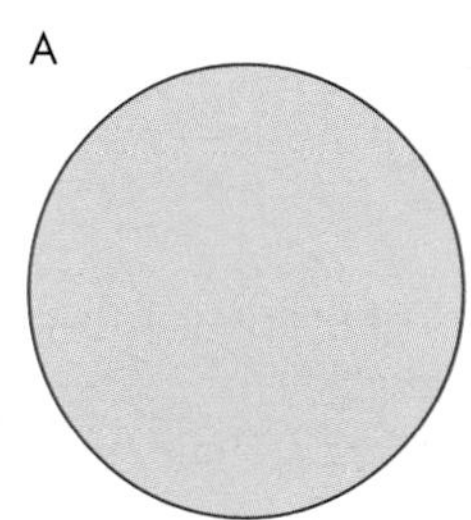

B

The radius of B is 12 cm.

The circumference of circle B is 50% longer than the circumference of circle A.

Work out the radius of circle A.

12 A

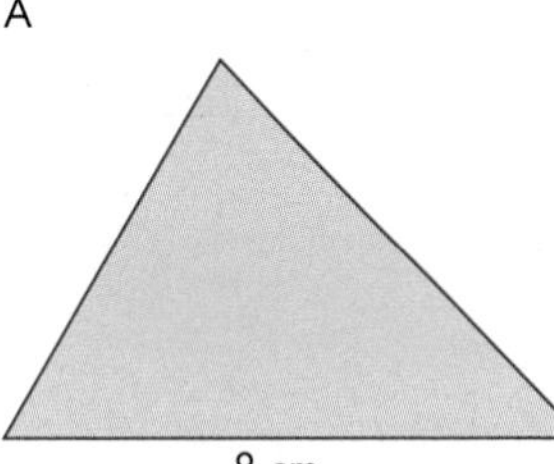

B

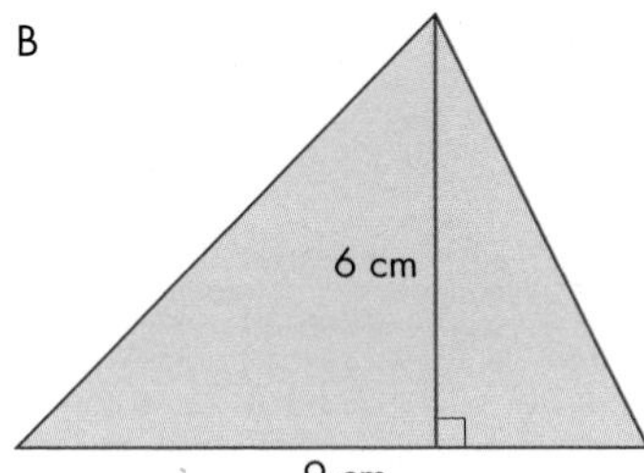

The area of triangle B is 35% more than triangle A.

Work out the height of triangle A.

FM 13 A company is in financial trouble. The workers are asked to take a 10% pay cut for each of the next two years.

a Rob works out that his pay in two years' time will be £1296 per month. How much is his pay now?

b Instead he offers to take an immediate pay cut of 14% and have his pay frozen at that level for two years. Has he made the correct decision?

AU 14 The population in a village is 30% of the size of the population in a neighbouring town.

a If both populations double, what is the population of the village as a percentage of the population of the town?

b If the population of the village stays the same but the population of the town doubles, what is the population of the village as a percentage of the population of the town?

PS 15 A man's salary was increased by 5% in one year and reduced by 5% in the next year. Is his final salary greater or less than the original one and by how many per cent?

PS 16 A woman's salary increased by 5% in one year and then increased the following year by 5% again.

Her new salary was £19 845.

How much was the increase, in pounds, in the first year?

PS 17 The VAT rate in Spain is 18%. A quick way of estimating the pre-VAT price of an item with VAT added is to divide by 6 and then multiply by 5. For example, if an item costs €360 including VAT, it cost approximately $(360 \div 6) \times 5 =$ €300 before VAT. Show that this gives an estimate to within €5 of the pre-VAT price for items costing up to €280.

PS 18 After a 6% increase followed by an 8% increase, the monthly salary of a chef was £1431. What was the original salary?

PS 19 Cassie invests some money at 4% interest per annum for five years. After five years, she had £1520.82 in the bank. How much did she invest originally?

AU 20 A teacher asked her class to work out the original price of a cooker that after a 12% increase cost £291.20.

This is Baz's answer: 12% of 291.20 = £34.94

Original price = 291.20 – 34.94 = 256.26 ≈ £260

When the teacher read out the answer Baz ticked his work as correct.

What errors has he made?

3.7 Ratio

This section will show you how to:
- simplify a ratio
- express a ratio as a fraction
- divide amounts into given ratios
- complete calculations from a given ratio and partial information

Key words
cancel
common units
ratio
simplest form

A **ratio** is a way of comparing the sizes of two or more quantities.

A ratio can be expressed in a number of ways. For example, if Joy is five years old and James is 20 years old, the ratio of their ages is:

	Joy's age : James's age	
which is:	5 : 20	
which simplifies to:	1 : 4	(dividing both sides by 5)

A ratio is usually given in one of these three ways.

Joy's age : James's age	or	5 : 20	or	1 : 4
Joy's age to James's age	or	5 to 20	or	1 to 4
$\frac{\text{Joy's age}}{\text{James's age}}$	or	$\frac{5}{20}$	or	$\frac{1}{4}$

Common units

When working with a ratio involving different units, *always convert them to a* **common unit**. A ratio can be simplified only when the units of each quantity are the same, because the ratio itself has no units. Once the units are the same, the ratio can be simplified or **cancelled**.

For example, the ratio of 125 g to 2 kg must be converted to the ratio of 125 g to 2000 g, so that you can simplify it.

	125 : 2000	
Divide both sides by 25:	5 : 80	
Divide both sides by 5:	1 : 16	The ratio 125 : 2000 can be simplified to 1 : 16.

EXAMPLE 16

Express 25 minutes : 1 hour as a ratio in its simplest form.

The units must be the same, so change 1 hour into 60 minutes.

25 minutes : 1 hour = 25 minutes : 60 minutes
= 25 : 60 — Cancel the units (minutes).
= 5 : 12 — Divide both sides by 5.

So 25 minutes : 1 hour simplifies to 5 : 12

Ratios as fractions

A ratio in its **simplest form** can be expressed as portions by expressing the whole numbers in the ratio as fractions with the same denominator (bottom number).

EXAMPLE 17

A garden is divided into lawn and shrubs in the ratio 3 : 2.

What fraction of the garden is covered by **a** lawn, **b** shrubs?

The denominator (bottom number) of the fraction comes from *adding the numbers in the ratio* (that is, 2 + 3 = 5).

a The lawn covers $\frac{3}{5}$ of the garden.

b The shrubs cover $\frac{2}{5}$ of the garden.

EXERCISE 3H

1 A length of wood is cut into two pieces in the ratio 3 : 7. What fraction of the original length is the longer piece?

2 Jack and Thomas find a bag of marbles which they divide between them in the ratio of their ages. Jack is 10 years old and Thomas is 15 years old. What fraction of the marbles did Jack get?

3 Dave and Sue share a pizza in the ratio of 2 : 3. They eat it all.

a What fraction of the pizza did Dave eat?

b What fraction of the pizza did Sue eat?

4 A camp site allocates space to caravans and tents in the ratio 7 : 3. What fraction of the total space is given to:

a the caravans **b** the tents?

5 Two sisters, Amy and Katie, share a packet of sweets in the ratio of their ages. Amy is 15 and Katie is 10. What fraction of the sweets does each sister get?

6 **a** The recipe for a fruit punch is 1.25 litres of fruit crush to 6.75 litres of lemonade. What fraction of the punch is each ingredient?

FM **b** A different recipe for fruit punch is 1 litre of fruit crush to 5 litres of lemonade.

Roy wants to make the fruit punch with the biggest proportion of fruit crush. Which recipe should he use?

PS **7** Three cows, Gertrude, Gladys and Henrietta produced milk in the ratio 2 : 3 : 4. Henrietta produced $1\frac{1}{2}$ more litres than Gladys. How much milk did the three cows produce altogether?

D

D

8 In a safari park at feeding time, the elephants, the lions and the chimpanzees are given food in the ratio 10 to 7 to 3. What fraction of the total food is given to:

a the elephants **b** the lions **c** the chimpanzees?

9 Three brothers, James, John and Joseph, share a huge block of chocolate in the ratio of their ages.

James is 20, John is 12 and Joseph is 8.

What fraction of the bar of chocolate does each brother get?

10 The recipe for a pudding is 125 g of sugar, 150 g of flour, 100 g of margarine and 175 g of fruit. What fraction of the pudding is each ingredient?

C

AU 11 June wins three-quarters of her bowls matches. She loses the rest.

What is the ratio of wins to losses?

PS 12 Three brothers share some cash.

The ratio of Mark's and David's share is 1 : 2.

The ratio of David's and Paul's share is 1 : 2.

What is the ratio of Mark's share to Paul's share?

PS 13 In a garden, the area is divided into lawn, vegetables and flowers in the ratio 3 : 2 : 1.

If one-third of the lawn is dug up and replaced by flowers what is the ratio of lawn : vegetables : flowers now?

Give your answer as a ratio in its simplest form.

Dividing amounts in a given ratio

To divide an amount in a given ratio, you first look at the ratio to see how many parts there are altogether.

For example 4 : 3 has four parts and three parts giving seven parts altogether. Seven parts is the whole amount.

One part can then be found by dividing the whole amount by 7. Three parts and four parts can then be worked out from one part.

EXAMPLE 18

Divide £28 in the ratio 4 : 3.

4 + 3 = 7 parts altogether
So 7 parts = £28
Divide by 7
1 part = £4
4 parts = 4 × £4 = £16 and 3 parts = 3 × £4 = £12

So £28 divided in the ratio 4 : 3 gives £16 and £12.

To divide an amount in a given ratio you can also use fractions.

Express the whole numbers in the ratio as fractions with the same common denominator.

Then multiply the amount by each fraction.

EXAMPLE 19

Divide £40 between Peter and Hitan in the ratio 2 : 3.

Changing the ratio to fractions gives:

$$\text{Peter's share} = \frac{2}{(2+3)} = \frac{2}{5}$$

$$\text{Hitan's share} = \frac{3}{(2+3)} = \frac{3}{5}$$

So, Peter receives £40 × $\frac{2}{5}$ = £16 and Hitan receives £40 × $\frac{3}{5}$ = £24

Note that whichever method you use, you should always check that the final values add up to the original amount: £16 + £12 = £28 and £16 + £24 = £40.

EXERCISE 3I

1 Divide the following amounts in the given ratios.

- **a** 400 g in the ratio 2 : 3
- **b** 280 kg in the ratio 2 : 5
- **c** 500 in the ratio 3 : 7
- **d** 1 km in the ratio 19 : 1
- **e** 5 hours in the ratio 7 : 5
- **f** £100 in the ratio 2 : 3 : 5
- **g** £240 in the ratio 3 : 5 : 12
- **h** 600 g in the ratio 1 : 5 : 6

2 The ratio of female to male members of Lakeside Gardening Club is 7 : 3.

The total number of members of the club is 250.

- **a** How many members are female?
- PS **b** What percentage of members are male?

3 A supermarket aims to stock branded goods and their own goods in the ratio 2 : 3.

They stock 500 kg of breakfast cereal.

- **a** What percentage of the cereal stock is branded?
- **b** How much of the cereal stock is their own?

4 The Illinois Department of Health reported that, for the years 1981 to 1992, when it tested a total of 357 horses for rabies, the ratio of horses with rabies to those without was 1 : 16.

How many of these horses had rabies?

C

C

FM **5** Being overweight increases the chances of an adult suffering from heart disease. A way to test whether an adult has an increased risk is shown below.

W and H refer to waist and hip measurements.
For women, there is increased risk when $W/H > 0.8$
For men, there is increased risk when $W/H > 1.0$

a Find whether the following people have an increased risk of heart disease.

Miss Mott: waist 26 inches, hips: 35 inches
Mrs Wright: waist 32 inches, hips: 37 inches
Mr Brennan: waist 32 inches, hips: 34 inches
Ms Smith: waist 31 inches, hips: 40 inches
Mr Kaye: waist 34 inches, hips: 33 inches

b Give three examples of waist and hip measurements that would suggest no risk of heart disease for a man, but would suggest a risk for a woman.

6 Rewrite the following scales as ratios, as simply as possible.

a 1 cm to 4 km **b** 4 cm to 5 km **c** 2 cm to 5 km
d 4 cm to 1 km **e** 5 cm to 1 km **f** 2.5 cm to 1 km

7 A map has a scale of 1 cm to 10 km.

a Rewrite the scale as a ratio in its simplest form.
b What is the actual length of a lake that is 4.7 cm long on the map?
c How long will a road be on the map if its actual length is 8 km?

8 A map has a scale of 2 cm to 5 km.

a Rewrite the scale as a ratio in its simplest form.
b How long is a path that measures 0.8 cm on the map?
c How long should a 12 km road be on the map?

9 The scale of a map is 5 cm to 1 km.

a Rewrite the scale as a ratio in its simplest form.
b How long is a wall that is shown as 2.7 cm on the map?
c The distance between two points is 8 km; how far will this be on the map?

10 A car is 240 miles from Manchester. A lorry is 180 miles from Manchester.

a Work out the ratio of the distances, giving your answer in its simplest form.
b Two hours later the ratio of the distances is exactly the same.
The car is 120 miles from Manchester.

How far is the lorry from Manchester?

PS **c** If the ratio of the distances stays the same for the entire journeys to Manchester, which vehicle, if either, arrives first?

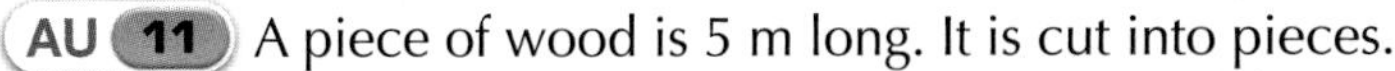

AU **11** A piece of wood is 5 m long. It is cut into pieces.
The lengths of the pieces are in the ratio 4 : 3 : 2 : 1.
The biggest piece is then cut in the ratio 4 : 1 so that there are now five pieces.

How long is the smallest piece?

12 You can simplify a ratio by changing it into the form 1 : n.

For example, 5 : 7 can be rewritten as $\frac{5}{5} : \frac{7}{5} = 1 : 1.4$

Rewrite each of the following in the form 1 : n.

a 5 : 8

b 4 : 13

c 8 : 9

d 25 : 36

e 5 : 27

f 12 : 18

g 5 hours : 1 day

h 4 hours : 1 week

i £4 : £5

Calculating with ratios when only part of the information is known

EXAMPLE 20

A fruit drink is made by mixing orange squash with water in the ratio 2 : 3.

How much water needs to be added to 5 litres of orange squash to make the drink?

2 parts are 5 litres.
Divide by 2.
1 part is 2.5 litres
3 parts = 2.5 litres × 3 = 7.5 litres

So 7.5 litres of water are needed to make the drink.

EXAMPLE 21

Two business partners, John and Ben, divided their total profit in the ratio 3 : 5. John received £2100. How much did Ben get?

John's £2100 was $\frac{3}{8}$ of the total profit. (Check you know why.)

$\frac{1}{8}$ of the total profit = £2100 ÷ 3 = £700

So, Ben's share, which was $\frac{5}{8}$ of the total, amounted to £700 × 5 = £3500

EXERCISE 3J

C

1 Derek, aged 15, and Ricki, aged 10, shared all the conkers they found in the woods in the same ratio as their ages. Derek had 48 conkers.

a Simplify the ratio of their ages.

b How many conkers did Ricki have?

c How many conkers did they find altogether?

2 Two types of crisps, plain and salt 'n' vinegar, were bought for a school party in the ratio 5 : 3. The school bought 60 packets of salt 'n' vinegar crisps.

a How many packets of plain crisps did they buy?

b How many packets of crisps altogether did they buy?

3 Robin is making a drink from orange juice and lemon juice in the ratio 9 : 1. If Robin has only 3.6 litres of orange juice, how much lemon juice does he need to make the drink?

4 When I picked my strawberries, I found some had been spoilt by snails. The rest were good. These were in the ratio 3 : 17. Eighteen of my strawberries had been spoilt by snails. How many good strawberries did I find?

5 A blend of tea is made by mixing Lapsang with Assam in the ratio 3 : 5. I have a lot of Assam tea but only 600 g of Lapsang. How much Assam do I need to make the blend using all the Lapsang?

FM **6** An old recipe to make pancakes says, "For every four ounces of flour, add two eggs and half a pint milk. This is enough for 10 pancakes".

Jamie wants to make two pancakes each for 15 people. He has 1 litre of milk.

Will he have enough milk? Explain how you decide.

7 The ratio of male to female spectators at ice hockey games is 4 : 5. At the Steelers' last match, 4500 men watched the match. What was the total attendance at the game?

8 'Proper tea' is made by putting milk and tea together in the ratio 2 : 9. How much proper tea can be made if you have 1 litre of milk?

9 A teacher always arranged the content of each of his lessons to Year 10 as 'teaching' and 'practising learnt skills' in the ratio 2 : 3.

a If a lesson lasted 35 minutes, how much teaching would he do?

b If he decided to teach for 30 minutes, how long would the lesson be?

10 A 'good' children's book is supposed to have pictures and text in the ratio 17 : 8. In a book I have just looked at, the pictures occupy 23 pages.

a Approximately how many pages of text should this book have to be deemed a 'good' children's book?

b What percentage of a 'good' children's book will be text?

11 Three business partners, Kevin, John and Margaret, put money into a business in the ratio 3 : 4 : 5. They shared any profits in the same ratio. Last year, Margaret made £3400 out of the profits. How much did Kevin and John make last year?

AU **12** The ratio of daffodils to tulips in a flower bed is 3 : 7.

Which of the following statements is true (T), false (F) or could be true (C). The first one has been done for you.

a There are 25 daffodils in the flower bed. **F**

b There are 140 flowers altogether in the flower bed

c The fraction of daffodils in the flower bed is $\frac{3}{7}$.

d The percentage of tulips in the flower bed is 70%.

e If half of the daffodils were dug up the ratio of daffodils to tulips would now be 3 : 14.

PS **13** In a factory, the ratio of female employees to male employees is 3 : 8. There are 85 more males than females.

How many females work in the factory?

PS **14** There is a group of boys and girls waiting for school buses. 25 girls get on the first bus. The ratio of boys to girls at the stop is now 3 : 2. 15 boys get on the second bus. There are now the same number of boys and girls at the bus stop.

How many students altogether were originally at the bus stop?

PS **15** A jar contains 100 cc of a mixture of oil and water in the ratio 1 : 4. Enough oil is added to make the ratio of oil to water 1 : 2.

How much water needs to be added to make the ratio of oil to water 1 : 3?

16 The soft drinks Cola, Orange Fizz and Zesto were bought for the school disco in the ratio 10 : 5 : 3. The school bought 80 cans of Orange Fizz.

a How much Cola did they buy?

b How much Zesto did they buy?

17 **a** Iqra is making a drink from lemonade, orange juice and ginger ale in the ratio 40 : 9 : 1. If Iqra has only 4.5 litres of orange juice, how much of the other two ingredients does she need to make the drink?

AU **b** Another drink made from lemonade, orange juice and ginger ale uses the ratio 10 : 2 : 1.

Which drink has a larger proportion of ginger ale, Iqra's or this one? Show how you work out your answer.

18 Bob is making concrete, using sand and cement in the ratio 3 : 1. He has three 25 kg bags of cement. How much sand will he need if he is to use all his cement?

PS **19** The ratio of my sister's age to my age is 10 : 9.
The ratio of my brother's age to my age is 29 : 27.
I am over 40 years old but under 70 years old.

What is my age?

GRADE BOOSTER

- **D** You can add and subtract fractions
- **C** You can calculate percentage increases and decreases
- **C** You can calculate with mixed numbers
- **C** You can work out compound interest problems
- **C** You can solve problems using ratio in appropriate situations
- **B** You can do reverse percentage problems
- **A** You can solve complex problems involving percentage increases and percentage decreases

What you should know now

- How to calculate with fractions
- How to do percentage problems
- How to divide any amount in a given ratio

1 Mrs Senior earns £320 per week. She is awarded a pay rise of 4%.

How much does she earn each week after the pay rise?

2 Five girls run a 100 metre race.

Their times are shown in the table.

Name	Amy	Bavna	Charlotte	Di	Ellie
Time (seconds)	49.0	45.5	51.3	44.7	48.1

a Write down the median time.

b The five girls run another 100 metre race.

They all reduce their times by 10%.

i Calculate Amy's new time.

ii Who won this race?

iii Who improved her time by the least amount of time?

3 Mr Shaw's bill for new tyres is £120 plus VAT. VAT is charged at 17.5%. What is his total bill?

4 Andy's salary is £24 000 per year.

He is paid the same amount each month.
He is given a pay rise of 10%.

Calculate his new **monthly** salary.

You **must** show your working. *(4 marks)*

AQA, November 2008, Paper 1 Higher, Question 3

5 Mr and Mrs Jones are buying a tumble dryer that normally costs £250. They save 12% in a sale. How much do they pay for the tumble dryer?

6 Work out the value of $\frac{3}{5} - \frac{3}{8}$

7 On Monday Joe drinks $2\frac{1}{3}$ pints of milk.
On Tuesday he drinks $1\frac{3}{4}$ pints of milk.

Work out the total amount of milk that Joe drinks on Monday and Tuesday. *(3 marks)*

AQA, June 2005, Paper 1 Intermediate, Question 13

8 Andy uses $\frac{3}{8}$ of a tin of creosote to creosote 2 m of fencing. What is the least number of tins he needs to creosote 10 m of fencing?

9 Pythagoras made a number of calculations trying to find an approximation for π.

Here are a few of the closest approximations

$\frac{22}{7}$ $\frac{54}{17}$ $\frac{221}{71}$ $\frac{312}{77}$

a Put these approximations into order of size, largest on the left, smallest of the right.

b Use your calculator to find which of the above is the closest approximation to π.

10 John has £2000 to invest.

He sees this advert.

SureFire Investments

Don't see your money go up in smoke!

Double your money in 10 years!

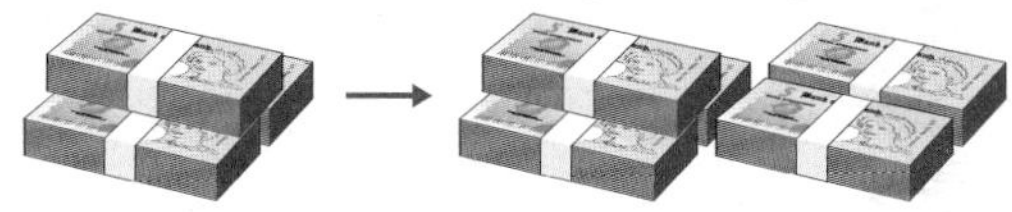

The average annual growth of our investment account is **7.2%**

Will John double his money in ten years with SureFire Investments?

You **must** show your working. *(4 marks)*

AQA, June 2006, Paper 2 Higher, Question 9

11 During 2003 the number of unemployed people in Barnsley fell from 2800 to 2576. What was the percentage decrease?

12 A painter has 50 litres of paint. Each litre covers 2.5 m^2.

The area to be painted is 98 m^2.

Estimate the percentage of paint used.

13 **a** Poppy the dog has two meals a day.

At each meal Poppy eats $\frac{2}{5}$ of a tin of dog food.

On Monday morning there are 5 tins of dog food in the cupboard. Is this enough dog food to feed Poppy for one week?

You must show your working.

b Work out $4\frac{2}{3} \div 1\frac{3}{4}$ *(3 marks)*

AQA, November 2006, Paper 1 Higher, Question 4

14 Zoe invests £6000 in a savings account that pays 3.5% compound interest per year.

How much does she have in the account after 6 years?

15 Simon weighed 3.7 kg when he was born.

One year later he weighed 10.9 kg.

Calculate the percentage increase in his weight. *(3 marks)*

AQA, November 2005, Module 3, Question 2

16 The house price index for a flat in Leeds was 190 in August 2006, compared with a base of 100 in April 2000.

a Write down the percentage increase in the price of flats in Leeds in that period. *(1 mark)*

b A flat cost £80 000 in April 2000.

What was its likely value in August 2006? *(2 marks)*

AQA, June 2008, Paper 1 Higher, Question 12

17 There are 126 people at a party.

The ratio of adults to children at the party is 1 : 6.

a How many adults and children are there? *(3 marks)*

b Nine more adults arrive.

Including these adults, what is the new ratio of adults to children?

Give your answer in the form 1 : k, where k is to be found. *(3 marks)*

AQA, June 2008, Module 3, Question 3

18 A bag contains 6 blue balls and 8 red balls.

Some more red balls are added. The ratio of blue balls to red balls can now be written in the form 1 : n, where n is an integer.

What is the smallest number of red balls that can be added? *(2 marks)*

AQA, Question 10, Specification A, Paper 2, June 2008

19 Helen weighed 100 kg. Her target was to weigh 70 kg or less.

Her weight decreased by 4% each month. Has she achieved her target after nine months?

You **must** show your working. *(3 marks)*

AQA, June 2008, Module 3 Higher, Question 7

20 Gotland is an island which forms part of Sweden. The area of Gotland is 3140 square kilometres.

This area is 0.8% of the total area of Sweden.

What is the total area of Sweden? *(3 marks)*

AQA, November 2005, Module 3 Higher, Question 4

21 Jack and Jill want to buy some towels.

A store displays the following signs.

January Sales
All towels 60% off
Normally £10
January sale price £4

Today Only
EXTRA
25% off the
January sale price

Jack: That is 85% off the original price

Jill: No, it is only 70% off the original price

Who is correct?

Explain your answer fully. *(2 marks)*

AQA, June 2006, Paper 2 Higher, Question 5

22 A leaking water tank loses 36% of its contents each day.

Isobel says that the tank will have lost over 90% of its original contents by the end of the fifth day.

Is Isobel correct?

You **must** explain your answer. *(3 marks)*

AQA, November 2007, Module 3 Higher, Question 9

23 The cost of bananas increased by 25% one week but then fell the following week back to the original price.

By what percentage did the cost of bananas fall in the following week?

24 110 men and women visit a cinema. There are 20% more men than women.

How many men are at the cinema?

Worked Examination Questions

FM **1** Kelly bought a television set. After a reduction of 15% in a sale, the one she bought cost her £319.60. What was the original price of the television set?

1 Multiplier is 0.85 or 85% is equivalent to £319.60

A 15% reduction is a multiplier of 0.85, or realising the sale price is 85% scores 1 mark for method.

£319.60 ÷ 0.85, or 100% is equivalent to £319.60 ÷ 85 × 100

Showing the correct calculation that will lead to the correct answer scores 1 mark for method.

= £376

£376 gets 1 mark for accuracy.

Total: 3 marks

PS **2** A plant in a greenhouse is 10 cm high. It increases its height by 15% each day. How many days does it take to double in height?

2 1.15 is the multiplier

Recognising the multiplier scores 1 mark.

$10 \times 1.15 = 11.5$ (one day)

10×1.15^2 (or 11.5×1.15) $= 13.225$

Getting to this stage scores 1 mark for method.

$10 \times 1.15^3 = 15.2$

$10 \times 1.15^4 = 17.5$

$10 \times 1.15^5 = 20.1$ and therefore it takes 5 days to double its height

Reaching the correct solution gets 1 mark for accuracy.

Total: 3 marks

PS AU **3** Decide whether $\frac{2}{3} + \frac{4}{5}$ is greater or less than $1\frac{7}{9}$. Show clearly how you decide.

3 The answer to part a is less than $1(\frac{33}{40})$ so the numerator $(\frac{2}{3} + \frac{4}{5})$ must be smaller than the denominator $(1\frac{7}{9})$.

or

$\frac{2}{3} + \frac{4}{5} = \frac{22}{15}$ from part a

$= \frac{198}{135}$

and $1\frac{7}{9} = \frac{240}{135}$

Any valid reason would be acceptable and scores 1 mark.

Total: 3 marks

3

Functional Maths
Understanding VAT

VAT means 'Value Added Tax'. You pay it when you buy goods or services and it is normally included in the price of the goods or services. The rate of VAT can vary from country-to-country and even product-to-product (for example, some goods are exempt from VAT, such as food and children's clothes).

In the economic recession of 2009, the British government reduced the VAT rate from 17.5% to 15%, to help stimulate economic recovery. This had practical implications for shops, which had to find the best way to include the VAT reduction in the price labels in their shops.

Your task

To accommodate the reduced VAT rate, many shops had to change their price labels overnight.

Write a report advising shops on how best to accommodate the change in VAT. Use mathematical evidence to support your advice.

Consider the points below in your report:

- Some shops took 2.5% from their displayed prices. Was this the correct thing to do? Use evidence to support your answer.
- Is there a quick calculation that shops could use to work out the new price at the till?
- Does the change in VAT rate affect shops' overall profit?
- Does the change in VAT have an effect on the prices displayed in the annual 'January sales'?

On 1st January 2010, the government reversed the VAT reduction.

- What should shopkeepers do now to come into line with the increased VAT rate?

Getting started

The price of a TV is £420 + VAT. What is its final price if VAT is charged at:

- 15%
- 17.5%
- 20%?

The price of a TV is £450 including VAT. What would the pre-VAT price have been if VAT is charged at:

- 15%
- 17.5%
- 20%?

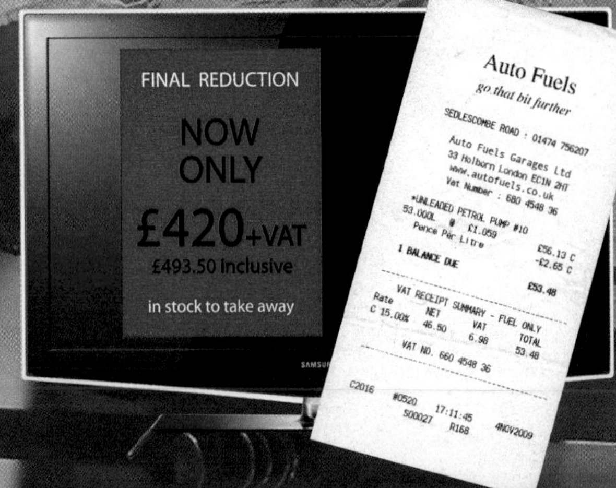

Extension

Research two case studies to find out how shops actually dealt with the changes in VAT. Identify the mathematics that the shops used and evaluate the approach that they took.

Why this chapter matters

We use proportion and speed as part of our everyday lives to help when dealing with facts or to compare two or more pieces of information.

Proportions are often used to compare sizes, speed is used to compare distances with the time taken to travel them.

Speed

What do you think of as a high speed?

On 16 August 2009 Usain Bolt set a new world record for the 100 m sprint of 9.58 seconds. This is an average speed of 23.3 mph.

The sailfish is the fastest fish and can swim at 68 mph.

The cheetah is the fastest land animal and can run at 75 mph.

The fastest bird is the swift, which can fly at 106 mph.

Ratio and proportion facts

- Russia is the largest country. Vatican City is the smallest country. The area of Russia is nearly 39 million times bigger than the area of Vatican City.
- Monaco has the most people per square mile. Mongolia has the least people per square mile. The ratio of the number of people per square mile in Monaco to the number of people in Mongolia is 10 800 : 1.
- Japan has the highest life expectancy. Sierra Leone has the lowest life expectancy. On average, people in Japan live over twice as long as people in Sierra Leone.
- Taiwan has the most mobile phones per 100 people (106.5). This is approximately four times more than in Thailand (26.04).
- About one-seventh of England is green-belt land.

This chapter is about comparing pieces of information. You can compare the speeds of Usain Bolt, the sailfish, the cheetah and the swift by answering questions such as: How much faster is a sailfish than Usain Bolt?

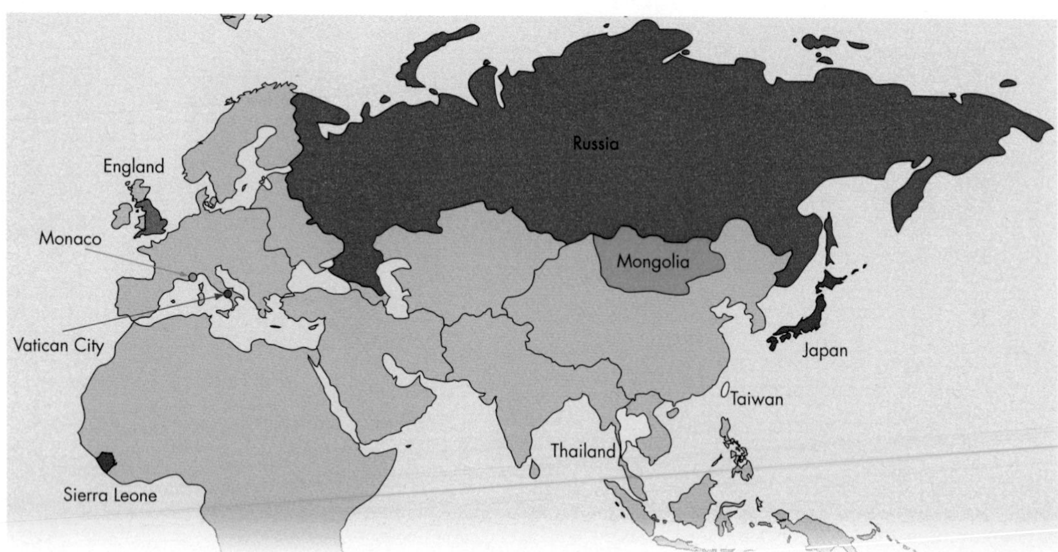

Chapter

Number: Proportions

1. Speed, time and distance
2. Direct proportion problems
3. Best buys
4. Density

This chapter will show you ...

- **D** how to solve problems involving direct proportion
- **D** how to compare prices of products
- **D** to **C** how to calculate speed
- **B** how to calculate density

Visual overview

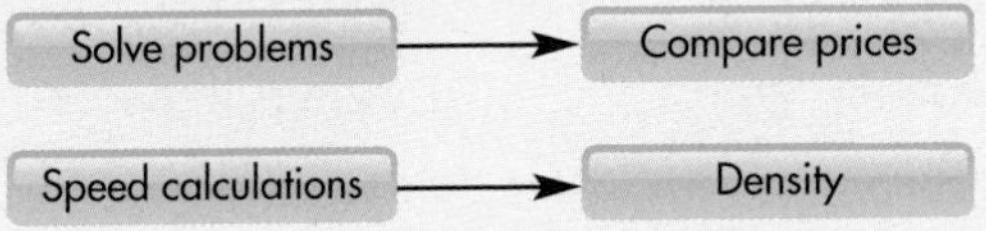

What you should already know

- Multiplication tables up to 10 × 10 **(KS3 level 4, GCSE grade G)**
- How to simplify fractions **(KS3 level 5, GCSE grade G)**
- How to find a fraction of a quantity **(KS3 level 4, GCSE grade F)**
- How to multiply and divide, with and without a calculator **(KS3 level 5, GCSE grade G)**

Quick check

1 Cancel the following fractions.

a $\frac{6}{10}$ **b** $\frac{4}{20}$ **c** $\frac{4}{12}$ **d** $\frac{32}{50}$ **e** $\frac{36}{90}$ **f** $\frac{18}{24}$ **g** $\frac{16}{48}$

2 Find the following quantities.

a $\frac{2}{5}$ of £30 **b** $\frac{3}{8}$ of £88 **c** $\frac{7}{10}$ of 250 litres **d** $\frac{5}{8}$ of 24 kg

e $\frac{2}{3}$ of 60 m **f** $\frac{5}{6}$ of £42 **g** $\frac{9}{20}$ of 300 g **h** $\frac{3}{10}$ of 3.5 litres

4.1 Speed, time and distance

This section will show you how to:

- recognise the relationship between speed, distance and time
- calculate average speed from distance and time
- calculate distance travelled from the speed and the time
- calculate the time taken on a journey from the speed and the distance

Key words
average
distance
speed
time

The relationship between **speed**, **time** and **distance** can be expressed in three ways:

$$\textit{speed} = \frac{\textit{distance}}{\textit{time}} \qquad \textit{distance} = \textit{speed} \times \textit{time} \qquad \textit{time} = \frac{\textit{distance}}{\textit{speed}}$$

In problems relating to speed, you usually mean **average** speed, as it would be unusual to maintain one exact speed for the whole of a journey.

The diagram will help you remember the relationships between distance (D), time (T) and speed (S).

D
S T

$$D = S \times T \qquad S = \frac{D}{T} \qquad T = \frac{D}{S}$$

EXAMPLE 1

Paula drove a distance of 270 miles in 5 hours. What was her average speed?

Paula's average speed $= \frac{\text{distance she drove}}{\text{time she took}} = \frac{270}{5} = 54$ miles per hour (mph)

EXAMPLE 2

Edith drove from Sheffield to Peebles for $3\frac{1}{2}$ hours at an average speed of 60 mph. How far is it from Sheffield to Peebles?

Since: *distance* = *speed* × *time*

the distance from Sheffield to Peebles is given by

$60 \times 3.5 = 210$ miles

Note: You need to change the time to a decimal number and use 3.5 (not 3.30).

EXAMPLE 3

Sean is going to drive from Newcastle upon Tyne to Nottingham, a distance of 190 miles. He estimates that he will drive at an average speed of 50 mph. How long will it take him?

$$\text{Sean's time} = \frac{\text{distance he covers}}{\text{his average speed}} = \frac{190}{50} = 3.8 \text{ hours}$$

Change the 0.8 hour to minutes by multiplying by 60, to give 48 minutes.

So, the time for Sean's journey will be 3 hours 48 minutes.

Remember: When you calculate a time and get a decimal answer, as in Example 3, *do not mistake* the decimal part for minutes. You must either:

- leave the time as a decimal number and give the unit as hours, or
- change the decimal part to minutes by multiplying it by 60 (1 hour = 60 minutes) and give the answer in hours and minutes.

EXERCISE 4A

D

HINTS AND TIPS

Remember to convert time to a decimal if you are using a calculator, for example, 8 hours 30 minutes is 8.5 hours.

1 A cyclist travels a distance of 90 miles in 5 hours. What was her average speed?

2 How far along a motorway would you travel if you drove at 70 mph for 4 hours?

3 I drive to Bude in Cornwall from Sheffield in about 6 hours. The distance from Sheffield to Bude is 315 miles. What is my average speed?

4 The distance from Leeds to London is 210 miles. The train travels at an average speed of 90 mph. If I catch the 9.30 am train in London, at what time should I expect to arrive in Leeds?

5 How long will an athlete take to run 2000 m at an average speed of 4 metres per second?

6 Copy and complete the following table.

	Distance travelled	Time taken	Average speed
a	150 miles	2 hours	
b	260 miles		40 mph
c		5 hours	35 mph
d		3 hours	80 km/h
e	544 km	8 hours 30 minutes	
f		3 hours 15 minutes	100 km/h
g	215 km		50 km/h

D

7 Eliot drove from Sheffield to Inverness, a distance of 410 miles, in 7 hours 45 minutes.

a Change the time 7 hours 45 minutes to a decimal.

b What was the average speed of the journey? Round your answer to 1 decimal place.

8 Colin drives home from his son's house in 2 hours 15 minutes. He says that he drives at an average speed of 44 mph.

a Change the 2 hours 15 minutes to a decimal.

b How far is it from Colin's home to his son's house?

9 The distance between Paris and Le Mans is 200 km. The express train between Paris and Le Mans travels at an average speed of 160 km/h.

a Calculate the time taken for the journey from Paris to Le Mans, giving your answer as a decimal number of hours.

b Change your answer to part a to hours and minutes.

C

FM 10 The distance between Sheffield and Land's End is 420 miles.

a What is the average speed of a journey from Sheffield to Land's End that takes 8 hours 45 minutes?

b If Sam covered the distance at an average speed of 63 mph, how long would it take him?

FM 11 A train travels at 50 km/h for 2 hours, then slows down to do the last 30 minutes of its journey at 40 km/h.

a What is the total distance of this journey?

b What is the average speed of the train over the whole journey?

FM 12 Jade runs and walks the 3 miles from home to work each day. She runs the first 2 miles at a speed of 8 mph, then walks the next mile at a steady 4 mph.

a How long does it take Jade to get to work?

b What is her average speed?

13 Change the following speeds to metres per second.

a 36 km/h **b** 12 km/h **c** 60 km/h

d 150 km/h **e** 75 km/h

HINTS AND TIPS

Remember that there are 3600 seconds in an hour and 1000 metres in a kilometre. So to change from km/h to m/s multiply by 1000 and divide by 3600.

14 Change the following speeds to kilometres per hour.

a 25 m/s **b** 12 m/s **c** 4 m/s **d** 30 m/s **e** 0.5 m/s

AU 15 A train travels at an average speed of 18 m/s.

a Express its average speed in km/h.

b Find the approximate time the train would take to travel 500 m.

c The train set off at 7.30 on a 40 km journey. At approximately what time will it reach its destination?

HINTS AND TIPS

To change from m/s to km/h multiply by 3600 and divide by 1000.

16 A cyclist is travelling at an average speed of 24 km/h.

a What is this speed in metres per second?

b What distance does he travel in 2 hours 45 minutes?

c How long does it take him to travel 2 km?

d How far does he travel in 20 seconds?

HINTS AND TIPS

To convert a decimal fraction of an hour to minutes, just multiply by 60.

PS **17** How much longer does it take to travel 100 miles at 65 mph than at 70 mph?

18 It takes me 20 minutes to walk from home to the bus station.

I catch the bus from the bus station to work each morning. My bus journey is 10 miles and usually takes 30 minutes. I can catch a bus at 20 minutes past the hour or 10 minutes to the hour. When I get off the bus it takes me 5 minutes to walk to work.

FM **a** What is the average speed of my bus?

PS **b** I have to be at work for 08.30. What time is the latest I can leave home to be at work on time?

4.2 Direct proportion problems

This section will show you how to:

- recognise and solve problems, using direct proportion

Key words
direct proportion
unit cost
unitary method

Suppose you buy 12 items that each cost the same. The total amount you spend is 12 times the cost of one item.

That is, the total cost is said to be in **direct proportion** to the number of items bought. The cost of a single item (the **unit cost**) is the constant factor that links the two quantities.

Direct proportion is not only concerned with costs. Any two related quantities can be in direct proportion to each other.

The best way to solve all problems involving direct proportion is to start by finding the single unit value. This method is called the **unitary method**, because it involves referring to a single *unit* value.

Remember: Before solving a direct proportion problem, think carefully about it to make sure that you know how to find the required single unit value.

EXAMPLE 4

If eight pens cost £2.64, what is the cost of five pens?

First, we need to find the cost of *one* pen. This is £2.64 ÷ 8 = £0.33

So, the cost of five pens is £0.33 × 5 = £1.65

EXAMPLE 5

Eight loaves of bread will make packed lunches for 18 people. How many packed lunches can be made from 20 loaves?

First, find how many lunches one loaf will make.

One loaf will make 18 ÷ 8 = 2.25 lunches

So, 20 loaves will make 2.25 × 20 = 45 lunches

EXERCISE 4B

D

1 If 30 matches weigh 45 g, what would 40 matches weigh?

2 Five bars of chocolate cost £2.90. Find the cost of nine bars.

3 Eight men can chop down 18 trees in a day. How many trees can 20 men chop down in a day?

4 Find the cost of 48 eggs when 15 eggs can be bought for £2.10.

5 Seventy maths textbooks cost £875.

a How much will 25 maths textbooks cost?

b How many maths textbooks can you buy for £100?

> **HINTS AND TIPS**
>
> **Remember** to work out the value of one unit each time. Always check that answers are sensible.

FM **6** A lorry uses 80 litres of diesel fuel on a trip of 280 miles.

a How much diesel would the same lorry use on a trip of 196 miles?

b How far would the lorry travel on a full tank (100 litres) of diesel?

FM **7** During the winter, I find that 200 kg of coal keeps my open fire burning for 12 weeks.

a If I want an open fire all through the winter (18 weeks), how much coal will I need?

b Last year I bought 150 kg of coal. For how many weeks did I have an open fire?

8 It takes a photocopier 16 seconds to produce 12 copies of a document. How long will it take to produce 30 copies?

D

9 A recipe for 12 biscuits uses:

200 g margarine   400 g sugar   500 g flour   300 g ground rice

a What quantities are needed for:

i 6 biscuits

ii 9 biscuits

iii 15 biscuits?

PS **b** What is the maximum number of biscuits I could make if I had just 1 kg of each ingredient?

AU **10** Peter the butcher sells sausages in packs of 6 for £2.30.

Paul the butcher sells sausages in packs of 10 for £3.50.

I have £10 to spend on sausages. If I want to buy as many sausages as possible from one shop, which shop should I use? Show your working.

C

PS **11** A shredding machine can shred 20 sheets of paper in 14 seconds. The bin has room for 1000 sheets of shredded paper.

How long will it take to fill the bin if the machine has to stop for 3 minutes after every 200 sheets to prevent overheating?

FM **12** Here is a recipe for making Yorkshire pudding.

Adjust this recipe to use it for two people.
Justify any decision you make.

Yorkshire pudding recipe (Serves 8)
125 g plain flour
235 ml whole milk
2 eggs
3 g salt
45 ml beef dripping or lard

FM **13** An aircraft has two fuel tanks, one in each wing.

The tanks each hold 40 litres when full.

The left tank is quarter full. The right tank is half full.

How much fuel is needed so that both tanks are three-quarters full?

4.3 Best buys

This section will show you how to:
- find the cost per unit weight
- find the weight per unit cost
- use the above to find which product is the cheaper

Key words
best buy
better value
value for money

When you wander around a supermarket and see all the different prices for the many different-sized packets, it is rarely obvious which are the **best buys**. However, with a calculator you can easily compare **value for money** by finding either:

the cost per unit weight *or* the weight per unit cost

To find:

- *cost per unit weight*, divide *cost by weight*
- *weight per unit cost*, divide *weight by cost.*

The next two examples show you how to do this.

EXAMPLE 6

A 300 g tin of cocoa costs £1.20.

Work out (a) the cost per gram (b) the number of grams per penny.

First change £1.20 to 120p. Then divide, using a calculator, to get:

cost per unit weight 120p ÷ 300 g = 0.4p per gram
weight per unit cost 300 g ÷ 120p = 2.5 g per penny

EXAMPLE 7

A supermarket sells two different-sized packets of Whito soap powder. The medium size contains 800 g and costs £1.60 and the large size contains 2.5 kg and costs £4.75. Which is the better buy?

Find the weight per unit cost for both packets.

Medium: 800 g ÷ 160p = 5 g per penny
Large: 2500 g ÷ 475p = 5.26 g per penny

From these it is clear that there is more weight per penny with the large size, which means that the large size is the better buy.

Sometimes it is easier to us a scaling method to compare prices to find the **better value**.

EXAMPLE 8

Small
Price £3.40

Large
Price £4.95

12 is a common factor of 24 and 36 so work out the cost of 12 fish fingers.

For the small box 12 fish fingers cost £3.40 ÷ 2 = £1.70

For the large box 12 fish fingers cost £4.95 ÷ 3 = £1.65

So the large box is better value.

EXAMPLE 9

Price £1.45

Price £1.20

30 is the least common multiple of 5 and 6 so work out the cost of 30 yoghurts.

For the six-pack the cost of 30 yoghurts is £1.45 × 5 = £7.25

For the five-pack the cost of 30 yoghurts is £1.20 × 6 = £7.20

So the five-pack is better value.

EXERCISE 4C

1 Compare the prices of the following pairs of products and state which, if either, is the better buy.

- **a** Chocolate bars: £2.50 for a 5-pack, £4.50 for a 10-pack
- **b** Eggs: £1.08 for 6, £2.25 for 12
- **c** Car shampoo: £4.99 for 2 litres, £2.45 for 1 litre
- **d** Dishwasher tablets: £7.80 for 24, £3.90 for 12
- **e** Carrots: 29p for 250 grams, 95p for 750 grams
- **f** Bread rolls: £1.39 for a pack of 6, £4.90 for a pack of 20
- **g** Juice: £2.98 for 2, £4 for 3

D

D

FM AU **2** Compare the following pairs of product and state which is the better buy, and why.

a Coffee: a medium jar which is 140 g for £1.10 or a large jar which is 300 g for £2.18

b Beans: a 125 g tin at 16p or a 600 g tin at 59p

c Flour: a 3 kg bag at 75p or a 5 kg bag at £1.20

d Toothpaste: a large tube which is 110 ml for £1.79 or a medium tube which is 75 ml for £1.15

e Frosted flakes: a large box which is 750 g for £1.64 or a medium box which is 500 g for £1.10

f Rice Crisps: a medium box which is 440 g for £1.64 or a large box which is 600 g for £2.13

g Shampoo: a bottle containing 400 ml for £1.15 or a bottle containing 550 ml for £1.60

FM **3** Julie wants to respray her car with yellow paint. In the local shop, she sees the following tins:

small tin: 350 ml at a cost of £1.79

medium tin: 500 ml at a cost of £2.40

large tin: 1.5 litres at a cost of £6.70

a What is the cost per litre of paint in the small tin?

b Which tin is offered at the lowest price per litre?

FM **4** Tisco's sells bottled water in three sizes.

a Work out the cost per litre of the 'handy' size.

b Which bottle is the best value for money?

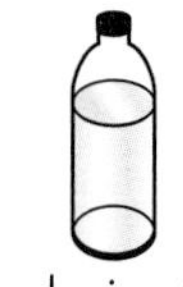
Handy size 40 cl
£0.38

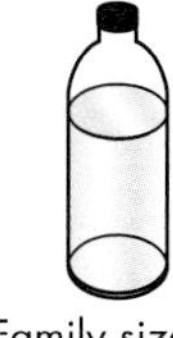
Family size 2 l
£0.98

Giant size 5 l
£2.50

PS **5** Two drivers are comparing the petrol consumption of their cars.

Ahmed says, "I get 320 miles on a tank of 45 litres".

Bashir says, "I get 230 miles on a tank of 32 litres".

Whose car is more economical?

PS **6** Mary and Jane are arguing about which of them is better at mathematics.

Mary scored 49 out of 80 on a test.

Jane scored 60 out of 100 on a test of the same standard.

Who is better at mathematics?

AU PS **7** Paula and Kelly are comparing their running times.

Paula completed a 10-mile run in 65 minutes.

Kelly completed a 10-km run in 40 minutes.

Given that 8 km are equal to 5 miles, which girl has the greater average speed?

4.4 Density

This section will show you how to:
- solve problems involving density

Key words
density
mass
volume

Density is the **mass** of a substance per unit **volume**, usually expressed in grams per cm^3. The relationship between the three quantities is:

$$density = \frac{mass}{volume}$$

You can remember this with a triangle similar to that for distance, speed and time.

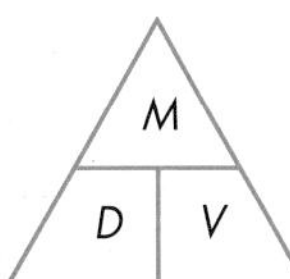

mass = *density* × *volume*

density = *mass* ÷ *volume*

volume = *mass* ÷ *density*

Note: Density is defined in terms of mass. The common metric units for mass are grams and kilograms. Try not to mix up mass with weight. The common metric unit for weight is the Newton. You may have learnt about the difference between mass and weight in science.

EXAMPLE 10

A piece of metal has a mass of 30 g and a volume of 4 cm^3. What is the density of the metal?

$$\text{Density} = \frac{\text{mass}}{\text{volume}}$$

$$= \frac{30}{4} = 7.5 \text{ g/cm}^3$$

EXAMPLE 11

What is the mass of a piece of rock which has a volume of 34 cm^3 and a density of 2.25 g/cm^3?

Mass = volume × density

= 34 × 2.25 = 76.5 g

EXERCISE 4D

B

1 Find the density of a piece of wood with a mass of 6 g and a volume of 8 cm^3.

2 Calculate the density of a metal if 12 cm^3 of it has a mass of 100 g.

3 Calculate the mass of a piece of plastic, 20 cm^3 in volume, if its density is 1.6 g/cm^3.

4 Calculate the volume of a piece of wood which has a mass of 102 g and a density of 0.85 g/cm^3.

5 Find the mass of a marble model, 56 cm^3 in volume, if the density of marble is 2.8 g/cm^3.

6 Calculate the volume of a liquid with a mass of 4 kg and a density of 1.25 g/cm^3.

7 Find the density of the material of a pebble which has a mass of 34 g and a volume of 12.5 cm^3.

8 It is estimated that the statue of Queen Victoria in Endcliffe Park, Sheffield, has a volume of about 4 m^3.

The density of the material used to make the statue is 9.2 g/cm^3. What is the estimated mass of the statue?

9 I bought a 50 kg bag of coal, and estimated the total volume of coal to be about 28 000 cm^3.

What is the density of coal, in g/cm^3?

10 A 1 kg bag of sugar has a volume of about 625 cm^3. What is the density of sugar in g/cm^3?

PS **11** Two statues look identical and both appear to be made out of gold. One of them is a fake.

The density of gold is 19.3 g/cm^3.

The statues each have a volume of approximately 200 cm^3.

The first statue has a mass of 5.2 kg.

The second statue has a mass of 3.8 kg.

Which one is the fake?

AU **12** A piece of metal has a mass of 345 g and a volume of 15 cm^3.

A different piece of metal has a mass of 400 g and a density of 25 g/cm^3.

Which piece of metal has the bigger volume and by how much?

FM **13** Two pieces of scrap metal are melted down to make a single piece of metal.

The first piece has a mass of 1.5 tonnes and a density of 7000 kg/m^3.

The second piece has a mass of 1 tonne and a density of 8000 kg/m^3.

Work out the total volume of the new piece.

GRADE BOOSTER

- **D** You can calculate distance from speed and time
- **D** You can calculate time from speed and distance
- **C** You can solve problems involving speed
- **B** You can solve problems involving density

What you should know now

- How to divide any amount into a given ratio
- The relationships between speed, time and distance
- How to do problems involving direct proportion
- How to compare the prices of products
- How to work out the density of materials

1 Two towns, A and B, are connected by a motorway of length 100 miles and a dual carriageway of length 80 miles as shown.

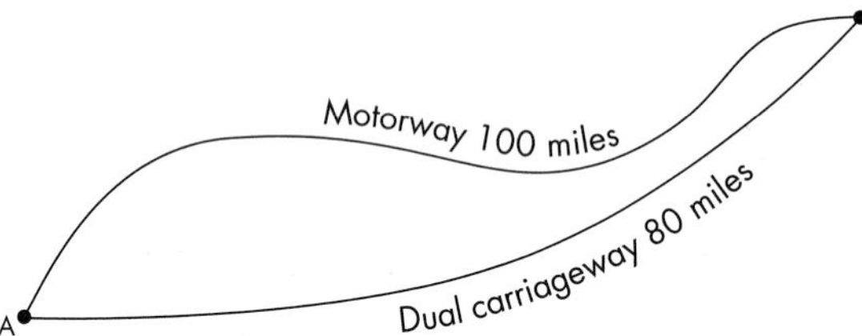

Jack travels from A to B along the motorway at an average speed of 60 mph.

Fred travels from A to B along the dual carriageway at an average speed of 50 mph.

What is the difference in time between the two journeys? Give your answers in minutes. *(4 marks)*

AQA, June 2005, Paper 2 Intermediate, Question 8

2 Blushing Pink paint is made by mixing red and white paint.

The ratio of red paint to white paint used is 3 : 1.

a What percentage of Blushing Pink is red paint? *(2 marks)*

b One day 27 000 litres of red paint are used to make Blushing Pink paint. *(3 marks)*

How many litres of Blushing Pink are made?

AQA, November 2008, Module 3 Higher, Question 1

3 Paula goes for a 30-minute run.

For the first 20 minutes she runs at an average speed of 9 miles per hour. In the next 10 minutes she runs a distance of 1 mile.

Work out the average speed for her 30-minute run.

Give your answer in miles per hour. *(4 marks)*

AQA, Question 3, Specification B, Module 3, Section A, November 2008

4 Bill and Ben buy £10 worth of lottery tickets. Ben pays £7 and Bill pays £3. They decide to share any prize in the ratio of the money they each paid.

a They win £350. How much does Bill get?

b What percentage of the £350 does Ben get?

5 A car produces 2.78 kg of carbon dioxide per hour when driven in a city. The car travels 30 miles in a city at an average speed of 20 mph.

How much carbon dioxide does the car produce during its journey? *(3 marks)*

AQA, November 2006, Paper 2, Question 2

6 Susan completes a journey in two stages. In stage 1 of her journey, she drives at an average speed of 80 km/h and takes 1 hour 45 minutes.

a How far does Susan travel in stage 1 of her journey?

b Altogether, Susan drives 190 km and takes a total time of 2 hours 15 minutes. What is her average speed, in km/h, in stage 2 of her journey?

7 Ruth, Sally and Jenny share £500 between them. Jenny receives the smallest amount of £120. The ratio of Sally's share to Jenny's share is 4 : 3.

Work out the ratio of Ruth's share to Sally's share. Give your answer in its simplest form.

AQA, June 2011 H1 Question 10

8 Two lanes of a motorway are being resurfaced. Each lane is 4 metres wide.

The speed limit through the road works is 80 km/h.
It takes 15 minutes to drive through the road works at this speed.

Calculate the area being resurfaced.
Give your answers in square kilometres.

AQA, June 2011 H2 Question 3

9 A chemist sells a brand of shampoo in two different sizes.

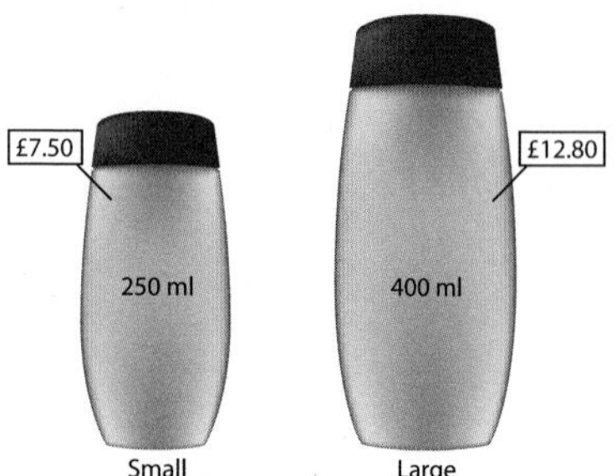

Which is the better value?
Show your working.

AQA, November 2010 H1 Question 4

Worked Examination Questions

1 To be on time, a train must complete a journey of 210 miles in 3 hours.

a Calculate the average speed of the train for the whole journey when it is on time.

b The train averages a speed of 56 mph over the first 98 miles of the journey. Calculate the average speed for the remainder of the journey so that the train arrives on time.

1 a Average speed = distance ÷ time

210 ÷ 3 = 70 mph

1 mark for method. 1 mark for correct answer.

Note that one question in the examination will ask you to state the units of your answer. This is often done with a speed question.

b 98 ÷ 56 = 1.75 which is 1 hour and 45 minutes.

First find out how long the train took to do the first 98 miles.

98 ÷ 56 gets 1 mark for method.
1.75 or 1 hour 45 minutes gets 1 mark for accuracy.

(210 − 98) ÷ (3 − 1.75) = 112 ÷ 1.25
= 89.6 mph

Now work out the distance still to be travelled (112 miles) and the time left (1 hour 15 minutes = 1.25 hours). Divide distance by time to get the average speed.

(210 − 98) ÷ (3 − 1.75) gets 1 mark for method even if there is an error in one of the figures.
89.6 or an answer following an arithmatic error gets 1 mark for accuracy.

Total: 6 marks

PS **2** Jonathan is comparing two ways to travel from his flat in London to his parents' house.

Tube, train and taxi

It takes 35 minutes to get to the railway station by tube in London. A train journey from London to Doncaster takes 1 hour 40 minutes. From Doncaster it is 15 miles by taxi at an average speed of 20 mph.

Car

The car journey is 160 miles at an average speed of 50 mph. Which is the slower journey, tube, train and taxi or car?

2 Time = distance ÷ speed = $\frac{15}{20}$

Work out the time taken by taxi.
This gets 1 mark for method.

= 0.75 hour (or 45 minutes)

This gets 1 mark for accuracy.

Total time = 35 minutes + 1 hour 40 minutes + 45 minutes

Work out the **total time** for tube, train and taxi.

= 3 hours

This is required to compare with the car.
This gets 1 mark.

Time = distance ÷ speed = 160 ÷ 50

= 3.2 hours (or 3 hours 12 minutes)

Work out the time taken by car.
This gets 1 mark for method and 1 mark for accuracy.

Car is 12 minutes slower.

State the conclusion following from your results. This gets 1 mark.

Total: 6 marks

4

Functional Maths
Organising your birthday dinner

To celebrate your birthday, you have decided to hold a large dinner party for your friends and family.

You have invited 15 people in total and so far you have had responses from eight people, all of whom can attend.

As the dinner party will be a big event for you, you want to begin preparing for it straight away. You decide to plan for several different themes, so that you can work out the price of the ingredients for each menu for your dinner and work out how much money you will need to spend.

Chilli pasta *Serves 2*

175 g pasta
40 ml olive oil
2 onions
4 cloves garlic
1 red jalapeno pepper
185 g yellow peppers, roasted
Basil

Ragu Bolognese *Serves 16*

450 g minced beef
450 g minced pork
1¼ kg pasta
30 ml olive oil
225 g chicken liver
2 onions
4 garlic cloves
150 g streaky bacon
400 g chopped tomatoes
200 g tomato purée
400 ml red wine
Basil

Wild mushroom tart *Serves 4*

275 g puff pastry
25 g butter
300 g wild mushrooms
25 g cheese
1 clove garlic
1 egg

Steak and kidney pudding *Serves 6*

450 g diced beef
150 g kidney
30 ml beef dripping
2 onions
40 g plain flour
Thyme
Bay leaves
Parsley
1 pint brown beef stock
175 g self-raising flour
75 g suet

Getting started

- Find the cost of 175 g of pasta, if you can buy 500 g for 70p.
- The butcher sells sausages in packs of eight for £2.50. How much would you pay for three sausages, if he will sell them individually?
- How many 300-g portions will 1.3 kg provide?
- Which is cheaper, 200 g of tomatoes for £1.90 or 350 g for £2.50?

Prices of ingredients	Quantity	Cost
Beef dripping	500 g	55p
Beef stock	12 cubes	98p
Minced beef	500 g	£2.89
Diced beef	1 kg	£5.20
Kidney	400 g	£1.20
Minced pork	500 g	£2.00
Chicken liver	400 g	99p
Streaky bacon	300 g	£2.00
Suet	200 g	65p
Puff pastry	500 g	79p
Butter	250 g	85p
Eggs	6	91p
Pasta	500 g	70p
Onion	1kg	75p
Jalapeno pepper	200 g jar	£1.25
Yellow pepper	each	80p
Plain flour	1.5 kg	43p
Self-raising flour	1.5 kg	43p
Cheese	1 kg	£5.12
Chopped tomatoes	400 g tin	55p
Tomato purée	200 g	33p
Wild mushrooms	125 g	£1.69
Olive oil	500 ml	£188
Garlic (8 cloves per bulb)	3 bulbs	89p
Thyme	16 g	68p
Bay leaves	6 grams	88p
Parsley	180g	79p
Basil	115 g	£2.20
Red wine	750 ml	£3.99

Your task

You have not chosen the main course of your dinner yet. To help you decide, you have chosen four of your favourite recipes.

You now need to look at the lists of ingredients and the price list, then work out what you will need for each one, and how much it will cost.

1 First, work on the assumption that no more than the eight guests, who have already confirmed that they are coming, will be able to attend your party.

  Work out the ingredients you will need for each recipe, and the cost of these ingredients (in total and per portion).

2 To make sure that you are able to cater for more guests, work out the ingredients and costs for each recipe for:
  - 10 guests
  - 15 guests.

3 You decide to set a budget for the ingredients for your main course and you choose £75 as a starting point.

  Evaluate how realistic this budget is, considering the number of guests that could attend and the level of choice you would like to offer, taking account that some will eat meat and some will be vegetarians.

  Remember: you must include yourself when you are working out the ingredients that you must buy and the potential cost.

Extension

Your friends and family will need to travel to get to your dinner party. Your friend Sam will set out at 4.30pm and travel 35 miles to your house; your cousin Charlie will set out at 4.00pm and travel 65 miles. If they both travel at the same speed, who will reach your house first? How fast will they each have to travel for both to reach your house at the same time?

Why this chapter matters

People have always needed to measure areas and volumes.

In everyday life, you will, for instance, need to find the area to work out how much carpet to buy to cover a floor; or you will need to find the volume to see how much water is needed to fill a swimming pool. You can do this quickly using formulae.

Measuring the world

From the earliest times, farmers have needed to know the area of their fields to see how many crops they could grow or animals they could support. One of the oldest units of area used in England was the acre, which was the amount of land that a man could plough in a day. When land is bought and sold, the cost depends on the area. That is true today, too, and a considerable part of the cost of a new house will be the cost of the land it stands on.

Volumes are important too. Volumes tell us how much space there is inside any structure. Whether it is a house, barn, aeroplane, car or office, the volume is important. Did you know, for example, that in England there is a regulation that governs the number of people who can use an office, and this is based on the volume of the room?

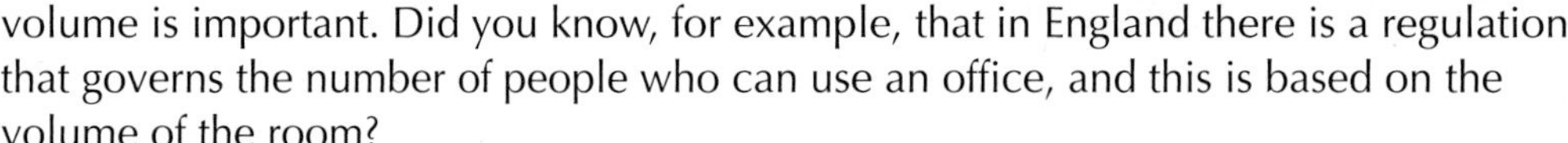

Volumes of containers for liquids also need to be measured. Think, for example, of a car fuel tank, the water tank in a building, your local swimming pool or a reservoir. It is important to be able to calculate the capacity of all these things.

So how do we measure areas and volumes? Some shapes and objects are easy. Others take more ingenuity and skill. In this chapter, you will learn formulae that can be used to calculate areas and volumes of different shapes, based on a few measurements. Many of these formulae were first worked out thousands of years ago. We know, for example, that Archimedes, perhaps the greatest mathematician who ever lived, discovered how to find the volume and surface area of a sphere over 2000 years ago. The fact that these formulae are still in use today shows how important they are.

Chapter 5

Geometry: Shapes

This chapter will show you ...

- **D** how to calculate the area of a trapezium
- **C** how to calculate the volume and surface area of a prism
- **B** how to calculate the volume and surface area of a cylinder
- **A** how to calculate the length of an arc and the area of a sector

Visual overview

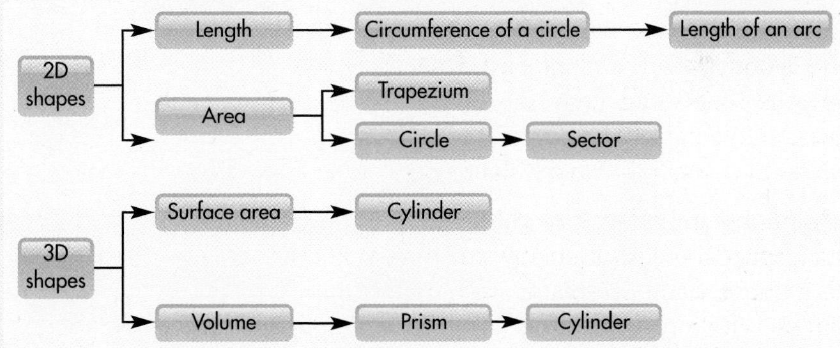

What you should already know

- The area of a rectangle is given by Area = length × width or $A = lw$ **(KS3 level 5, GCSE grade F)**
- The area of a triangle is given by Area = $\frac{1}{2}$ × base × height or $A = \frac{1}{2}bh$ **(KS3 level 6, GCSE grade D)**
- The area of a parallelogram is given by Area = base × height or $A = bh$ **(KS3 level 6, GCSE grade E)**
- The circumference of a circle is given by $C = \pi d$, where d is the diameter of the circle **(KS3 level 6, GCSE grade D)**
- The area of a circle is given by $A = \pi r^2$, where r is the radius of the circle **(KS3 level 6, GCSE grade D)**
- The volume of a cuboid is given by Volume = length × width × height or $V = lwh$ **(KS3 level 6, GCSE grade E)**

continued

- The common metric units to measure area, volume and capacity are shown in this table **(KS3 level 6, GCSE grade E)**

Area	Volume	Capacity
$100 \text{ mm}^2 = 1 \text{ cm}^2$	$1000 \text{ mm}^3 = 1 \text{ cm}^3$	$1000 \text{ cm}^3 = 1$ litre
$10\,000 \text{ cm}^2 = 1 \text{ m}^2$	$1\,000\,000 \text{ cm}^3 = 1 \text{ m}^3$	$1 \text{ m}^3 = 1000$ litres

Quick check

1 Find the areas of the following shapes.

a

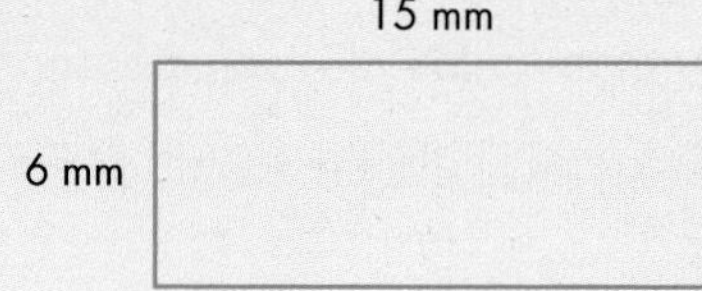

b

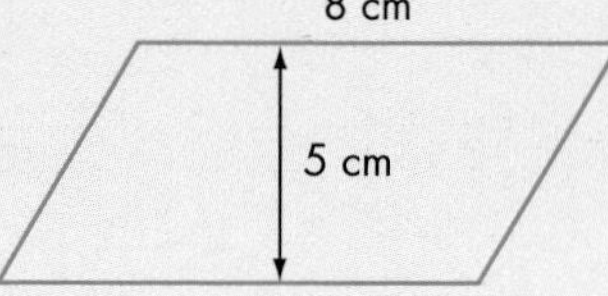

c

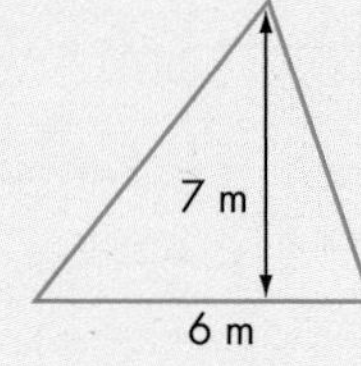

2 Find the volume of this cuboid.

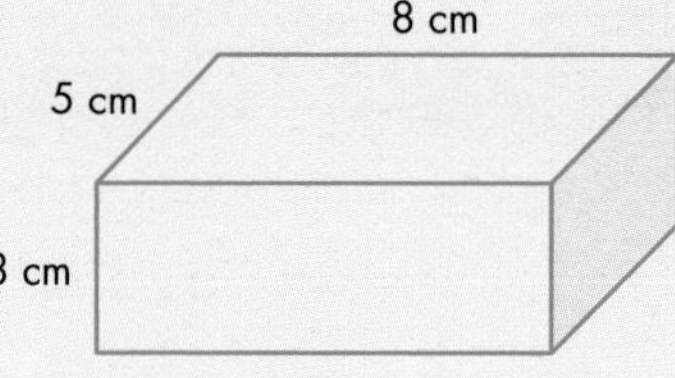

5.1 Circumference and area of a circle

This section will show you how to:

- calculate the circumference and area of a circle

Key words

π
area
circumference

EXAMPLE 1

Calculate the **circumference** of the circle. Give your answer to 3 significant figures.

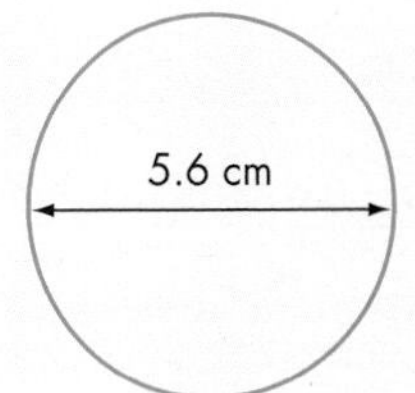

$C = \pi d$

$= \pi \times 5.6$ cm

$= 17.6$ cm (to 3 significant figures)

EXAMPLE 2

Calculate the **area** of the circle. Give your answer in terms of π.

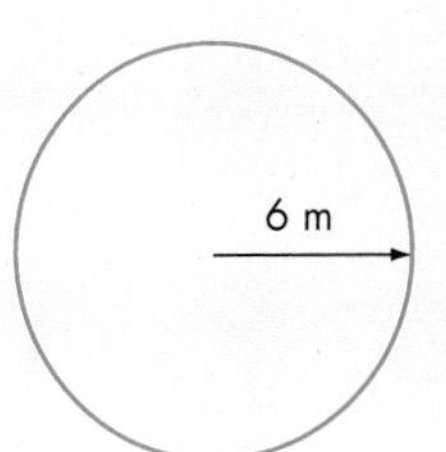

$A = \pi r^2$

$= \pi \times 6^2$ m^2

$= 36\pi$ m^2

EXERCISE 5A

1 Copy and complete the following table for each circle. Give your answers to 1 decimal place.

	Radius	Diameter	Circumference	Area
a	4.0 cm			
b	2.6 m			
c		12.0 cm		
d		3.2 m		

2 Find the circumference of each of the following circles. Give your answers in terms of π.

a Diameter 5 cm **b** Radius 4 cm **c** Radius 9 m **d** Diameter 12 cm

3 Find the area of each of the following circles. Give your answers in terms of π.

a Radius 5 cm **b** Diameter 12 cm **c** Radius 10 cm **d** Diameter 1 m

D

AU 4 A rope is wrapped eight times around a capstan (a cylindrical post), the diameter of which is 35 cm. How long is the rope?

PS 5 The roller used on a cricket pitch has a radius of 70 cm.

A cricket pitch has a length of 20 m. How many complete revolutions does the roller make when rolling the pitch?

6 The diameter of each of the following coins is as follows.

1p: 2.0 cm, 2p: 2.6 cm, 5p: 1.7 cm, 10p: 2.4 cm

Calculate the area of one face of each coin. Give your answers to 1 decimal place.

7 The distance around the outside of a large pipe is 2.6 m. What is the diameter of the pipe?

AU 8 What is the total perimeter of a semicircle of diameter 15 cm?

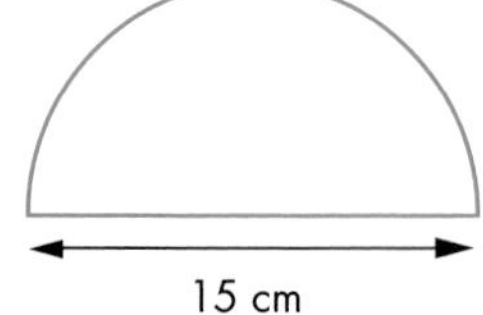

FM 9 A restaurant sells two sizes of pizzas. The diameters are 24 cm and 30 cm. The restaurant claims that the larger size is 50% bigger.

Your friend disagrees and wants to complain to the local trading standards officer. What would you advise? Give a reason for your answer.

10 Calculate the area of each of these shapes, giving your answers in terms of π.

a

12 cm

b

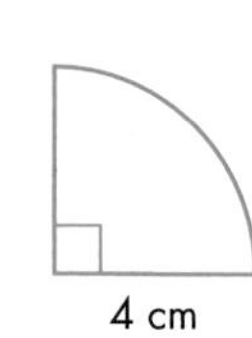

AU 11 Calculate the area of the shaded part of the diagram, giving your answer in terms of π.

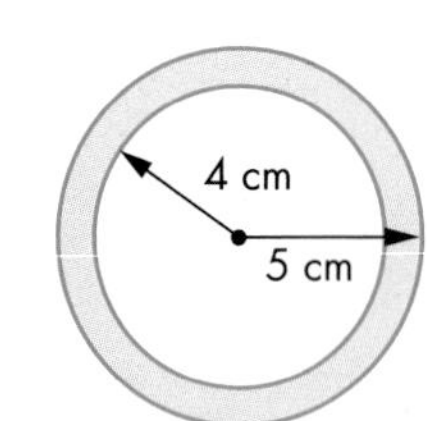

12 This is the plan of a large pond with a gravel path all around it. What area needs to be covered with gravel?

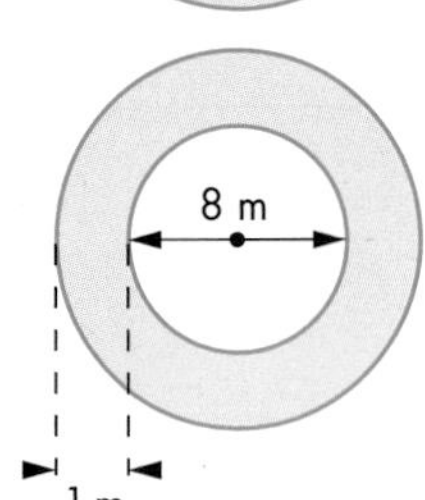

13 A tree in Sequoia National Park in USA is considered to be the largest in the world. It has a circumference at the base of 31.3 m. Would the base of the tree fit inside your classroom?

AU 14 The wheel of a bicycle has a diameter of 70 cm. The bicycle travels 100 m.

How many complete revolutions does the wheel make?

5.2 Area of a trapezium

This section will show you how to:
- find the area of a trapezium

Key words
trapezium

The area of a **trapezium** is calculated by finding the average of the lengths of its parallel sides and multiplying this by the perpendicular distance between them.

$A = \frac{1}{2}(a + b)h$

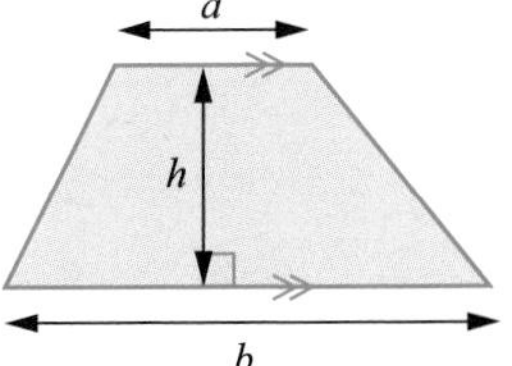

EXAMPLE 3

Find the area of the trapezium ABCD.

$A = \frac{1}{2}(4 + 7) \times 3 \text{ cm}^2$

$= 16.5 \text{ cm}^2$

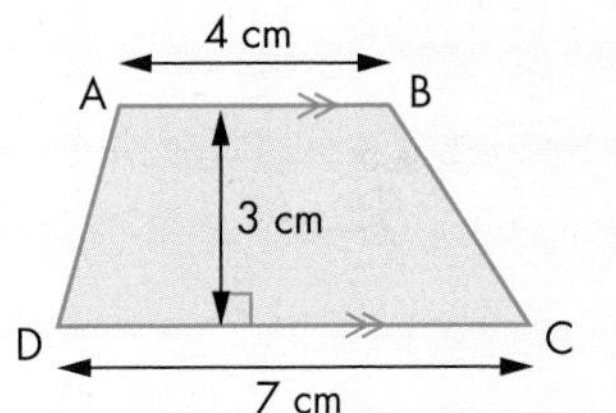

EXERCISE 5B

1 Copy and complete the following table for each trapezium.

	Parallel side 1	Parallel side 2	Vertical height	Area
a	8 cm	4 cm	5 cm	
b	10 cm	12 cm	7 cm	
c	7 cm	5 cm	4 cm	
d	5 cm	9 cm	6 cm	
e	3 m	13 m	5 m	
f	4 cm	10 cm		42 cm^2
g	7 cm	8 cm		22.5 cm^2
h	6 cm		5 cm	40 cm^2

2 Calculate the perimeter and the area of each of these trapeziums.

a

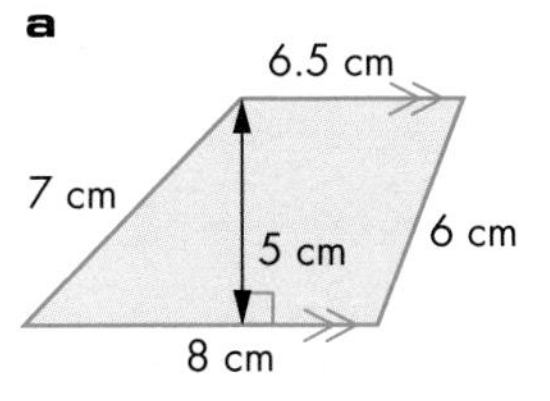

b

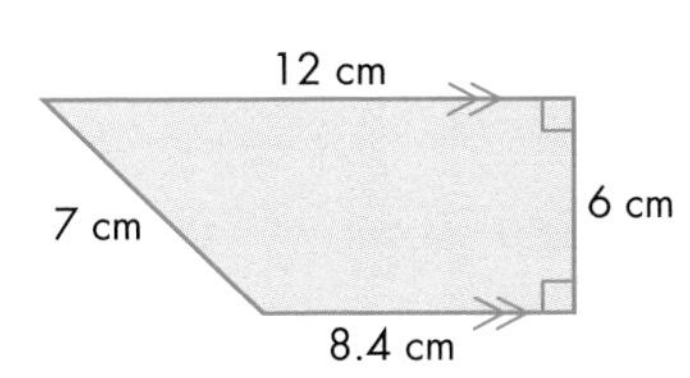

c

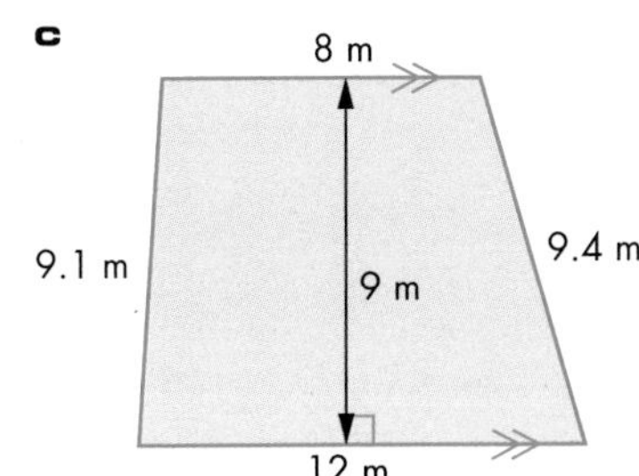

D

D

AU **3** How does this diagram show that the area of a trapezium is $\frac{1}{2}(a + b)h$?

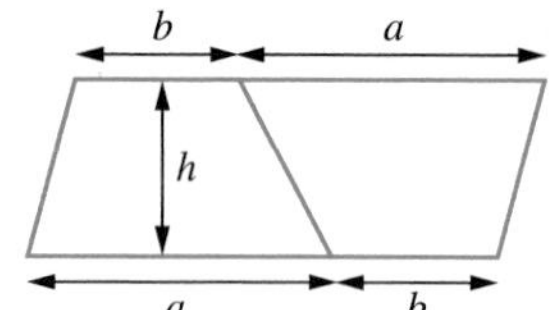

PS **4** Find the area of each part of this picture frame.

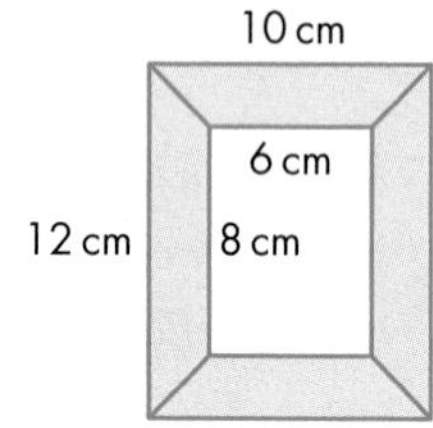

C

5 Calculate the area of each of these compound shapes.

a

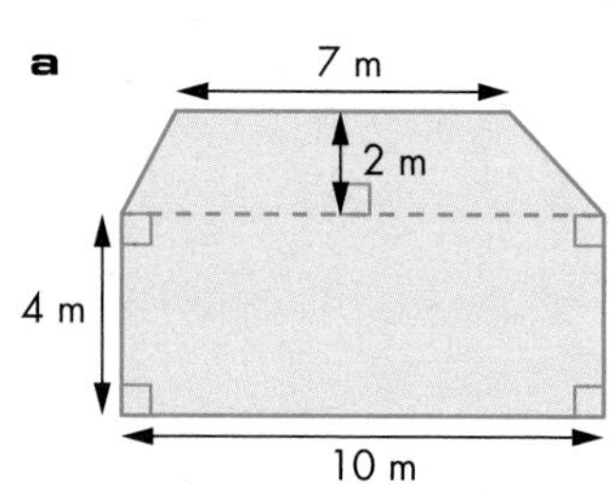

b

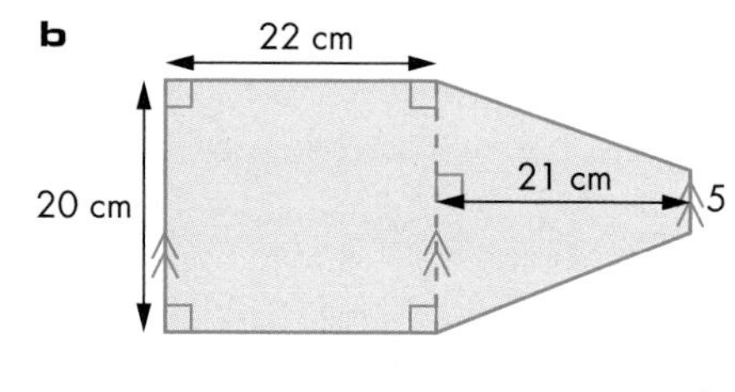

c

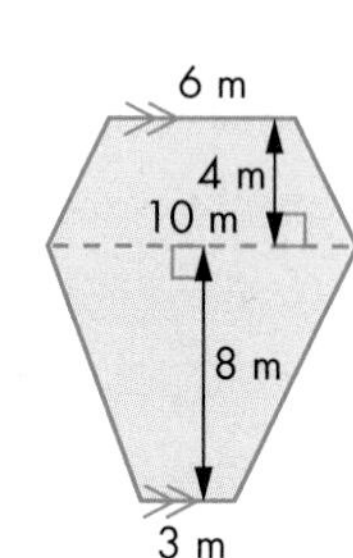

AU **6** Calculate the area of the shaded part in this diagram.

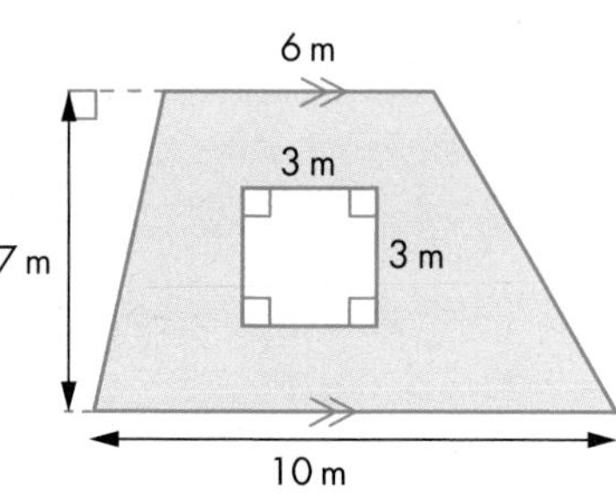

FM **7** This is a sketch of a shed with four walls and a sloping roof.

A one-litre can of wood-protection paint will cover 10 m^2.

How many one-litre cans do you need to put two coats of preservative on each of the four walls?

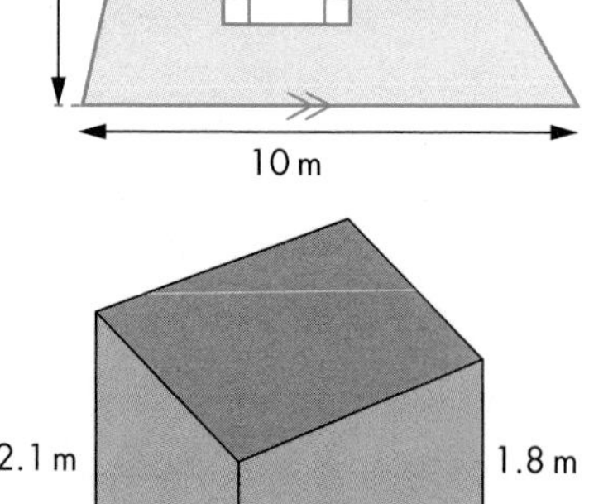

AU **8** What percentage of this shape has been shaded?

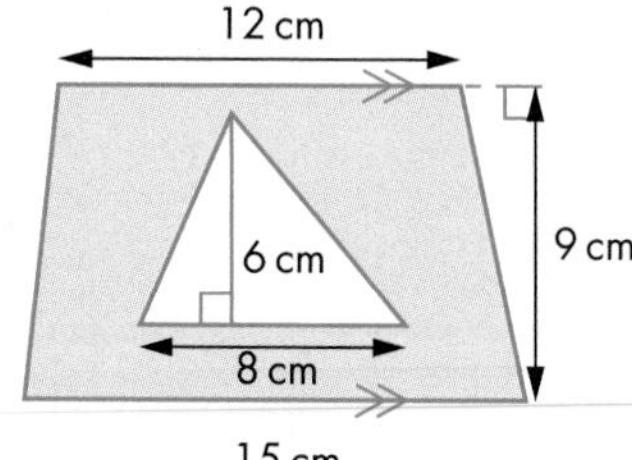

AU **9** The shape of most of Egypt (see map) roughly approximates to a trapezium. The north coast is about 900 km long, the south boundary is about 1100 km long and the distance from north to south is about 1100 km.

What is the approximate area of this part of Egypt?

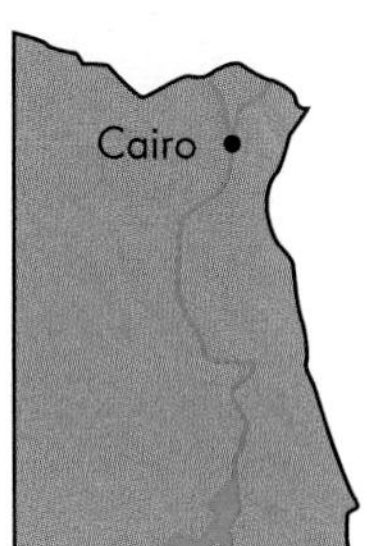

10 The diagram shows an isosceles trapezium.

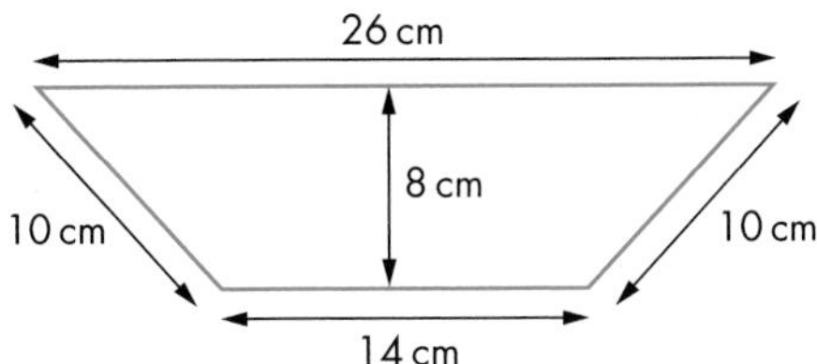

Calculate its area.

C

5.3 Sectors

This section will show you how to:
- calculate the length of an arc and the area of a sector

Key words
arc
sector
subtend

A **sector** is part of a circle, bounded by two radii of the circle and one of the **arcs** formed by the intersections of these radii with the circumference.

The angle **subtended** at the centre of the circle by the arc of a sector is known as the angle of the sector.

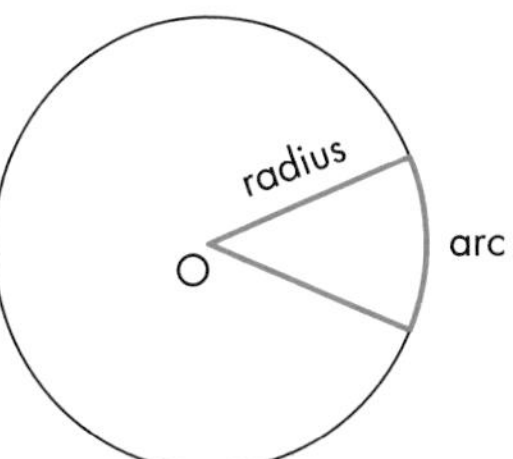

When a circle is divided into only two sectors, the larger one is called the major sector and the smaller one is called the minor sector.

Likewise, their arcs are called the major arc and the minor arc respectively.

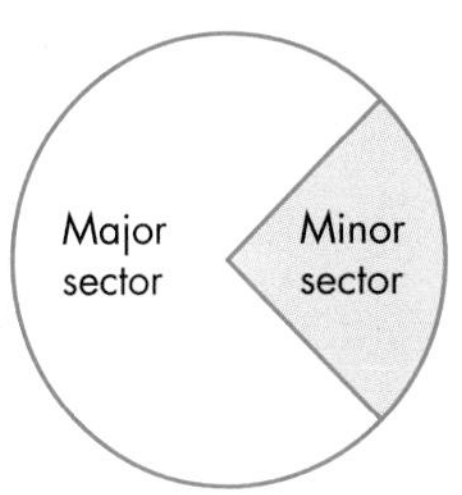

Length of an arc and area of a sector

A sector is a fraction of the whole circle, the size of the fraction being determined by the size of angle of the sector. The angle is often written as θ, a Greek letter pronounced *theta*. For example, the sector shown in the diagram represents the fraction $\frac{\theta}{360}$.

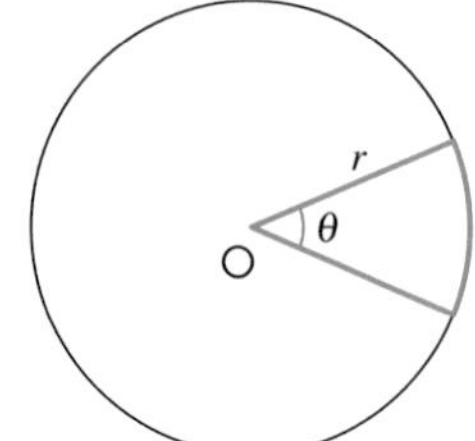

This applies to both its arc length and its area. Therefore,

$$\text{arc length} = \frac{\theta}{360} \times 2\pi r \quad \text{or} \quad \frac{\theta}{360} \times \pi d$$

$$\text{sector area} = \frac{\theta}{360} \times \pi r^2$$

EXAMPLE 4

Find the arc length and the area of the sector in the diagram.

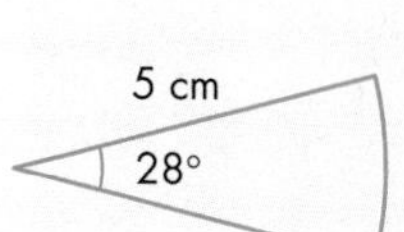

The sector angle is 28° and the radius is 5 cm. Therefore,

$$\text{arc length} = \frac{28}{360} \times \pi \times 2 \times 5 = 2.4 \text{ cm (1 decimal place)}$$

$$\text{sector area} = \frac{28}{360} \times \pi \times 5^2 = 6.1 \text{ cm}^2 \text{ (1 decimal place)}$$

EXERCISE 5C

A

1 For each of these sectors, calculate: **i** the arc length **ii** the sector area.

a

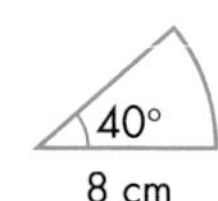

b

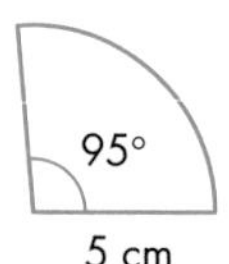

c

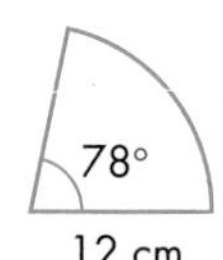

d

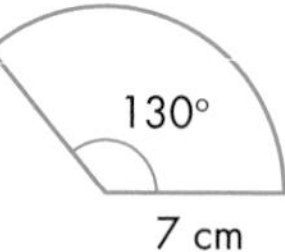

2 Calculate the arc length and the area of a sector whose arc subtends an angle of 60° at the centre of a circle with a diameter of 12 cm. Give your answer in terms of π.

3 Calculate the total perimeter of each of these sectors.

a

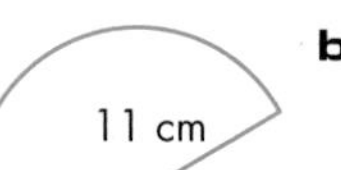

b

22°
8.5 cm

4 Calculate the area of each of these sectors.

a

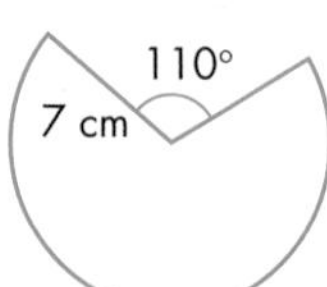

b

50°
8 cm

5 O is the centre of a circle of radius 12.5 cm.

Calculate the length of the arc ACB.

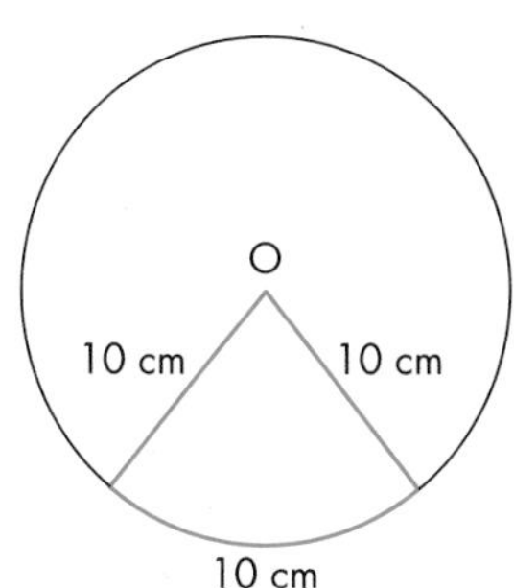

PS 6 **a** Calculate the angle of the minor sector of this circle. Give your answer in terms of π.

b Angles are sometime measured in radians.

The angle you found in part **a** is equal to one radian.

By comparing the sector to an equilateral triangle of side 10 cm, explain why one radian must be a bit less than 60°.

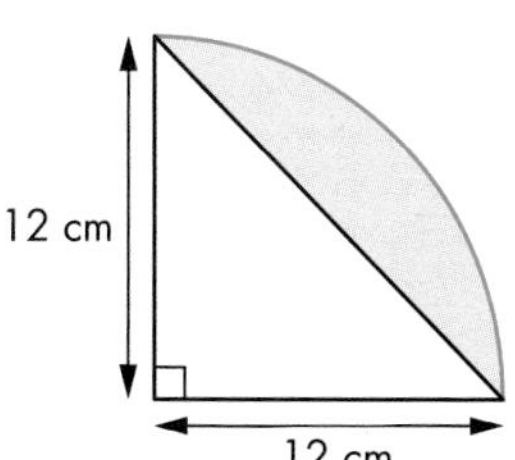

AU 7 The diagram shows a quarter of a circle. Calculate the area of the shaded shape, giving your answer in terms of π.

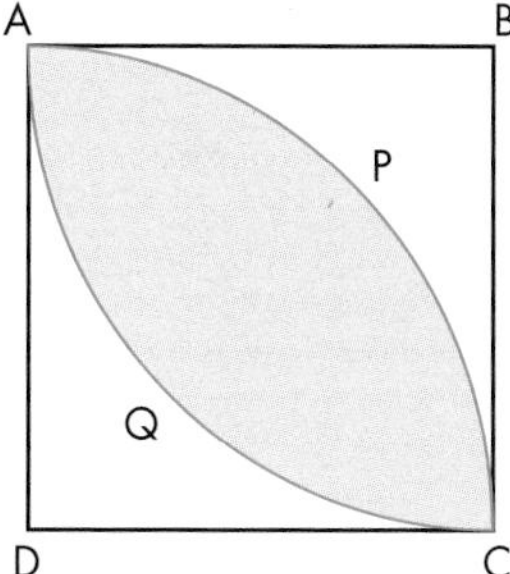

PS 8 ABCD is a square of side length 8 cm. APC and AQC are arcs of the circles with centres D and B. Calculate the area of the shaded part.

A B P Q D C

FM 9 Antique clocks are powered by a pendulum which swings from side to side.

The pendulum of an old clock is 90 cm long.

It swings from side to side through an angle of 10°.

How wide must the clock case be made so that the pendulum can swing freely?

10 Find:

a the perimeter

b the area of this shape.

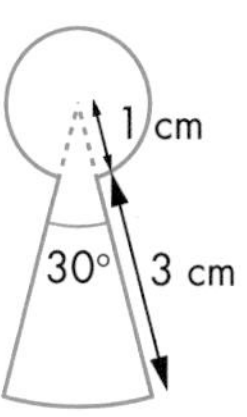

5.4 Volume of a prism

This section will show you how to:
- calculate the volume of a prism

Key words
cross-section
prism

A **prism** is a 3D shape which has the same **cross-section** running all the way through it.

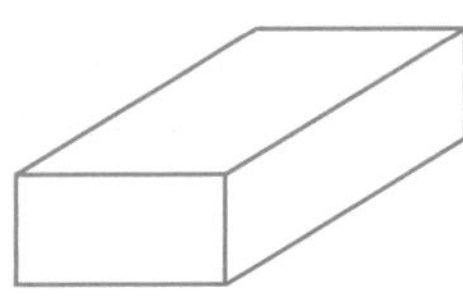

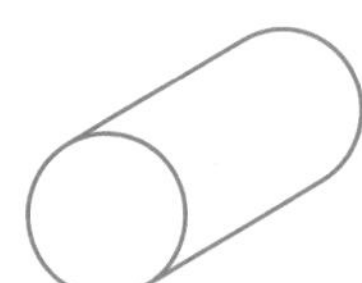

Name:	Cuboid	Triangular prism	Cylinder
Cross-section:	Rectangle	Isosceles Triangle	Circle

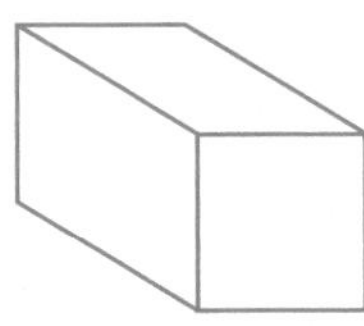

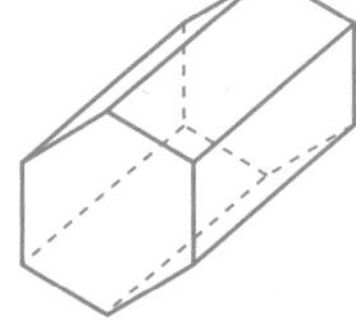

Name:	Cuboid	Hexagonal prism
Cross-section:	Square	Regular Hexagon

The volume of a prism is found by multiplying the area of its cross-section by the length of the prism (or height if the prism is stood on end).

That is, volume of prism = area of cross-section × length **or** $V = Al$

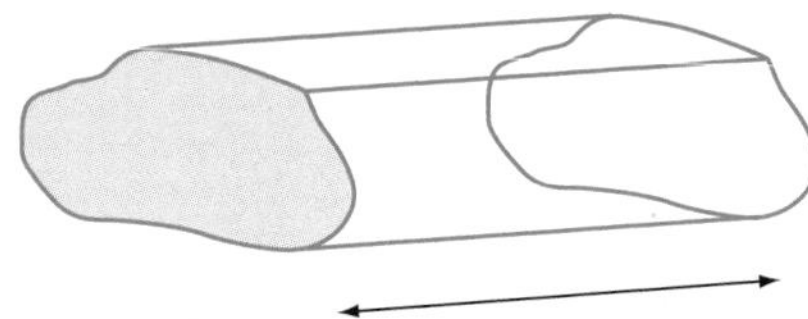

EXAMPLE 5

Find the volume of the triangular prism.

The area of the triangular cross-section $= A = \frac{5 \times 7}{2} = 17.5 \text{ cm}^2$

The volume is the area of its cross-section × length $= Al$

$= 17.5 \times 9 = 157.5 \text{ cm}^3$

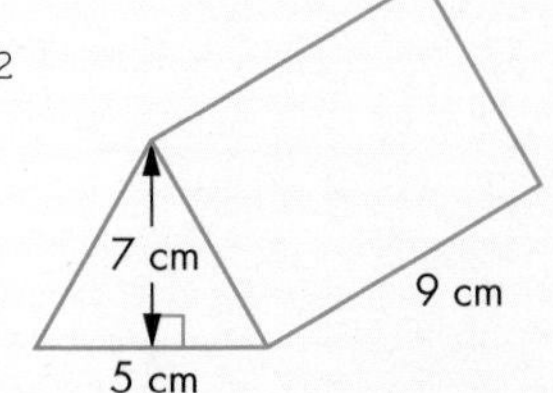

EXERCISE 5D 

1 For each prism shown:

i calculate the area of the cross-section **ii** calculate the volume.

a

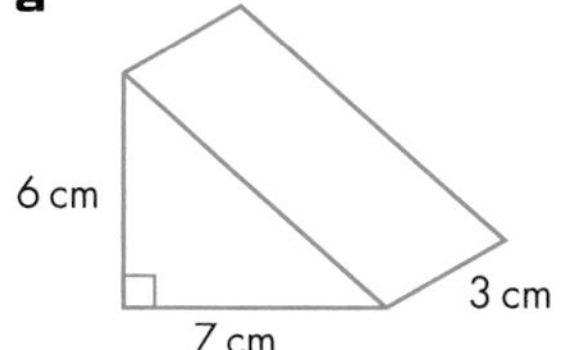

b

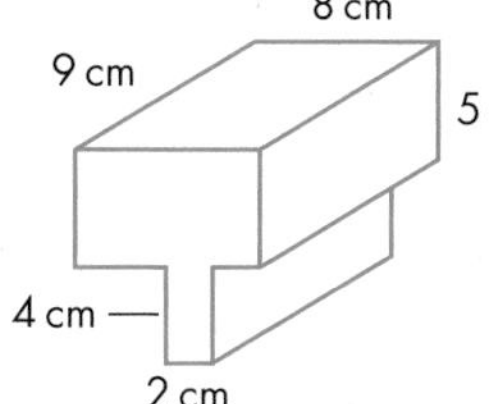

c

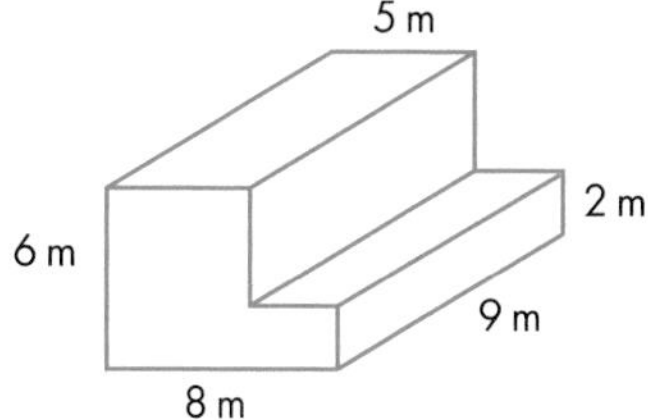

2 Calculate the volume of each of these prisms.

a

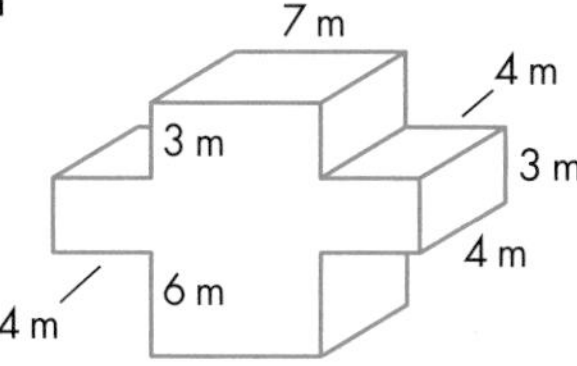

b

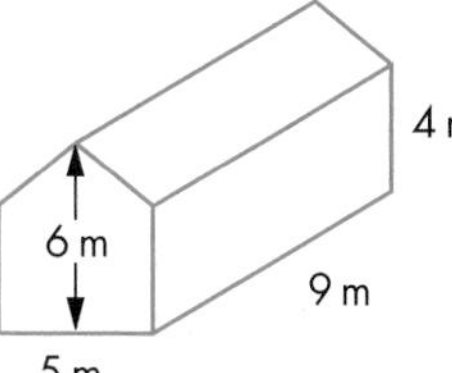

c

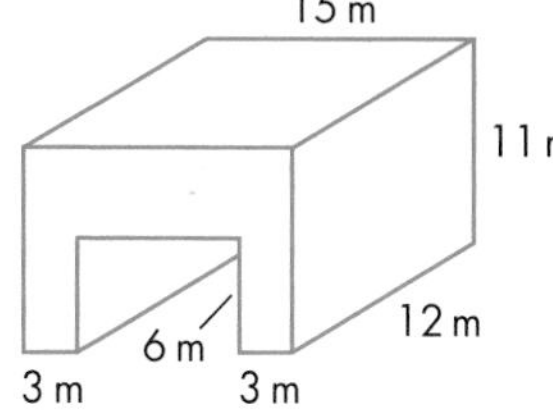

FM 3 A swimming pool is 10 m wide and 25 m long.

It is 1.2 m deep at one end and 2.2 m deep at the other end. The floor slopes uniformly from one end to the other.

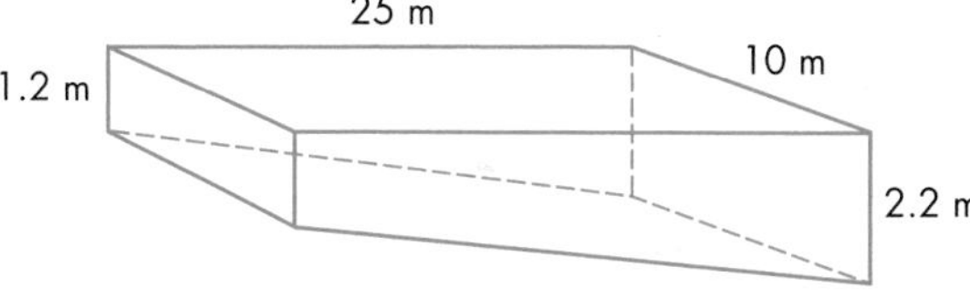

a Explain why the shape of the pool is a prism.

b The pool is filled with water at a rate of 2 m^3 per minute. How long will it take to fill the pool?

PS 4 A conservatory is in the shape of a prism. Calculate the volume of air inside the conservatory with the dimensions shown in the diagram.

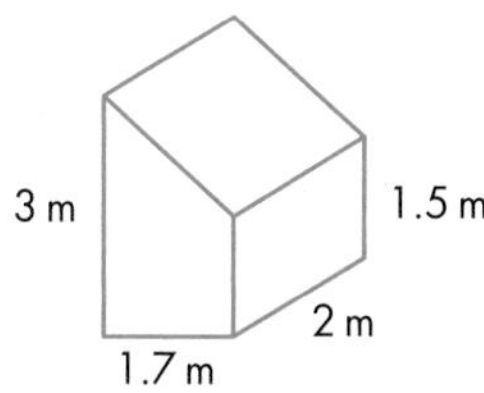

5 Each of these prisms has a uniform cross-section in the shape of a right-angled triangle.

a Find the volume of each prism. **b** Find the total surface area of each prism.

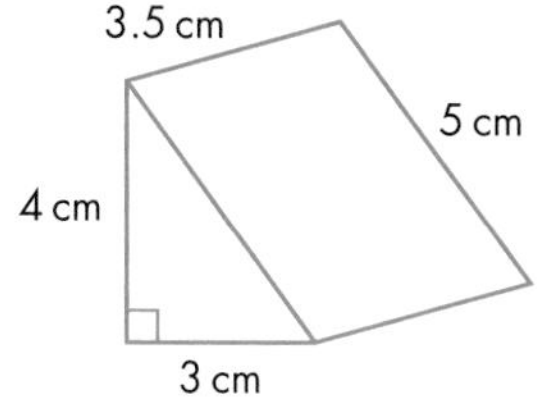

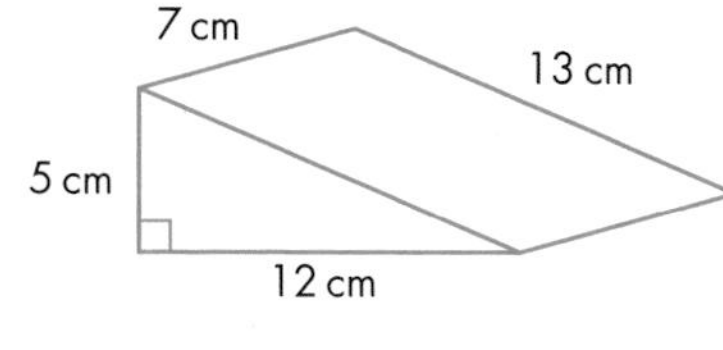

B

AU 6 The top and bottom of the container shown here are the same size, both consisting of a rectangle, 4 cm by 9 cm, with a semicircle at each end. The depth is 3 cm.

Find the volume of the container.

PS 7 In 2009 the sculptor Anish Kapoor exhibited a work called *Svayambh* at the Royal Academy in London. It was a block of red wax in the shape of a prism.

The cross-section was in the shape of an arched entrance.

It was 8 m long and weighed 30 tonnes. It slowly travelled through the galleries on a track.

Calculate the volume of wax used.

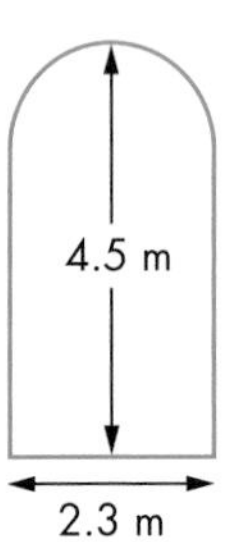

AU 8 A horse trough is in the shape of a semicircular prism as shown.

What volume of water will the trough hold when it is filled to the top? Give your answer in litres.

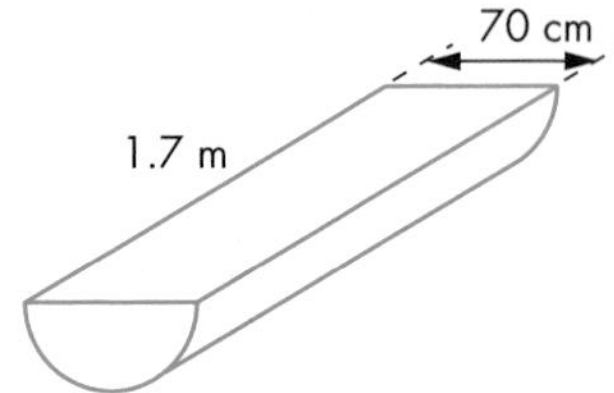

FM 9 The dimensions of the cross-section of a girder (in the shape of a prism), 2 m in length, are shown on the diagram. The girder is made of iron. 1 cm^3 of iron weighs 79 g.

What is the mass of the girder?

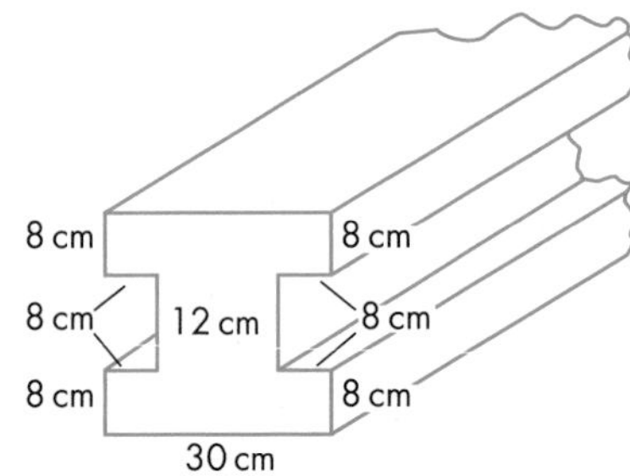

AU 10 Calculate the volume of this prism.

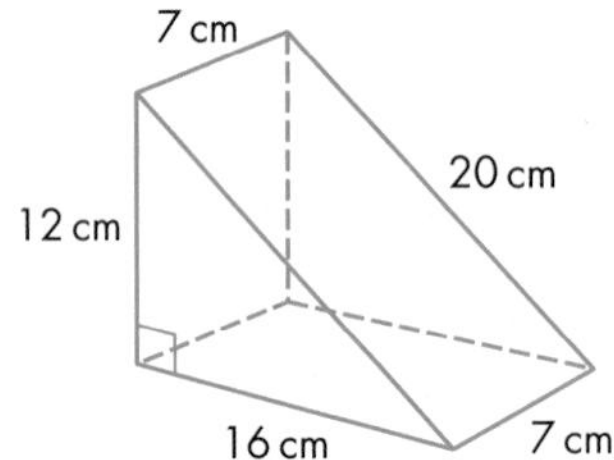

5.5 Cylinders

This section will show you how to:
- calculate the volume and surface area of a cylinder

Key words
cylinder
surface area
volume

Volume

Since a **cylinder** is an example of a prism, its **volume** is found by multiplying the area of one of its circular ends by the height.

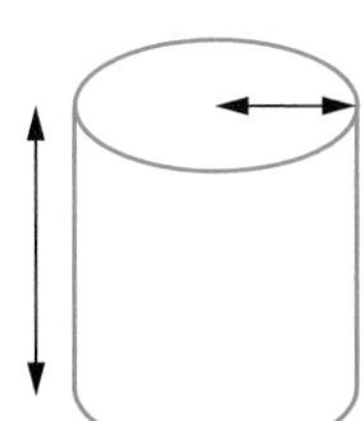

That is, volume = $\pi r^2 h$

where r is the radius of the cylinder and h is its height or length.

EXAMPLE 6

What is the volume of a cylinder having a radius of 5 cm and a height of 12 cm?

Volume = area of circular base × height
$= \pi r^2 h$
$= \pi \times 5^2 \times 12 \text{ cm}^3$
$= 942 \text{ cm}^3$ (3 significant figures)

Surface area

The total **surface area** of a cylinder is made up of the area of its curved surface plus the area of its two circular ends.

The curved surface area, when opened out, is a rectangle with length equal to the circumference of the circular end.

curved surface area = circumference of end × height of cylinder
$= 2\pi rh$ **or** πdh

area of one end $= \pi r^2$

Therefore, total surface area = $2\pi rh + 2\pi r^2$ **or** $\pi dh + 2\pi r^2$

EXAMPLE 7

What is the total surface area of a cylinder with a radius of 15 cm and a height of 2.5 m?

First, you must change the dimensions to a *common unit*. Use centimetres in this case.

Total surface area = $\pi dh + 2\pi r^2$
$= \pi \times 30 \times 250 + 2 \times \pi \times 15^2 \text{ cm}^2$
$= 23\,562 + 1414 \text{ cm}^2$
$= 24\,976 \text{ cm}^2$
$= 25\,000 \text{ cm}^2$ (3 significant figures)

EXERCISE 5E

B

1 For the cylinders below find:

i the volume **ii** the total surface area.

Give your answers to 3 significant figures.

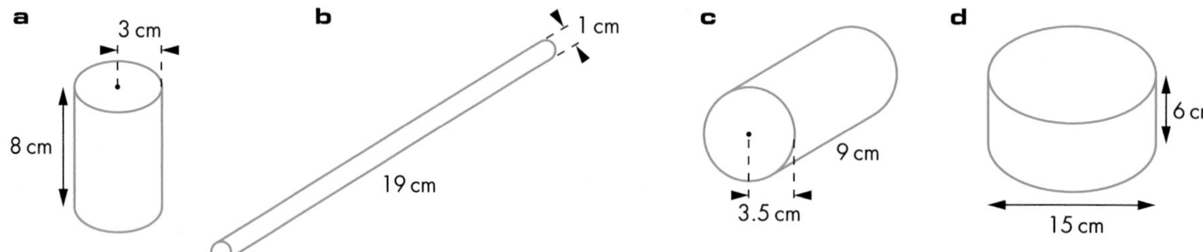

2 For each of these cylinder dimensions find:

i the volume **ii** the curved surface area.

Give your answers in terms of π.

a Base radius 3 cm and height 8 cm **b** Base diameter 8 cm and height 7 cm

c Base diameter 12 cm and height 5 cm **d** Base radius of 10 m and length 6 m

AU 3 The diameter of a marble, cylindrical column is 60 cm and its height is 4.2 m. The cost of making this column is quoted as £67.50 per cubic metre. What is the estimated total cost of making the column?

AU 4 Find the mass of a solid iron cylinder 55 cm high with a base diameter of 60 cm. 1 cm^3 iron has a mass of 7.9 g.

5 A solid cylinder has a diameter of 8.4 cm and a height of 12.0 cm. Calculate the volume of the cylinder.

8.4 cm
12 cm

FM 6 A cylindrical food can has a height of 10.5 cm and a diameter of 7.4 cm.

What can you say about the size of the paper label around the can?

7 A cylindrical container is 65 cm in diameter. Water is poured into the container until it is 1 m deep. How much water is in the container? Give your answer in litres.

FM 8 A drinks manufacturer wishes to market a new drink in a can. The quantity in each can must be 330 ml.

Suggest a suitable height and diameter for the can.

You might like to look at the dimensions of a real drinks can.

9 A cylindrical can of soup has a diameter of 7 cm and a height of 9.5 cm. It is full of soup, which weighs 625 g. What is the density of the soup?

B

A*

AU **10** A metal bar, 1 m long and with a diameter of 6 cm, has a mass of 22 kg. What is the density of the metal from which the bar is made?

PS **11** Wire is commonly made by putting hot metal through a hole in a plate.

What length of wire of diameter 1 mm can be made from a 1 cm cube of metal?

FM **12** The engine size of a car is measured in litres. This tells you the total capacity of the cylinders in which the pistons move up and down. For example, in a 1.6 litre engine with four cylinders, each cylinder will have a capacity of 0.4 litres.

Cylinders of a particular size can be long and thin or short and fat; they will give the engine different running characteristics.

In a racing car, the diameter can be approximately twice the length. This means the engine will run at very high revs.

Suggest possible dimensions for a 0.4 litre racing car cylinder.

PS **13** Cylinder A has radius r and height h. It has a volume of 100π cm^3.

Cylinders B, C and D have dimensions as shown.

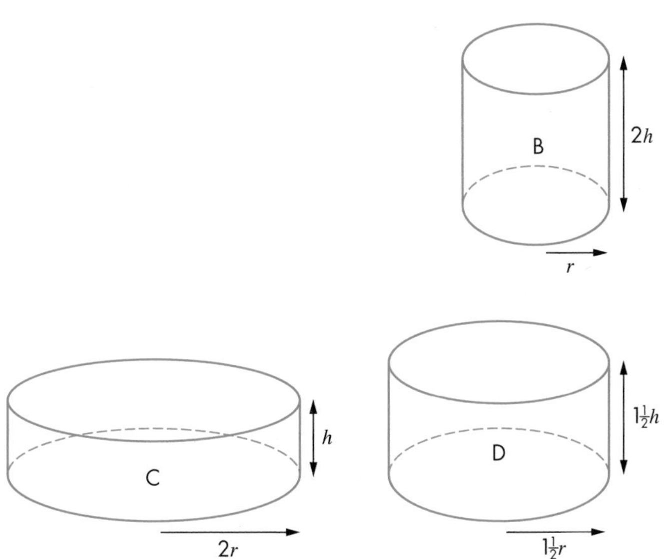

Arrange the three cylinders in order of volume, starting with the smallest.

You must show your working.

PS **14** A cube of side r and a cylinder of radius, r and height h have the same volume.

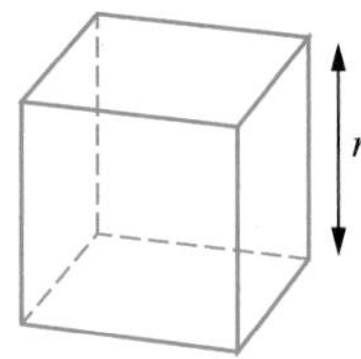

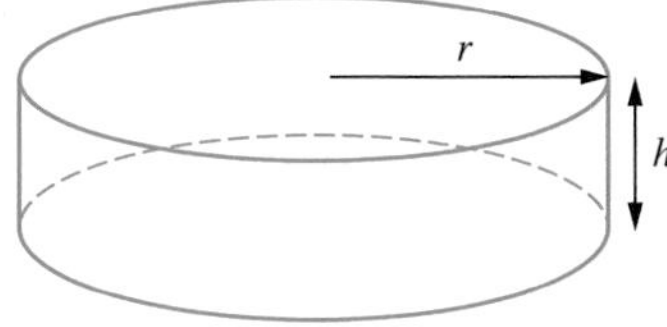

Show that h is approximately $0.32r$.

SUMMARY

GRADE BOOSTER

D You can calculate the circumference and area of a circle

D You can calculate the area of a trapezium

C You can calculate the volume of prisms and cylinders

B You can calculate the length of an arc and the area of a sector

What you should know now

- For a sector of radius r and angle θ:

  Arc length $= \frac{\theta}{360} \times 2\pi r$ or $\frac{\theta}{360} \times \pi d$

  Area of a sector $= \frac{\theta}{360} \times \pi r^2$

- The area of a trapezium is given by:

  $A = \frac{1}{2}(a + b)h$

  where h is the vertical height, and a and b are the lengths of the two parallel sides
- The volume of a prism is given by $V = Al$, where A is the cross-section area and l is the length of the prism
- The volume of a cylinder is given by $V = \pi r^2 h$, where r is the radius and h is the height or length of the cylinder
- The curved surface area of a cylinder is given by $S = 2\pi rh$, where r is the radius and h is the height or length of the cylinder

1

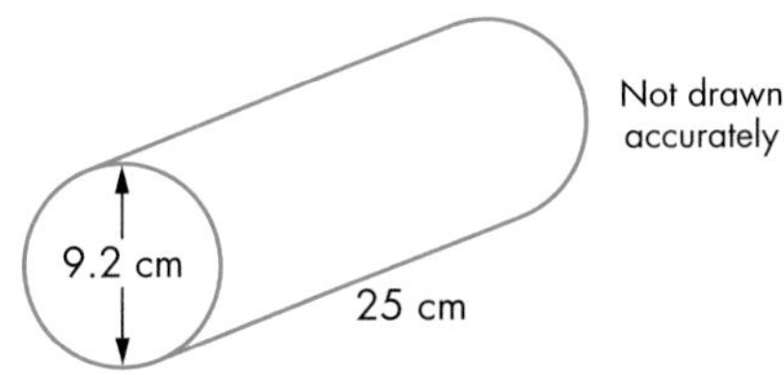

a Calculate the area of one end of the cylinder. *(2 marks)*

b Calculate the **total** surface area of the cylinder. You **must** show your working. *(3 marks)*

AQA, June 2007, Module 5, Paper 2 Higher, Question 3

2 **a** Cylinder A has a height of 5 cm and a diameter of 16 cm.

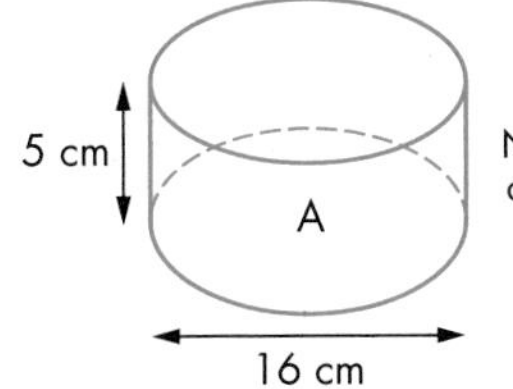

Calculate the volume of the cylinder A.

Give your answer in terms of π.

State the units of your answer. *(4 marks)*

b Cylinder B has a height of 20 cm and a radius of r cm. *(3 marks)*

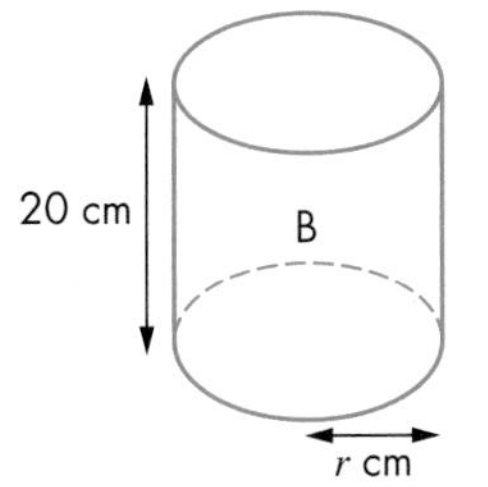

Cylinder B has the same volume as cylinder A.

Calculate the value of r. *(3 marks)*

AQA, May 2008, Paper 1 Higher, Question 14

3

A solid cube of side 25 cm has a circular hole cut through vertically.

The circle has a diameter of 14 cm.

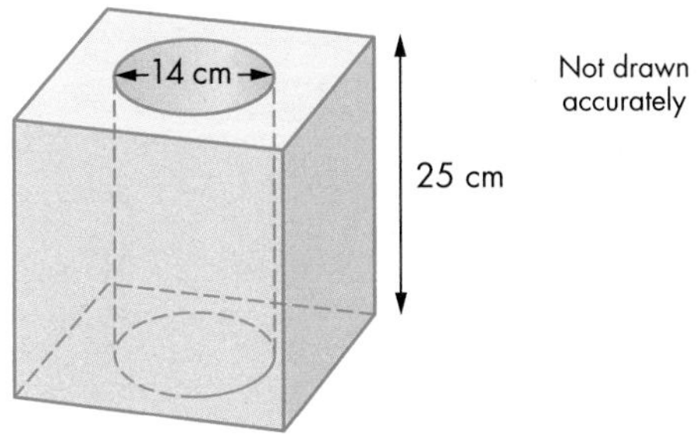

Calculate the volume remaining. *(4 marks)*

AQA, November 2008, Paper 2 Higher, Question 17

4 AB is a minor arc of a circle of radius 5.2 m.

Angle AOB = 100°

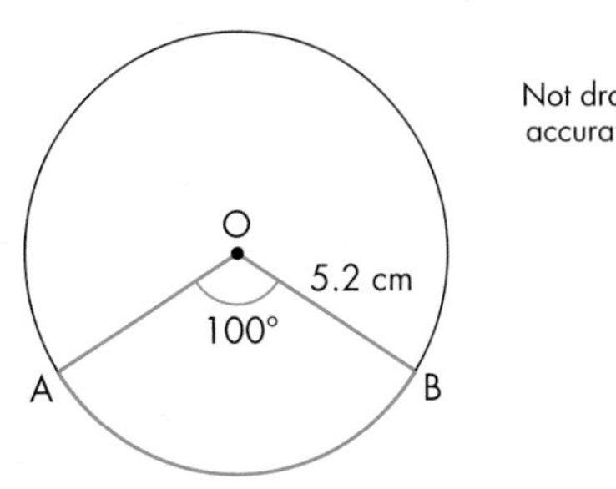

Calculate the length of the minor arc AB. *(3 marks)*

AQA, June 2007, Module 5, Paper 2 Higher, Question 11

5 The diagram shows a prism.

The cross-section of the prism is a sector of a circle of radius 12 cm.

The angle of the sector is 60°.

The prism is 20 cm long.

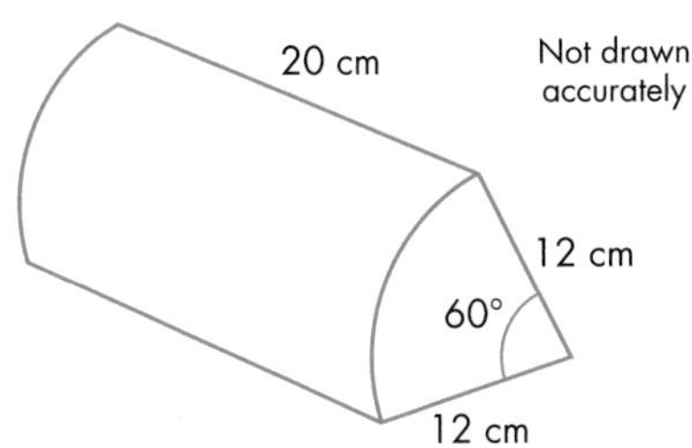

Calculate the volume of the prism.

Give your answer in terms of π. *(4 marks)*

AQA, November 2008, Paper 1 Higher, Question 21

6

A solid sphere of radius 3 cm just fits inside a hollow cone of radius 6 cm and height 8 cm.

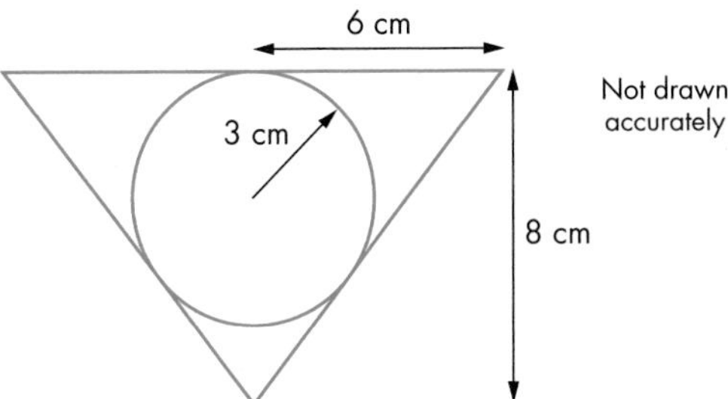

Calculate the fraction of the volume of the cone taken up by the sphere.

You **must** show your working. *(3 marks)*

AQA, November 2007, Paper 2 Higher, Question 18

Worked Examination Questions

AU **1** The diagram shows a pepper pot. The pot consists of a cylinder and a hemisphere. The cylinder has a diameter of 5 cm and a height of 7 cm.

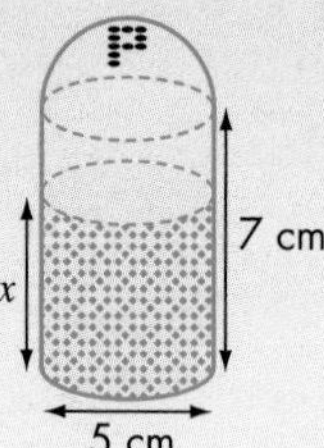

The pepper takes up half the total volume of the pot. Find the depth of pepper in the pot marked x in the diagram.

1 Volume of pepper pot

$= \pi r^2 h + \frac{2}{3}\pi r^3$ — This gets 1 method mark for setting up equation.

$= \pi \times 2.5^2 \times 7 + \frac{2}{3} \times \pi \times 2.5^3$ cm^3 — This gets 1 method mark for correct substitutions.

$= 170.2$ cm^3 — This gets 1 accuracy mark for correct answers.

So volume of pepper = 85.1 cm^3

Therefore,

$\pi r^2 x = 85.1$ cm^3 — This gets 1 method mark for setting up equation.

and $x = \dfrac{85.1}{\pi \times 2.5^2} = 4.3$ cm (1 decimal place) — This gets 1 method mark for correct rearrangement and 1 accuracy mark for correct answer.

Total: 6 marks

Worked Examination Questions

FM **2** Aluminium craft wire is available in different diameters. One manufacturer sells a 5 metre coil of wire with a diameter of 1.6 mm for £5.00.

What volume of aluminium is that?

2 Volume = $\pi \times r^2 \times h = \pi \times 0.8^2 \times 5000$

$= 10\,000 \text{ mm}^3$ to 2 sf

Total: 4 marks

Find the volume of a cylinder with a length of 5 m and a diameter of 1.6 mm. This gets 1 mark for method.
The units need to be the same. In millimetres, the length is 5000mm. This gets 1 mark for method.
This gets 1 mark for accuracy for correct substitution.

This gets 1 mark for accuracy and rounding answer to 2 sf.
The answer could also be given in cm^3 (10 or 10.1).
Rounding off to 3 sf would also be acceptable (10 100).
Note that the £5.00 is not used in the question. Sometimes extraneous information is included to test if you can pick out relevant information and ignore irrelevant information.

PS **3** Three balls of diameter 8.2 cm just fit inside a cylindrical container.

What is the internal volume of the container?

Give your answer to a sensible degree of accuracy.

3 The diameter of the cylinder is 8.2 cm and the height is 24.6 cm.

The internal volume = $\pi \times 4.1^2 \times 24.6$

$= 1299.13\ldots$

$= 1300 \text{ cm}^3$ to 2 sf

Total: 4 marks

This gets 1 mark for method.

This gets 1 mark for method, 1 mark for accuracy and 1 mark for sensible rounding.

5

Functional Maths
Fitting a carpet

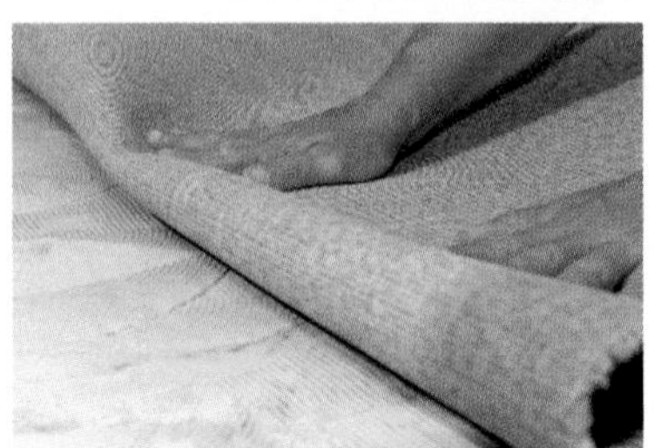

Steve is building an extension in his house to accommodate an extra family room. He will want to lay a carpet in the new room.

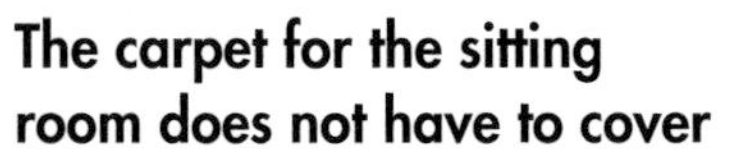

The carpet for the sitting room does not have to cover the whole room and can only be bought in whole metre quantities. Your task is to look at the information below and advice Steve about what size the carpet he should buy.

Your task:

Using all the information on these pages, advise Steve on the size of carpet he should buy.

Begin by thinking about the following points:

- What shape Steve's new room will be – think of at least **two** potential shapes for Steve's new room and plan the carpet size for each
- The dimensions of each room shape
- The area of each room that could be covered by the carpet

Remember: Keep it simple to start with and use realistic measurements.

Discuss your idea with a partner, making sure that you have taken into account all the criteria that Steve has give you. Write down the assumptions that you have made and then find the total carpet size that each of your potential rooms will require.

Steve's criteria

Steve has given you the following criteria:

- The carpet must cover at least half of the floor area but not more than three-quarters of the floor area.
- Due to roll sizes, the carpet can only be purchased in whole-metre widths and lengths, for example 2 m by 3 m.
- Steve can only tell you that his room has a total floor area of $30m^2$.

Extension

Steve will need to add skirting board to his room once the carpet is laid. How much skirting board will he need for each of your room designs?

Why this chapter matters

If you were asked to circle one of these to describe mathematics, which would it be?

Art Science Sport Language

In fact, you could circle them all.

Mathematics is important in **art**. The *Mona Lisa,* probably the most famous painting in the world, uses the proportions of the 'Golden Ratio' (approximately 1.618). This 'Golden Ratio' as shown by the red rectangles marked on this copy of the painting, is supposed to be particularly attractive to the human eye.

Obviously, you cannot do much **science** without using mathematics. In 1962, a *Mariner* space probe went off course and had to be destroyed because someone had used a wrong symbol in a mathematical formula that was part of its programming.

Is mathematics used in **sport**? There are national and international competitions each year that use mathematics. For example, there is a world Sudoku championship each year, and university students compete in the Mathematics Olympiad each year.

But perhaps the most important description in the list above is maths as a **language**.

Mathematics is the only universal language. If you write the equation $3x = 9$, it will be understood by people in all countries.

Algebra is the way that the language of mathematics is expressed.

Algebra comes from the Arabic *al-jabr* which means something similar to 'completion'. It was used in a book written in 820AD by a Persian mathematician called al-Khwarizmi.

The use of symbols developed until the middle of the 17th century, when René Descartes introduced what is regarded as the basis of the algebra we use today.

Chapter

Algebra: Expressions and equations 1

1 Basic algebra

2 Factorisation

3 Solving linear equations

4 Setting up equations

5 Trial and Improvement

This chapter will show you ...

- **D** how to manipulate basic algebraic expressions by multiplying terms together, expanding brackets and collecting like terms
- **C** how to factorise linear expressions
- **C** how to set up and solve linear equations
- **B** how to solve linear equations

Visual overview

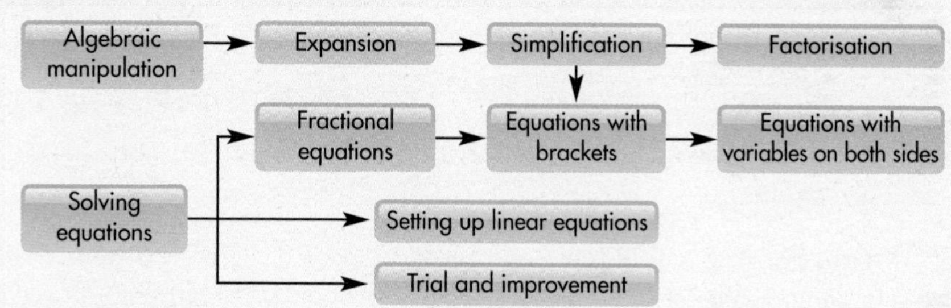

What you should already know

- The basic language of algebra **(KS3 level 5, GCSE grade E)**
- How to collect together like terms **(KS3 level 5, GCSE grade E)**
- How to multiply together terms such as $2m \times 3m$ **(KS3 level 5, GCSE grade E)**

Quick check

1 Expand the following.

a $2(x + 6)$ **b** $4(x - 3)$ **c** $6(2x - 1)$

2 Simplify the following.

a $4y + 2y - y$ **b** $3x + 2 + x - 5$ **c** $2(x + 1) - 3(x + 2)$

3 Simplify the following.

a $3 \times 2x$ **b** $4y \times 2y$ **c** $c^2 \times 2c$

4 Solve the following equations

a $2x + 4 = 6$ **b** $3x - 5 = 4$ **c** $\frac{x}{3} + 2 = 5$

d $\frac{x}{2} = 4$ **e** $\frac{x}{3} = 8$ **f** $\frac{2x}{5} = 6$

6.1 Basic algebra

This section will remind you how to:

- apply the rules of algebra
- simplify algebraic expressions by multiplying terms
- simplify algebraic expressions by collecting **like terms**
- **expand and simplify** brackets
- factorise expressions
- **substitute** numbers into expressions and formulae

Key words

brackets
coefficient
constant term
equation
expand
expand and simplify
expression
factor
factorisation
formula
identity
like terms
simplify
substitute
terms
variable

Remind yourself of some words used in algebra that you need to know.

Variable: This is what the letters used to represent numbers are called. They can take on any value so they 'vary'.

Coefficient: This is the number in front of a letter, so in $2x$ the coefficient of x is 2.

Expression: This is any combination of letters and numbers. For example, $2x + 4y$ and $\frac{p-6}{5}$ are expressions.

Equations: You will have met these in Book 1. These contain, as the name suggests, an equals sign and at least one variable. The important fact is that a value can be found for the variable. This is called *solving the equation.*

Formula: You may already have seen many formulae (the plural of formula). These are like equations in that they contain an equals sign, but there is more than one variable and they are rules for working out things such as area or the cost of taxi fares.

For example, $V = x^3$, $A = \frac{1}{2}bh$ and $C = 3 + 4m$ are formulae.

Identity: This looks like a formula, but the important fact about an identity is that it is true for all values, whether numerical or algebraic. For example, $5n \equiv 2n + 3n$ and $(x + 1)^2 \equiv x^2 + 2x + 1$ are identities. Note that the special sign $\equiv$ is used in an identity.

Terms: These are the separate parts of expressions, equations, formulae and identities. In $3x + 2y - 7$, there are three terms, $3x$, $+ 2y$ and -7. Expressions inside **brackets** are treated as a single term in calculations, although they can be multiplied out in expansions.

Constant term: This is any single number in an expression or equation, so in $x^2 + 3x + 7$, the constant term is 7.

EXAMPLE 1

Expand **a** $3(2x + 7)$ **b** $2x(3x - 4y)$

a $3 \times 2x + 3 \times 7 = 6x + 21$

b $2x \times 3x - 2x \times 4y = 6x^2 - 8xy$

Substitution

EXAMPLE 2

Find the value of $3x^2 - 5$ when **a** $x = 3$ **b** $x = -4$

Whenever you **substitute** a number for a variable in an expression always put the value in **brackets** before working it out. This will avoid errors in calculation, especially with negative numbers.

a When $x = 3$ $3(3)^2 - 5 = 3 \times 9 - 5 = 27 - 5 = 22$

b When $x = -4$ $3(-4)^2 - 5 = 3 \times 16 - 5 = 48 - 5 = 43$

EXAMPLE 3

Find the value of $L = a^2 - 8b^2$ when $a = -6$ and $b = \frac{1}{2}$.

Substitute for the letters.

$L = (-6)^2 - 8(\frac{1}{2})^2$

$L = 36 - 8 \times \frac{1}{4} = 36 - 2 = 34$

Note: If you do not use brackets and write -6^2, this could be wrongly evaluated as -36.

EXERCISE 6A

1 Find the value of $4b + 3$ when: **a** $b = 2.5$ **b** $b = -1.5$ **c** $b = \frac{1}{2}$

2 Evaluate $\frac{x}{3}$ when: **a** $x = 6$ **b** $x = 24$ **c** $x = -30$

3 Find the value of $\frac{12}{y}$ when: **a** $y = 2$ **b** $y = 4$ **c** $y = -6$

4 Evaluate $3w - 4$ when: **a** $w = -1$ **b** $w = -2$ **c** $w = 3.5$

5 Find the value of $\frac{24}{y}$ when: **a** $x = -5$ **b** $x = \frac{1}{2}$ **c** $x = \frac{3}{4}$

6 Where $P = \frac{5w - 4y}{w + y}$, find P when:

a $w = 3$ and $y = 2$ **b** $w = 6$ and $y = 4$ **c** $w = 2$ and $y = 3$

7 Where $A = b^2 + c^2$, find A when:

a $b = 2$ and $c = 3$ **b** $b = 5$ and $c = 7$ **c** $b = -1$ and $c = -4$

8 Where $A = \frac{180(n - 2)}{n + 5}$, find A when:

a $n = 7$ **b** $n = 3$ **c** $n = -1$

D

9 Where $Z = \frac{y^2 + 4}{4 + y}$, find Z when:

a $y = 4$ **b** $y = -6$ **c** $y = -1.5$

FM 10 A taxi company uses the following rule to calculate their fares.

Fare = £2.50 plus 50p per kilometre.

a How much is the fare for a journey of 3 km?

b Farook pays £9.00 for a taxi ride. How far was the journey?

c Maisy knows that her house is 5 miles from town. She has £5.50 left in her purse after a night out. Has she got enough for a taxi ride home?

FM 11 A holiday cottage costs £150 per day to rent.
A group of friends decide to rent the cottage for seven days.

a Which of the following formulae would represent the cost per day if there are n people in the group and they share the cost equally?

$\frac{150}{n}$ $\frac{150}{7n}$ $\frac{1050}{n}$ $\frac{150n}{7}$

b Eventually 10 people go on the holiday.
When they get the bill, they find that there is a discount for a seven-day rental. After the discount, they each find it costs them £12.50 less than they expected.

How much does a seven-day rental cost?

HINTS AND TIPS

To check your choice in part **a**, make up some numbers and try them in the formulae. For example, take $n = 5$.

AU 12 Kaz knows that x, y and z have the values 2, 8 and 11, but she does not know which variable has which value.

a What is the maximum value that the expression $2x + 6y - 3z$ could have?

b What is the minimum value that the expression $5x - 2y + 3z$ could have?

HINTS AND TIPS

You can just try all combinations but, if you think for a moment, the $6y$ term obviously has to be the biggest, and this will give you a clue to the other terms.

FM 13 The formula for the electricity bill each quarter in a household is £7.50 + £0.07 per unit.
A family uses 6720 units in a quarter.

a How much is their total bill?

b The family pay a direct debit of £120 per month towards their electricity costs.

By how much will they be in credit or debit after the quarter?

AU 14 x and y are different positive whole numbers.
Choose values for x and y so that the expression

$5x + 3y$

a evaluates to an odd number

b evaluates to a prime number.

You will need to remember the prime numbers, 2, 3, 5, 7, 11, 13, 17, 19, …

C

PS 15 The formula for the area, A, of a rectangle with length l and width w is $A = lw$.

The formula for the area, T, of a triangle with base b and height h is $T = \frac{1}{2}bh$.

Find values of l, w, b and h so that $A = T$.

AU 16 **a** p is an odd number and q is an even number.

Say if the following expressions are odd or even.

i $p + q$ **ii** $p^2 + q$ **iii** $2p + q$ **iv** $p^2 + q^2$

b x, y and z are all odd numbers.

Write an expression using x, y and z so that the value of the expression is always even.

There are many answers for part (b) and part (a).

PS 17 A formula for the cost of delivery, in pounds, of orders from a do-it-yourself warehouse is:

$$D = 2M - \frac{C}{5}$$

where D is the cost of the delivery, M is the distance in miles from the store and C is the cost of the goods to be delivered.

a How much is the delivery cost when $M = 30$ and $C = 200$?

b Bob buys goods worth £300 and lives 10 miles from the store.

i The formula gives that the cost of delivery is a negative value. What is this value?

ii Explain why Bob will not get a rebate from the store.

c Martha buys goods worth £400. She calculates that her cost of delivery will be zero.

What is the greatest distance Martha could live from the store?

18 Say if each of the following is an expression (E), equation (Q), formula (F) or identity (I).

A: $2x - 5$

B: $s = \sqrt{A}$

C: $2(x + 3) \equiv 2x + 6$

D: $2x - 3 = 1$

AU 19 Marvin hires a car for the day for £40. He wants to know how much it costs him for each mile he drives.

Petrol is 98p per litre and the car does 10 miles per litre.

Marvin works out the following formula for the cost per mile, C in pounds, for M miles driven:

$$C = 0.098 + \frac{40}{M}$$

a Explain each term of the formula.

b How much will it cost per mile if Marvin drives 200 miles that day?

HINTS AND TIPS

Use the information in the question in your explanation.

Expansion

In mathematics, to '**expand**' usually means 'multiply out'. For example, expressions such as $3(y + 2)$ and $4y^2(2y + 3)$ can be expanded by multiplying them out.

Remember that there is an invisible multiplication sign between the outside number and the opening bracket. So $3(y + 2)$ is really $3 \times (y + 2)$ and $4y^2(2y + 3)$ is really $4y^2 \times (2y + 3)$.

You expand by multiplying *everything inside* the brackets by what is outside the brackets.

So in the case of the two examples above,

$$3(y + 2) = 3 \times (y + 2) = 3 \times y + 3 \times 2 = 3y + 6$$

$$4y^2(2y + 3) = 4y^2 \times (2y + 3) = 4y^2 \times 2y + 4y^2 \times 3 = 8y^3 + 12y^2$$

Look at these next examples of expansion, which show clearly how the term outside the brackets has been multiplied with the terms inside them.

$2(m + 3) = 2m + 6$ $\qquad$ $y(y^2 - 4x) = y^3 - 4xy$

$3(2t + 5) = 6t + 15$ $\qquad$ $3x^2(4x + 5) = 12x^3 + 15x^2$

$m(p + 7) = mp + 7m$ $\qquad$ $-3(2 + 3x) = -6 - 9x$

$x(x - 6) = x^2 - 6x$ $\qquad$ $-2x(3 - 4x) = -6x + 8x^2$

$4t(t^3 + 2) = 4t^4 + 8t$ $\qquad$ $3t(2 + 5t - p) = 6t + 15t^2 - 3pt$

Note: The signs change when a negative quantity is outside the brackets. For example,

$a(b + c) = ab + ac$ $\qquad$ $a(b - c) = ab - ac$

$-a(b + c) = -ab - ac$ $\qquad$ $-a(b - c) = -ab + ac$

$-(a - b) = -a + b$ $\qquad$ $-(a + b - c) = -a - b + c$

Note: A minus sign on its own in front of the brackets is actually –1, so:

$$-(x + 2y - 3) = -1 \times (x + 2y - 3) = -1 \times x + -1 \times 2y + -1 \times -3 = -x - 2y + 3$$

The effect of a minus sign outside the brackets is to change the sign of everything inside the brackets.

EXERCISE 6B

D

1 Expand these expressions.

a $2(3 + m)$ **b** $5(2 + l)$ **c** $3(4 - y)$ **d** $4(5 + 2k)$

e $3(2 - 4f)$ **f** $2(5 - 3w)$ **g** $5(2k + 3m)$ **h** $4(3d - 2n)$

i $t(t + 3)$ **j** $k(k - 3)$ **k** $4t(t - 1)$ **l** $2k(4 - k)$

m $4g(2g + 5)$ **n** $5h(3h - 2)$ **o** $y(y^2 + 5)$ **p** $h(h^3 + 7)$

q $k(k^2 - 5)$ **r** $3t(t^2 + 4)$ **s** $3d(5d^2 - d^3)$ **t** $3w(2w^2 + t)$

u $5a(3a^2 - 2b)$ **v** $3p(4p^3 - 5m)$ **w** $4h^2(3h + 2g)$ **x** $2m^2(4m + m^2)$

2 The local supermarket is offering £1 off a large tin of biscuits. Morris wants five tins.

a If the price of one tin is £t, which of the expressions below represents how much it will cost Morris to buy five tins?

$5(t - 1)$ $\quad$ $5t - 1$ $\quad$ $t - 5$ $\quad$ $5t - 5$

b Morris has £20 to spend. Will he have enough money for five tins? Show working to justify your answer.

AU 3 Dylan wrote the following.

$3(5x - 4) = 8x - 4$

Dylan has made two mistakes.

Explain the mistakes that Dylan has made.

> **HINTS AND TIPS**
>
> It is not enough to give the right answer. You must try to explain why Dylan wrote 8 for 3 × 5 instead of 15.

PS 4 The expansion $2(x + 3) = 2x + 6$ can be shown by this diagram.

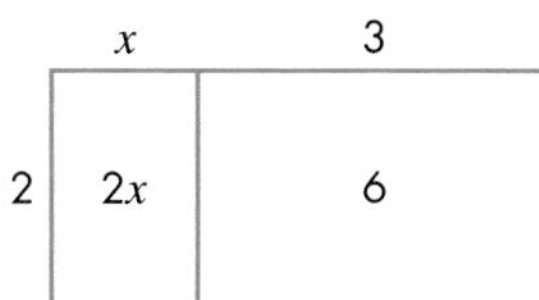

a What expansion is shown in this diagram?

3 | $6y$ | 9

b Write down an expansion that is shown on this diagram.

$12z$ | 8

Simplification

Simplification is the process whereby an expression is written down as simply as possible, with any **like terms** being combined. Like terms are terms that have the same letter(s) raised to the same power and can differ only in their numerical **coefficients** (numbers in front). For example,

m, $3m$, $4m$, $-m$ and $76m$ are all like terms in m

t^2, $4t^2$, $7t^2$, $-t^2$, $-3t^2$ and $98t^2$ are all like terms in t^2

pt, $5tp$, $-2pt$, $7pt$, $-3tp$ and $103pt$ are all like terms in pt

Note: All the terms in tp are also like terms to all the terms in pt.

When simplifying an expression, you can only add or subtract like terms. For example,

$4m + 3m = 7m$	$3y + 4y + 3 = 7y + 3$	$4h - h = 3h$
$2t^2 + 5t^2 = 7t^2$	$2m + 6 + 3m = 5m + 6$	$7t + 8 - 2t = 5t + 8$
$3ab + 2ba = 5ab$	$5k - 2k = 3k$	$10g - 4 - 3g = 7g - 4$

Expand and simplify

When two brackets are expanded there are often like terms that can be collected together. Algebraic expressions should always be simplified as much as possible.

EXAMPLE 4

$$3(4 + m) + 2(5 + 2m) = 12 + 3m + 10 + 4m = 22 + 7m$$

EXAMPLE 5

$$3t(5t + 4) - 2t(3t - 5) = 15t^2 + 12t - 6t^2 + 10t = 9t^2 + 22t$$

EXERCISE 6C

D

1 Simplify these expressions.

a $4t + 3t$ **b** $3d + 2d + 4d$ **c** $5e - 2e$ **d** $3t - t$

e $2t^2 + 3t^2$ **f** $6y^2 - 2y^2$ **g** $3ab + 2ab$ **h** $7a^2d - 4a^2d$

AU **2** Find the missing terms to make these equations true.

a $4x + 5y + \ldots - \ldots = 6x + 3y$

b $3a - 6b - \ldots + \ldots = 2a + b$

PS **3** ABCDEF is an 'L' shape.
AB = DE = x
AF = $3x - 1$ and EF = $2x + 1$

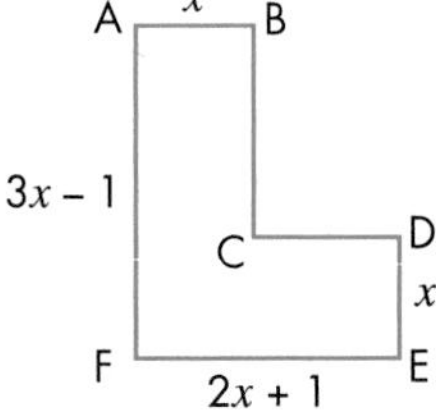

a Explain why the length BC = $2x - 1$

b Find the perimeter of the shape in terms of x.

c If $x = 2.5$ cm what is the perimeter of the shape?

HINTS AND TIPS

Make sure your explanation uses expressions. Do not try to explain in words alone.

C

4 Expand and simplify.

a $3(4 + t) + 2(5 + t)$ **b** $5(3 + 2k) + 3(2 + 3k)$

c $4(3 + 2f) + 2(5 - 3f)$ **d** $5(1 + 3g) + 3(3 - 4g)$

5 Expand and simplify.

a $4(3 + 2h) - 2(5 + 3h)$ **b** $5(3g + 4) - 3(2g + 5)$

c $5(5k + 2) - 2(4k - 3)$ **d** $4(4e + 3) - 2(5e - 4)$

HINTS AND TIPS

Be careful with minus signs. For example, $-2(5e - 4) = -10e + 8$

6 Expand and simplify.

a $m(4 + p) + p(3 + m)$

b $k(3 + 2h) + h(4 + 3k)$

c $4r(3 + 4p) + 3p(8 - r)$

d $5k(3m + 4) -2m(3 - 2k)$

7 Expand and simplify.

a $t(3t + 4) + 3t(3 + 2t)$

b $2y(3 + 4y) + y(5y - 1)$

c $4e(3e - 5) - 2e(e - 7)$

d $3k(2k + p) - 2k(3p - 4k)$

8 Expand and simplify.

a $4a(2b + 3c) + 3b(3a + 2c)$

b $3y(4w + 2t) + 2w(3y - 4t)$

c $5m(2n - 3p) - 2n(3p - 2m)$

d $2r(3r + r^2) - 3r^2(4 - 2r)$

FM 9 A two-carriage train has f first-class seats and $2s$ standard-class seats.

A three-carriage train has $2f$ first-class seats and $3s$ standard-class seats.

On a weekday, 5 two-carriage trains and 2 three-carriage trains travel from Hull to Liverpool.

a Write down an expression for the total number of first-class and standard-class seats available during the day.

b On average in any day, half of the first-class seats are used at a cost of £60.
On average in any day, three-quarters of the standard-class seats are used at a cost of £40.

How much money does the rail company earn in an average day on this route?
Give your answer in terms of f and s.

c $f = 15$ and $s = 80$. It costs the rail company £30 000 per day to operate this route.
How much profit do they make on an average day?

HINTS AND TIPS

There is more than one answer. You don't have to give them all.

AU 10 Fill in whole-number values so that the following expansion is true.

$3(\ldots x + \ldots y) + 2(\ldots x + \ldots y) = 11x + 17$

PS 11 A rectangle with sides 5 and $3x + 2$ has a smaller rectangle with sides 3 and $2x - 1$ cut from it.

Work out the remaining area.

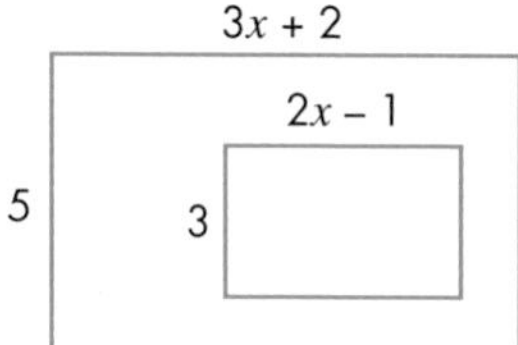

HINTS AND TIPS

Write out the expression for the difference between the two rectangles and then work it out.

6.2 Factorisation

This section will show you how to:
- factorise an algebraic expression

Key words
common factor
factorisation

Factorisation is the opposite of expansion. It puts an expression back into the brackets it may have come from.

In factorisation, you have to look for the **common factors** in *every* term of the expression.

EXAMPLE 6

Factorise each expression.

a $6t + 9m$ **b** $6my + 4py$ **c** $5k^2 - 25k$ **d** $10a^2b - 15ab^2$

a First look at the numerical coefficients 6 and 9.

These have a common factor of 3.

Then look at the letters, t and m.

These do not have any common factors as they do not appear in both terms.

The expression can be thought of as $3 \times 2t + 3 \times 3m$, which gives the factorisation:

$6t + 9m = 3(2t + 3m)$

Note: You can always check a factorisation by expanding the answer.

b First look at the numbers.

These have a common factor of 2. m and p do not occur in both terms but y does, and is a common factor, so the factorisation is:

$6my + 4py = 2y(3m + 2p)$

c 5 is a common factor of 5 and 25 and k is a common factor of k^2 and k.

$5k^2 - 25k = 5k(k - 5)$

d 5 is a common factor of 10 and 15, a is a common factor of a^2 and a, b is a common factor of b and b^2.

$10a^2b - 15ab^2 = 5ab(2a - 3b)$

Note: If you multiply out each answer, you will get the expressions you started with.

EXERCISE 6D

1 Factorise the following expressions.

a $6m + 12t$
b $9t + 3p$
c $8m + 12k$
d $4r + 8t$
e $mn + 3m$
f $5g^2 + 3g$
g $4w - 6t$
h $3y^2 + 2y$
i $4t^2 - 3t$
j $3m^2 - 3mp$
k $6p^2 + 9pt$
l $8pt + 6mp$
m $8ab - 4bc$
n $5b^2c - 10bc$
o $8abc + 6bed$
p $4a^2 + 6a + 8$
q $6ab + 9bc + 3bd$
r $5t^2 + 4t + at$
s $6mt^2 - 3mt + 9m^2t$
t $8ab^2 + 2ab - 4a^2b$
u $10pt^2 + 15pt + 5p^2t$

FM **2** Three friends have a meal together. They each have a main meal costing £6.75 and a dessert costing £3.25.

Chris says that the bill will be $3 \times 6.75 + 3 \times 3.25$.

Mary says that she has an easier way to work out the bill as $3 \times (6.75 + 3.25)$.

a Explain why Chris' and Mary's methods both give the correct answer.

b Explain why Mary's method is better.

c What is the total bill?

3 Factorise the following expressions where possible. List those that do not factorise.

a $7m - 6t$
b $5m + 2mp$
c $t^2 - 7t$
d $8pt + 5ab$
e $4m^2 - 6mp$
f $a^2 + b$
g $4a^2 - 5ab$
h $3ab + 4cd$
i $5ab - 3b^2c$

AU **4** Three students are asked to factorise the expression $12m - 8$. These are their answers.

Aidan	Bernice	Craig
$2(6m - 4)$	$4(3m - 2)$	$4m(3 - \frac{2}{m})$

All the answers are accurately factorised, but only one is the normally accepted answer.

a Which student gave the correct answer?

b Explain why the other two students' answers are not acceptable as correct answers.

AU **5** Explain why $5m + 6p$ cannot be factorised.

PS **6** Alvin has correctly factorised the top and bottom of an algebraic fraction and cancelled out the terms to give a final answer of $2x$. Unfortunately some of his work has had coffee spilt on it. What was the original fraction?

$$\frac{4x\ldots}{2\ldots} = \frac{4\ldots}{2(x-3)} = 2x$$

6.3 Solving linear equations

This section will remind you how to:

- solve linear equations by rearrangement

Key words
do the same to both sides
inverse operations
rearrangement
solution
variable

Fractional equations

Work through these examples to remind yourself how to solve equations.

Remember: to solve equations, you can **do the same to both sides**, use **inverse operations** or **rearrangement** to find the **solution** for the **variable**.

EXAMPLE 7

Solve the following equations.

a $\frac{y}{5} + 7 = 4$ **b** $\frac{z + 4}{3} = 5$

a Subtract 7 from both sides: $\frac{y}{5} = -3$

Multiply both sides by 5: $y = -15$

Check: $-15 \div 5 + 7 = -3 + 7 = 4$ ✓

b Multiply both sides by 3: $z + 4 = 15$

Subtract 4 from both sides: $z = 11$

Check: $(11 + 4) \div 3 = 15 \div 3 = 5$ ✓

Don't forget to check your answer in the original equation.

EXAMPLE 8

Solve this equation. $\frac{3x}{4} - 3 = 1$

First add 3 to both sides: $\frac{3x}{4} = 4$

Now multiply both sides by 4: $3x = 16$

Now divide both sides by 3: $x = \frac{16}{3} = 5\frac{1}{3}$

Check: $\frac{3 \times 5\frac{1}{3}}{4} - 3 = \frac{16}{4} - 3 = 4 - 3 = 1$

EXERCISE 6E

1 Solve these equations.

a $\frac{f}{5} + 2 = 8$ **b** $\frac{w}{3} - 5 = 2$ **c** $\frac{x}{8} + 3 = 12$

d $\frac{5t}{4} + 3 = 18$ **e** $\frac{3y}{2} - 1 = 8$ **f** $\frac{2x}{3} + 5 = 12$

g $\frac{t}{5} + 3 = 1$ **h** $\frac{x + 3}{2} = 5$ **i** $\frac{t - 5}{2} = 3$

j $\frac{3x + 10}{2} = 8$ **k** $\frac{2x + 1}{3} = 5$ **l** $\frac{5y - 2}{4} = 3$

m $\frac{6y + 3}{9} = 1$ **n** $\frac{2x - 3}{5} = 4$ **o** $\frac{5t + 3}{4} = 1$

AU 2 The solution to the equation $\frac{2x - 3}{5} = 3$ is $x = 9$.

Make up *two* more *different* equations of the form $\frac{ax \pm b}{5} = d$,

for which the answer is also 3, where a, b, c and d are positive whole numbers.

AU 3 A teacher asked her class to solve the equation $\frac{2x + 4}{5} = 6$.

Amanda wrote:	Betsy wrote:
$2x + 4 = 6 \times 5$	$\frac{2x}{5} = 6 + 4$
$2x + 4 - 4 = 30 - 4$	$2x = 6 + 4 + 5$
$2x = 26$	$2x = 15$
$2x \div 2 = 26 \div 2$	$2x - 2 = 15 - 2$
$x = 13$	$x = 13$

When the teacher read out the correct answer of 13, both students ticked their work as correct.

a Which student used the correct method?

b Explain the mistakes the other student made.

D

C

Brackets

When we have an equation that contains **brackets**, we first must multiply out the brackets and then solve the resulting equation.

EXAMPLE 9

Solve $5(x + 3) = 25$

First multiply out the brackets to get:

$5x + 15 = 25$

Rearrange: $5x = 25 - 15 = 10$

Divide by 5: $\frac{5x}{5} = \frac{10}{5}$

$x = 2$

EXAMPLE 10

Solve $3(2x - 7) = 15$

Multiply out the brackets to get:

$6x - 21 = 15$

Add 21 to both sides: $6x = 36$

Divide both sides by 6: $x = 6$

EXERCISE 6F

D

1 Solve each of the following equations. Some of the answers may be decimals or negative numbers. Remember to check that each answer works for its original equation. Use your calculator if necessary.

a $2(x + 5) = 16$ **b** $5(x - 3) = 20$

c $3(t + 1) = 18$ **d** $4(2x + 5) = 44$

e $2(3y - 5) = 14$ **f** $5(4x + 3) = 135$

g $4(3t - 2) = 88$ **h** $6(2t + 5) = 42$

i $2(3x + 1) = 11$ **j** $4(5y - 2) = 42$

k $6(3k + 5) = 39$ **l** $5(2x + 3) = 27$

m $9(3x - 5) = 9$ **n** $2(x + 5) = 6$ **o** $5(x - 4) = -25$

p $3(t + 7) = 15$ **q** $2(3x + 11) = 10$ **r** $4(5t + 8) = 12$

HINTS AND TIPS

Once the brackets have been expanded the equations become straightforward. Remember to multiply *everything* inside the brackets with what is outside.

D

AU 2 Fill in values for a, b and c so that the answer to this equation is $x = 4$.

$a(bx + 3) = c$

PS 3 My son is x years old. In five years' time, I will be twice his age and both our ages will be multiples of 10.
The sum of our ages will be between 50 and 100.
How old am I now?

HINTS AND TIPS

Set up an equation and put it equal to 60, 70, 80, etc. Solve the equation and see if the answer fits the conditions.

Equations with the variable on both sides

When a letter (or variable) appears on both sides of an equation, it is best to use the '**do the same to both sides**' method of **solution**, and collect all the terms containing the letter on the left-hand side of the equation. But when there are more of the letters on the right-hand side, it is easier to turn the equation round. When an equation contains brackets, they must be multiplied out first.

EXAMPLE 11

Solve this equation. $5x + 4 = 3x + 10$

There are more xs on the left-hand side, so leave the equation as it is.

Subtract $3x$ from both sides: $2x + 4 = 10$

Subtract 4 from both sides: $2x = 6$

Divide both sides by 2: $x = 3$

EXAMPLE 12

Solve this equation. $2x + 3 = 6x - 5$

There are more xs on the right-hand side, so turn the equation round.

$6x - 5 = 2x + 3$

Subtract $2x$ from both sides: $4x - 5 = 3$

Add 5 to both sides: $4x = 8$

Divide both sides by 4: $x = 2$

EXAMPLE 13

Solve this equation. $3(2x + 5) + x = 2(2 - x) + 2$

Multiply out both brackets: $6x + 15 + x = 4 - 2x + 2$

Simplify both sides: $7x + 15 = 6 - 2x$

There are more xs on the left-hand side, so leave the equation as it is.

Add $2x$ to both sides: $9x + 15 = 6$

Subtract 15 from both sides: $9x = -9$

Divide both sides by 9: $x = -1$

EXERCISE 6G

D

1 Solve each of the following equations.

a $2x + 3 = x + 5$ **b** $5y + 4 = 3y + 6$

c $4a - 3 = 3a + 4$ **d** $5t + 3 = 2t + 15$

e $7p - 5 = 3p + 3$ **f** $6k + 5 = 2k + 1$

g $4m + 1 = m + 10$ **h** $8s - 1 = 6s - 5$

> **HINTS AND TIPS**
>
> **Remember:** 'Change sides, change signs'. Show all your working. Rearrange *before* you simplify. If you try to do these at the same time you could get it wrong.

PS 2 Terry says:

I am thinking of a number. I multiply it by 3 and subtract 2.

June says:

I am thinking of a number. I multiply it by 2 and add 5.

Terry and June find that they both thought of the same number and both got the same final answer.

What number did they think of?

> **HINTS AND TIPS**
>
> Set up equations; put them equal and solve.

C

3 Solve each of the following equations.

a $2(d + 3) = d + 12$ **b** $5(x - 2) = 3(x + 4)$

c $3(2y + 3) = 5(2y + 1)$ **d** $3(h - 6) = 2(5 - 2h)$

e $4(3b - 1) + 6 = 5(2b + 4)$ **f** $2(5c + 2) - 2c = 3(2c + 3) + 7$

AU 4 Explain why the equation $3(2x + 1) = 2(3x + 5)$ cannot be solved.

> **HINTS AND TIPS**
>
> Expand the brackets and collect terms on one side as usual. What happens?

PS 5 Wilson has eight coins of the same value and seven pennies.

Chloe has 11 coins of the same value as those that Wilson has and she also has five pennies.

Wilson says, "If you give me one of your coins and four pennies, we will have the same amount of money."

What is the value of the coins that Wilson and Chloe have?

> **HINTS AND TIPS**
>
> Call the coin x and set up the equations, e.g. Wilson has $8x + 7$, and then take one x and 4 from Chloe and add one x and 4 to Wilson. Then put the equations equal and solve.

AU 6 Explain why there are an infinite number of solutions to the equation:

$$2(6x + 9) = 3(4x + 6)$$

6.4 Setting up equations

This section will remind you how to:

- set up equations from given information, and then solve them

Key words
do the same to both sides
equation
rearrange
solve

Equations are used to represent situations, so that you can **solve** real-life problems. Many real-life problems can be solved by setting them up as linear equations. You can **do the same to both sides** or use **rearrangement** to **solve** the problem.

EXAMPLE 14

A milkman sets off from the dairy with eight crates of milk, each containing b bottles.

He delivers 92 bottles to a large factory and finds that he has exactly 100 bottles left on his milk float. How many bottles were in each crate?

The equation is:

$8b - 92 = 100$

$8b = 192$ (Add 92 to both sides.)

$b = 24$ (Divide both sides by 8.)

EXAMPLE 15

The rectangle shown has a perimeter of 40 cm.

Find the value of x.

$3x + 1$

$x + 3$

The perimeter of the rectangle is:

$3x + 1 + x + 3 + 3x + 1 + x + 3 = 40$

This simplifies to: $8x + 8 = 40$

Subtract 8 from both sides: $8x = 32$

Divide both sides by 8: $x = 4$

EXERCISE 6H

Set up an equation to represent each situation described below. Then solve the equation. Remember to check each answer.

FM **1** A man buys a daily paper from Monday to Saturday for d pence. He buys a Sunday paper for £1.80. His weekly paper bill is £7.20.

What is the price of his daily paper?

D

D

2 The diagram shows a rectangle.

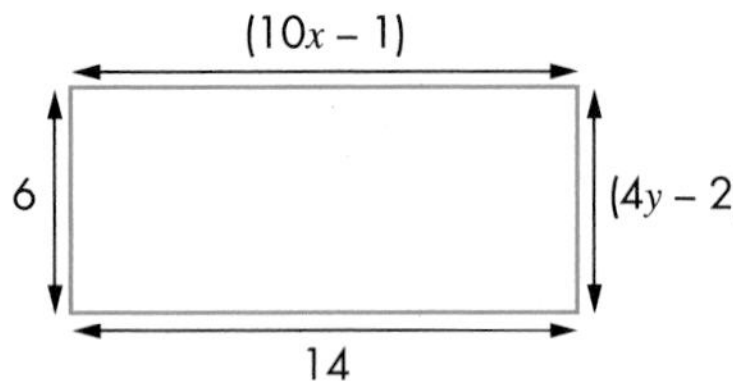

a What is the value of x?

b What is the value of y?

HINTS AND TIPS

Use the letter x for the variable unless you are given a letter to use. Once the equation is set up solve it by the methods shown above.

PS 3 In this rectangle, the length is 3 cm more than the width. The perimeter is 12 cm.

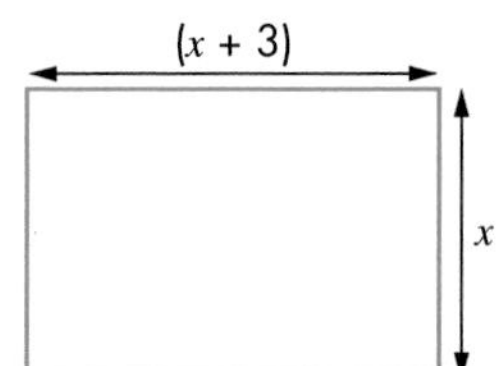

a What is the value of x?

b What is the area of the rectangle?

4 Mary has two bags, each of which contains the same number of sweets. She eats four sweets. She then finds that she has 30 sweets left. How many sweets were there in each bag to start with?

FM 5 A carpet costs £12.75 per square metre.

The shop charges £35 for fitting. The final bill was £137.

How many square metres of carpet were fitted?

FM 6 Moshin bought eight garden chairs. When he got to the till he used a £10 voucher as part payment. His final bill was £56.

a Set this problem up as an equation, using c as the cost of one chair.

b Solve the equation to find the cost of one chair.

FM 7 This diagram shows the traffic flow through a one-way system in a town centre.

Cars enter at A and at each junction the fractions show the proportion of cars that take each route.

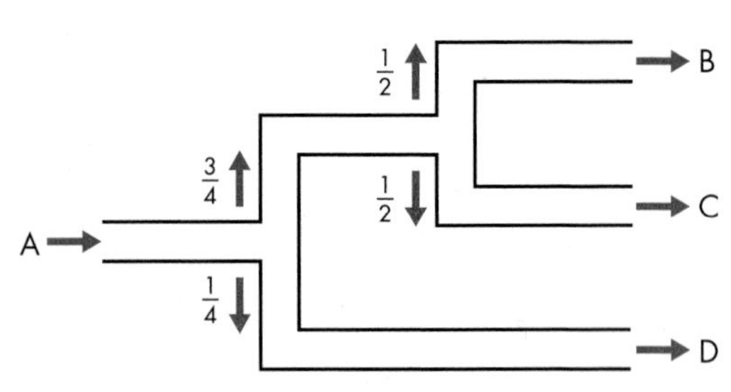

a 1200 cars enter at A. How many come out of each of the exits, B, C and D?

b If 300 cars exit at B, how many cars entered at A?

c If 500 cars exit at D, how many exit at B?

D

FM **8** A rectangular room is 3 m longer than it is wide.
The perimeter is 16 m.

Carpet costs £9.00 per square metre. How much will it cost to carpet the room?

HINTS AND TIPS

Set up an equation to work out the length and width, then calculate the area.

C

PS **9** A boy is Y years old. His father is 25 years older than he is. The sum of their ages is 31. How old is the boy?

PS **10** Another boy is X years old. His sister is twice as old as he is. The sum of their ages is 27. How old is the boy?

11 The diagram shows a square.

Find x if the perimeter is 44 cm.

$(4x - 1)$

PS **12** Max thought of a number. He then multiplied his number by 3. He added 4 to the answer. He then doubled that answer to get a final value of 38. What number did he start with?

13 The angles of a triangle are $2x$, $x + 5°$ and $x + 35°$.

a Write down an equation to show this.

b Solve your equation to find the value of x.

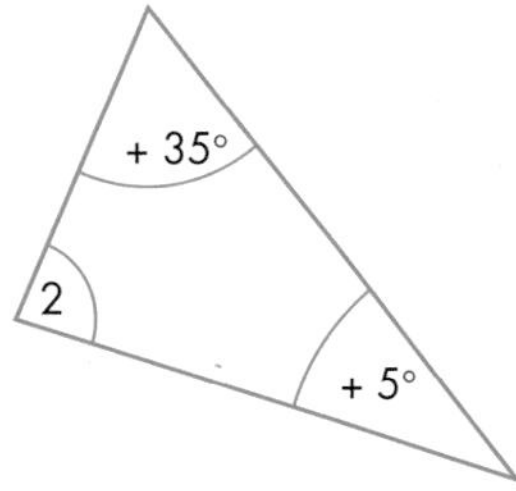

FM **14** Five friends went for a meal in a restaurant. The bill was £x.
They decided to add a £10 tip and split the bill between them.

Each person paid £9.50.

a Set this problem up as an equation.

b Solve the equation to work out the bill before the tip was added.

AU **15** The diagram shows two number machines that perform the same operations.

a Starting with an input value of 7, work through the left-hand machine to get the output.

b Find an input value that gives the same value for the output.

c Write down the algebraic expressions in the right-hand machine for an input of n. (The first operation has been filled in for you.)

d Set up an equation for the same input and output and show each step in solving the equation to get the answer in part **b**.

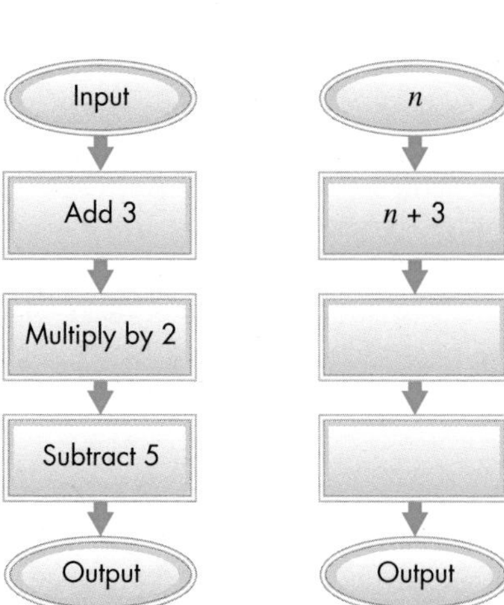

C

PS **16** A teacher asked her class to find three angles of a triangle that were consecutive even numbers.

Tammy wrote: $x + x + 2 + x + 4 = 180$

$$3x + 6 = 180$$

$$3x = 174$$

$$x = 58$$

So the angles are 58°, 60° and 62°.

The teacher then asked the class to find four angles of a quadrilateral that are consecutive even numbers.

Can this be done? Explain your answer.

HINTS AND TIPS

Do the same type of working as Tammy did for a triangle. Work out the value of x. What happens?

FM **17** Mary has a large and a small bottle of cola. The large bottle holds 50 cl more than the small bottle.

From the large bottle she fills four cups and has 18 cl left over.

From the small bottle she fills three cups and has 1 cl left over.

How much cola does each bottle hold?

HINTS AND TIPS

Set up equations for both using x as the amount of pop in a cup. Put them equal but remember to add 50 to the small bottle equation to allow for the difference. Solve for x, then work out how much is in each bottle.

6.5 Trial and improvement

This section will show you how to:

- estimate the answers to some questions that do not have exact solutions, using the method of trial and improvement

Key words

comment
decimal place
guess
trial and improvement

Certain equations cannot be solved exactly. However, a close enough solution to such an equation can be found by the **trial-and-improvement** method. (Sometimes this is wrongly called the trial-and-error method.)

The idea is to keep trying different values in the equation to take it closer and closer to the 'true' solution. This step-by-step process is continued until a value is found that gives a solution that is close enough to the accuracy required.

The trial-and-improvement method is the way in which computers are programmed to solve equations.

EXAMPLE 16

Solve the equation $x^3 + x = 105$, giving the solution correct to 1 **decimal place.**

Step 1 You must find the two consecutive whole numbers between which x lies. You do this by intelligent guessing.

Try $x = 5$: $125 + 5 = 130$ Too high – next trial needs to be much smaller.

Try $x = 4$: $64 + 4 = 68$ Too low.

So now you know that the solution lies between $x = 4$ and $x = 5$.

Step 2 You must find the two consecutive 1-decimal-place numbers between which x lies. Try 4.5, which is halfway between 4 and 5.

This gives $91.125 + 4.5 = 95.625$ Too small.

Now attempt to improve this by trying 4.6.

This gives $97.336 + 4.6 = 101.936$ Still too small.

Try 4.7 which gives 108.523. This is too high.

So the solution is between 4.6 and 4.7.

It looks as though 4.7 is closer but there is a very important final step.

Step 3 Now try the value that is halfway between the two 1-decimal-place values. In this case it is 4.65.

This gives 105.194 625.

This means that 4.6 is nearer the actual solution than 4.7.

Never assume that the one-decimal-place number that gives the closest value to the solution is the answer.

The diagram on the right shows why this is.

The approximate answer is $x = 4.6$ to 1 decimal place.

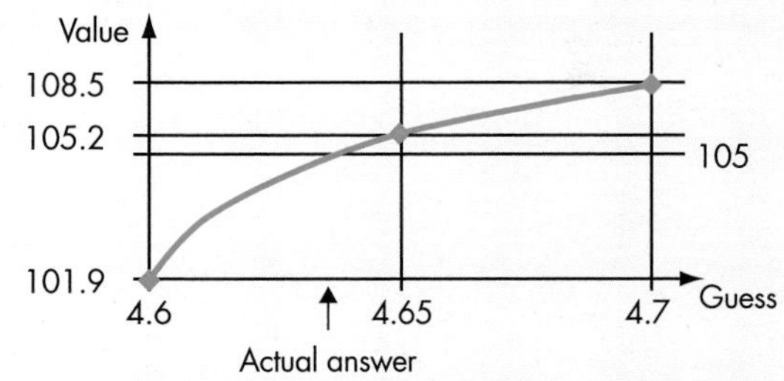

The best way to answer this type of question is to set up a table to show working. You will need three columns: **guess** (the trial), the equation to be solved and a **comment** – whether the value of the equation is too high or too low.

Guess	$x^3 + x$	Comment
4	68	Too low
5	130	Too high
4.5	95.625	Too low
4.6	101.936	Too low
4.7	108.523	Too high
4.65	105.194 625	Too high

EXERCISE 6I

C

1 Find the two consecutive *whole numbers* between which the solution to each of the following equations lies.

a $x^2 + x = 24$ **b** $x^3 + 2x = 80$ **c** $x^3 - x = 20$

2 Copy and complete the table by using trial and improvement to find an approximate solution to:

$x^3 + 2x = 50$

Give your answer correct to 1 decimal place.

Guess	$x^3 + 2x$	Comment
3	33	Too low
4	72	Too high

3 Copy and complete the table by using trial and improvement to find an approximate solution to:

$x^3 - 3x = 40$

Give your answer correct to 1 decimal place.

Guess	$x^3 - 3x$	Comment
4	52	Too high

4 Use trial and improvement to find an approximate solution to:

$2x^3 + x = 35$

Give your answer correct to 1 decimal place.

You are given that the solution lies between 2 and 3.

HINTS AND TIPS

Set up a table to show your working. This makes it easier for you to show method and the examiner to mark.

5 Use trial and improvement to find an exact solution to:

$4x^2 + 2x = 12$

Do not use a calculator.

6 Find a solution to each of the following equations, correct to 1 decimal place.

a $2x^3 + 3x = 35$ **b** $3x^3 - 4x = 52$ **c** $2x^3 + 5x = 79$

PS 7 A rectangle has an area of 100 cm^2. Its length is 5 cm longer than its width.

a Show that, if x is the width, then $x^2 + 5x = 100$.

b Find, correct to 1 decimal place, the dimensions of the rectangle.

8 Use trial and improvement to find a solution to the equation $x^2 + x = 40$.

FM 9 Rob is designing a juice carton to hold $\frac{1}{2}$ litre (500 cm^3).

He wants the sides of the base in the ratio 1 : 2.

He wants the height to be 8 cm more than the shorter side of the base.

Use trial and improvement to find the dimensions of the carton.

HINTS AND TIPS

Call the length of the side with 'ratio 1' x, write down the other two sides in terms of x and then write down an equation for the volume = 500.

GRADE BOOSTER

- **D** You can expand linear brackets
- **D** You can substitute numbers into expressions
- **D** You can factorise simple linear expresions
- **D** You can solve simple linear equations which include the variable inside brackets
- **D** You can solve linear equations where the variable occurs in the numerator of a fraction
- **D** You can solve linear equations where the variable appears on both sides of the equals sign.
- **D** You can manipulate algebraic expressions
- **C** You can expand and simplify expressions
- **C** You can solve linear fractional equations
- **C** You can solve linear equations where brackets have to be expanded
- **C** You can set up and solve linear equations from practical and real-life situations
- **C** You can solve non-linear equations, using trial and improvement

What you should know now

- How to manipulate and simplify algebraic expressions, including those with linear brackets
- How to factorise linear expressions
- How to solve all types of linear equations

EXAMINATION QUESTIONS

1 Solve the equation $\frac{1}{2}x - 5 = \frac{1}{4}x + 3$ *(3 marks)*

AQA, June 2005, Paper 1 Intermediate, Question 20 (a)

2 **a** **i** Expand and simplify $(y + 5)(y - 1)$ *(2 marks)*

ii When y is an odd number, explain why $(y + 5)(y - 1)$ is an even number. *(1 mark)*

b Factorise $2xy - 6y^2$ *(2 marks)*

AQA, November 2006, Paper 1 Higher, Questions 8(c) and (d),

3 Expand and simplify $(x - 3)(x + 4)$ *(2 marks)*

AQA, June 2008, Paper 2 Higher, Question 11

4 **a** Draw arrows to join each item on the left with its correct description on the right.

One of them has been done for you.

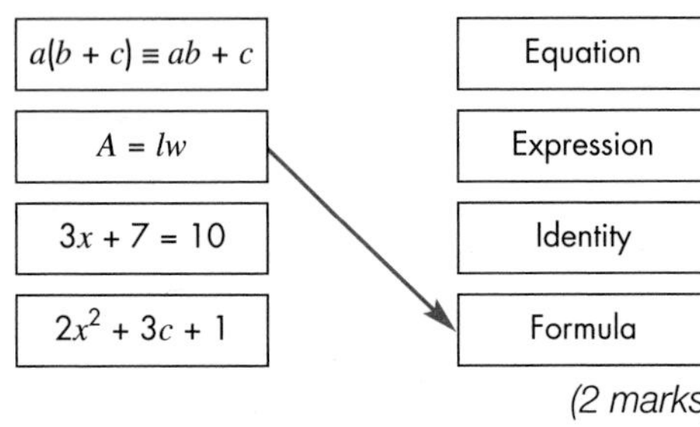

(2 marks)

b A two-stage operation is shown:

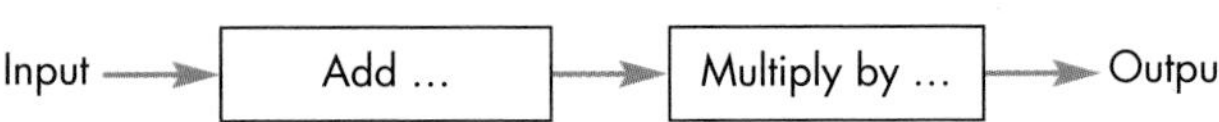

Fill values in the boxes so that when the input is an odd number, the output is also an odd number. *(2 marks)*

AQA, November 2008, Paper 2 Higher, Question 9

5 Solve the equations.

a $\frac{17 - x}{3} = 4.5$

b $2(y - 3) = 5 - 3y$

c $3(2z - 1) + 4(z + 3) = 5(2z - 1) + 4(3z - 1)$

6 **a** Simplify:

i $y^7 \times y^2$ *(1 mark)*

ii $y^7 \div y^2$ *(1 mark)*

iii $(y^7)^2$ *(1 mark)*

b **i** If $y = -1$, which answer in part **a** is positive? *(1 mark)*

ii If $y = 0.5$, which answer in part **a** has the greatest value? *(1 mark)*

AQA, November 2005, Paper 1 Higher, Question 8

7  Solve the equation:

$4(x + 3) = 9(x - 2)$ *(3 marks)*

AQA, June 2006, Paper 2 Higher, Question 10

8 Factorise:

$5x^2 + 20x$ *(1 mark)*

AQA, June 2006, Paper 2 Higher, Question 15

9 **a** Solve

$\frac{x}{4} + 1 = 6$ *(2 marks)*

b Solve

$\frac{4}{y + 1} = 3$ *(2 marks)*

c Factorise fully $6ab^2 - 2ab$ *(2 marks)*

AQA, June 2007, Paper 1 Higher, Question 14

10  Expand and simplify

$4(2x + 1) - 3(x - 4)$

AQA, November 2011 F2 Question 25

11  Solve

$6x - 1 = 2x + 4$

AQA, November 2011 H2 Question 11

12 **a** Solve

$4(3w - 7) = 32$

b Solve

$\frac{26 - y}{5} = 4$

AQA, November 2011 H1 Question 8

13  Use trial and improvement to find a solution to the equation

$x^3 + 6x = 29$

Continue the table of results

Give your solution to 1 decimal place.

x	$x^3 + 6x$	Comment
2	20	Too small

AQA, June 2011 H2 Question 12

14 This triangle is equilateral.

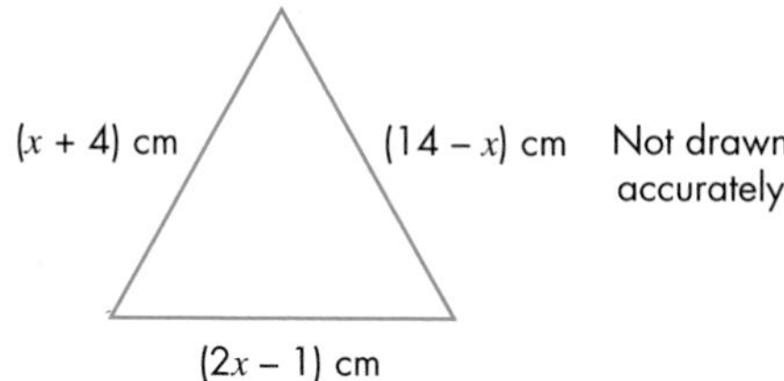

a Work out the value of x. *(3 marks)*

b Hence, or otherwise, work out the perimeter of the triangle.

AQA, June 2011 H1 Question 7

15 **a** Factorise

$x^2 + 7x$ *(1 mark)*

b Expand

$5(3x + 8)$ *(1 mark)*

c Expand and simplify

$3(2x + 1) - 2(x - 3)$ *(2 marks)*

AQA, November 2010 H2 Question 5

Worked Examination Questions

1 a Expand and simplify $6(x - 2) - 4(x - 5)$

b Solve the equation $6x - 7 = 2x + 11$

1 a $6x - 12 - 4x + 20$ — Multiply out each bracket. This will get 1 method mark even if you get one term wrong.
If you get all 4 terms correct (especially the last term) you get 1 accuracy mark

$6x - 4x - 12 + 20$

$2x + 8$ — Collect the like terms together. Do the x terms separately from the number terms. This will get 1 accuracy mark but this is a follow through so you can still get this if you collect your terms together.

b $6x - 2x = 11 + 7$ — Move the x terms to the left hand side and the number terms to the right hand side. This gets 1 method mark. It is better not to try to evaluate the expressions at this stage as you could make a mistake.

$4x = 18$ — Collect the x and number terms together. This gets 1 method mark

$x = 4.5$ — Divide both sides by 4 to get the answer. This gets 1 method mark

Total: 6 marks

2 a Expand $5(x - 3)$

b Factorise $x^2 + 3x$

c Solve the equation $\frac{x - 3}{2} = 3x + 4$

2 a $5x - 15$ — Multiply both terms inside the bracket by the number outside. This is 1 mark each term but the minus sign must be seen between them.

b $x(x + 3)$ — The only term common to both terms is x. Check by expanding the answer. $x \times x + 3 \times x = x^2 + 3x$. This is 1 mark.

c $x - 3 = 2(3x + 4)$ — Cross-multiply by 2. This is 1 method mark.

$x - 3 = 6x + 8$

$6x + 8 = x - 3$ — Expand the bracket and as there are more x terms on the right swap sides. This is 1 accuracy mark.

$6x - x = -3 - 8$ — Rearrange by taking all x terms to the left and number terms to the right. Then collect terms together. This is 1 method mark.

$5x = -11$

$x = -2.2$ — Divide both side by 5 to get the answer. This is 1 accuracy mark.

Total: 7 marks

Worked Examination Questions

3 A rectangle has sides of $(3x + 1)$ cm and $(2x - 3)$ cm.

The perimeter is 21 cm.

$3x + 1$

$2x - 3$

Work out the value of x.

3 $3x + 1 + 2x - 3 = 5x - 2$ — Adding up the two sides (or all four sides to get $10x - 4$) would get 1 method mark

$5x - 2 = 10.5$ — Putting the total of the two sides = 10.5 or $10x - 4 = 21$ would get a method mark.

$5x = 12.5$ — Rearranging the equation would get 1 method mark.

$x = 2.5$ — The correct answer would get 1 accuracy mark.

Total: 4 marks

4 The cost of a newspaper is x pence.

The cost of a magazine is £2.10 more than the newspaper.

The cost of 2 magazines is the same as the cost of 8 newspapers.

a Show clearly that $2x + 420 = 8x$

b Solve $2x + 420 = 8x$

4 a $2(x + 210) = 8x$ — As the price of the newspaper is in pence you must convert £2.10 to pence. The equation above shows that 2 magazines = 8 newspapers. This gets 1 mark. It isn't necessary to expand this as the answer is already given.

b $420 = 6x$ — Take $2x$ from both sides. This gets 1 method mark.

$x = 70$ — Divide by 6 to get the answer. This is 1 accuracy mark.

Total: 3 marks

Worked Examination Questions

5 One solution of $x^3 + 3x = 120$ is between 4 and 5.
Use trial and improvement to find the solution.
Give your answer to one decimal place.

5 Set up a table.

x	$x^3 + 3x$	Comment
4	76	Too low
5	140	Too high
4.5	104.615	Too low
4.6	111.136	Too low
4.7	117.923	Too low
4.8	124.992	Too high
4.75	121.422	Too high

It isn't necessary to test 4 and 5 as these are given in the question but it helps to check you are using your calculator correctly

Testing any value correctly between 4 and 5 will get 1 method mark

Testing 1 decimal place values that 'bracket' the answer gets 1 method mark

Testing the 'halfway' value between the two 1 dp values and stating the answer is essential to get the accuracy mark

Total: 3 marks

Worked Examination Questions

6 Alf is x years old.
Bill is 4 years younger than Alf.
Chloe is twice as old as Bill.
The total of their ages is 48.
Form an equation in x and use it work out Alf's age.

6 $x - 4$ and $2(x - 4)$

Writing down either of the ages of Bill or Chloe will get 1 mark. Writing them both down will get 2 marks.

$x + x - 4 + 2(x - 4) = 48$

Adding all the ages and putting them equal to 48 will get 1 method mark.

$4x - 12 = 48$

Simplifying the expression on the left will get 1 accuracy mark.

$x = 15$

Solving the equation to get the correct answer will get 1 accuracy mark.

Functional Maths
Temperature scales

There are three temperature scales in use today, Fahrenheit, Celsius and Kelvin.

The Fahrenheit scale uses 32°F degrees for the temperature at which pure water freezes and 212°F for the temperature at which pure water boils. It was devised by a German physicist, Daniel Gabriel Fahrenheit in the 18th century. It was in general use in English-speaking countries until the 1970s. Many older British people still think in terms of Fahrenheit temperatures so it is often quoted in newspapers.

The Celsius scale is the one that is most commonly used everyday. It is based on 0°C for the temperature at which water freezes and 100°C for the temperature at which water boils. It is also called the centigrade scale. It was invented in 1742 by the Swedish astronomer Anders Celsius.

The Kelvin scale is based on 0K for the lowest possible temperature that is achievable (Called **absolute zero**) and 273K for the temperature at which water freezes. It is used by scientists and was named for the British physicist William Thomson, Baron Kelvin.

Your task

Draw three parallel number lines. Label them °F, °C and K.

Starting with 0 at the bottom of the Kelvin scale label equivalent points on each scale for absolute zero, the freezing point of water and the boiling point of water.

Look up some other significant temperature, such as normal body temperature, the temperature at which paper burns, the temperature at which lead melts and add these to each scale.

The relationship between Celsius (C) and Fahrenheit (F) is given by:

$$C = \frac{5}{9}(F - 32)$$

The relationship between Kelvin (K) and Celsius (C) is given by:

$$K = C + 273$$

Use these formulae to check that your temperatures on the scales you wrote down are correct.

Find a relationship between Kelvin (K) and Fahrenheit (F).

Use a newspaper or the internet to find an up to date weather map of Britain showing the maximum temperatures.

Change these to a different scale.

Look up other temperature scales on the internet (there are at least two others).

Which scale has the most practical use? Explain your choice.

Why this chapter matters

Mathematicians are interested in the way that numbers work, rather than just looking at specific calculations. Using letters to represent numbers, they can use formulae and solve the very complicated equations that occur in modern technology. Without the use of algebra, mankind would not have developed aircraft or walked on the Moon.

2000 BC The Babylonians discover that the ratio of the circumference of a circle to its diameter is approximately 3.125. This is the first time an approximation to π is used.

The Chinese made the first reference to negative numbers.

100 AD Heron of Alexandria did some work that led to the need for the square root of a negative number but this was dismissed for many centuries as something that was impossible.

1500 AD Italian mathematicians came across formulae that could only be solved if the square root of –1 was used.

1550 AD Rafael Bombelli was the first mathematician to introduce the notation $\sqrt{-1} = \mathrm{i}$.

1618 AD When doing work on logarithms, the Scottish mathematician John Napier published a list of 'natural logarithms' that were based on a number with a value of about 2.718, although this number was not quoted.

1680 AD A Swiss mathematician, Jacob Bernoulli, found the value 2.71828… which is the limit of $(1 + \frac{1}{n})^n$. (Try this on your calculator with a big number for n. If $n = 1000$, then $1.001^{1000} = 2.71692\ldots$)

1727 AD Another Swiss mathematician, Leonhard Euler, gave the value the letter e.

1748 AD Euler publishes his famous formula $\ln_e(\cos x + \mathrm{i}\sin x) = \mathrm{i}x$ which can also be expressed as $e^{\mathrm{i}\pi} = -1$.

Why is Euler's formula so important? Any advanced culture, be it human or alien, would need a counting system that included a unit of 1. It would also need a symbol to record zero. Circles occur all over the Universe, so the ratio π would be familiar. The same is true for e, which is a constant that occurs naturally. So Euler's formula really is universal. Many people even think it proves the existence of God.

However, other people think that the fact that five of the most important numbers in mathematics, 0, 1, e, i and π, all occur in such a neat formula is pure coincidence.

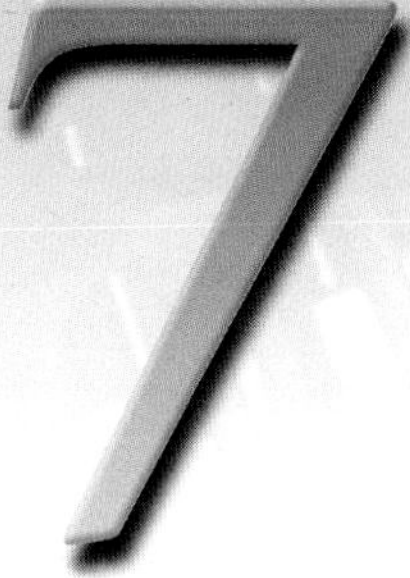

Algebra: Expressions and equations 2

1 Simultaneous equations

2 Rearranging formulae

This chapter will show you ...

B how to solve simultaneous equations

A how to rearrange formulae

Visual overview

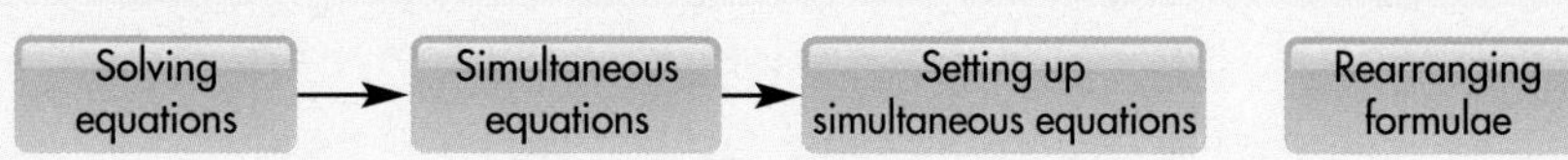

What you should already know

- The basic language of algebra **(KS3 level 5, GCSE grade E)**
- How to collect together like terms **(KS3 level 5, GCSE grade E)**
- How to multiply together terms such as $2m \times 3m$ **(KS3 level 5, GCSE grade E)**

Quick check

1 Expand the following.

a $2(x + 6)$ **b** $4(x - 3)$ **c** $6(2x - 1)$

2 Simplify the following.

a $4y + 2y - y$ **b** $3x + 2 + x - 5$ **c** $2(x + 1) - 3(x + 2)$

3 Simplify the following.

a $3 \times 2x$ **b** $4y \times 2y$ **c** $c^2 \times 2c$

4 Solve the following equations

a $2x + 4 = 6$ **b** $3x - 5 = 4$ **c** $\frac{x}{3} + 2 = 5$

d $\frac{x}{2} = 4$ **e** $\frac{x}{3} = 8$ **f** $\frac{2x}{5} = 6$

7.1 Simultaneous equations

This section will show you how to:

- solve simultaneous linear equations in two variables

Key words
balance the coefficients
check
coefficient
eliminate
simultaneous equations
substitute
variable

A pair of **simultaneous equations** is exactly that — two equations (usually linear) for which you want the same solution, and which you therefore *solve together*. For example,

$x + y = 10$ has many solutions:

$x = 2, y = 8$ $\quad$ $x = 4, y = 6$ $\quad$ $x = 5, y = 5$ …

and $2x + y = 14$ has many solutions:

$x = 2, y = 10$ $\quad$ $x = 3, y = 8$ $\quad$ $x = 4, y = 6$ …

But only *one* solution, $x = 4$ and $y = 6$, satisfies both equations at the same time.

Elimination method

Here, you solve simultaneous equations by the *elimination method*. There are six steps in this method. **Step 1** is to **balance the coefficients** of one of the **variables**. **Step 2** is to **eliminate** this variable by adding or subtracting the equations. **Step 3** is to solve the resulting linear equation in the other variable. **Step 4** is to **substitute** the value found back into one of the previous equations. **Step 5** is to solve the resulting equation. **Step 6** is to **check** that the two values found satisfy the original equations.

EXAMPLE 1

Solve the equations: $6x + y = 15$ and $4x + y = 11$

Label the equations so that the method can be clearly explained.

$6x + y = 15$ $\quad$ (1)
$4x + y = 11$ $\quad$ (2)

Step 1: Since the y-term in both equations has the same **coefficient** there is no need to balance them.

Step 2: Subtract one equation from the other. (Equation (1) minus equation (2) will give positive values.)

(1) − (2) $\quad$ $2x = 4$

Step 3: Solve this equation: $\quad$ $x = 2$

EXAMPLE 1 (continued)

Step 4: Substitute $x = 2$ into one of the original equations. (Usually the one with smallest numbers involved.)

So substitute into: $4x + y = 11$

which gives: $8 + y = 11$

Step 5: Solve this equation: $y = 3$

Step 6: Test the solution in the original equations. So substitute $x = 2$ and $y = 3$ into $6x + y$, which gives $12 + 3 = 15$ and into $4x + y$, which gives $8 + 3 = 11$. These are correct, so you can confidently say the solution is $x = 2$ and $y = 3$.

EXAMPLE 2

Solve these equations.

$$5x + y = 22 \quad (1)$$
$$2x - y = 6 \quad (2)$$

Step 1: Both equations have the same y-coefficient but with *different* signs so there is no need to balance them.

Step 2: As the signs are different, *add* the two equations, to eliminate the y-terms.

(1) + (2) $7x = 28$

Step 3: Solve this equation: $x = 4$

Step 4: Substitute $x = 4$ into one of the original equations, $5x + y = 22$,

which gives: $20 + y = 22$

Step 5: Solve this equation: $y = 2$

Step 6: Test the solution by putting $x = 4$ and $y = 2$ into the original equations, $2x - y$, which gives $8 - 2 = 6$ and $5x + y$ which gives $20 + 2 = 22$. These are correct, so the solution is $x = 4$ and $y = 2$.

Substitution method

This is an alternative method. Which method you use depends very much on the coefficients of the variables and the way that the equations are written in the first place. There are five steps in the substitute method.

Step 1 is to rearrange one of the equations into the form $y = \ldots$ or $x = \ldots$.

Step 2 is to substitute the right-hand side of this equation into the other equation in place of the variable on the left-hand side.

Step 3 is to expand and solve this equation.

Step 4 is to substitute the value into the $y = \ldots$ or $x = \ldots$ equation.

Step 5 is to check that the values work in both original equations.

EXAMPLE 3

Solve the simultaneous equations: $y = 2x + 3$, $3x + 4y = 1$

Because the first equation is in the form $y = \ldots$ it suggests that the substitution method should be used.

Again label the equations to help with explaining the method.

$y = 2x + 3$ (1)

$3x + 4y = 1$ (2)

Step 1: As equation (1) is in the form $y = \ldots$ there is no need to rearrange an equation.

Step 2: Substitute the right-hand side of equation (1) into equation (2) for the variable y.

$$3x + 4(2x + 3) = 1$$

Step 3: Expand and solve the equation. $3x + 8x + 12 = 1$, $11x = -11$, $x = -1$

Step 4: Substitute $x = -1$ into $y = 2x + 3$: $y = -2 + 3 = 1$

Step 5: Test the values in $y = 2x + 3$ which gives $1 = -2 + 3$ and $3x + 4y = 1$, which gives $-3 + 4 = 1$. These are correct so the solution is $x = -1$ and $y = 1$.

EXERCISE 7A

B

1 Solve these simultaneous equations.

In question **1** parts **a** to **i** the coefficients of one of the variables are the same so there is no need to balance them. Subtract the equations when the identical terms have the same sign. Add the equations when the identical terms have opposite signs. In parts **j** to **l** use the substitution method.

a $4x + y = 17$
$2x + y = 9$

b $5x + 2y = 13$
$x + 2y = 9$

c $2x + y = 7$
$5x - y = 14$

d $3x + 2y = 11$
$2x - 2y = 14$

e $3x - 4y = 17$
$x - 4y = 3$

f $3x + 2y = 16$
$x - 2y = 4$

g $x + 3y = 9$
$x + y = 6$

h $2x + 5y = 16$
$2x + 3y = 8$

i $3x - y = 9$
$5x + y = 11$

j $2x + 5y = 37$
$y = 11 - 2x$

k $4x - 3y = 7$
$x = 13 - 3y$

l $4x - y = 17$
$x = 2 + y$

PS 2 In this sequence, the next term is found by multiplying the previous term by a and then adding b. a and b are positive whole numbers.

3 14 47 … …

a Explain why $3a + b = 14$

b Set up another equation in a and b.

c Solve the equations to solve for a and b.

d Work out the next two terms in the sequence.

Balancing coefficients in one equation only

You were able to solve all the pairs of equations in Exercise 5H, question **1** simply by adding or subtracting the equations in each pair, or just by substituting without rearranging. This does not always happen. The next examples show what to do when there are no identical terms to begin with, or when you need to rearrange.

EXAMPLE 4

Solve these equations. $3x + 2y = 18$ (1)

$2x - y = 5$ (2)

Step 1: Multiply equation (2) by 2. There are other ways to balance the coefficients but this is the easiest and leads to less work later. With practice, you will get used to which will be the best way to balance the coefficients.

$2 \times (2)$ $4x - 2y = 10$ (3)

Label this equation as number (3).

Be careful to multiply every term and not just the y-term, it sometimes helps to write:

$2 \times (2x - y = 5) \Rightarrow 4x - 2y = 10$ (3)

Step 2: As the signs of the y-terms are opposite, add the equations.

(1) + (3) $7x = 28$

Be careful to add the correct equations. This is why labelling them is useful.

Step 3: Solve this equation: $x = 4$

Step 4: Substitute $x = 4$ into any equation, say $2x - y = 5 \Rightarrow 8 - y = 5$

Step 5: Solve this equation: $y = 3$

Step 6: Check: (1), $3 \times 4 + 2 \times 3 = 18$ and (2), $2 \times 4 - 3 = 5$, which are correct so the solution is $x = 4$ and $y = 3$.

EXAMPLE 5

Solve the simultaneous equations: $3x + y = 5$ (1)

$5x - 2y = 10$ (2)

Step 1: Multiply the first equation by 2: $6x + 2y = 10$ (3)

Step 2: Add (1) + (3): $11x = 22$

Step 3: Solve: $x = 2$

Step 4: Substitute back: $3 \times 2 + y = 5$

Step 5: Solve: $y = -1$

Step 6: Check: (1) $3 \times 2 - 1 = 5$ and (2) $5 \times 2 - 2x -1 = 10 + 2 = 12$, which are correct.

EXERCISE 7B

B

1 Solve parts **a** to **c** by the substitution method and the rest by first changing one of the equations in each pair to obtain identical terms, and then adding or subtracting the equations to eliminate those terms.

a $5x + 2y = 4$
$4x - y = 11$

b $4x + 3y = 37$
$2x + y = 17$

c $x + 3y = 7$
$2x - y = 7$

d $2x + 3y = 19$
$6x + 2y = 22$

e $5x - 2y = 26$
$3x - y = 15$

f $10x - y = 3$
$3x + 2y = 17$

g $3x + 5y = 15$
$x + 3y = 7$

h $3x + 4y = 7$
$4x + 2y = 1$

i $5x - 2y = 24$
$3x + y = 21$

k $5x - 2y = 4$
$3x - 6y = 6$

l $2x + 3y = 13$
$4x + 7y = 31$

m $3x - 2y = 3$
$5x + 6y = 12$

AU **2** **a** Mary is solving the simultaneous equations $4x - 2y = 8$ and $2x - y = 4$.

She finds a solution of $x = 5$, $y = 6$ which works for both equations.

Explain why this is not a unique solution.

b Max is solving the simultaneous equations $6x + 2y = 9$ and $3x + y = 7$.

Why is it impossible to find a solution that works for both equations?

Balancing coefficients in both equations

There are also cases where *both* equations have to be changed to obtain identical terms. The next example shows you how this is done.

Note: The substitution method is not suitable for these types of equations as you end up with fractional terms.

EXAMPLE 6

Solve these equations. $4x + 3y = 27$ (1)

$5x - 2y = 5$ (2)

Both equations have to be changed to obtain identical terms in either x or y. However, you can see that if you make the y-coefficients the same, you will add the equations. This is always safer than subtraction, so this is obviously the better choice. We do this by multiplying the first equation by 2 (the y-coefficient of the other equation) and the second equation by 3 (the y-coefficient of the other equation).

Step 1: (1) × 2 or $2 \times (4x + 3y = 27) \Rightarrow 8x + 6y = 54$ (3)

(2) × 3 or $3 \times (5x - 2y = 5) \Rightarrow 15x - 6y = 15$ (4)

Label the new equations (3) and (4).

Step 2: Eliminate one of the variables: (3) + (4) $23x = 69$

Step 3: Solve the equation: $x = 3$

Step 4: Substitute into equation (1): $12 + 3y = 27$

Step 5: Solve the equation: $y = 5$

Step 6: Check: (1), $4 \times 3 + 3 \times 5 = 12 + 15 = 27$, and (2), $5 \times 3 - 2 \times 5 = 15 - 10 = 5$, which are correct so the solution is $x = 3$ and $y = 5$.

EXERCISE 7C

1 Solve the following simultaneous equations.

a $2x + 5y = 15$
$3x - 2y = 13$

b $2x + 3y = 30$
$5x + 7y = 71$

c $2x - 3y = 15$
$5x + 7y = 52$

d $3x - 2y = 15$
$2x - 3y = 5$

e $5x - 3y = 14$
$4x - 5y = 6$

f $3x + 2y = 28$
$2x + 7y = 47$

g $2x + y = 4$
$x - y = 5$

h $5x + 2y = 11$
$3x + 4y = 8$

i $x - 2y = 4$
$3x - y = -3$

j $3x + 2y = 2$
$2x + 6y = 13$

k $6x + 2y = 14$
$3x - 5y = 10$

l $2x + 4y = 15$
$x + 5y = 21$

m $3x - y = 5$
$x + 3y = -20$

n $3x - 4y = 4.5$
$2x + 2y = 10$

o $x - 5y = 15$
$3x - 7y = 17$

B

B

PS 2 Here are four equations.

A: $5x + 2y = 1$
B: $4x + y = 9$
C: $3x - y = 5$
D: $3x + 2y = 3$

Here are four sets of (x, y) values.

$(1, -2)$, $(-1, 3)$, $(2, 1)$, $(3, -3)$

Match each pair of (x, y) values to a pair of equations.

HINTS AND TIPS

You could solve each possible set of pairs but there are six to work out. Alternatively you can substitute values into the equations to see which work.

A

AU 3 Find the area of the triangle enclosed by these three equations.

$y - x = 2$ $\quad\quad$ $x + y = 6$ $\quad\quad$ $3x + y = 6$

AU 4 Find the area of the triangle enclosed by these three equations.

$x - 2y = 6$ $\quad\quad$ $x + 2y = 6$ $\quad\quad$ $x + y = 3$

HINTS AND TIPS

Find the point of intersection of each pair of equations, plot the points on a grid and use any method to work out the area of the resulting triangle.

7.2 Rearranging formulae

This section will show you how to:

- rearrange formulae, using the same methods as for solving equations

Key words

expression
rearrange
subject
transpose
variable

The **subject** of a formula is the **variable** (letter) in the formula which stands on its own, usually on the left-hand side of the equals sign. For example, x is the subject of each of the following equations.

$x = 5t + 4$ $\quad\quad$ $x = 4(2y - 7)$ $\quad\quad$ $x = \frac{1}{t}$

To change the existing subject to a different variable, you have to **rearrange** (**transpose**) the formula to get that variable on the left-hand side. You do this by using the same rules as for solving equations. Move the terms concerned from one side of the equals sign to the other. The main difference is that when you solve an equation each step gives a numerical value. When you rearrange a formula each step gives an algebraic **expression**.

EXAMPLE 7

Make m the subject of this formula. $T = m - 3$

Move the 3 so that the m is on its own. $T + 3 = m$

Reverse the formula. $m = T + 3$

EXAMPLE 8

From the formula $P = 4t$, express t in terms of P.

(This is another common way of asking you to make t the subject.)

Divide both sides by 4: $\frac{P}{4} = \frac{4t}{4}$

Reverse the formula: $t = \frac{P}{4}$

EXAMPLE 9

From the formula $C = 2m^2 + 3$, make m the subject.

Move the 3 so that the $2m^2$ is on its own $C - 3 = 2m^2$

Divide both sides by 2: $\frac{C-3}{2} = \frac{2m^2}{2}$

Reverse the formula: $m^2 = \frac{C-3}{2}$

Take the square root on both sides: $m = \sqrt{\frac{C-3}{2}}$

EXERCISE 7D

C

HINTS AND TIPS

Remember about inverse operations, and the rule 'change sides, change signs'.

1. $T = 3k$ Make k the subject.
2. $X = y - 1$ Express y in terms of X.
3. $Q = \frac{p}{3}$ Express p in terms of Q.
4. $A = 4r + 9$ Make r the subject.
5. $W = 3n - 1$ Make n the subject.
6. $p = m + t$ **a** Make m the subject. **b** Make t the subject.
7. $g = \frac{m}{v}$ Make m the subject.

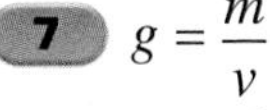

8. $t = m^2$ Make m the subject.
9. $C = 2\pi r$ Make r the subject.
10. $A = bh$ Make b the subject.

C

11 $P = 2l + 2w$ Make l the subject.

12 $m = p^2 + 2$ Make p the subject.

FM 13 The formula for converting degrees Fahrenheit to degrees Celsius is $C = \frac{5}{9}(F - 32)$.

a Show that when $F = -40$, C is also equal to -40.

b Find the value of C when $F = 68$.

c Use this flow diagram to establish the formula for converting degrees Celsius to degrees Fahrenheit.

FM 14 Kieran notices that the price of five cream buns is 75p more than the price of nine mince pies.
Let the price of a cream bun be x pence and the price of a mince pie be y pence.

a Express the cost of one mince pie, y, in terms of the price of a cream bun, x.

b If the price of a cream bun is 60p, how much is a mince pie?

> **HINTS AND TIPS**
> Set up a formula, using the first sentence of information, then rearrange it.

PS 15 Distance, speed and time are connected by the formula:

distance = speed × time.

A delivery driver drove 126 km in 1 hour and 45 minutes. On the return journey, he was held up at some road works so his average speed decreased by 9 km per hour.

How long was he held up at the road works?

> **HINTS AND TIPS**
> Work out the average speed for the first journey, then work out the average speed for the return journey.

B

16 $v = u + at$ **a** Make a the subject. **b** Make t the subject.

17 $A = \frac{1}{4}\pi d^2$ Make d the subject.

18 $W = 3n + t$ **a** Make n the subject. **b** Express t in terms of n and W.

19 $x = 5y - w$ **a** Make y the subject. **b** Express w in terms of x and y.

20 $k = 2p^2$ Make p the subject.

21 $v = u^2 - t$ **a** Make t the subject. **b** Make u the subject.

22 $k = m + n^2$ **a** Make m the subject. **b** Make n the subject.

23 $T = 5r^2$ Make r the subject.

24 $K = 5n^2 + w$ **a** Make w the subject. **b** Make n the subject.

GRADE BOOSTER

- **B** You can solve two simultaneous linear equations
- **B** You can rearrange more complicated formulae
- **A** You can set up and solve two simultaneous equations from a practical problem

What you should know now

- How to set up and/or solve a pair of linear simultaneous equations
- How to rearrange formula where the subject appears once

EXAMINATION QUESTIONS

1 Solve the simultaneous equations:

$5x + 6y = 28$

$x + 3y = 2$

You must show your working.

Do not use trial and improvement. *(3 marks)*

AQA, June 2007, Paper 1 Higher, Question 12

2 ABC is an isosceles triangle.

The lengths, in cm, of the sides are
AB = $4a + 3$, BC = $2b + 5$ and AC = $2a + b$

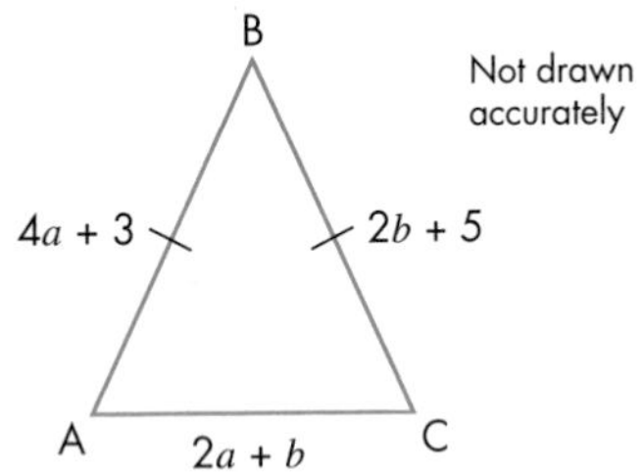

a AB = BC

Show that $2a - b = 1$ *(2 marks)*

b The perimeter of the triangle is 32 cm.

Find the values of a and b. *(4 marks)*

AQA, November 2005, Paper 2, Question 14

3 In this 'magic' triangle each side has a total of 24

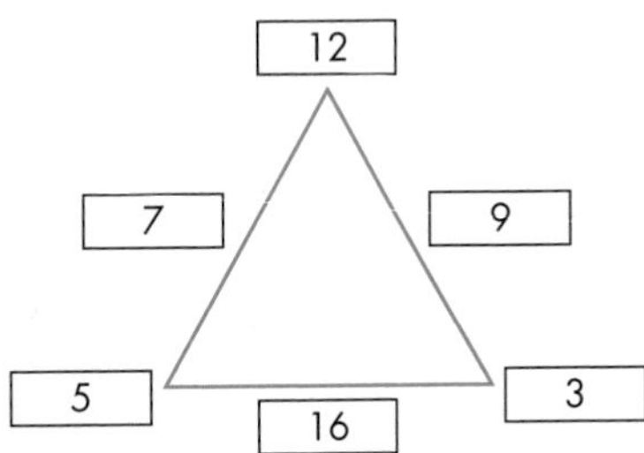

Here is another 'magic' triangle in which the sum of the three expressions on each of the three sides is the same.

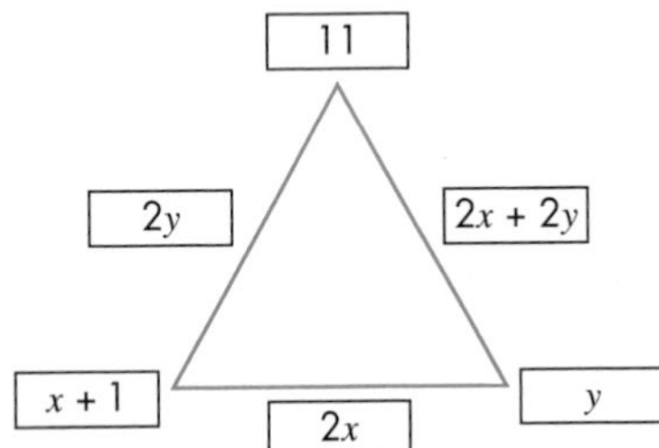

a By considering the left hand side and the right hand side, show that $x + y = 1$ *(2 marks)*

b By considering the left hand side and the bottom of the triangle, show that $2x - y = 11$ (2 marks)

c Solve the simultaneous equations

$x + y = 1$

$2x - y = 11$ *(2 marks)*

You **must** show your working.

d Complete the 'magic' triangle. *(1 mark)*

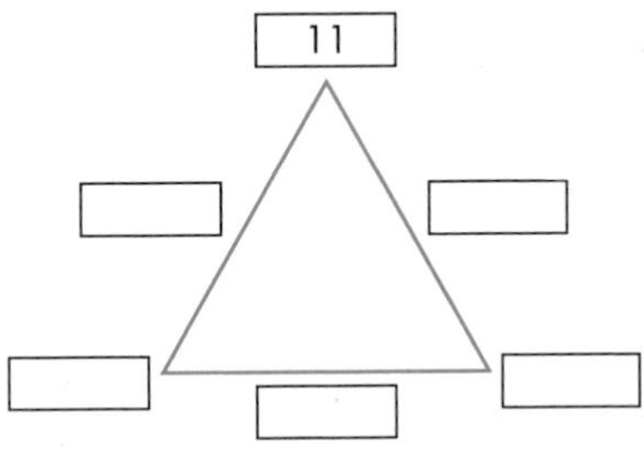

AQA, November 2008, Paper 1, Question 16

4 Amy and Jo are buying pens and rulers.

Amy buys three pens and two rulers for 75p.
Jo buys four pens and one ruler for 65p.

Work out the cost of a pen and the cost of a ruler.

You **must** show your working.
Do **not** use trial and improvement.

AQA, November 2011 H1 Question 11

5 Rearrange the formula

$$A = \frac{1}{2}(a + b)h$$

to make b the subject.

AQA, June 2011 H2 Question 15b

6 Make x the subject of the formula

$y = 3x + 7$

AQA, June 2011 H1 Question 13b

7 Solve the simultaneous equations

$3x - 2y = 7$

$x = y + 3$

Do **not** use trial and improvement.
You **must** show your working.

AQA, June 2011 H1 Question 15

8 **a** Make x the subject of

$$y = \frac{x}{w} - t$$

b Solve the simultaneous equations

$$2y = x + 6$$

$$y = 2x - 3$$

You **must** show your working.
Do **not** use trial and improvement.

AQA, November 2010 H1 Question 12

9 Rearrange the formula

$$n = p^2 + 6$$

to make p the subject

AQA, June 2010 H2 Question 15c

10 Solve the simultaneous equations

$$3x - 2y = 9$$

$$x + 4y = 10$$

You **must** show your working.
Do **not** use trial and improvement.

AQA, November 2009 H1 Question 16

Worked Examination Questions

1 $4x + 3y = 6$

$3x - 2y = 13$

Solve these simultaneous equations algebraically. Show your method clearly.

1 $4x + 3y = 23$ (1)

$3x - 2y = 13$ (2)

Label the equations and decide on the best way to get the coefficients of one variable the same.

(1) × 2 $8x + 6y = 46$ (3)

(2) × 3 $9x - 6y = 39$ (4)

(3) + (4) $17x = 85$

Making the y coefficients the same will be the most efficient way as the resulting equations will be added.

This gets 1 mark for method.

$x = 5$

Substitute into (1) $20 + 3y = 23$

$y = 1$

Solve the resulting equation. Substitute into one of the original equations. Work out the other value.

$4 \times 5 + 3 \times 1 = 23$ ✓

$3 \times 5 - 2 \times 1 = 13$ ✓

Check that these values work in the original equations.

Total: 6 marks

2 Nicky did a 22 km hill race. She ran x km to the top of the hill at an average speed of 8 km/h. She then ran y kilometres down the hill at an average speed of y km/h. She finished the race in 2 hours and 10 minutes.

Find out how long it took Nicky to get to the top of the hill.

2 $x + y = 22$

Set up two simultaneous equations using the information given. This scores 1 mark for method.

$\frac{x}{8} + \frac{y}{15} = 2\frac{1}{6}$

$15x + 8y = 260$

Multiply the second equation by 120 (LCM of 15, 8 and 6). This scores 1 mark for accuracy.

$8x + 8y = 176$

Balance the coefficients and subtract to eliminate y. This scores 1 mark for method.

$7x = 84$

$x = 12$

Time = $12 \div 8$ = 1 hour 30 minutes

Solve for x (A1) and work out the time using distance ÷ speed. This scores 1 mark for accuracy.

Total: 5 marks

Worked Examination Questions

3 **a** Rearrange $y = 5x - 3$ to make x the subject

b Rearrange $c^2 = a^2 + b^2$ to make a the subject

3 a $y + 3 = 5x$ — Just as when you are solving an equation you have to get x on its own. Taking −3 across the equal sign and making it into + 3 will get 1 method mark.

$x = \dfrac{y + 3}{5}$ — Dividing both sides by 5 will get the final answer. This is 1 accuracy mark. It is important to include the x = as this is a vital part of the rearranged formula.

Total: 2 marks

b $a^2 = c^2 - b^2$ — Start by taking b^2 from both sides. This is 1 method mark.

$a = \sqrt{(c^2 - b^2)}$ — Square root both sides. This will give the final answer which gets 1 accuracy mark.

Total: 2 marks

4 This table shows the costs of teas and coffees.

Number of teas	Number of coffees	Total cost
2	3	£8.75
4	1	£7.75
5	2	£11.15
1	6	£13.15
3	5	£14.10

Use the table to work out the cost of tea and coffee.

4 $2T + 3C = 875$ −(1)

$4T + C = 775$ −(2)

Setting up a pair of simultaneous equations will get 1 method mark. There are many ways of setting up the pair of equations. You could add the 2nd and 4th rows to get $5T + 7C = 2090$ which balances the teas with the 3rd row.

(1) x 2 $4T + 6C = 1750$ −(3)

Multiply one or both equation to get the same number of teas or coffees will get 1 method mark.

(3) − (1) $5C = 975$

$C = 195$

Eliminating one variable and solving the equation for the other variable will get 1 accuracy mark.

Substitute into (2)

$4T + 195 = 775$

$T = 145$

Tea = £1.45, Coffee = £1.95

Substituting back into one of the original equations and solving for the other variable gets 1 accuracy mark.

Total: 4 marks

Functional Maths
Walking using Naismith's rule

Many people go walking each weekend. It is good exercise and can be a very enjoyable pastime.

When walkers set out they often try to estimate the length of time the walk will take. There are many factors that could influence this, but one rule that can help in estimating how long the walk will take is Naismith's rule.

Naismith's rule

Naismith's rule is a rule of thumb that you can use when planning a walk by calculating how long it will take. The rule was devised by William Naismith, a Scottish mountaineer, in 1892.

The basic rule is:

Allow 1 hour for every 3 miles (5 km) forward, plus $\frac{1}{2}$ hour for every 1000 feet (300 m) ascent (height).

Getting started

Before you begin your main task, you may find it useful to fill in the following table to practise using Naismith's rule.

Can you use algebra to display the rule?

Day	Distance (km)	Height (m)	Time (m)
1	16	250	
2	18	0	
3	11	340	
4	13	100	
5	14	120	

Now, in small groups think about:

- What kind of things influence the speed at which you walk?
- Do different types of routes make people walk at different rates?
- If there is a large group of people will they all walk at the same rate?

Use all the ideas you have just discussed as you move on to your main task.

Your task

You are going to compare data to see if Naismith's rule is still a useful way to work out how much time to allow for different walks.

The table on the right shows the actual times taken by a school group as they did five different walks in five days. Use this information to work out the following

1 If the group had started at the same times and had the same breaks how long would the group have taken each day, according to Naismith's rule?

2 Do you think Naismith's rule is still valid today? Explain your reasons.

3 If your friend was going to climb Ben Nevis, setting out at 11.30 am, would you advise them to do the walk? You will need to research the distance and climb details of the pathway up Ben Nevis, in order to advise them fully.

Day	Distance (km)	Height (m)	Time (minutes)	Time (hours/minutes)	Start	Breaks	Finish
1	16	250	265	4h 15m	10.00	2h	4:15 pm
2	18	0	270	4h 30m	10.00	1h 30m	4:00 pm
3	11	340	199	3h 19m	09.30	2h 30m	3:19 pm
4	13	100	205	3h 15m	10.30	2h 30m	4:15 pm
5	14	120	222	3h 42m	10.30	2h 30m	4:42 pm

Extension

Produce a report that compares and contrasts the walking times for some of Britain's most famous walks, such as Ben Nevis, Snowdon, Helvellyn and the Pennine Way. Your report should contain realistic guidance on how best to approach these long walks, including:

- suggested day-by-day plans supported by mathematical evidence
- starting times
- places to rest
- how the walks will vary for walkers of different fitness levels
- how weather conditions could affect the walk and precautions that should be taken.

Using this information, evaluate how similar the walks are, and which walk would be the toughest for an average walker to complete.

Why this chapter matters

Ask any adult to name a mathematician. It is likely that, if they know any at all, the first name they think of will be Pythagoras. Could this be because this mathematical result is the only one – of those taught in schools – that has a person's name attached to it?

When a builder wants to build two walls at right angles to one another, how will he make sure the angle between them is 90°? The chances are that he will use a 3–4–5 triangle. What the builder may not know is that exactly the same technique has been used by builders for thousands of years, as far back as the construction of the pyramids of Egypt and maybe even before that.

The 3–4–5 rule works because it is a special case of Pythagoras' theorem.

Pythagoras' was a Greek who lived about 2500 years ago but he was certainly not the first person to discover the rule, although it has been named after him. We have written evidence that the theorem was know in ancient Mesopotamia, China and India and it was probably discovered independently at different times in different parts of the world.

This Chinese visual proof is based on one from a book called *Zhou Bi Suan Jing*, compiled between 500 and 200BC.

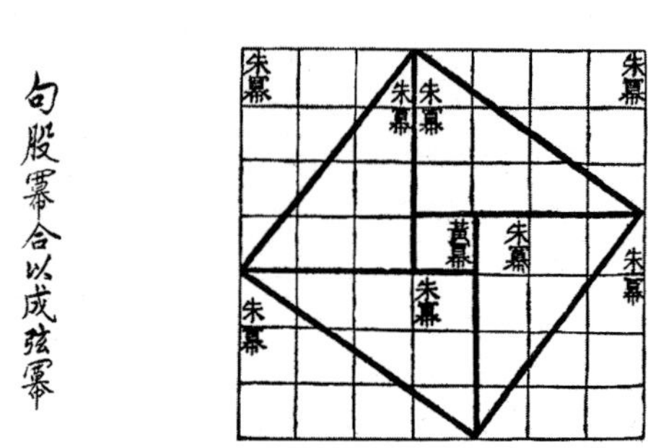

Pythagoras' theorem is interesting in its own right but it is also very important as a tool that is used in many areas of science, engineering and technology.

Mathematical trivia 1: Pythagoras' theorem has appeared in an episode of *The Simpsons*, when it was quoted incorrectly by Homer.

Mathematical trivia 2: In the year 2000, Uganda issued a 2000 shilling coin in the shape of a right-angled triangle. It had an image of Pythagoras and a statement of his theorem on one side.

Mathematical trivia 3: In 1940 Elisha Scott Loomis published a book containing 256 different proofs of Pythagoras' theorem. One of them was written by James Garfield, president of the United States, in 1881.

Chapter

Geometry: Pythagoras and trigonometry

This chapter will show you ...

- **C** how to use Pythagoras' theorem in right-angled triangles
- **C** how to solve problems using Pythagoras' theorem
- **B** how to use Pythagoras' theorem in three dimensions

Visual overview

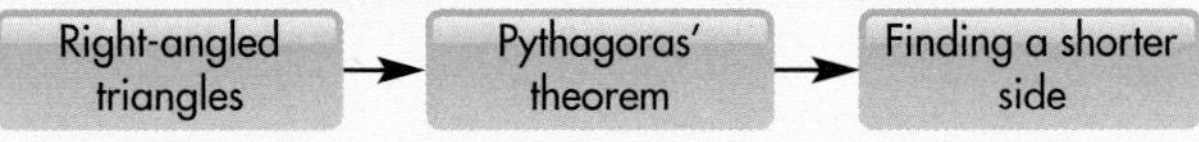

What you should already know

- how to find the square and square root of a number **(KS3 level 5, GCSE grade F)**
- how to round numbers to a suitable degree of accuracy **(KS3 level 6, GCSE grade E)**

Quick check

Use your calculator to evaluate the following, giving your answers to one decimal place.

1 2.3^2

2 15.7^2

3 0.78^2

4 $\sqrt{8}$

5 $\sqrt{260}$

6 $\sqrt{0.5}$

8.1 Pythagoras' theorem

This section will show you how to:

- calculate the length of the hypotenuse in a right-angled triangle

Key words
hypotenuse
Pythagoras' theorem

Pythagoras, who was a philosopher as well as a mathematician, was born in 580BC, on the island of Samos in Greece. He later moved to Crotona (Italy), where he established the Pythagorean Brotherhood, which was a secret society devoted to politics, mathematics and astronomy. It is said that when he discovered his famous theorem, he was so full of joy that he showed his gratitude to the gods by sacrificing a hundred oxen.

Consider squares being drawn on each side of a right-angled triangle, with sides 3 cm, 4 cm and 5 cm.

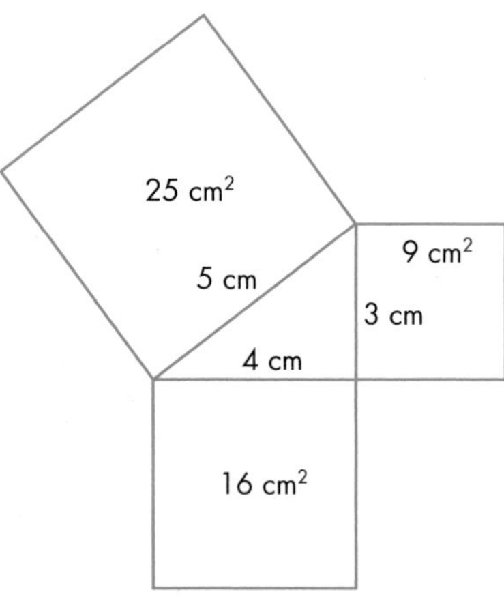

The longest side is called the **hypotenuse** and is always opposite the right angle.

Pythagoras' theorem can then be stated as follows:

For any right-angled triangle, the area of the square drawn on the hypotenuse is equal to the sum of the areas of the squares drawn on the other two sides.

The form in which most of your parents would have learnt the theorem when they were at school – and which is still in use today – is as follows:

In any right-angled triangle, the square of the hypotenuse is equal to the sum of the squares of the other two sides.

Pythagoras' theorem is more usually written as a formula:

$$c^2 = a^2 + b^2$$

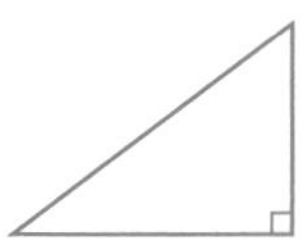

Remember that Pythagoras' theorem can only be used in right-angled triangles.

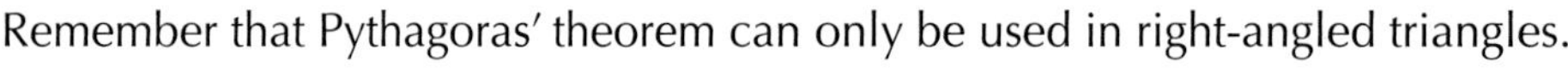

Finding the hypotenuse

EXAMPLE 1

Find the length of the hypotenuse, marked x on the diagram.

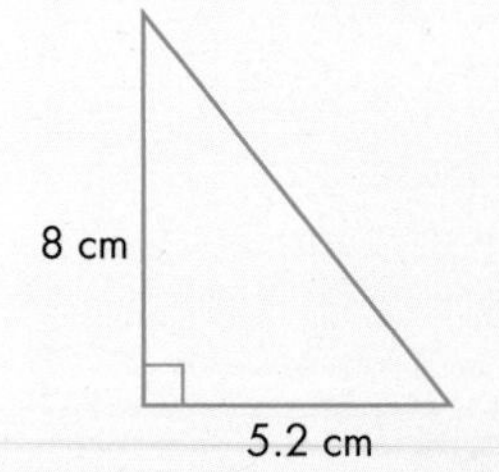

Using Pythagoras' theorem gives:

$$x^2 = 8^2 + 5.2^2 \text{ cm}^2$$
$$= 64 + 27.04 \text{ cm}^2$$
$$= 91.04 \text{ cm}^2$$

So $x = \sqrt{91.04} = 9.5$ cm (1 decimal place)

EXERCISE 8A

For each of the following triangles, calculate the length of the hypotenuse, x, giving your answers to 1 decimal place.

HINTS AND TIPS

In these examples you are finding the hypotenuse. The squares of the two short sides are added in every case.

1

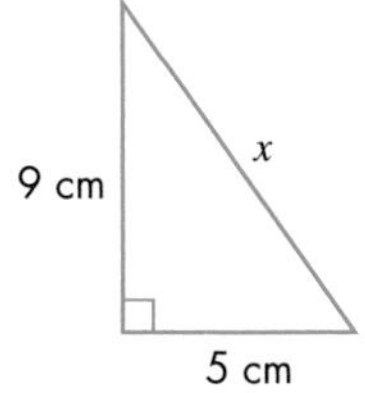

2

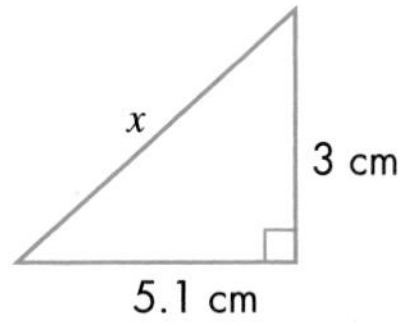

3

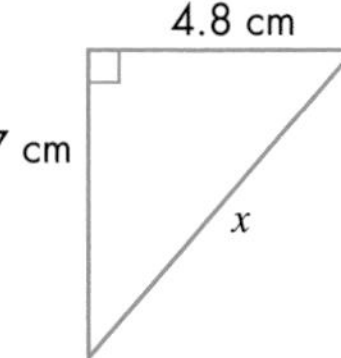

4

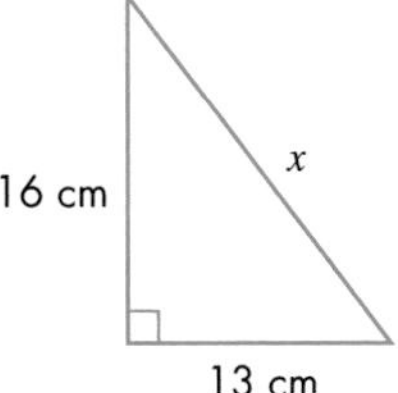

5

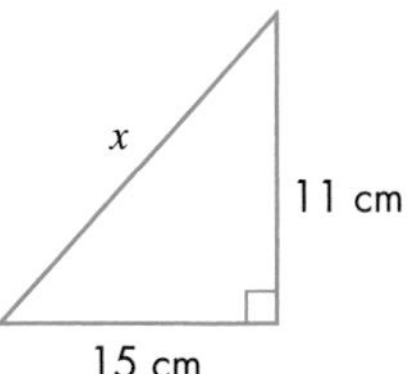

6

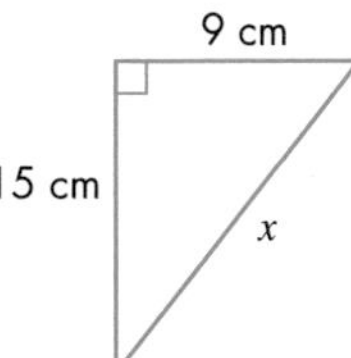

7

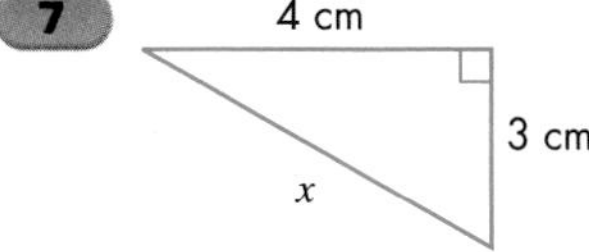

8

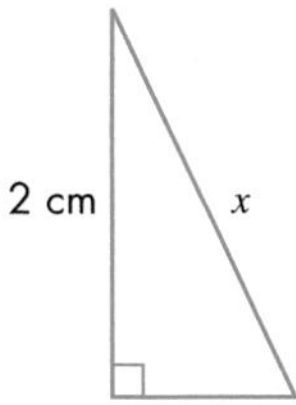

9

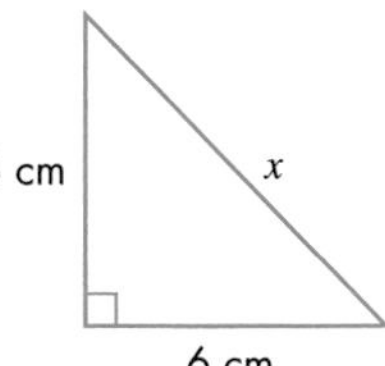

PS 10 How does this diagram show that Pythagoras' theorem is true?

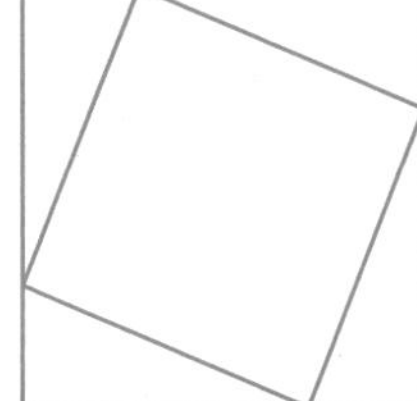

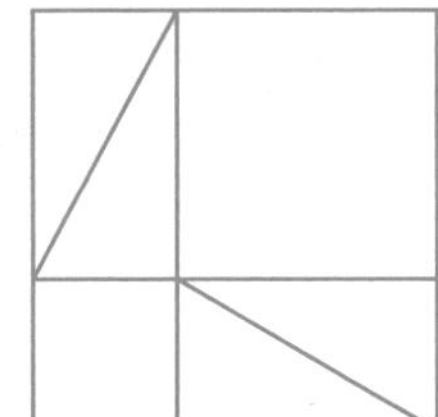

8.2 Finding a shorter side

This section will show you how to:
- calculate the length of a shorter side in a right-angled triangle

Key words
Pythagoras' theorem

By rearranging the formula for **Pythagoras' theorem**, the length of one of the shorter sides can easily be calculated.

$$c^2 = a^2 + b^2$$

So, $a^2 = c^2 - b^2$ or $b^2 = c^2 - a^2$

EXAMPLE 2

Find the length x.

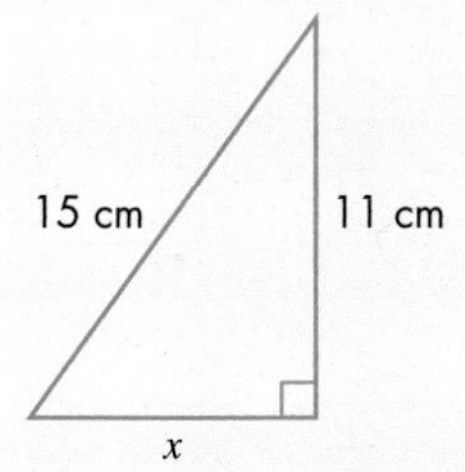

x is one of the shorter sides.

So using Pythagoras' theorem gives:

$$x^2 = 15^2 - 11^2 \text{ cm}^2$$
$$= 225 - 121 \text{ cm}^2$$
$$= 104 \text{ cm}^2$$

So $x = \sqrt{104} = 10.2$ cm (1 decimal place)

EXERCISE 8B

C

1 For each of the following triangles, calculate the length x, giving your answers to 1 decimal place.

HINTS AND TIPS

In these examples you are finding a short side. The square of the other short side is subtracted from the square of the hypotenuse in every case.

a

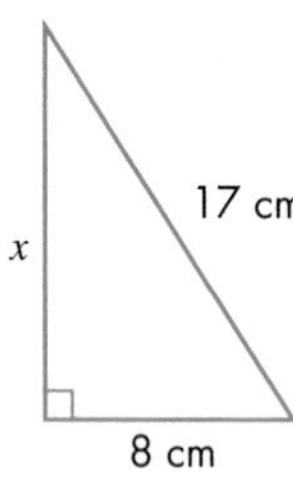

b

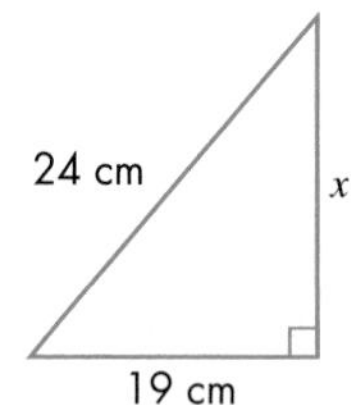

c

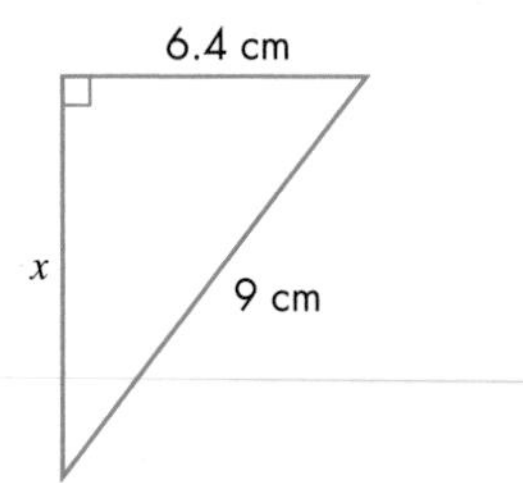

d

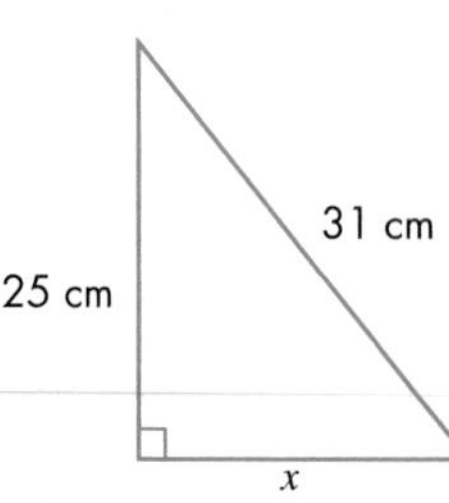

C

2 For each of the following triangles, calculate the length x, giving your answers to 1 decimal place.

a

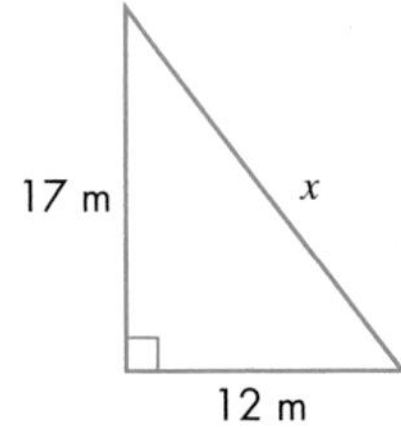

b

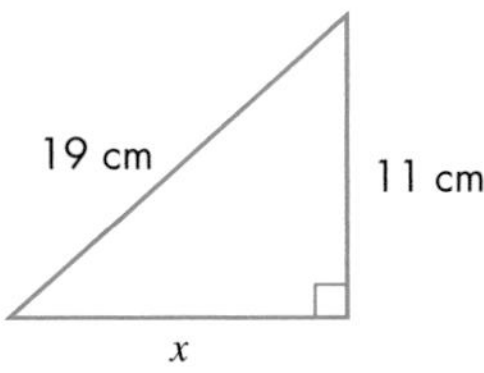

c

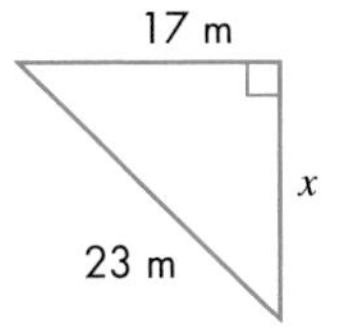

d

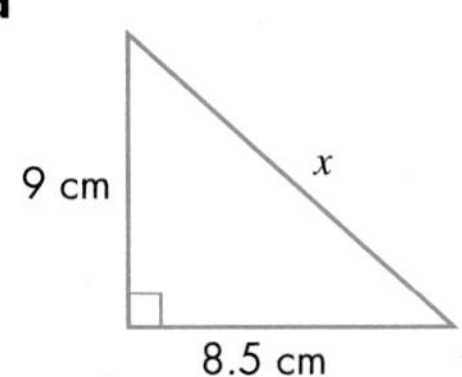

HINTS AND TIPS

These examples are a mixture. Make sure you combine the squares of the sides correctly. If you get it wrong in an examination, you get no marks.

3 For each of the following triangles, find the length marked x.

a

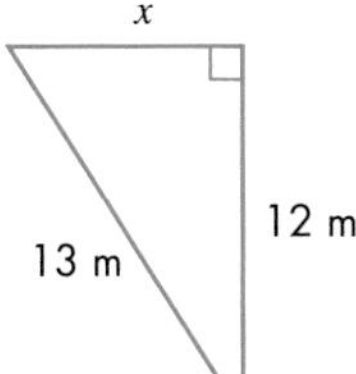

b

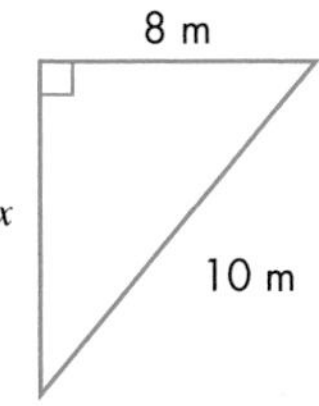

c

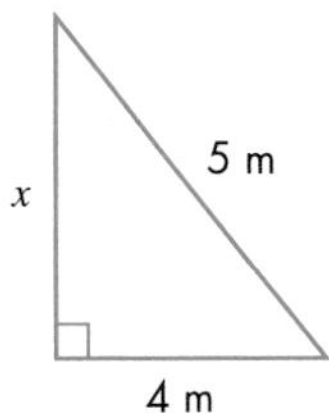

d

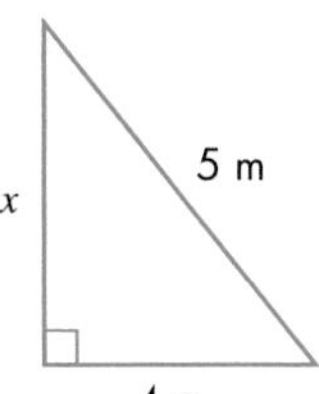

PS 4 In question **3** you found sets of three numbers which satisfy $a^2 + b^2 = c^2$.

Can you find any more?

5 Calculate the value of x.

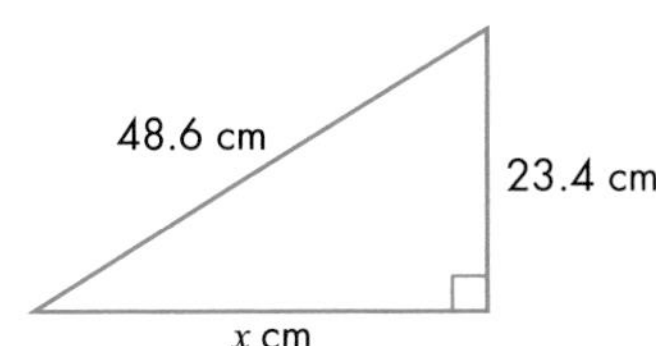

8.3 Applying Pythagoras' theorem in real situations

This section will show you how to:

- solve problems using Pythagoras' theorem

Key words
isosceles triangle
Pythagoras' theorem

Pythagoras' theorem can be used to solve certain practical problems. When a problem involves two lengths only, follow these steps.

- Draw a diagram for the problem that includes a right-angled triangle.
- Look at the diagram and decide which side has to be found: the hypotenuse or one of the shorter sides. Label the unknown side x.
- If it is the hypotenuse, square both numbers, add the squares and take the square root of the sum.
- If it is one of the shorter sides, square both numbers, subtract the squares and take the square root of the difference.
- Finally, round the answer to a suitable degree of accuracy.

EXAMPLE 3

A plane leaves Manchester airport heading due east. It flies 160 km before turning due north. It then flies a further 280 km and lands. What is the distance of the return flight if the plane flies straight back to Manchester airport?

First, sketch the situation.

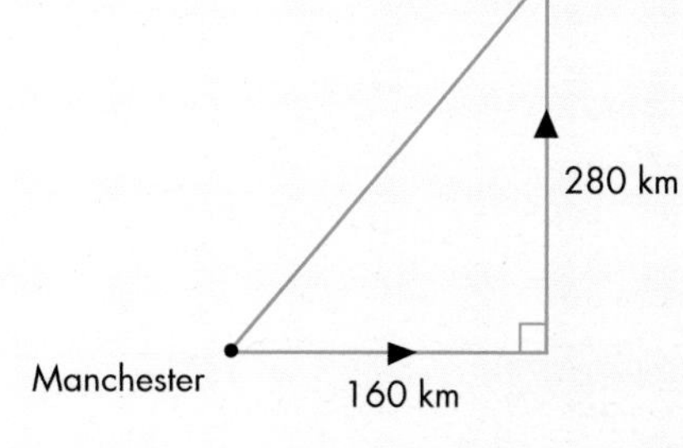

Using Pythagoras' theorem gives:

$$x^2 = 160^2 + 280^2 \text{ km}^2$$
$$= 25\,600 + 78\,400 \text{ km}^2$$
$$= 104\,000 \text{ km}^2$$

So $x = \sqrt{104104000} = 322$ km
(3 significant figures)

Remember the following tips when solving problems.

- Always sketch the right-angled triangle you need. Sometimes, the triangle is already drawn for you but some problems involve other lines and triangles that may confuse you. So identify which right-angled triangle you need and sketch it separately.
- Label the triangle with necessary information, such as the length of its sides, taken from the question. Label the unknown side x.
- Set out your solution as in Example 3. Avoid short cuts, since they often cause errors. You gain marks in your examination for clearly showing how you are applying Pythagoras' theorem to the problem.
- Round your answer to a suitable degree of accuracy.

EXERCISE 8C 

FM 1 A ladder, 12 m long, leans against a wall. The ladder reaches 10 m up the wall. The ladder is safe if the foot of the ladder is about 2.5 m away from the wall. Is this ladder safe?

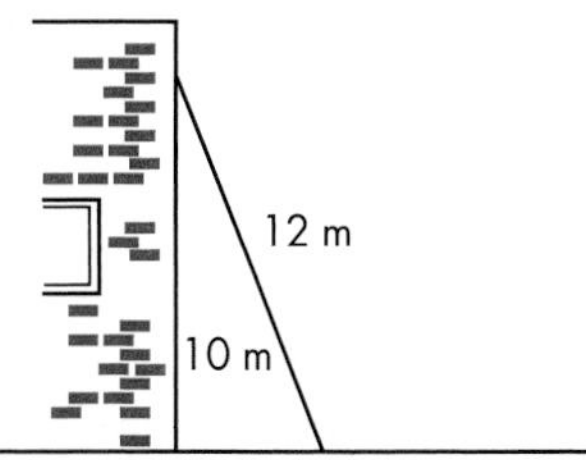

2 A model football pitch is 2 m long and 0.5-m wide. How long is the diagonal?

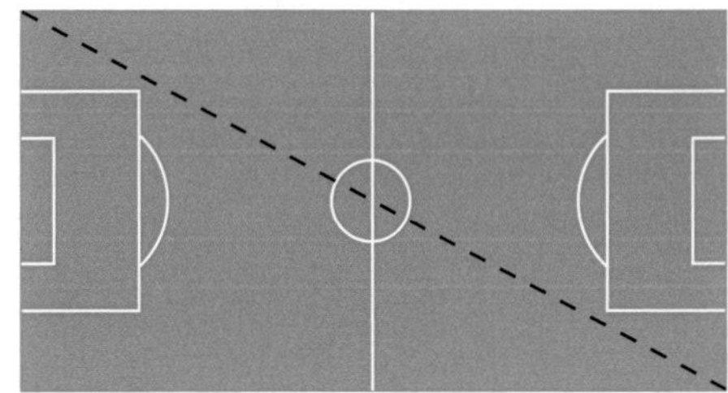

PS 3 How long is the diagonal of a square with a side of 8 m?

AU 4 A ship going from a port to a lighthouse steams 15 km east and 12 km north. The journey takes 1 hour. How much time would be saved by travelling directly to the lighthouse in a straight line?

FM 5 Some pedestrians want to get from point X on one road to point Y on another. The two roads meet at right angles.

Instead of following the roads, they decide to follow a footpath which goes directly from X to Y.

How much shorter is this route?

6 A mast on a sailboat is strengthened by a wire (called a stay), as shown on the diagram. The mast is 10 m tall and the stay is 11 m long. How far from the base of the mast does the stay reach?

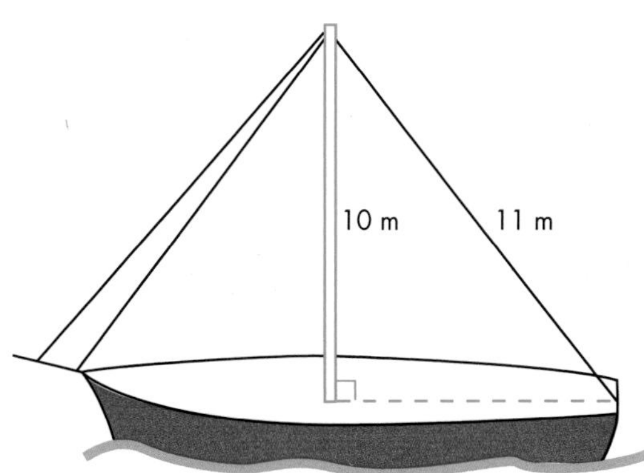

B

FM 7 A ladder, 4 m long, is put up against a wall.

a How far up the wall will it reach when the foot of the ladder is 1 m away from the wall?

b When it reaches 3.6 m up the wall, how far is the foot of the ladder away from the wall?

AU 8 A pole, 8 m high, is supported by metal wires, each 8.6 m long, attached to the top of the pole. How far from the foot of the pole are the wires fixed to the ground?

AU 9 A and B are two points on a coordinate grid. They have coordinates (13, 6) and (1, 1). How long is the line that joins them?

FM 10 The regulation for safe use of ladders states that: *the foot of a 5.00 m ladder must be placed between 1.20 m and 1.30 m from the foot of the wall.*

a What is the maximum height the ladder can safely reach up the wall?

b What is the minimum height the ladder can safely reach up the wall?

AU 11 Is the triangle with sides 7 cm, 24 cm and 25 cm a right-angled triangle? Give a reason for your answer.

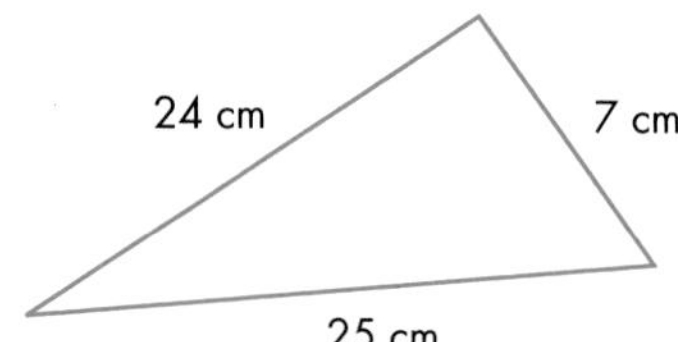

PS 12 A 4 m long ladder is leaning against a wall. The foot of the ladder is 1 m from the wall. The foot of the ladder is not securely held and slips 20 cm further away from the wall.

How far does the top of the ladder move down the wall?

PS 13 The diagonal of a rectangle is 10 cm. What can you say about the perimeter of the rectangle?

Pythagoras' theorem and isosceles triangles

This section shows you how to to use Pythagoras' theorem in isosceles triangles.

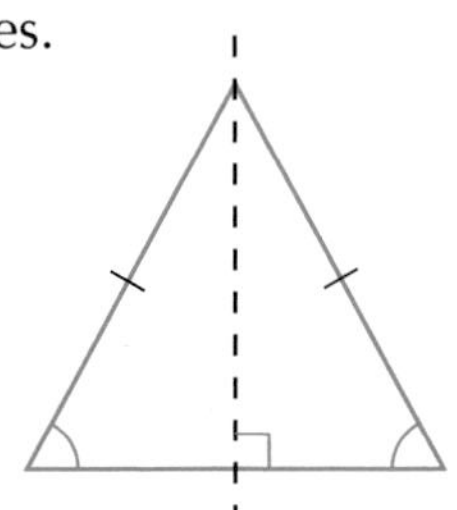

Every **isosceles triangle** has a line of symmetry that divides the triangle into two congruent right-angled triangles. So when you are faced with a problem involving an isosceles triangle, be aware that you are quite likely to have to split that triangle down the middle to create a right-angled triangle which will help you to solve the problem.

EXAMPLE 4

Calculate the area of this triangle.

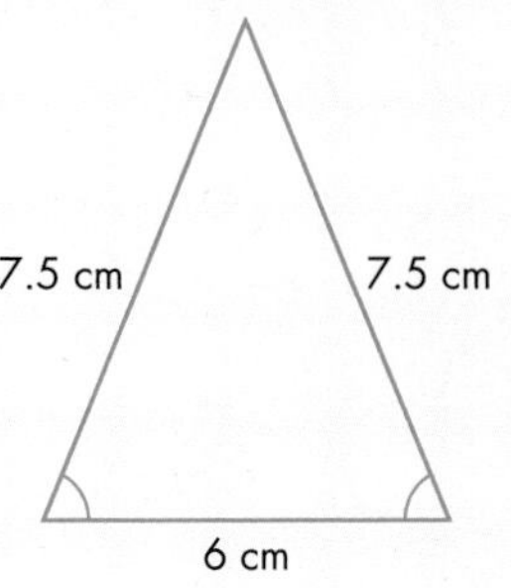

It is an isosceles triangle and you need to calculate its height to find its area.

First split the triangle into two right-angled triangles to find its height.

Let the height be x.

Then, using Pythagoras' theorem,

$$x^2 = 7.5^2 - 3^2 \text{ cm}^2$$
$$= 56.25 - 9 \text{ cm}^2$$
$$= 47.25 \text{ cm}^2$$

So $x = \sqrt{47.25}$ cm

$x = 6.87$ cm

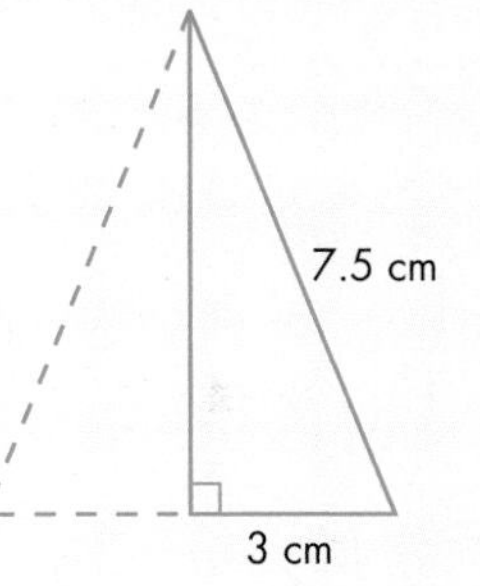

Keep the accurate figure in the calculator memory.

The area of the triangle is $\frac{1}{2} \times 6 \times 6.87$ cm^2 (from the calculator memory), which is 20.6 cm^2 (1 decimal place).

EXERCISE 8D

1 Calculate the areas of these isosceles triangles.

a

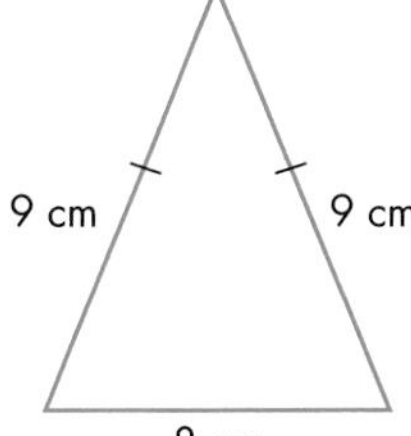

b

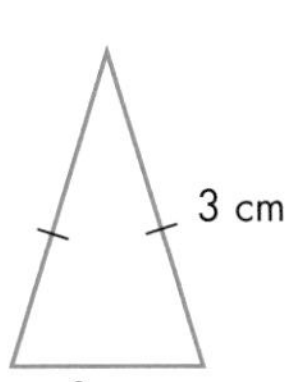

c

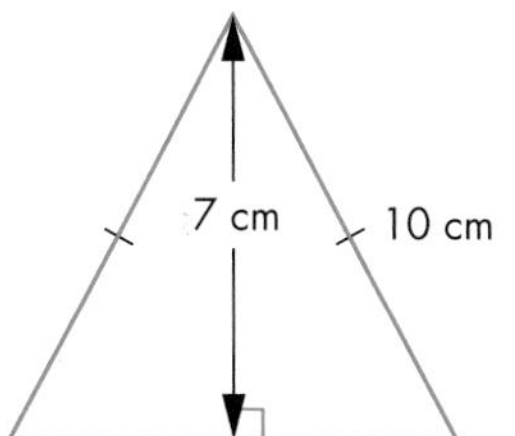

2 Calculate the area of an isosceles triangle whose sides are 8 cm, 8 cm and 6 cm.

PS **3** Calculate the area of an equilateral triangle of side 6 cm.

PS **4** An isosceles triangle has sides of 5 cm and 6 cm.

a Sketch the two different isosceles triangles that fit this data.

b Which of the two triangles has the greater area?

5 **a** Sketch a regular hexagon, showing all its lines of symmetry.

b Calculate the area of the hexagon if its side is 8 cm.

B

B

PS **6** Calculate the area of a hexagon of side 10 cm.

PS **7** These isosceles triangles have the same perimeter.

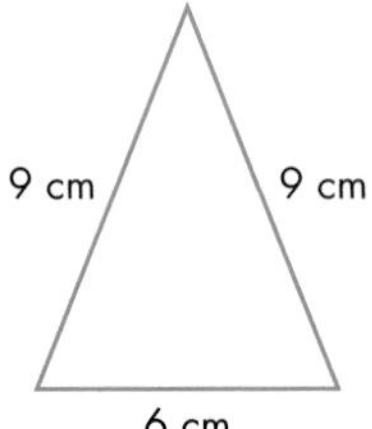

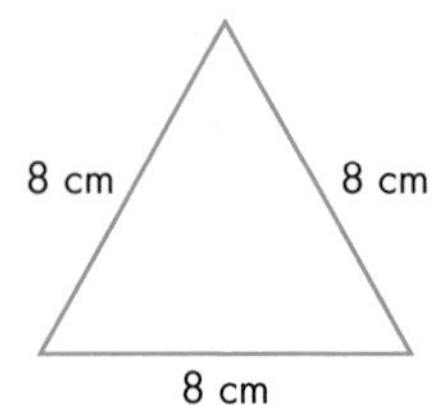

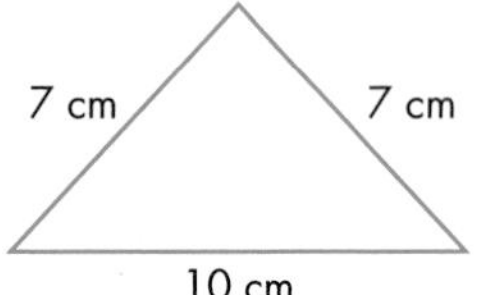

a Do the three triangles have the same area?

b Can you find an isosceles triangle with the same perimeter but a larger area?

c Can you generalise your findings?

FM **8** A piece of land is in the shape of an isosceles triangle with sides 6.5 m, 6.5 m and 7.4 m.

So that it can be sown with the correct quantity of grass seed to make a lawn, you have been asked to calculate the area.

What is the area of the land?

9 The diagram shows an isosceles triangle ABC.

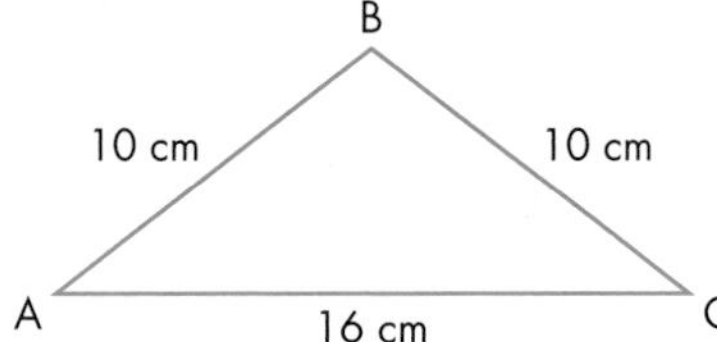

Calculate the area of triangle ABC.

State the units of your answer.

10 Calculate the lengths marked x in these isosceles triangles.

a

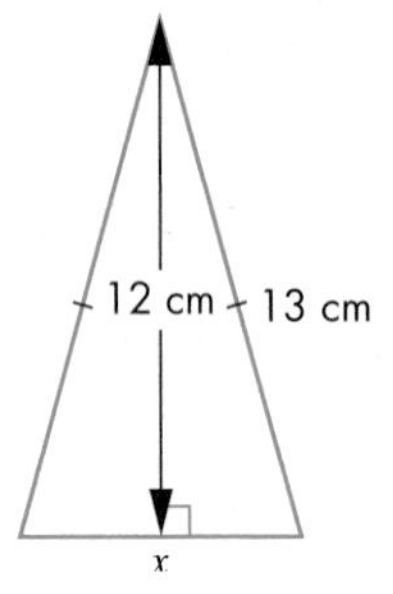

b

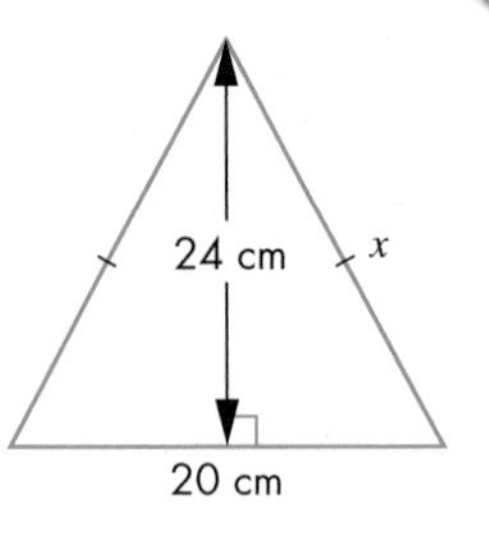

AU **c**

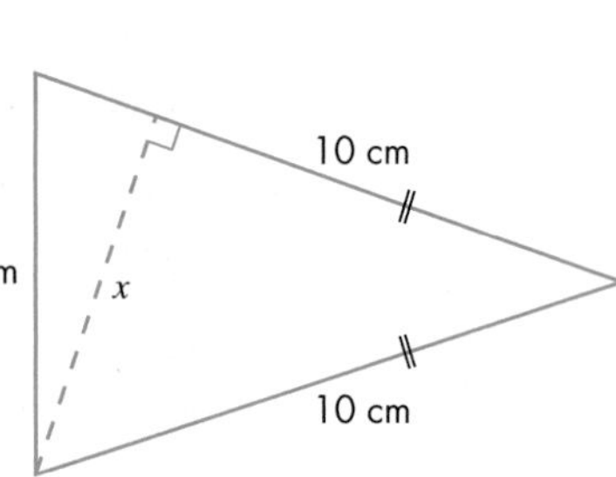

HINTS AND TIPS

Find the area first.

8.4 Pythagoras' theorem in three dimensions

This section will show you how to:
- use Pythagoras' theorem in problems involving three dimensions

Key words
3D
Pythagoras' theorem

This section shows you how to solve problems in **3D** using **Pythagoras' theorem**.

In your GCSE examinations, there may be questions which involve applying Pythagoras' theorem in 3D situations. Such questions are usually accompanied by clearly-labelled diagrams, which will help you to identify the lengths needed for your solutions.

You deal with these 3D problems in exactly the same way as 2D problems.

- Identify the right-angled triangle you need.
- Redraw this triangle and label it with the given lengths and the length to be found, usually x or y.
- From your diagram, decide whether it is the hypotenuse or one of the shorter sides which has to be found.
- Solve the problem, rounding to a suitable degree of accuracy.

EXAMPLE 5

What is the longest piece of straight wire that can be stored in this box measuring 30 cm by 15 cm by 20 cm?

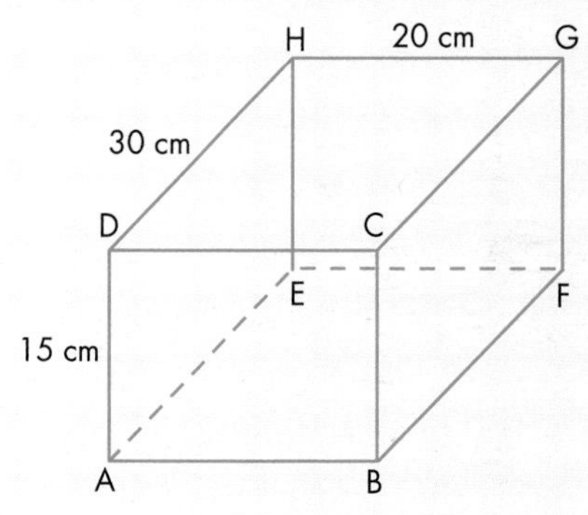

The longest distance across this box is any one of the diagonals AG, DF, CE or HB.

Let us take AG.

First, identify a right-angled triangle containing AG and draw it.

This gives a triangle AFG, which contains two lengths you do not know, AG and AF.

Let AG = x and AF = y

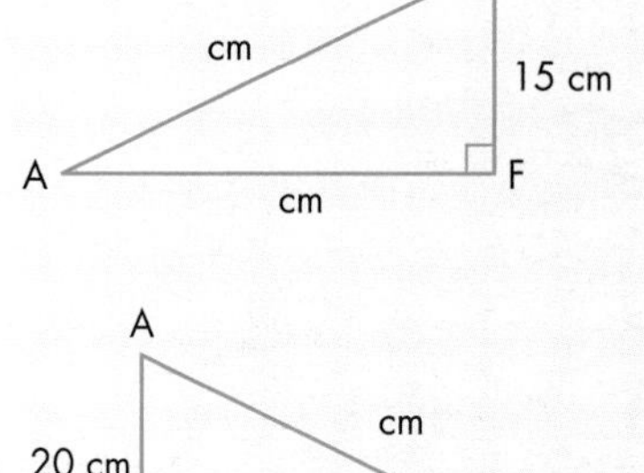

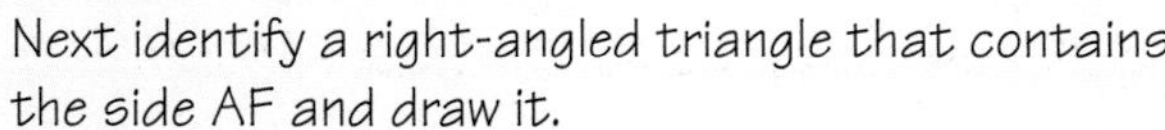

Next identify a right-angled triangle that contains the side AF and draw it.

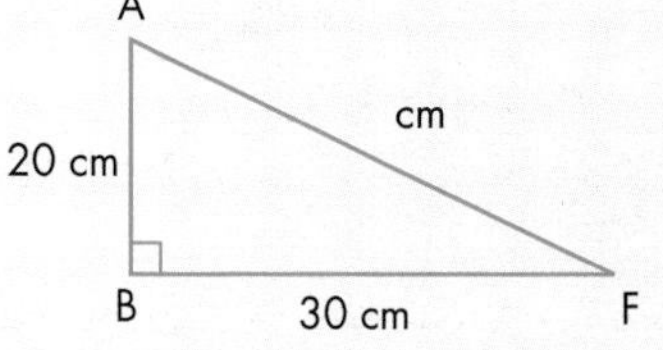

This gives a triangle ABF. You can now find AF.

By Pythagoras' theorem,

$y^2 = 30^2 + 20^2 \text{ cm}^2$

$y^2 = 1300 \text{ cm}^2$ (there is no need to find y)

EXAMPLE 5 (continued)

Now find AG using triangle AFG.

By Pythagoras' theorem,

$x^2 = y^2 + 15^2\ \text{cm}^2$

$x^2 = 1300 + 225 = 1525\ \text{cm}^2$

So $x = 39.1$ cm (1 decimal place)

So, the longest straight wire that can be stored in the box is 39.1 cm.

Note that in any cuboid with sides a, b and c, the length of a diagonal is given by:

$\sqrt{(a^2 + b^2 + c^2)}$

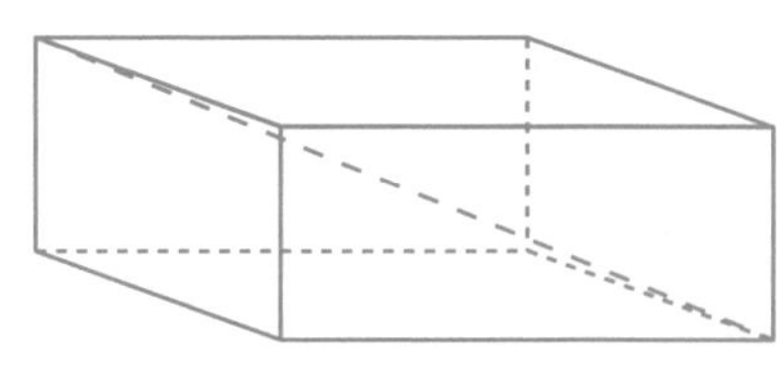

EXERCISE 8E

A **1** A box measures 8 cm by 12 cm by 5 cm.

a Calculate the lengths of the following.

i AC **ii** BG **iii** BE

b Calculate the diagonal distance BH.

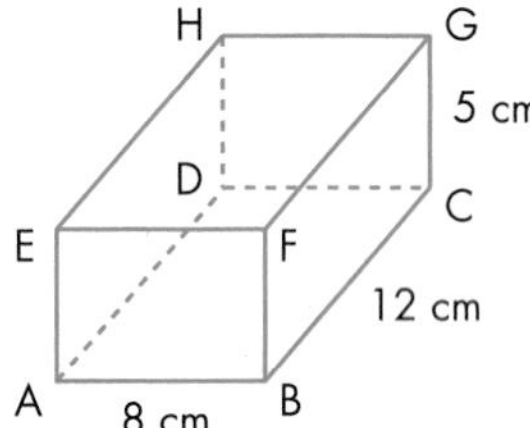

AU **2** A garage is 5 m long, 3 m wide and 3 m high. Can a 7 m long pole be stored in it?

AU **3** Spike, a spider, is at the corner S of the wedge shown in the diagram. Fred, a fly, is at the corner F of the same wedge.

a Calculate the shortest distance Spike would have to travel to get to Fred if she used the edges of the wedge.

b Calculate the distance Spike would have to travel across the face of the wedge to get directly to Fred.

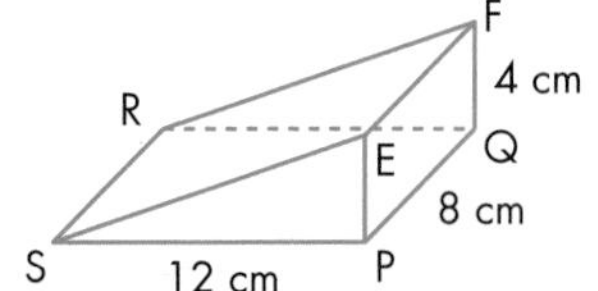
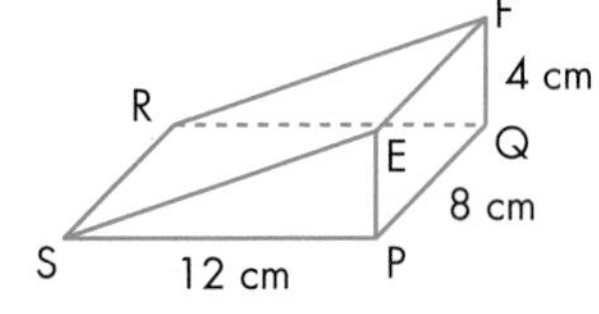

AU **4** Fred is now at the top of a baked-beans can and Spike is directly below him on the base of the can. To catch Fred by surprise, Spike takes a diagonal route round the can. How far does Spike travel?

HINTS AND TIPS

Imagine the can opened out flat.

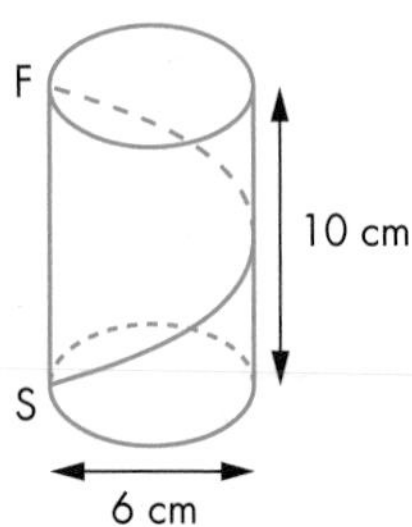

A

FM 5 A corridor is 3 m wide and turns through a right angle, as in the diagram.

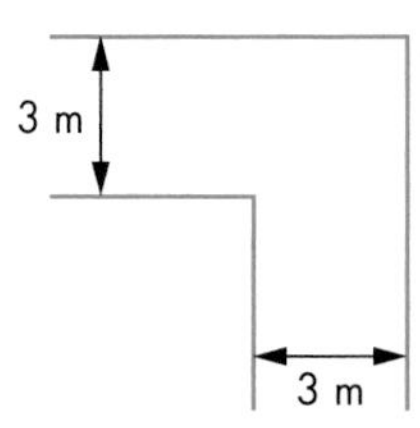

a What is the longest pole that can be carried along the corridor horizontally?

b If the corridor is 3 m high, what is the longest pole that can be carried along in any direction?

PS 6 If each side of a cube is 10 cm long, how far will it be from one corner of the cube to the opposite one?

AU 7 A pyramid has a square base of side 20 cm and each sloping edge is 25 cm long.

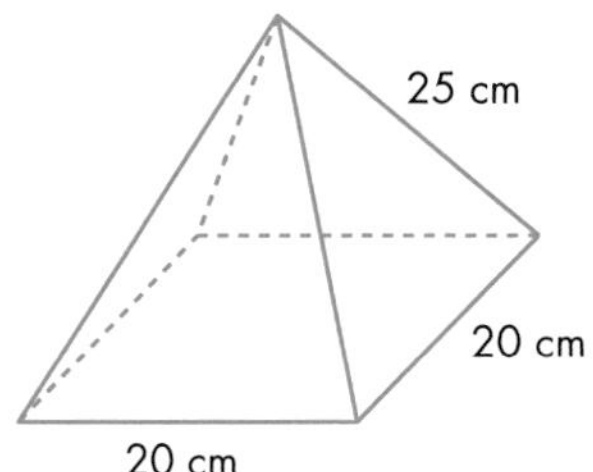

How high is the pyramid?

A*

AU 8 The diagram shows a square-based pyramid with base length 8 cm and sloping edges 9 cm. M is the midpoint of the side AB, X is the midpoint of the base, and E is directly above X.

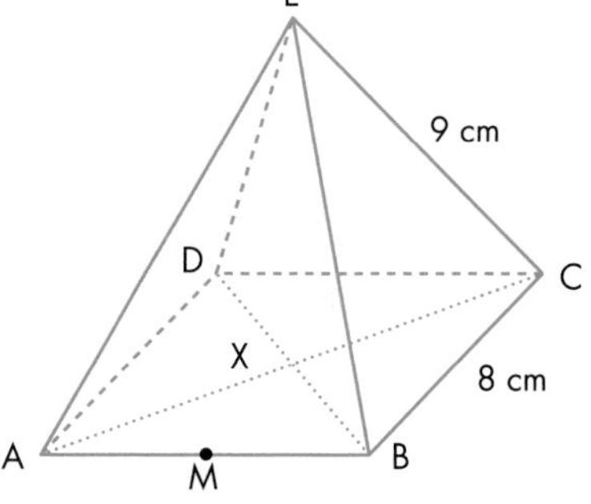

a Calculate the length of the diagonal AC.

b Calculate EX, the height of the pyramid.

c Using triangle ABE, calculate the length EM.

9 The diagram shows a cuboid with sides of 40 cm, 30 cm and 22.5 cm. M is the midpoint of the side FG. Calculate (or write down) these lengths, giving your answers to 3 significant figures if necessary.

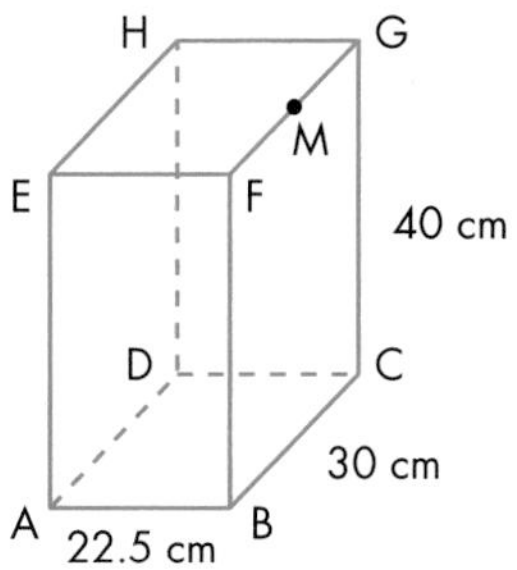

a AH **b** AG **c** AM **d** HM

GRADE BOOSTER

- **C** You can use Pythagoras' theorem in right-angled triangles
- **C** You can solve problems in 2D using Pythagoras' theorem
- **B** You can solve problems in 3D using Pythagoras' theorem

What you should know now

- How to use Pythagoras' theorem
- How to solve problems using Pythagoras' theorem

1 The diagram shows a solid triangular prism made of wood.

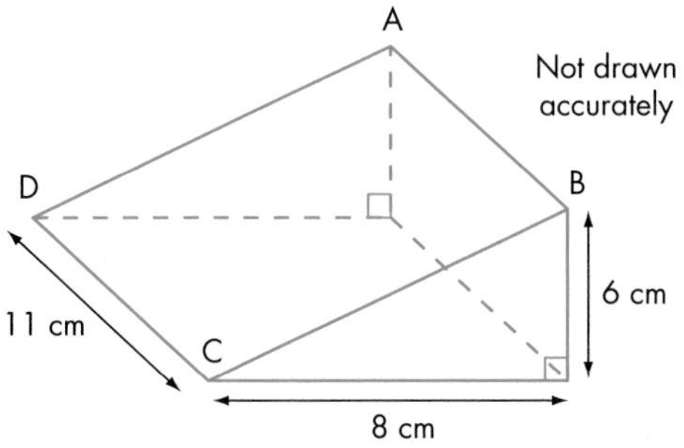

Work out the area of the rectangle ABCD.

AQA, November 2011 H1 Question 9 (modified)

2 Calculate the length, x cm, in the triangle below.

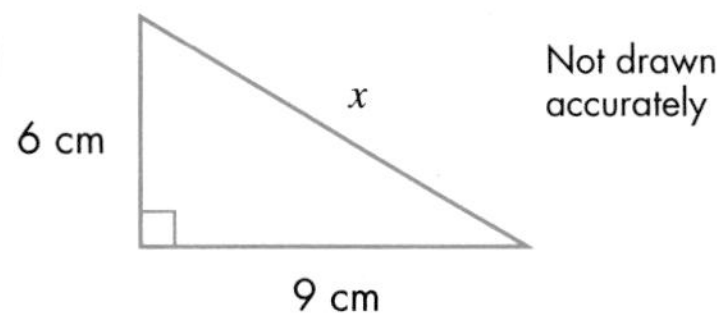

AQA, June 2008 H2 Question 12

3 The diagram shows an isosceles triangle ABC.

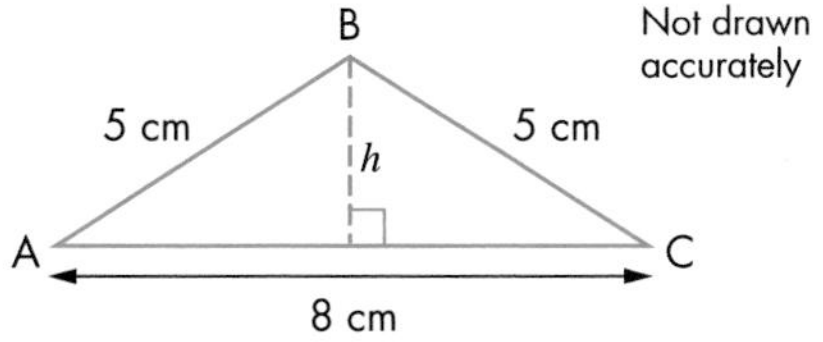

Calculate the area of the triangle ABC.

Show your working.

State the units of your answer. *(6 marks)*

AQA, June 2007, Module 5, Paper 1 Higher, Question 3

4 Calculate the length x.

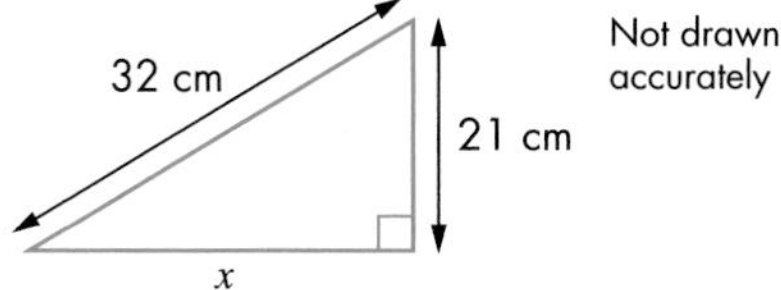

5 The diagram shows a cuboid ABCDEFGH with sides of 4 cm, 5 cm and 12 cm.

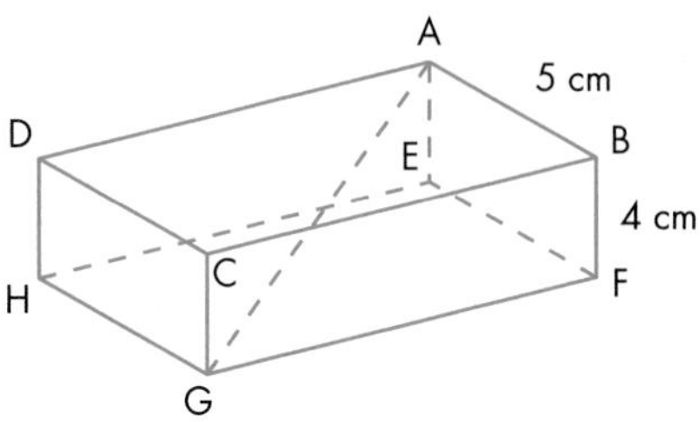

Calculate the length of the diagonal AG.

AQA, June 2009 H2 Question 21b

6 Triangle PQR is isosceles.

PQ = PR = 6 cm and QR = 4 cm.

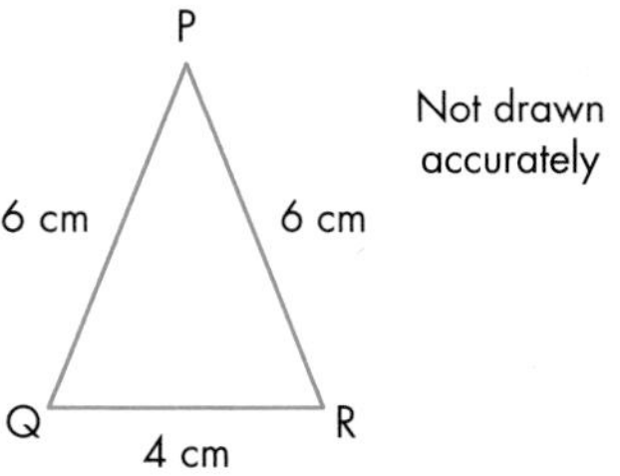

Show that the area of the triangle is $8\sqrt{2}$ cm^2.

AQA, June 2009 H1 Question 25a

7 The circle with centre X, has a radius of 5 cm.
The circle, with centre Y, has a radius of 3 cm.
The circles touch externally.
The circles have a common tangent AB.

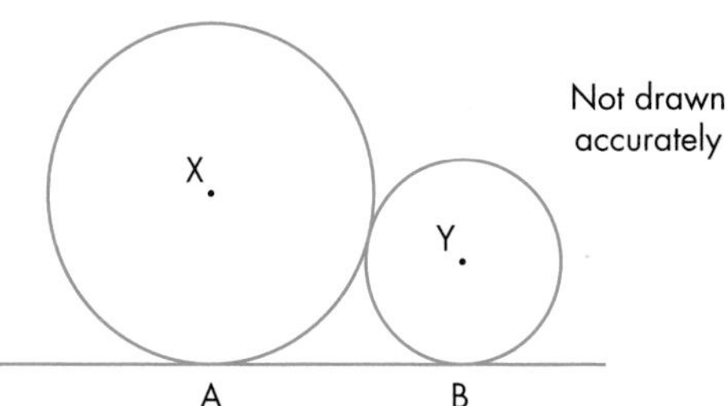

a Explain why ABYX is a trapezium.

b Show that AB = 7.75 cm to 3 significant figures.

AQA, November 2008 H2 Question 19

Worked Examination Questions

1 a Find the length x in this triangle.

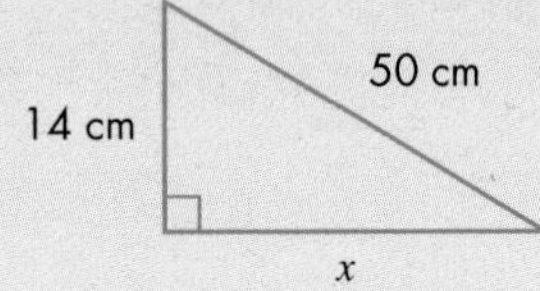

b Two towns A and B are linked by a two roads that intersect at a right angle as shown.

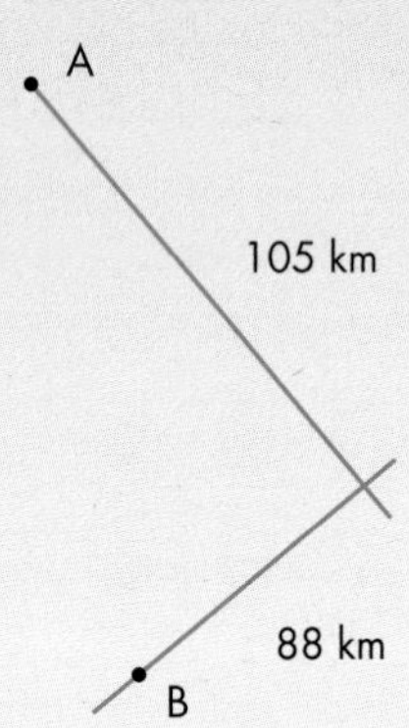

A new straight road is to be built directly from A to B.
How much distance will be saved by the new road when travelling from A to B?

1 a $50^2 - 14^2$ — This gets 1 method mark.

$\sqrt{2304}$ — This gets 1 method mark.

48 — This gets 1 accuracy mark.

Total: 3 marks

b $105^2 + 88^2$ — This gets 1 method mark.

$\sqrt{18769}$ — This gets 1 method mark.

137 — This gets 1 accuracy mark.

56 — This gets 1 accuracy mark.

Total: 4 marks

Worked Examination Questions

2 a ABC is a right-angled triangle. AC = 19 cm and AB = 9 cm.

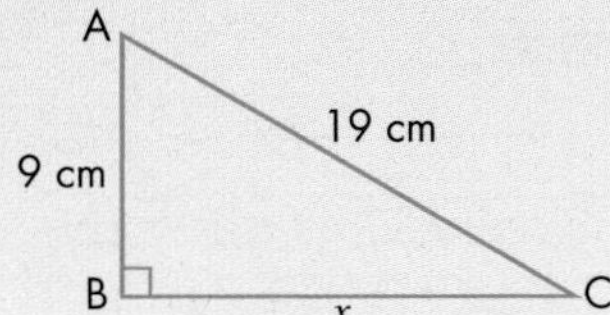

Calculate the length of BC.

b PQR is a right-angled triangle. PQ = 11 cm and QR = 24 cm.

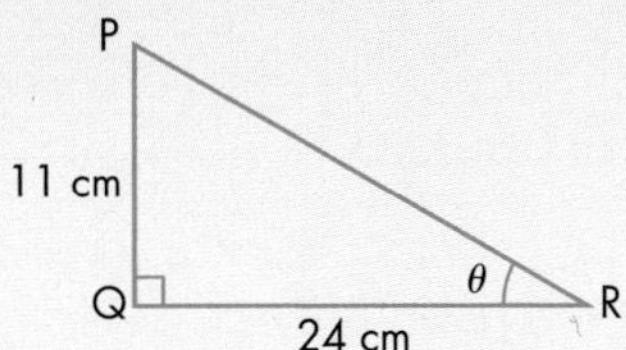

Calculate the size of angle PRQ.

c ABC and ACD are right-angled triangles. AD is parallel to BC.

AB = 12 cm, BC = 5 cm and AD = 33.8 cm.

Calculate the length of DC.

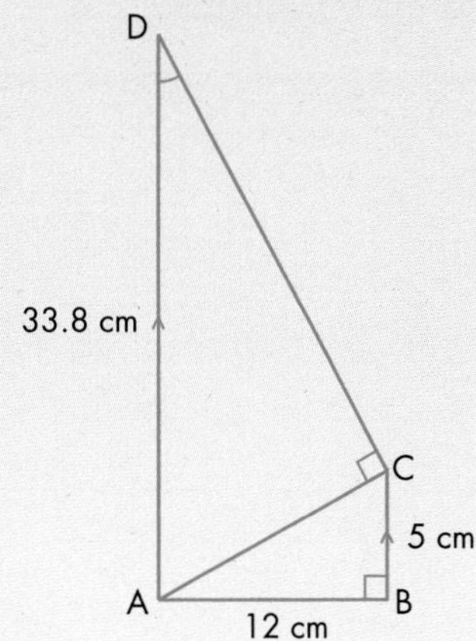

2 a Let BC = x

By Pythagoras' theorem

$x^2 = 19^2 - 9^2\ \text{cm}^2$ — This gets 1 mark for method.

$= 280\ \text{cm}^2$

So $x = \sqrt{280}$ — This gets 1 mark for method.

$= 16.7$ cm (3 sf) — This gets 1 mark for accuracy.

b Let $\angle PRQ = \theta$

So $\tan\theta = \frac{11}{24}$ — This gets 1 mark for method.

$\theta = \tan^{-1}\frac{11}{24} = 24.6°$ (1 dp) — This gets 1 mark for accuracy.

c AC = 13 — This is a 5, 12, 13 triangle which you should know. This is 1 mark.

$DC^2 = 33.8^2 - 13^2$

$DC = \sqrt{973.44}$ — Apply Pythagoras to triangle ADC. This is 2 method marks.

DC = 31.2 — This gets one mark for accuracy.

Total: 9 marks

8 Problem Solving
Map work using Pythagoras

When you are out walking on the hills it can be very useful to be able to estimate various distances that you have to cover. Of course, you can just use the scale on the map, but another way is by using Pythagoras' theorem with small right-angled triangles.

Getting started

- You know that the **square** of 2 is $2^2 = 4$.
  Write down the **square** of each of these numbers.
  5 0.1 0.4 0.03
  Think of a number that has a square between 0.1 and 0.001.
- You know that the **square root** of 4 is $\sqrt{4} = 2$.
  Now write down the **square root** of each of these numbers.
  16 81 0.01 0.025
  Think of a number that has a square root between 50 and 60.
- On a set of coordinate axes, draw the points A(1, 2) and B(4, 6).
  What is the distance from point A to point B?

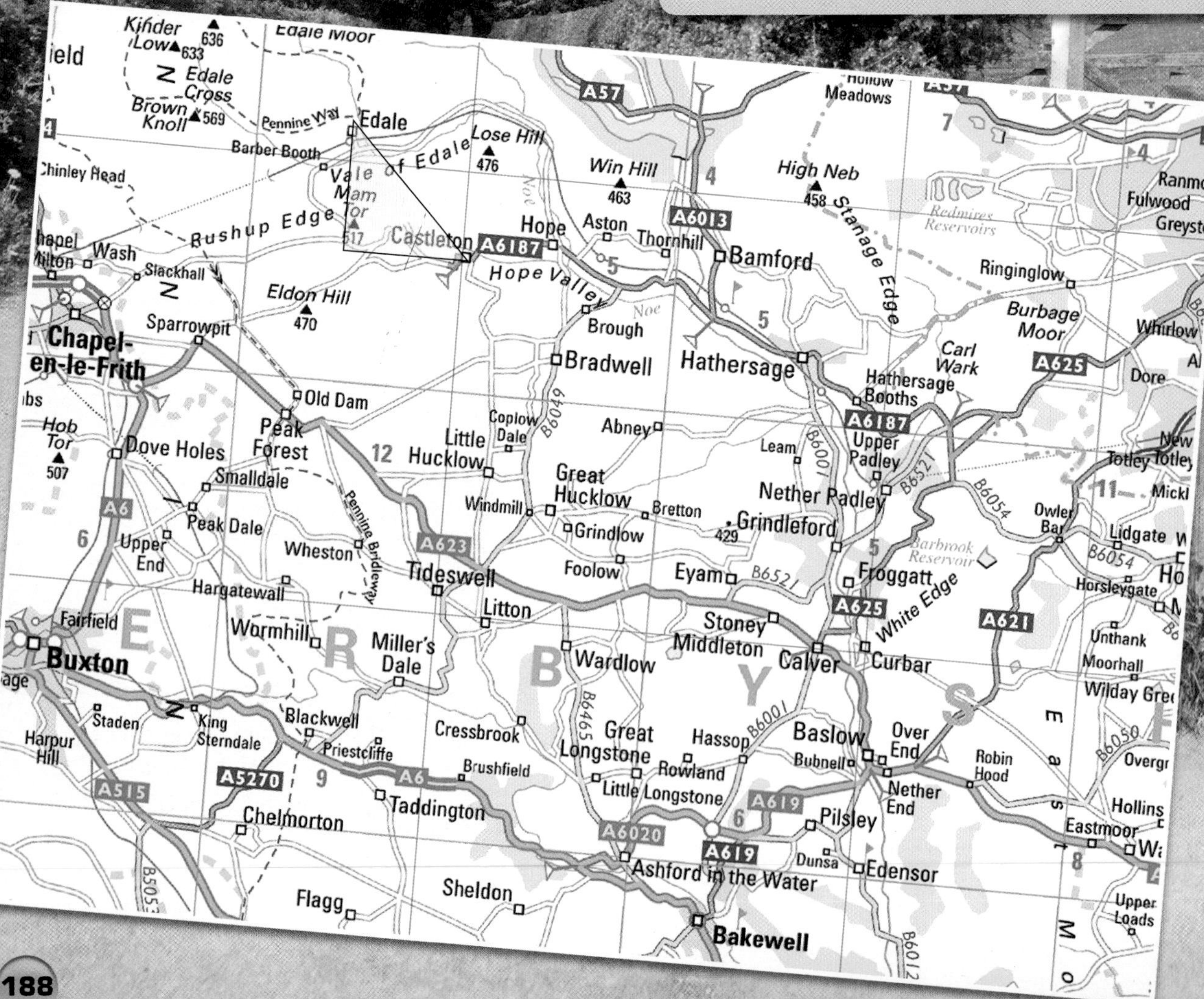

Your task

Freya and Chris often go out walking in the Peak District. On the left is a copy of the map they use. Use this map extract to complete these tasks.

1 Write five questions similar to the examples given on the right. Swap them with the person next to you.

   Answer each other's questions, making sure you show your working clearly.

   Now swap again and mark each other's answers. Give constructive feedback.

2 Plan a walk with a circular route that is between 20 and 35 km long.

   If the average person walks at approximately 4.5 miles per hour, estimate the time it would take to complete your route.

Example

Freya and Chris were at Edale. They wanted to know the rough distance to Castleton. Freya decided to set herself a maths problem, using Pythagoras' theorem.

She looked at the map and imagined the yellow right-angled triangle.

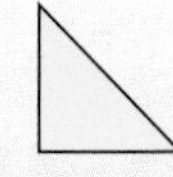

Using the fact that each square on the map represents an area 5 km by 5 km, she estimated each small side of the triangle to be 3 km.

Then she applied Pythagoras' theorem.

$3^2 + 3^2 = 9 + 9 = 18$

On the hillside, without a calculator, she estimated the square root of 18 to be just over 4, giving a distance of 4 km.

Another day, Freya and Chris were at Hucklow and wanted to know the distance to Hathersage.

Use Freya's method to estimate the distance from Hucklow to Hathersage.

Additional information

When working in distances, you need to work in either miles and other imperial units or kilometres and other metric units.

To change between these units there are some key conversion facts. Either use a textbook or the internet to find these.

Why this chapter matters

It is essential to understand angles. They help us to construct everything, from a building to a table. So angles literally shape our world.

Ancient civilisations used **right angles** in surveying and in constructing buildings. The ancient Greeks used the right angle to describe relationships between other angles. However, not everything can be measured in right angles. There is a need for a smaller, more useful unit. The ancient Babylonians chose a unit angle that led to the development of the **degree**, which is what we still use today.

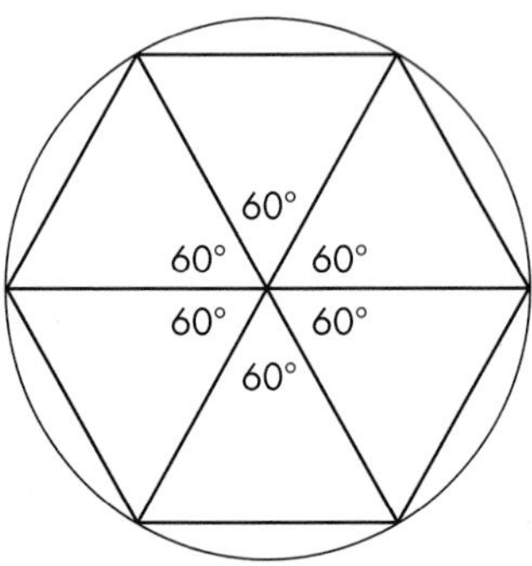

Most historians think that the ancient Babylonians thought of the 'circle' of the year as consisting of 360 days. This is not a bad approximation, given the crudeness of the ancient astronomical tools and often having to measure small angles with the naked eye. Mathematical historians believe that the ancient Babylonians knew that the side of a **regular hexagon** inscribed in a circle is equal to the **radius** of the circle. This may have led to the division of the full circle (360 days) into six equal parts, each part consisting of 60 days. They divided one angle of an **equilateral triangle** into 60 equal parts, now called degrees, then further subdivided a degree into 60 equal parts, called **minutes**, and a minute into 60 equal parts, called **seconds**.

Although many historians believe this is why 60 was the base of the Babylonian system of angle measurement, others think there was a different reason. The number 60 has many **factors**. Work with fractional parts of the whole (60) is greatly simplified, because 2, 3, 4, 5, 6, 10, 12, 15, 20 and 30 are all factors of 60.

Modern measurement of angles

Modern surveyors use theodolites for measuring angles.

A theodolite can be used for measuring both horizontal and vertical angles. It is a key tool in surveying and engineering work, particularly on inaccessible ground, but theodolites have been adapted for other specialised purposes in fields such as meteorology and rocket-launching technology. A modern theodolite comprises a movable telescope mounted within two perpendicular axes – the horizontal and the vertical axis. When the telescope is pointed at a desired object, the angle of each of these axes can be measured with great precision, typically on the scale of arcseconds. (There are 3600 **arcseconds** in 1°.)

Modern theodolite

Chapter 9

Geometry: Angles

1 Special triangles and quadrilaterals

2 Angles in polygons

This chapter will show you ...

to E D how to find angles in triangles and quadrilaterals

to D C how to find interior and exterior angles in polygons

Visual overview

2D shapes → Triangles → Quadrilaterals → Polygons → Interior and exterior angles

What you should already know

- The three interior angles of a triangle add up to 180°. So, $a + b + c = 180°$
  (KS3 level 5, GCSE grade E)

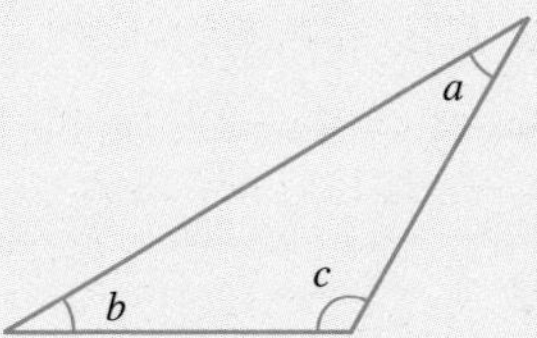

- The four interior angles of a quadrilateral quadrilateral add up to 360°.
  So, $a + b + c + d = 360°$
  (KS3 level 5, GCSE grade E)

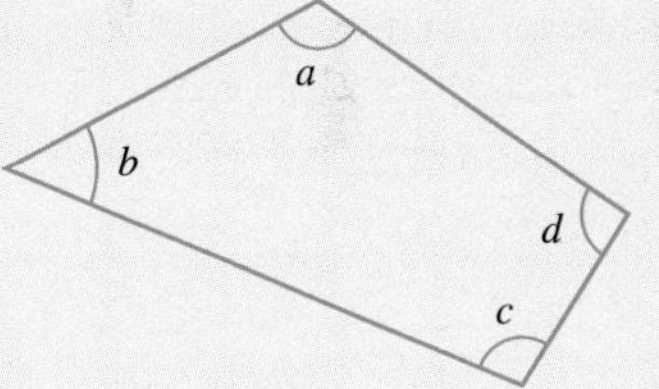

- Angles in parallel lines

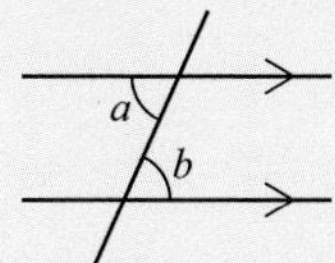

a and b are equal
a and b are alternate angles

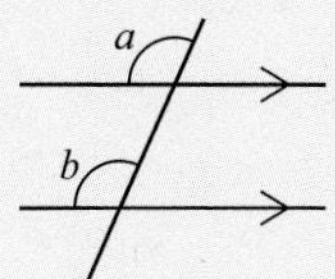

a and b are equal
a and b are corresponding angles

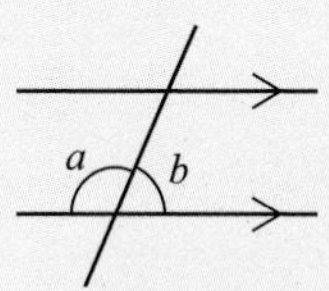

$a + b = 180°$
a and b are allied angles

(KS3 level 6, GCSE grade D)

continued

Quick check

Find the sizes of the lettered angles in these diagrams.

1

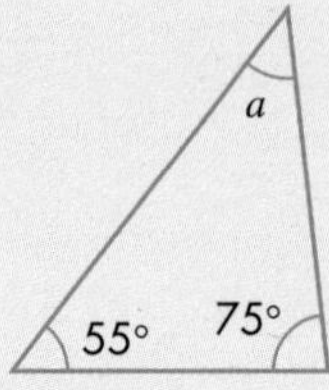

2

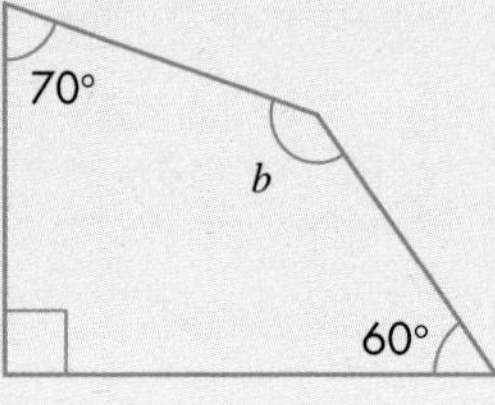

3

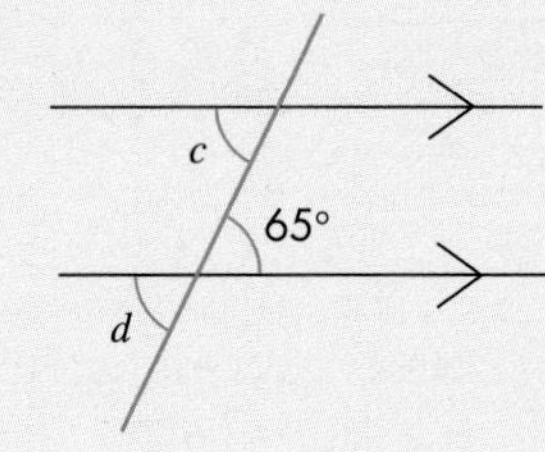

9.1 Special triangles and quadrilaterals

This section will show you how to:

- work out the sizes of angles in triangles and quadrilaterals

Key words
equilateral triangle
isosceles triangle
kite
parallelogram
rhombus
trapezium

Special triangles

An **equilateral triangle** is a triangle with all its sides equal.

Therefore, all three interior angles are 60°.

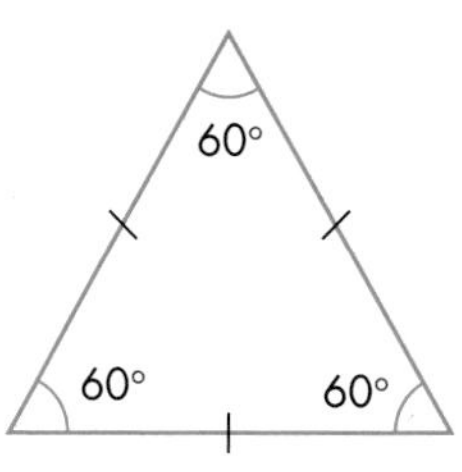

An **isosceles triangle** is a triangle with two equal sides, and therefore with two equal angles.

Notice how to mark the equal sides and equal angles.

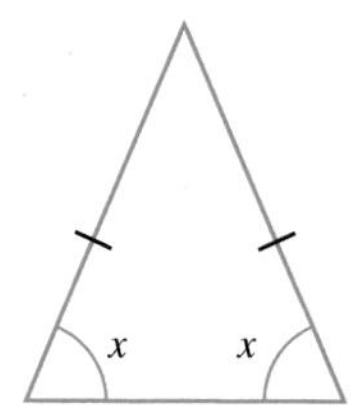

EXAMPLE 1

Find the size of the angle marked a in the triangle.

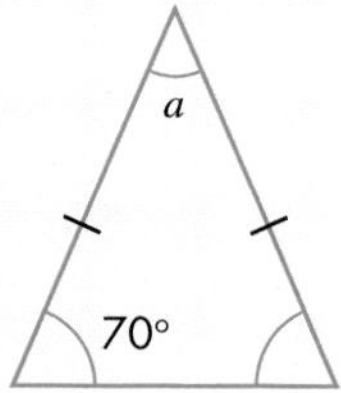

The triangle is isosceles, so both base angles are 70°.

So $a = 180° - (70° + 70°) = 180° - 140° = 40°$

Special quadrilaterals

A **parallelogram** has opposite sides that are parallel.

Its opposite sides are equal. Its diagonals bisect each other. Its opposite angles are equal: that is, ∠A = ∠C and ∠B = ∠D

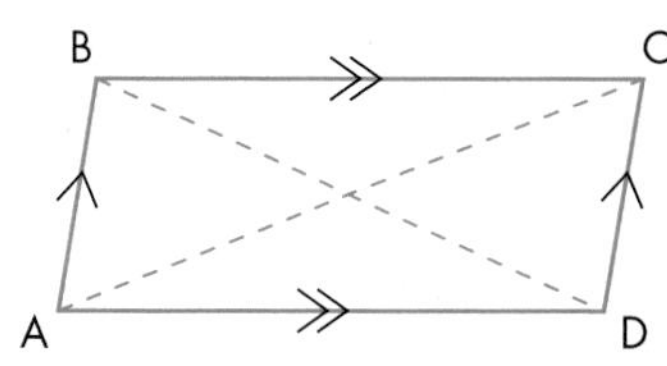

A **rhombus** is a parallelogram with all its sides equal.

Its diagonals bisect each other at right angles. Its diagonals also bisect the angles at the vertices.

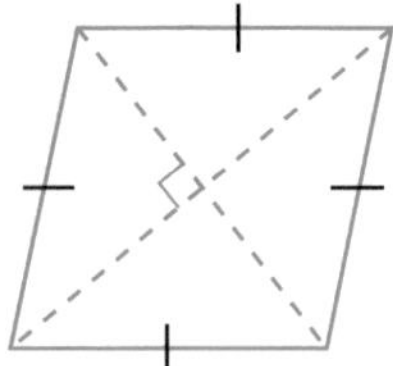

A **kite** is a quadrilateral with two pairs of equal adjacent sides.

Its longer diagonal bisects its shorter diagonal at right angles. The opposite angles between the sides of different lengths are equal.

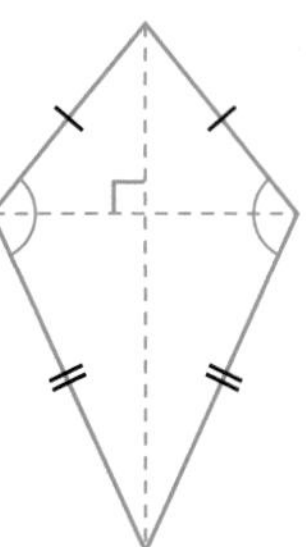

A **trapezium** has two parallel sides.

The sum of the interior angles at the ends of each non-parallel side is 180°: that is, $\angle A + \angle D = 180°$ and $\angle B + \angle C = 180°$

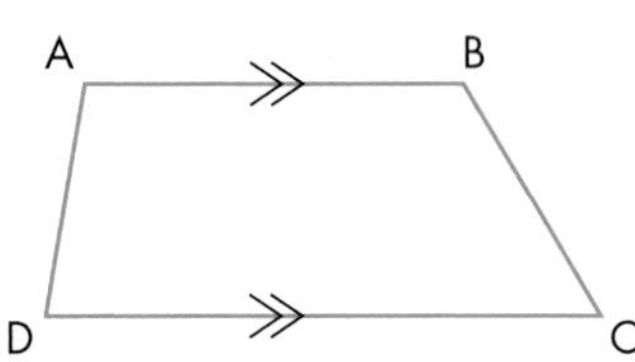

EXAMPLE 2

Find the size of the angles marked x and y in this parallelogram.

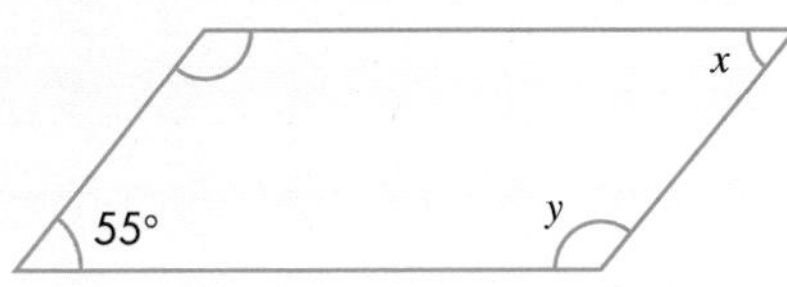

$x = 55°$ (opposite angles are equal) and $y = 125°$ ($x + y = 180°$)

EXERCISE 9A

1 Calculate the sizes of the lettered angles in each triangle.

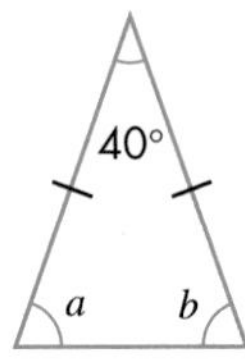

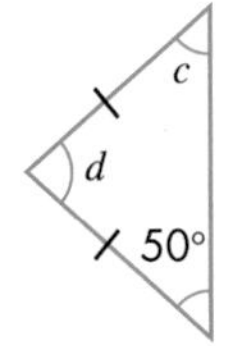

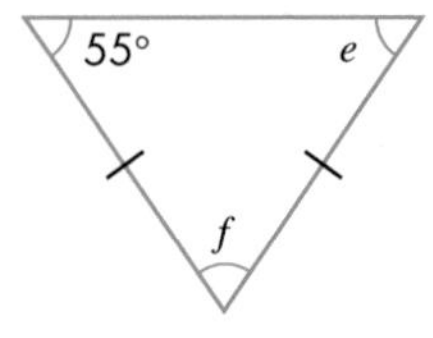

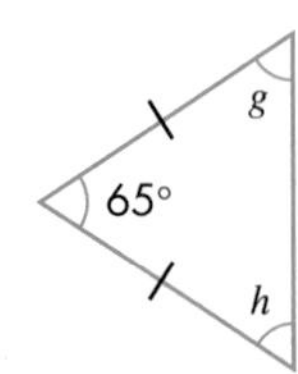

PS 2 An isosceles triangle has an angle of 50°. Sketch the two different possible triangles that match this description, showing what each angle is.

3 Find the sizes of the missing angles in these quadrilaterals.

a

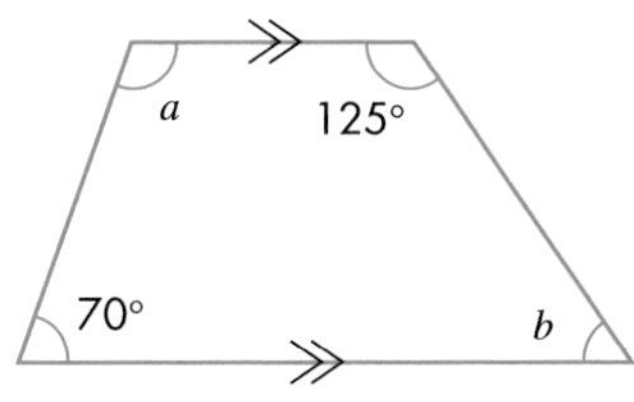

b

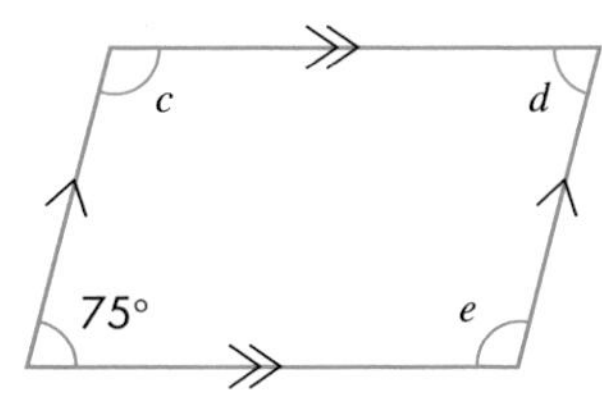

c

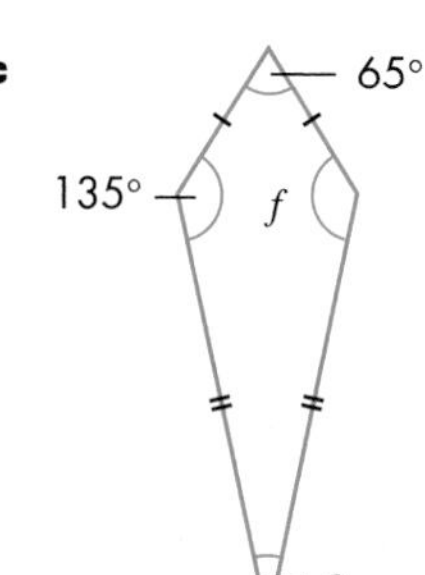

d

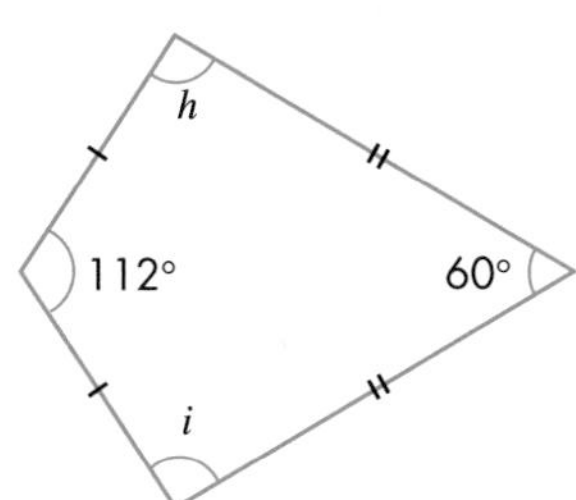

e

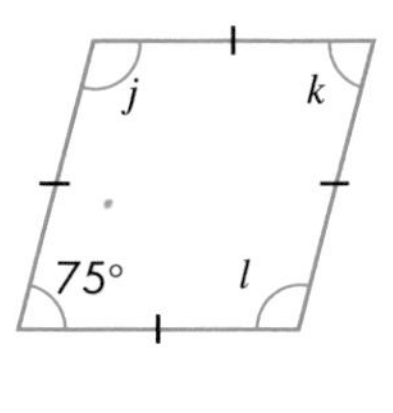

f

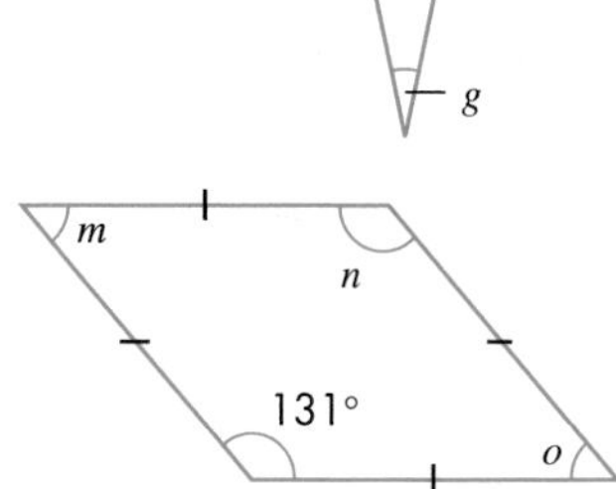

AU 4 The three angles of an isosceles triangle are $2x$, $x - 10$ and $x - 10$. What is the actual size of each angle?

5 Calculate the sizes of the lettered angles in these diagrams.

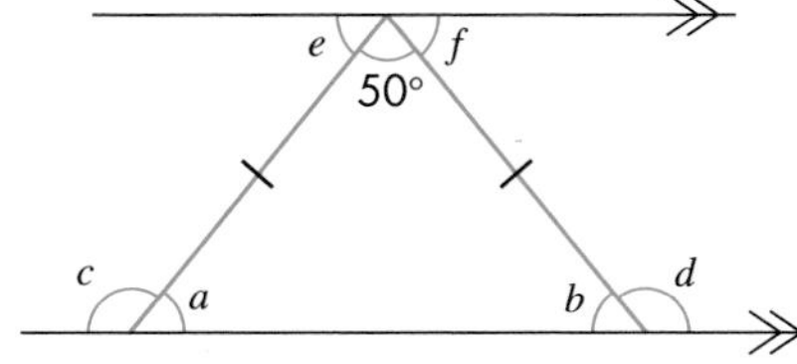

6 Calculate the values of x and y in each of these quadrilaterals.

a

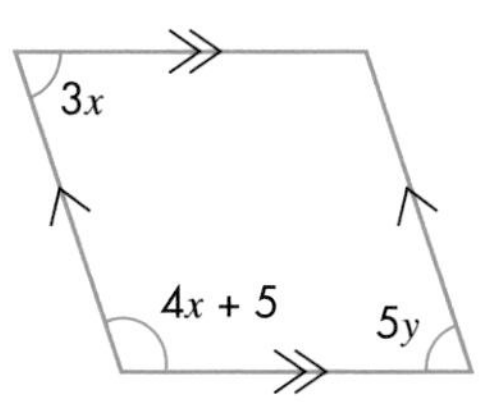

b

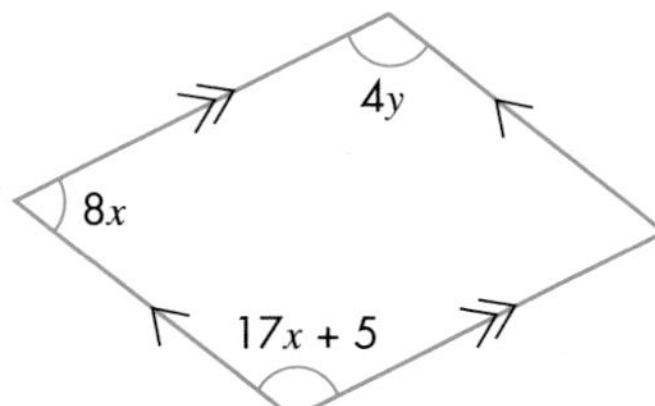

c

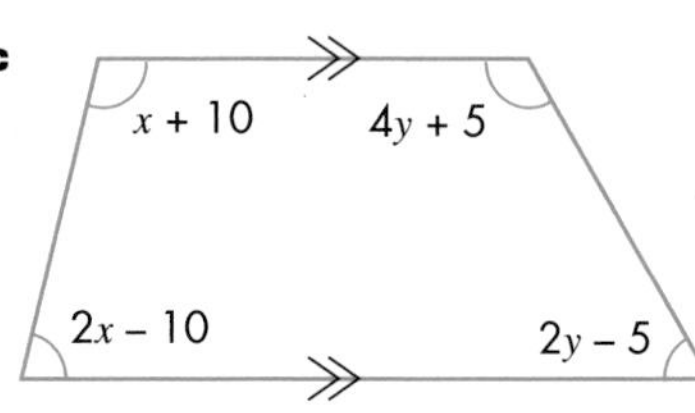

PS 7 Find the value of x in each of these quadrilaterals and hence state what type of quadrilateral it could be.

a A quadrilateral with angles $x + 10$, $x + 20$, $2x + 20$, $2x + 10$

b A quadrilateral with angles $x - 10$, $2x + 10$, $x - 10$, $2x + 10$

c A quadrilateral with angles $x - 10$, $2x$, $5x - 10$, $5x - 10$

d A quadrilateral with angles $4x + 10$, $5x - 10$, $3x + 30$, $2x + 50$

C

8 The diagram shows a parallelogram ABCD.

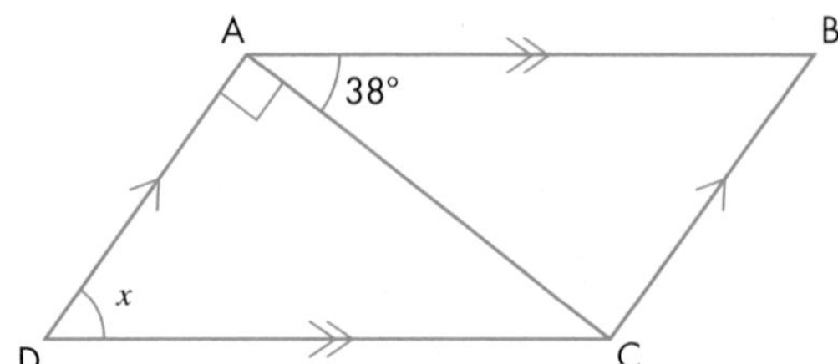

Work out the size of angle x, marked on the diagram.

9 Dani is making a kite and wants Angle C to be half of angle A.

Work out the size of angles B and D.

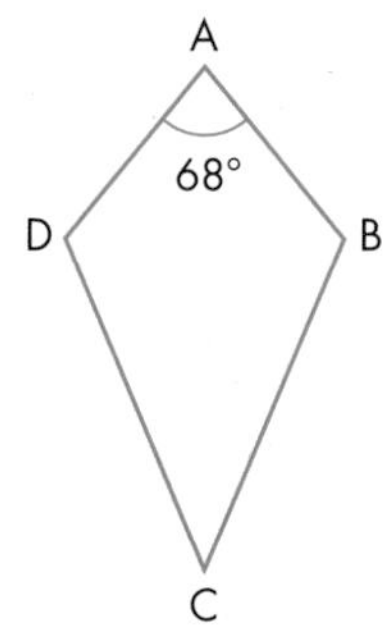

PS 10 This quadrilateral is made from two isosceles triangles.
They are both the same size.

Find the value of y in terms of x.

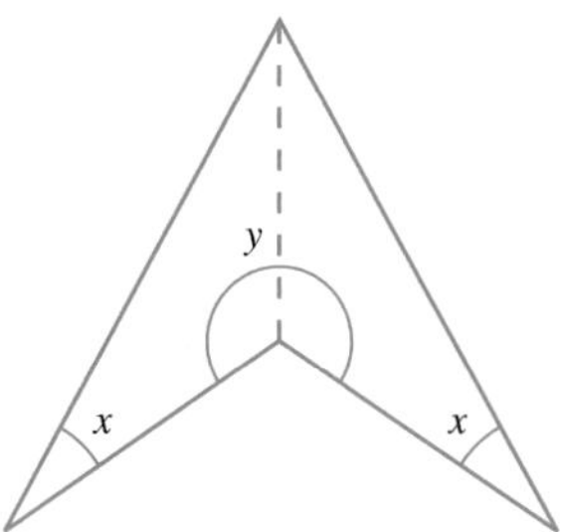

AU 11 The diagram shows a quadrilateral ABCD.

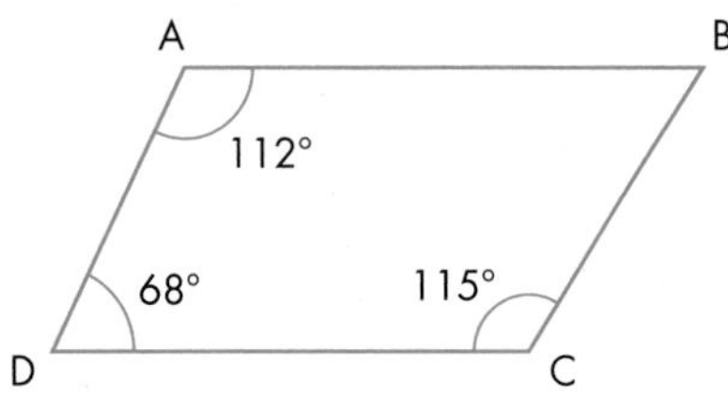

a Calculate the size of angle B.

b What special name is given to the quadrilateral ABCD?

Explain your answer.

9.2 Angles in polygons

This section will show you how to:

- work out the sizes of interior angles and exterior angles in a polygon

Key words

decagon, exterior angle, heptagon, hexagon, interior angle, nonagon, octagon, pentagon, polygon, regular polygon

A **polygon** has two kinds of angles.

- **Interior angles** are angles made by adjacent sides of the polygon and lying inside the polygon.
- **Exterior angles** are angles lying on the outside of the polygon, so that the interior angle + the exterior angle = 180°.

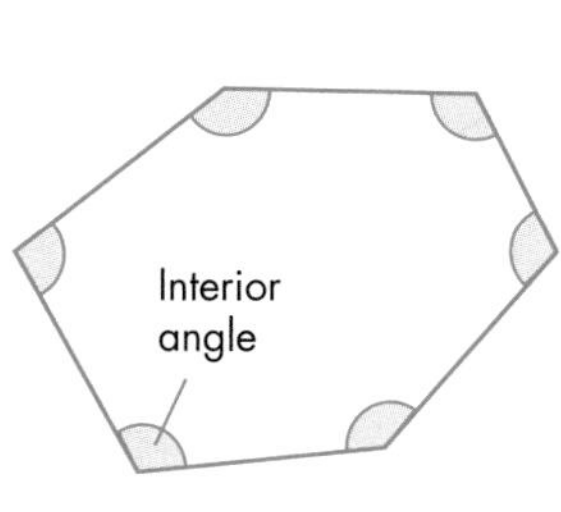

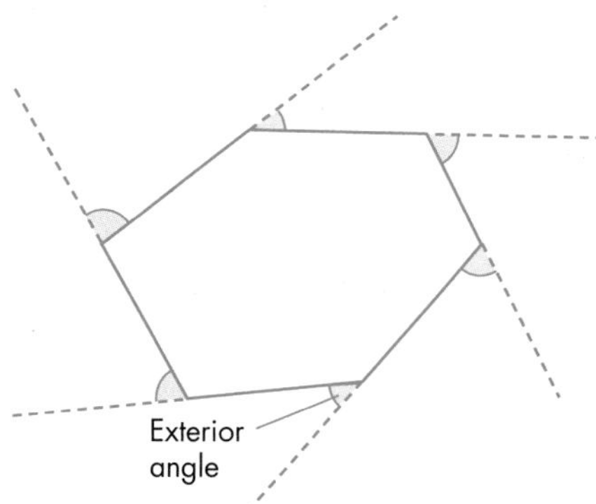

The *exterior* angles of *any* polygon add up to 360°.

Interior angles

You can find the sum of the interior angles of any polygon by splitting it into triangles.

Quadrilateral

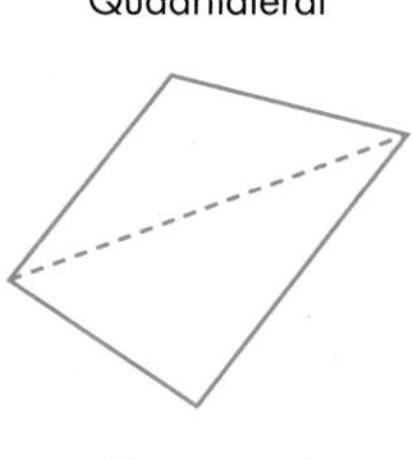

Two triangles

Pentagon

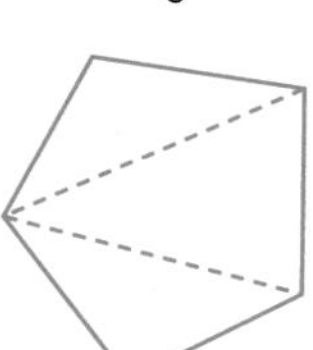

Three triangles

Hexagon

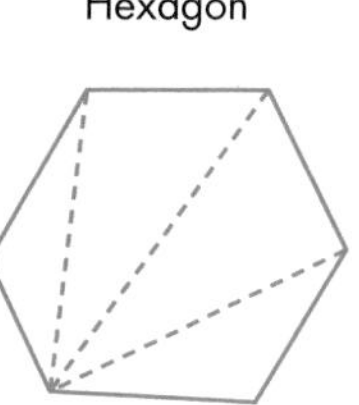

Four triangles

Heptagon

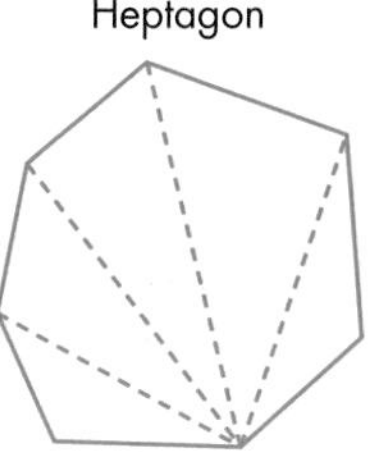

Five triangles

Since you already know that the angles in a triangle add up to 180°, you find the sum of the interior angles in a polygon by multiplying the number of triangles in the polygon by 180°, as shown in this table.

Shape	Name	Sum of interior angles
4-sided	Quadrilateral	2 × 180° = 360°
5-sided	**Pentagon**	3 × 180° = 540°
6-sided	**Hexagon**	4 × 180° = 720°
7-sided	**Heptagon**	5 × 180° = 900°
8-sided	**Octagon**	6 × 180° = 1080°
9-sided	**Nonagon**	7 × 180° = 1260°
10-sided	**Decagon**	8 × 180° = 1440°

As you can see from the table, for an n-sided polygon, the sum of the interior angles, S, is given by the formula:

$S = 180(n - 2)°$

Exterior angles

As you can see from the diagram, the sum of an exterior angle and its adjacent interior angle is 180°.

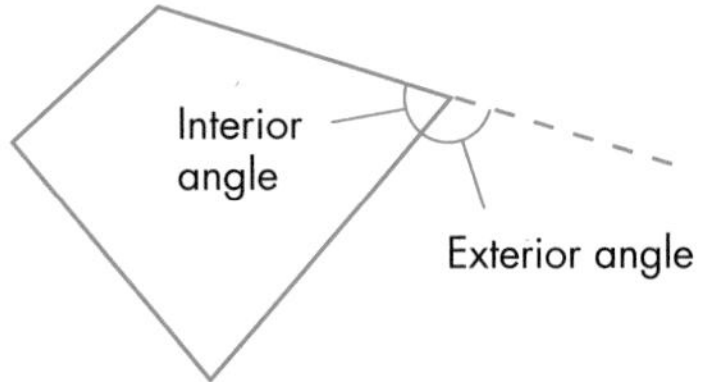

Regular polygons

A polygon is regular if all its interior angles are equal and all its sides have the same length. This means that all the exterior angles are also equal.

Here are two simple formulae for calculating the interior and the exterior angles of **regular polygons**.

The exterior angle, E, of a regular n-sided polygon is $E = \frac{360°}{n}$

The interior angle, I, of a regular n-sided polygon is $I = 180° - E = 180° - \frac{360°}{n}$

This can be summarised in the following table.

Regular polygon	Number of sides	Size of each exterior angle	Size of each interior angle
Square	4	90°	90°
Pentagon	5	72°	108°
Hexagon	6	60°	120°
Heptagon	7	$51\frac{3}{7}°$	$128\frac{4}{7}°$
Octagon	8	45°	135°
Nonagon	9	40°	140°
Decagon	10	36°	144°
n-sided	n	$\frac{360°}{n}$	$180° - \frac{360°}{n}$

EXAMPLE 3

Find the exterior angle, x, and the interior angle, y, for this regular octagon.

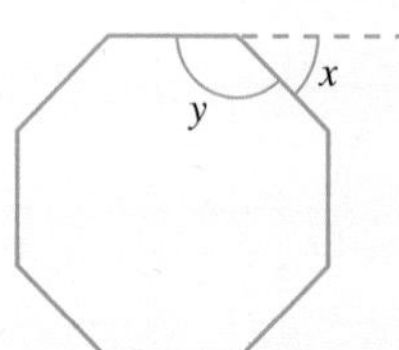

$x = \frac{360°}{8} = 45°$ and $y = 180° - 45° = 135°$

EXERCISE 9B

1 Calculate the sum of the interior angles of polygons with these numbers of sides.

a 10 sides **b** 15 sides **c** 100 sides **d** 45 sides

2 Calculate the size of the interior angle of regular polygons with these numbers of sides.

a 12 sides **b** 20 sides **c** 9 sides **d** 60 sides

3 Find the number of sides of polygons with these interior angle sums.

a 1260° **b** 2340° **c** 18 000° **d** 8640°

4 Find the number of sides of regular polygons with these exterior angles.

a 24° **b** 10° **c** 15° **d** 5°

5 Find the number of sides of regular polygons with these interior angles.

a 150° **b** 140° **c** 162° **d** 171°

6 Calculate the size of the unknown angle in each of these polygons.

a

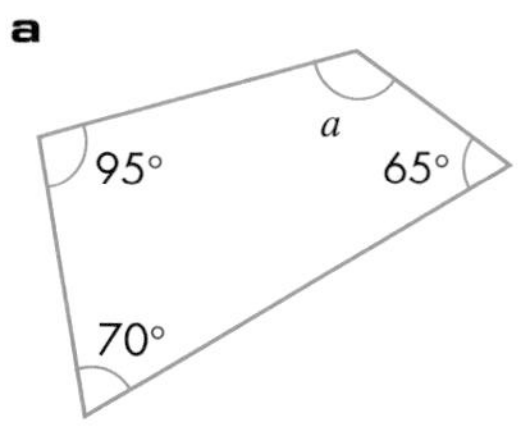

b

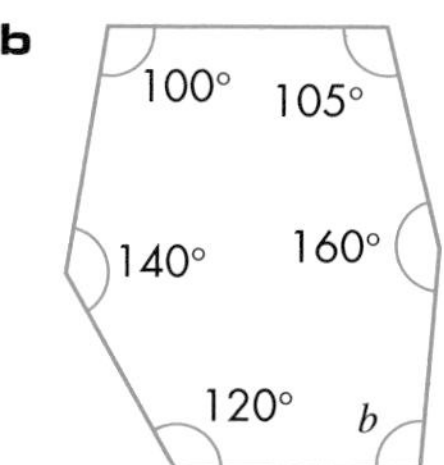

c

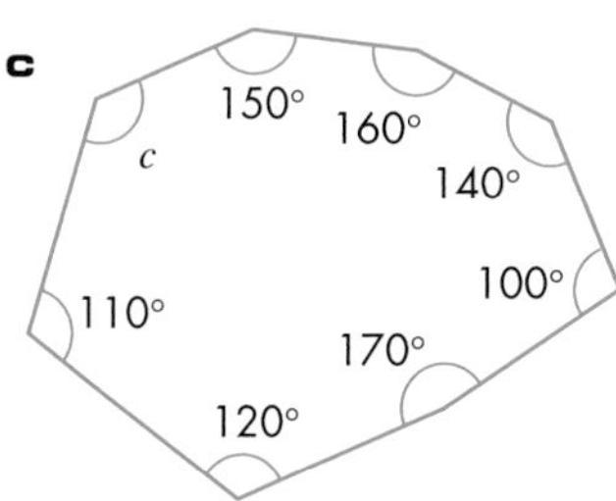

7 Find the value of x in each of these polygons.

a

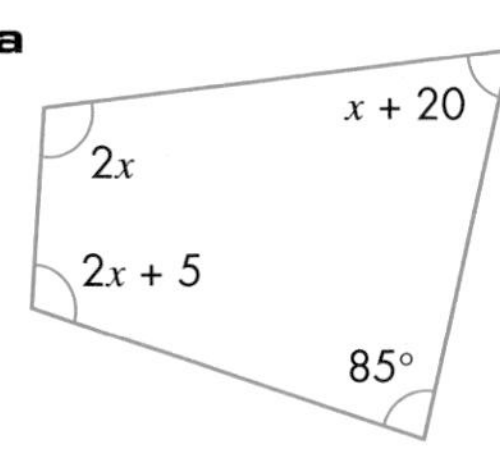

b

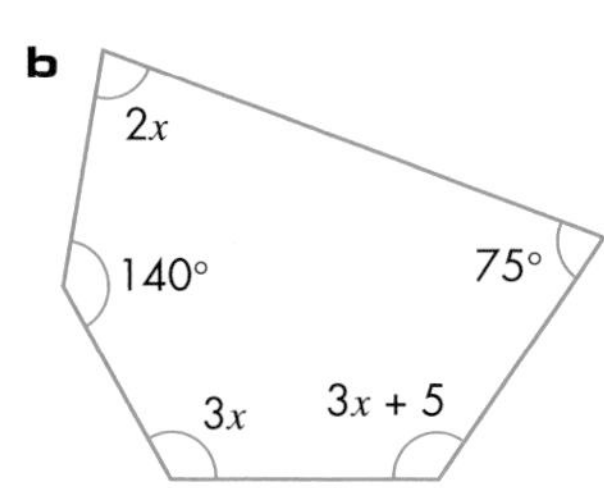

c

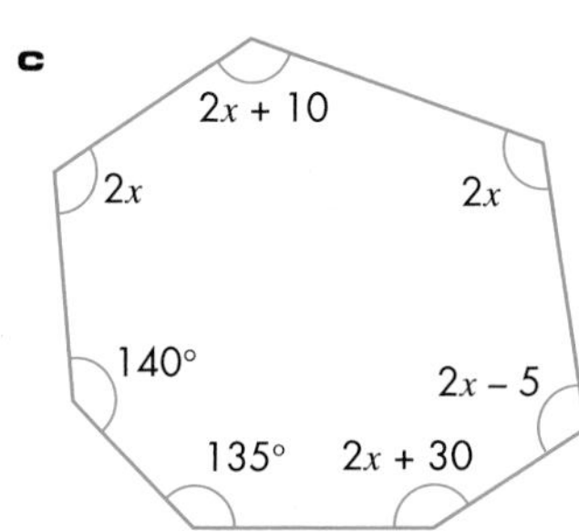

B

AU **8** What is the name of the regular polygon in which the interior angles are twice its exterior angles?

9 In a triangle ABC, the angles are in the ratio A : B : C = 5 : 1 : 3.

Work out the size of the largest angle.

HINTS AND TIPS

Angles in a triangle add up to 180°

10 The diagram shows a quadrilateral ABCD.

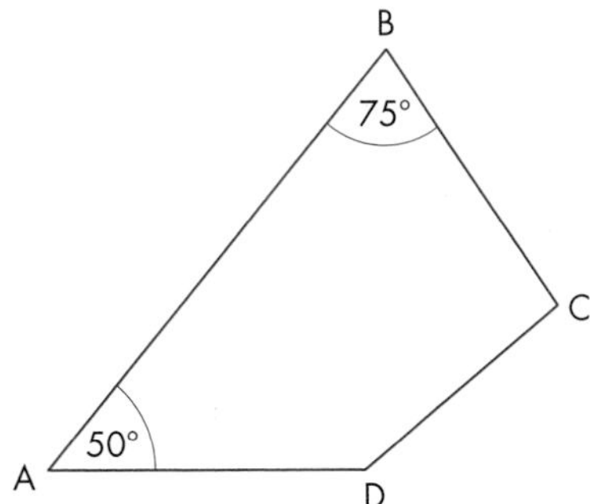

HINTS AND TIPS

Angles in a quadrilateral add up to 360°

The ratio of angle C to angle D is 2 : 3.

Work out the size of angle D.

PS **11** Wesley measured all the interior angles in a polygon. He added them up to make 991°, but he had missed out one angle.

a What type of polygon did Wesley measure? **b** What is the size of the missing angle?

12 **a** In the triangle ABC, angle A is 42° and angle B is 67°.

i Calculate the value of angle C.

ii What is the value of the exterior angle at C?

iii What connects the exterior angle at C with the sum of the angles at A and B?

b Prove that any exterior angle of a triangle is equal to the sum of the two opposite interior angles.

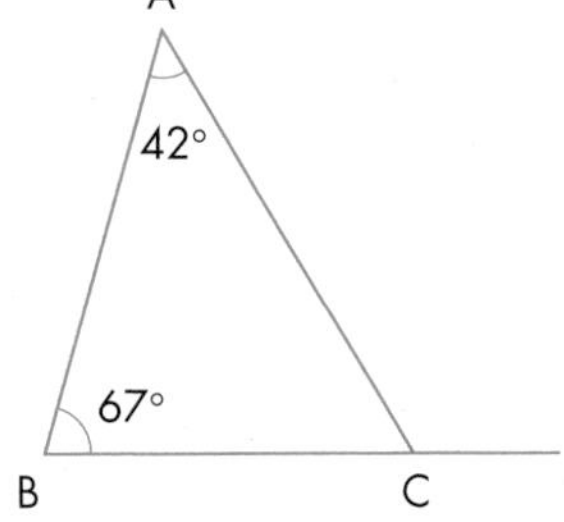

AU **13** Two regular pentagons are placed together.

Work out the value of a.

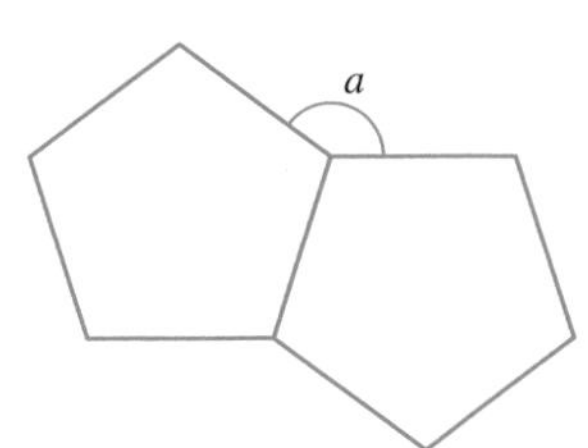

14 A joiner is making tables so that the shape of each one is half a regular hexagon, as shown in the diagram. He needs to know the size of each angle on the table top. What are the sizes of the angles?

PS **15** This star shape has 10 sides that are equal in length.

Each reflex interior angle is 200°.

Work out the size of each acute interior angle.

> **HINTS AND TIPS**
>
> Find the sum of the interior angles of a decagon first.

16 The diagram shows part of a regular polygon.

Each interior angle is 144°.

a What is the size of each exterior angle of the polygon?

b How many sides does the polygon have?

17 The interior angle to the exterior angle of a regular polygon is in the ratio 3 : 1.

Work out the number of sides of the polygon.

PS **18** A square is drawn inside a regular hexagon.

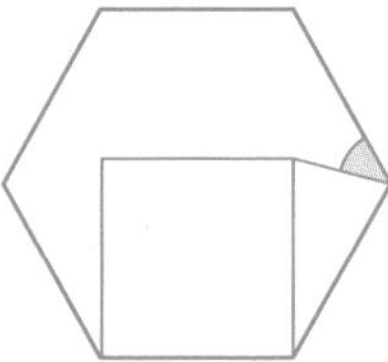

Calculate the size of the angle shown.

SUMMARY

GRADE BOOSTER

D You can find angles in triangles and quadrilaterals

C You can find interior angles and exterior angles in polygons

---

What you should know now

- How to find angles in any triangle or in any quadrilateral
- How to calculate interior and exterior angles in polygons

1 **a** Explain why the sum of the angles in any quadrilateral is 360°. *(2 marks)*

b A quadrilateral has one right angle.

The other angles are $2x$, $3x - 12$ and $x - 6$.

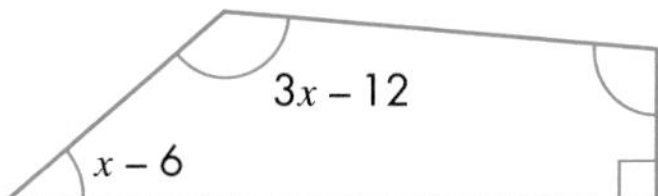

i Write down an equation in terms of x. *(1 mark)*

ii Solve your equation and find the size of the largest angle in the quadrilateral. *(3 marks)*

AQA, June 2007, Paper 1 Higher, Question 3

2 **a** The diagram shows a regular hexagon.

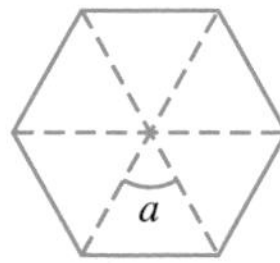

i Show clearly why angle $a = 60°$ *(1 mark)*

ii Work out the sum of the interior angles of a hexagon. *(2 marks)*

b P, Q, R and S are four vertices of a regular polygon. Each exterior angle is 18°.

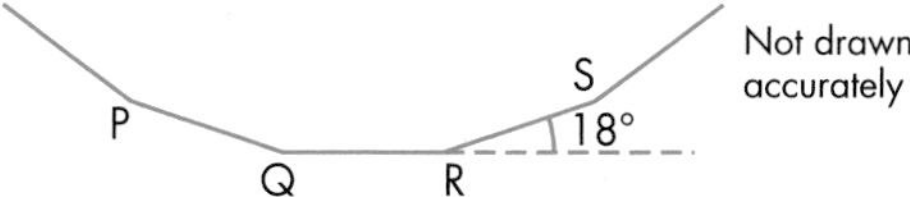

Work out the number of sides of the polygon.

AQA, November 2011 F2 Question 24

3 This is a regular pentagon.

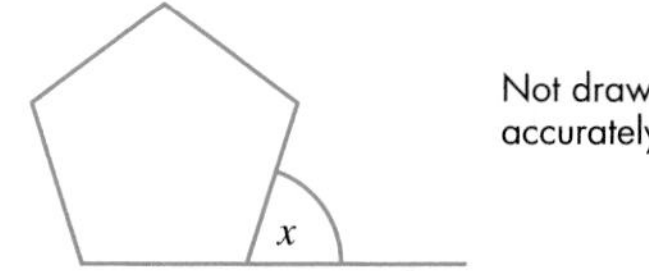

Work out the value of the exterior angle, marked x on the diagram.

AQA, June 2011 H2 Question 1

4 ABC is an isosceles triangle.

Calculate the value of the angle x.

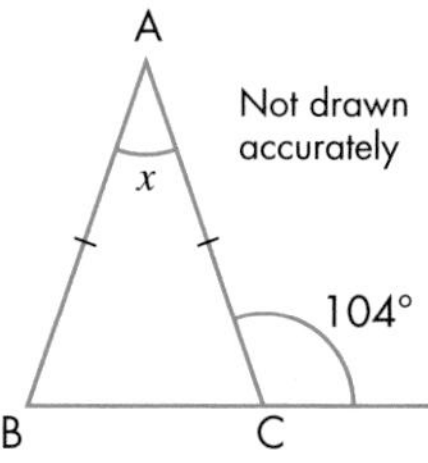

AQA, November 2010 H2 Question 4

5 On the diagram PQ is parallel to RS.

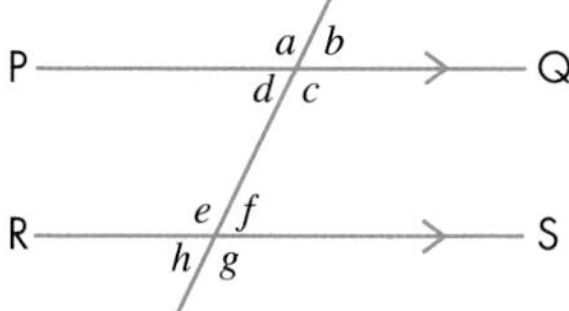

a Which angle is vertically opposite to angle a? *(1 mark)*

b Which angle is alternate to angle f? *(1 mark)*

c Which angle is corresponding to angle c? *(1 mark)*

AQA, November 2010 H1 Question 2

6 **a** The diagram shows a regular pentagon.

One side has been extended.

Which **one** of these statements is true?

A The exterior angle of a regular pentagon is equal to 360° ÷ 5 = 72°.

B The interior angle of a regular pentagon is equal to 360° ÷ 5 = 72°.

C The exterior angle of a regular pentagon is equal to 360° – 72° = 288°.

D The interior angle of a regular pentagon is equal to 360° – 72° = 288°.

(1 mark)

b The diagram shows two identical regular pentagons, touching, inside a rectangle.

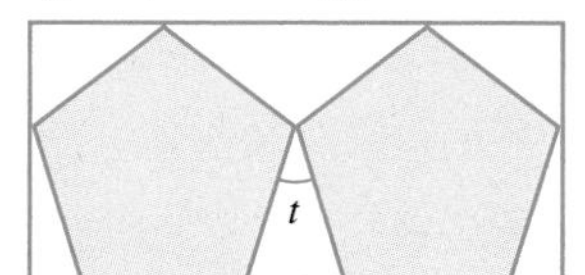

Work out the value of t. *(2 marks)*

AQA, May 2009, Module 5, Paper 1 Higher, Question 8(a)

7 In the diagram AB is parallel to CD.

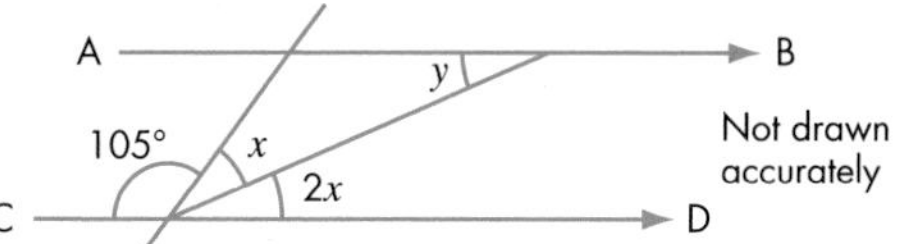

a Work out the value of x? *(2 marks)*

b Work out the value of y?
Give a reason for your answer

AQA, June 2010 H1 Question 7

Worked Examination Questions

1 Naomi has a collection of tiles that are all **regular polygons**.

The sides of all the tiles are the same length.

She says that a **regular hexagon** will fit exactly between **a square** and a **regular octagon**.

Explain why she is wrong.

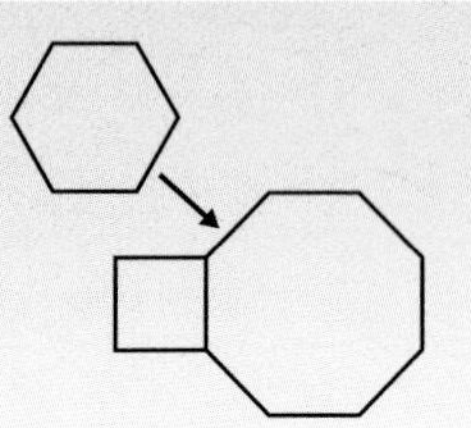

1 Interior angle of the square = 90°
Interior angle of the regular octagon = 135°
Interior angle of the regular hexagon = 120°

This gets 1 mark for all three correct.

90° + 135° + 120° = 345°

This gets 1 method mark for finding the total.

So the regular hexagon does not fit exactly as the three angles do not add up to 360°.

This gets 1 accuracy mark for the reason.

Total: 3 marks

2 ABCD is a trapezium

a work out the size of angle x

b Work out the size of angle y

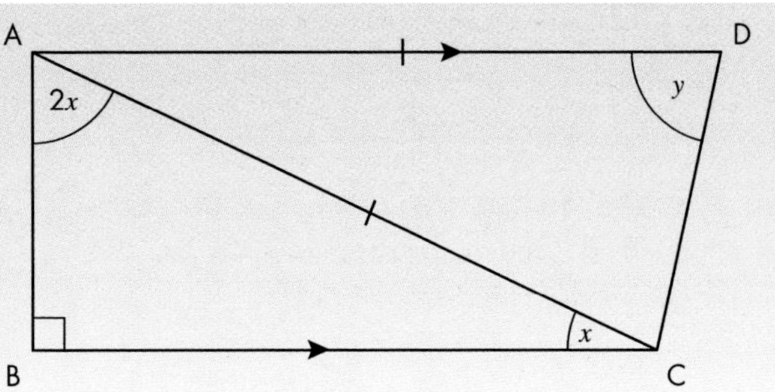

2 a $2x + x = 90$

ABC is a right angled triangle so $2x$ and x must add up to 90. This gets 1 method mark

$x = 30$

The correct answer gets 1 accuracy mark.

b Angle CAD = 30

This angle can be calculated in 2 ways. BAD = 90 or CAD is alternate to ACB. This gets 1 mark.

$y = (180 - 30) \div 2$

As triangle ACD is isosceles this will calculate the value of angle CDA. This gets 1 method mark.

$y = 75°$

The correct answer gets 1 accuracy mark.

Total: 5 marks

Worked Examination Questions

PS **3** The diagrams show a trapezium and a parallelogram.

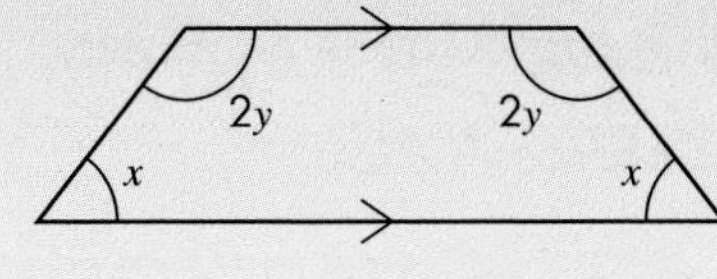

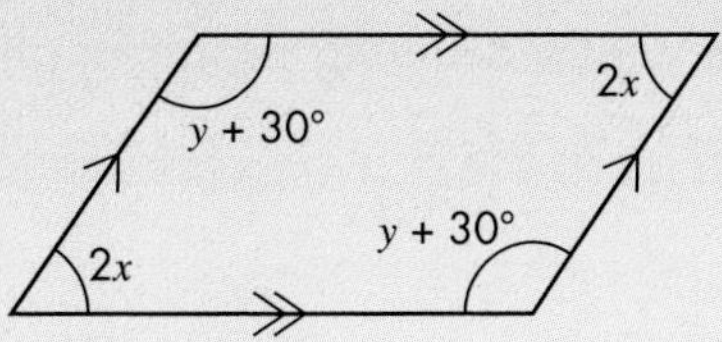

a Use the trapezium to explain why $x + 2y = 180°$.

b Use the parallelogram to form another equation in terms of x and y.

c Work out the values of x and y.

3 a The interior angles at either end of the trapezium add up to 180°

This gets 1 mark.

or you could say the two angles are allied angles

This gets 1 mark.

b $2x + y + 30° = 180°$ (allied angles)

This simplifies to $2x + y = 150°$

c You now have two equations to solve simultaneously.

$x + 2y = 180°$ (1)

$2x + y = 150°$ (2)

$2x + 4y = 360°$ (3) = (1) × 2

(3) − (2)

This gets 1 method mark for attempt to solve equations.

$3y = 210°$ so $y = 70°$

This gets 1 accuracy mark for one correct angle.

Substitute into (1) $x + 140° = 180°$, so $x = 40°$

This gets 1 accuracy mark for the other angle.

Total: 3 marks

9

Functional Maths
Tessellations

A tessellation is a pattern of shapes that fit together to leave no gaps.

This activity is best done with a drawing program although it can be done on paper.

Note: the instructions refer to the drawing facilities within Microsoft Word™.

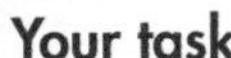

Your task

Step 1: First draw any quadrilateral. For example a Kite.

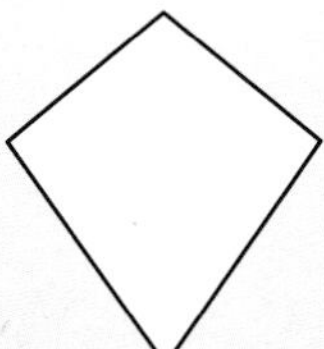

Step 2: Mark the midpoints of each side.

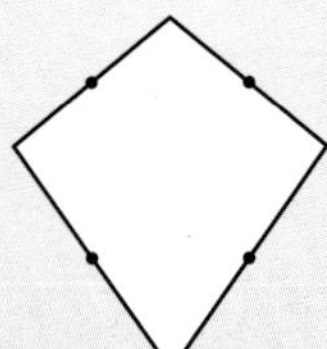

Step 3: Now draw any shape on half of two of the sides. [Curve, Scribble or Line]

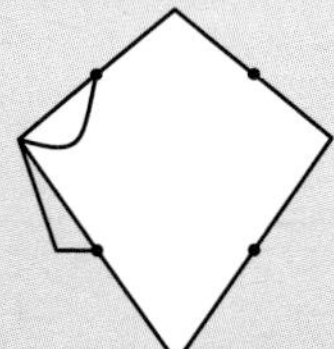

Step 4: Rotate these shapes around the midpoint and draw the rotation on the other half of the side. [Duplicate, Flip Vertical, Flip Horizontal]

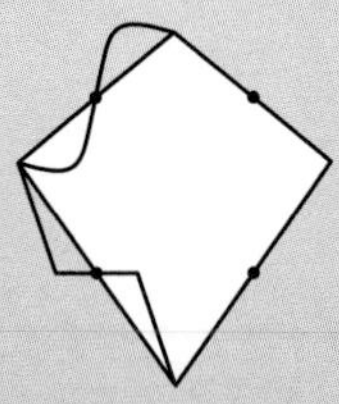

Step 5: Reflect these onto the other two sides [Group, Flip Horizontal]

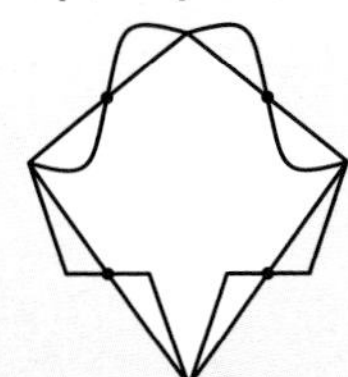

Step 6: Remove the original kite and midpoints. [Delete]

You now have a shape that will tessellate.

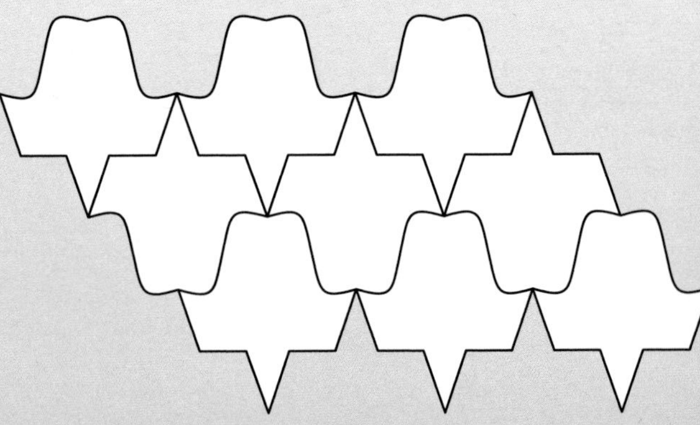

To make it into an Escher type of tessellation, draw something inside the original shape. [You can find clip art on the internet or scan in a picture.]

For example

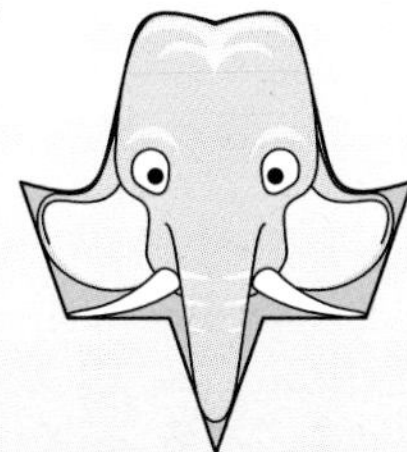

Do the same picture in a different colour and Flip it vertically.

Putting these together gives an Escher type of tessellation.

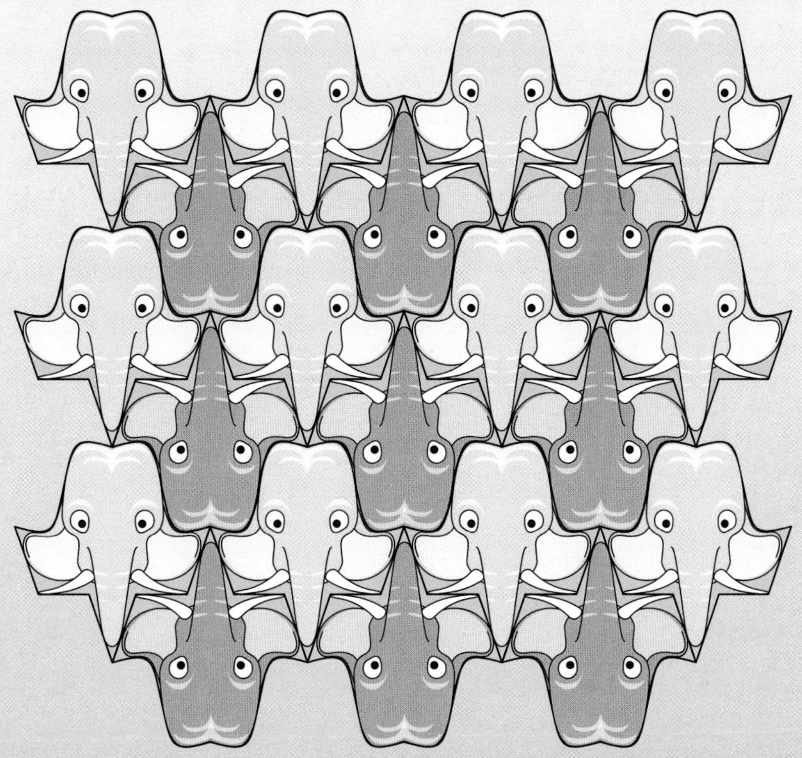

Why this chapter matters

For anything, from a house to a landscape gardening project, designers need to construct plans accurately, to be sure that everything will fit together properly. This will also give the people putting it together a blueprint to work from.

The need for accurate drawings is clear in bridge construction. Bridge engineers are responsible for producing practical bridge designs to meet the requirements of their employers. For example, a bridge intended to carry traffic over a newly constructed railway needs to be strong enough to bear the weight of the traffic and stable enough to counteract the effects of the moving traffic and strong winds. The designers produce a blueprint that has all the measurements, including heights, weights and angles, clearly marked on it. Construction workers then use this blueprint to build the bridge to the exact specifications set by the designers and engineers.

Generally, the construction workers work on both ends of the bridge at the same time, meeting in the middle. The blueprints are therefore essential for making sure that the bridge is safe and that the bridge meets in the middle.

Accurately-drawn blueprints were essential in the construction of the Golden Gate Bridge, which crosses the San Francisco Bay. When it was constructed in the 1930s, it was the longest suspension bridge in the world. The bridge engineers (who included Joseph Strauss and Charles Alton Ellis) had to draw precise blueprints to make sure that they had all the information necessary to build this innovative bridge and that it would be built correctly.

By contrast, a bridge built at a stadium for the Maccabiah Games in Israel was built without proper planning and without accurate blueprints. This led to the bridge collapsing soon after its construction in 1997, killing four athletes and injuring 64 people.

Just like a bridge engineer, you must be accurate in your construction, working with a freshly-sharpened pencil and a good pair of compasses, measuring and drawing angles carefully and drawing construction lines as faintly as possible.

Geometry: Constructions

This chapter will show you ...

- **D** how to construct triangles
- **C** how to bisect a line and an angle
- **C** how to construct perpendiculars
- **C** how to define a locus
- **C** how to solve locus problems

Visual overview

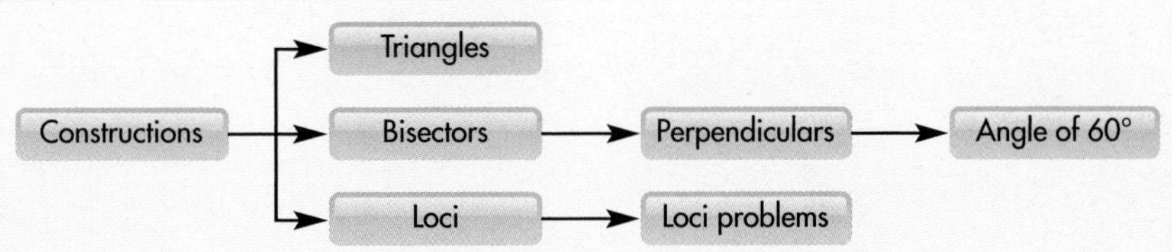

What you should already know

- How to measure lines and angles **(KS3 level 5, GCSE grade F)**
- How to use scale drawings **(KS3 level 5, GCSE grade E)**

Quick check

1 Measure the following lines.

a ______________________

b ____________________________

c ____

2 Measure the following angles.

a

b

10.1 Constructing triangles

This section will show you how to:

- construct triangles, using compasses, a protractor and a straight edge

Key words
angle
compasses
construct
side

There are three ways of **constructing** a triangle. Which one you use depends on what information you are given about the triangle.

When carrying out geometric constructions, always use a sharp pencil (preferably grade 2H rather than HB) to give you thin, clear lines. These may be called faint or feint lines. The examiner will be marking your construction and will be looking for accuracy, which requires fine, clean lines and points as small as you can make them, while ensuring they are clearly visible.

All three sides known

EXAMPLE 1

Construct a triangle with **sides** that are 5 cm, 4 cm and 6 cm long.

- **Step 1:** Draw the longest side as the base. In this case, the base will be 6 cm, which you draw using a ruler. (The diagrams in this example are drawn at half-size.)

- **Step 2:** Deal with the second longest side, in this case the 5 cm side. Open the **compasses** to a radius of 5 cm (the length of the side), place the point on one end of the 6 cm line and draw a short faint arc, as shown here.

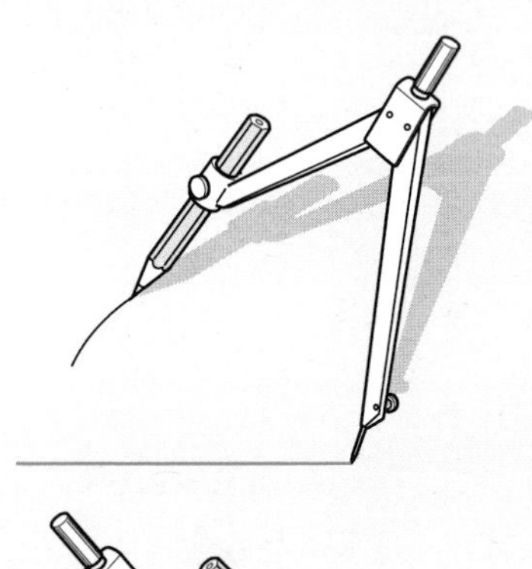

- **Step 3:** Deal with the shortest side, in this case the 4 cm side. Open the compasses to a radius of 4 cm, place the point on the other end of the 6 cm line and draw a second short faint arc to intersect the first arc, as shown here.

- **Step 4:** Complete the triangle by joining each end of the base line to the point where the two arcs intersect.

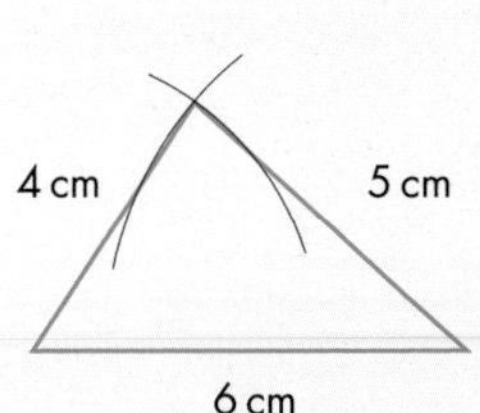

Note: The arcs are construction lines and so must be left in to show the examiner how you constructed the triangle.

Two sides and the included angle known

EXAMPLE 2

Draw a triangle ABC, in which AB is 6 cm, BC is 5 cm and the included **angle** ABC is 55°. (The diagrams in this example are drawn at half-size.)

- **Step 1:** Draw the longest side, AB, as the base. Label the ends of the base A and B.

A ——————————— B

- **Step 2:** Place the protractor along AB with its centre on B and make a point on the diagram at the 55° mark.

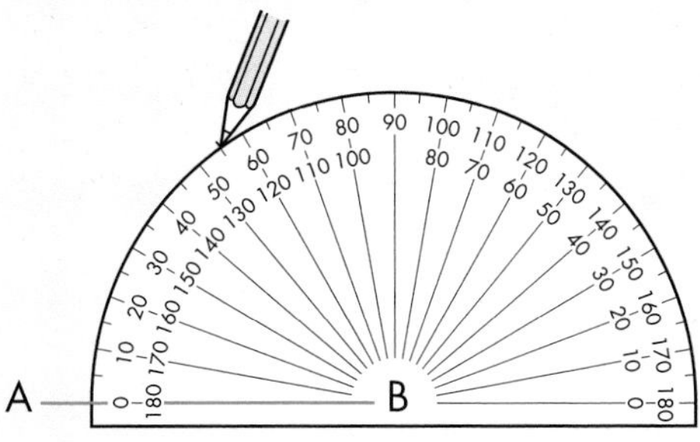

- **Step 3:** Draw a *faint* line from B through the 55° point. From B, using a pair of compasses, measure 5 cm along this line.
- Label the point where the arc cuts the line as C.

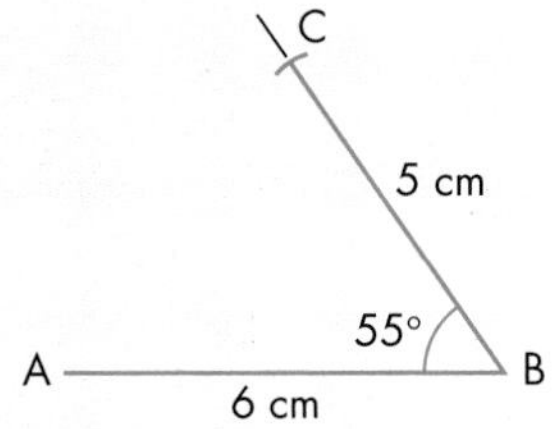

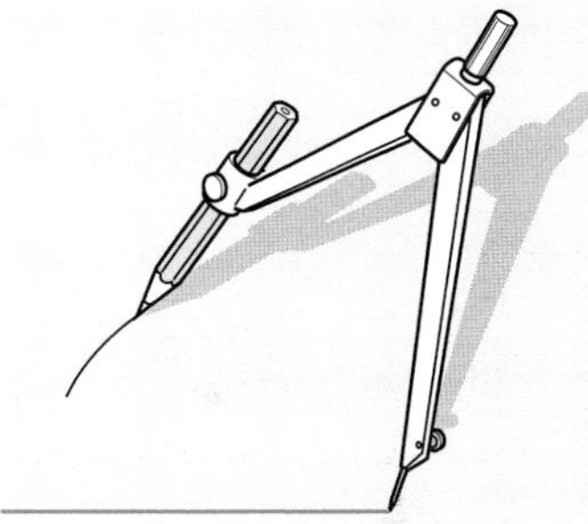

- **Step 4:** Join A and C to complete the triangle.

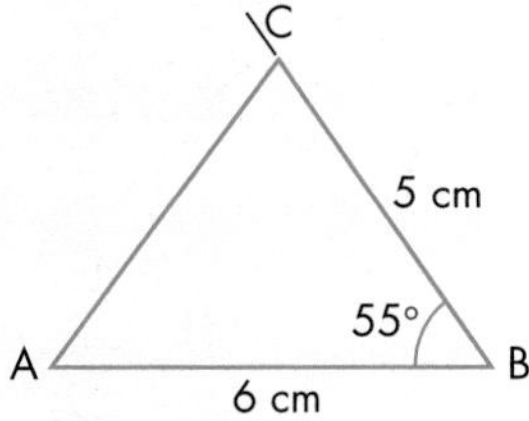

Note: Remember to use clean, sharp lines so that the examiner can see how the triangle has been constructed.

Two angles and a side known

When you know two angles of a triangle, you also know the third.

EXAMPLE 3

Draw a triangle ABC, in which AB is 7 cm, angle BAC is 40° and angle ABC is 65°.

- **Step 1:** As before, start by drawing the base, which here has to be 7 cm. Label the ends A and B.

- **Step 2:** Centre the protractor on A and mark the angle of 40°. Draw a clear, clean line from A through this point.

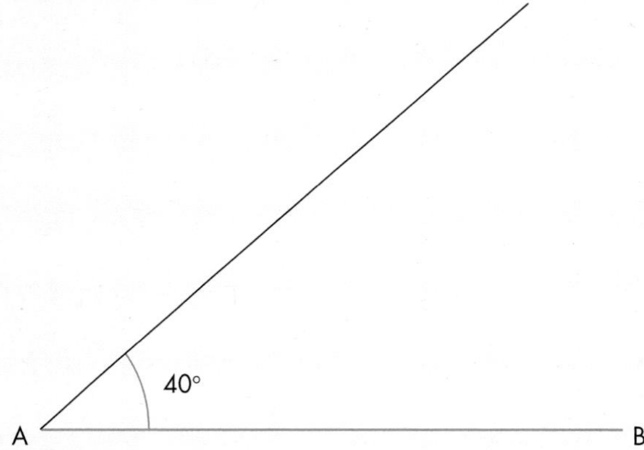

- **Step 3:** Centre the protractor on B and mark the angle of 65°. Draw a clear, clean line from B through this point, to intersect the 40° line drawn from A. Label the point of intersection as C.

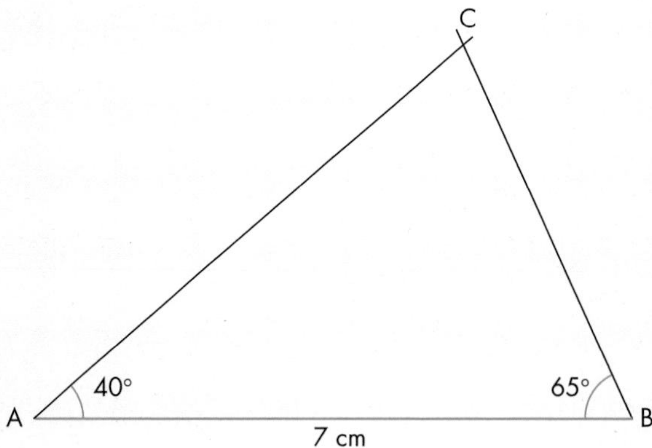

EXERCISE 10A

> **HINTS AND TIPS**
> Always make a sketch if one is not given in the question.

1 Draw the following triangles accurately and measure the sides and angles not given in the diagram.

a

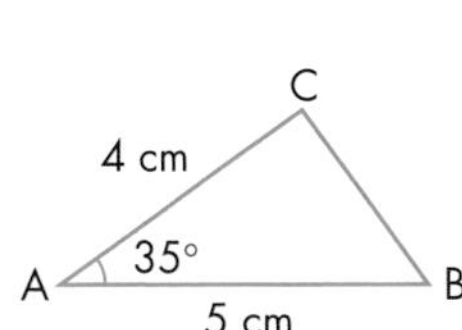

b

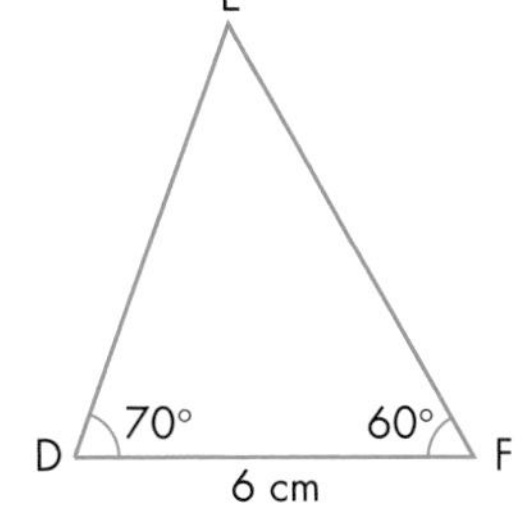

c

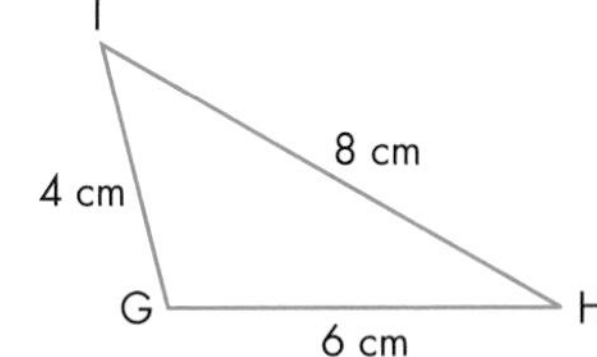

d

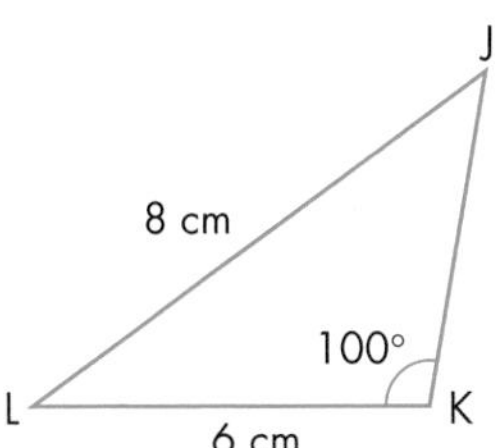

e

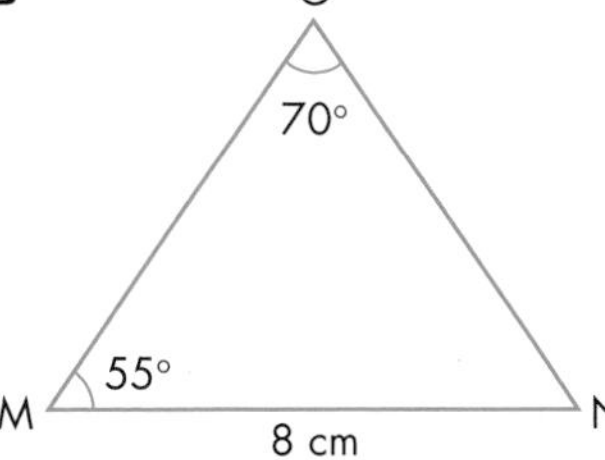

f

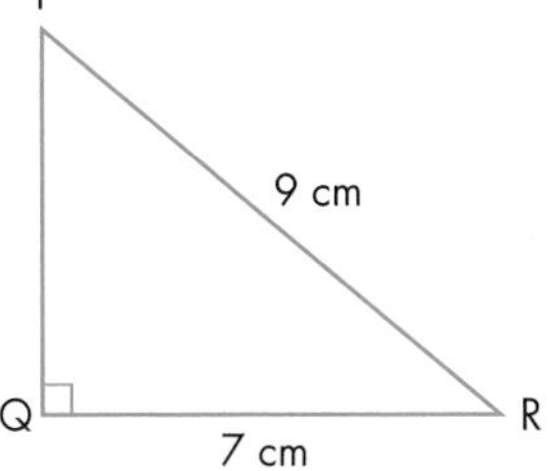

2 **a** Draw a triangle ABC, where AB = 7 cm, BC = 6 cm and AC = 5 cm.

b Measure the sizes of ∠ABC, ∠BCA and ∠CAB.

> **HINTS AND TIPS**
> Sketch the triangle first.

3 Draw an isosceles triangle that has two sides of length 7 cm and the included angle of 50°.

a Measure the length of the base of the triangle.

b What is the area of the triangle?

4 A triangle ABC has ∠ABC = 30°, AB = 6 cm and AC = 4 cm. There are two different triangles that can be drawn from this information.

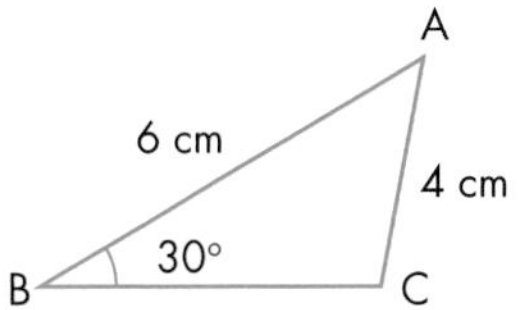

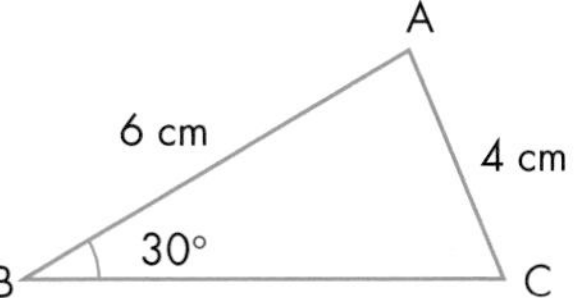

What are the two different lengths that BC can be?

5 Construct an equilateral triangle of side length 5 cm.

a Measure the height of the triangle.

b What is the area of this triangle?

D

6 Construct a parallelogram with sides of length 5 cm and 8 cm and with an angle of 120° between them.

8 cm
120°
5 cm
h

a Measure the height of the parallelogram.

b What is the area of the parallelogram?

7 Groundsmen painting white lines on a sports field may use a knotted rope, like the one shown below.

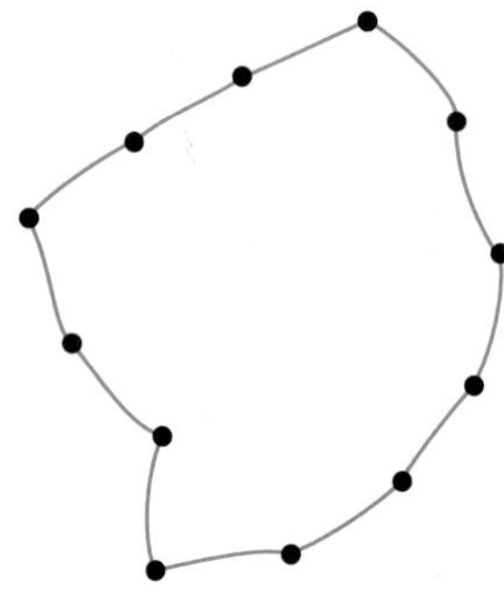

It has 12 equally-spaced knots.
It can be laid out to give a triangle, like this.

It will always be a right-angled triangle. This helps the groundsmen to draw lines perpendicular to each other.

Here are two more examples of such ropes.

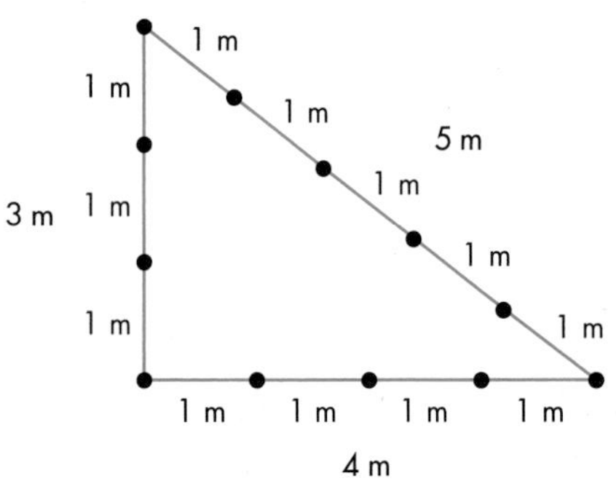

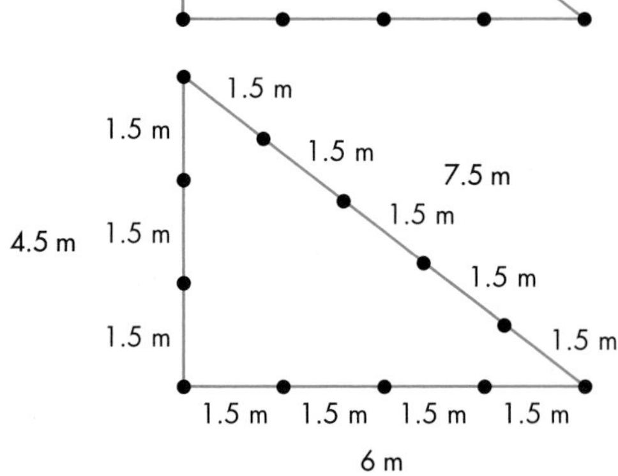

a Show, by constructing each of the above triangles (use a scale of 1 cm : 1 m), that each is a right-angled triangle.

b Choose a different triangle that you think might also be right-angled. Use the same knotted-rope idea to check.

PS 8 Construct the triangle with the largest area which has a total perimeter of 12 cm.

AU 9 Anil says that, as long as he knows all three angles of a triangle, he can draw it. Explain why Anil is wrong.

10.2 Bisectors

This section will show you how to:
- construct the bisectors of lines and angles
- construct angles of 60° and 90°

Key words
angle bisector
bisect
line bisector
perpendicular bisector

To **bisect** means to divide in half. So a bisector divides something into two equal parts.

- A **line bisector** divides a straight line into two equal lengths.
- An **angle bisector** is the straight line that divides an angle into two equal angles.

To construct a line bisector

It is usually more accurate to construct a line bisector than to measure its position (the midpoint of the line).

- **Step 1:** Here is a line to bisect.
- **Step 2:** Open your compasses to a radius of about three-quarters of the length of the line. Using each end of the line as a centre, and without changing the radius of your compasses, draw two intersecting arcs.

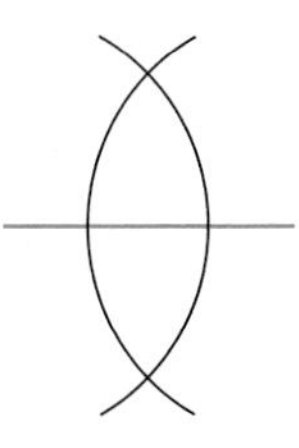

- **Step 3:** Join the two points at which the arcs intersect. This line is the **perpendicular bisector** of the original line.

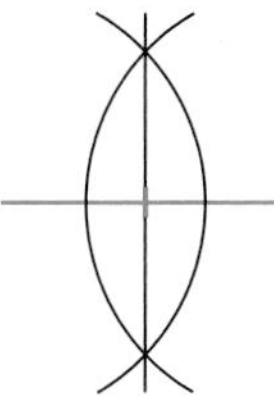

To construct an angle bisector

It is much more accurate to construct an angle bisector than to measure its position.

- **Step 1:** Here is an angle to bisect.
- **Step 2:** Open your compasses to any reasonable radius that is less than the length of the lines forming the angle. If in doubt, go for about 3 cm. With the vertex of the angle as centre, draw an arc through both lines.

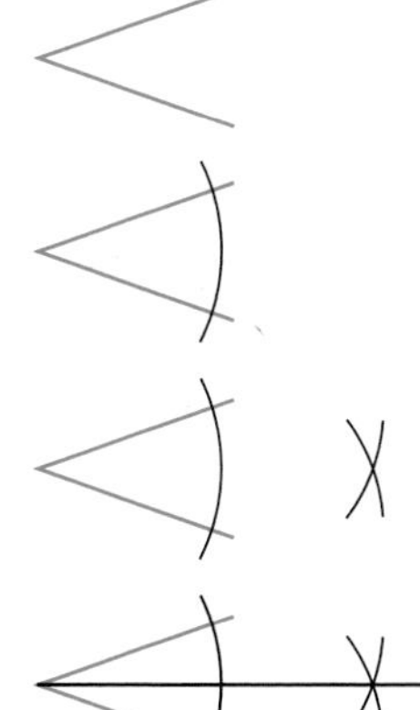

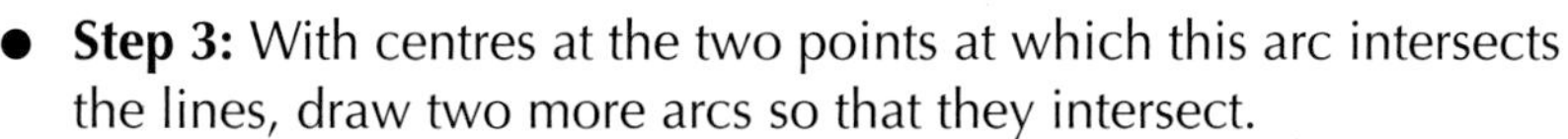

- **Step 3:** With centres at the two points at which this arc intersects the lines, draw two more arcs so that they intersect.
- **Step 4:** Join the point at which these two arcs intersect to the vertex of the angle.

This line is the angle bisector.

To construct an angle of 60°

It is more accurate to construct an angle of 60° than to measure and draw it with a protractor.

- **Step 1:** Draw a line and mark a point on it.
- **Step 2:** Open the compasses to a radius of about 4 cm.
  Using the point as the centre, draw an arc that crosses the line and extends almost above the point.

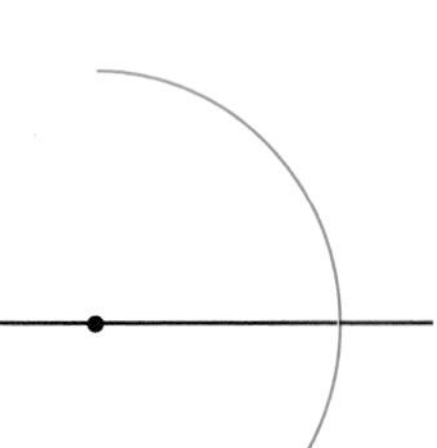

- **Step 3:** Keep the compasses set to the same radius.
  Using the point where the first arc crosses the line as a centre, draw another arc that intersects the first one.

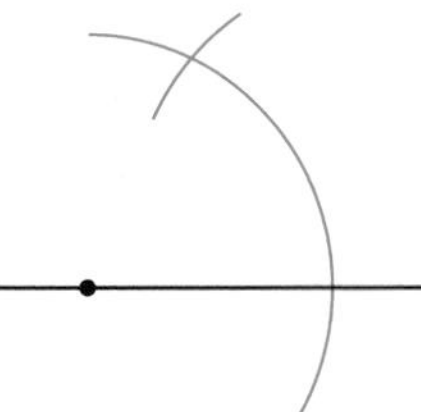

- **Step 4:** Join the original point to the point where the two arcs intersect.

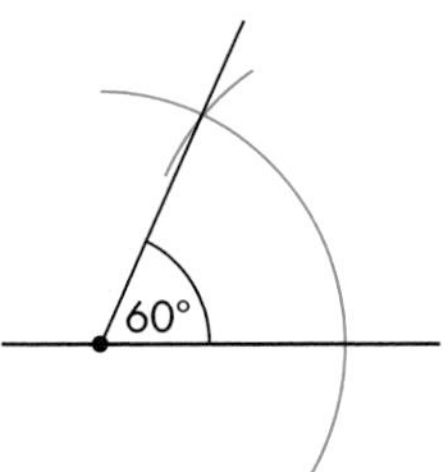

- **Step 5:** Use a protractor to check that the angle is 60°.

To construct a perpendicular from a point on a line (an angle of 90°)

This construction will produce a perpendicular from a point A on a line.

- Open your compasses to about 2 or 3 cm.
  With point A as centre, draw two short arcs to intersect the line at each side of the point.

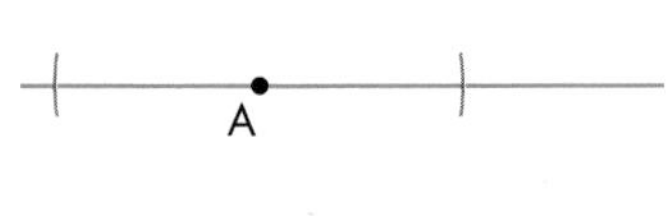

- Now extend the radius of your compasses to about 4 cm. With centres at the two points at which the arcs intersect the line, draw two arcs to intersect at X above the line.

- Join AX.

  AX is perpendicular to the line.

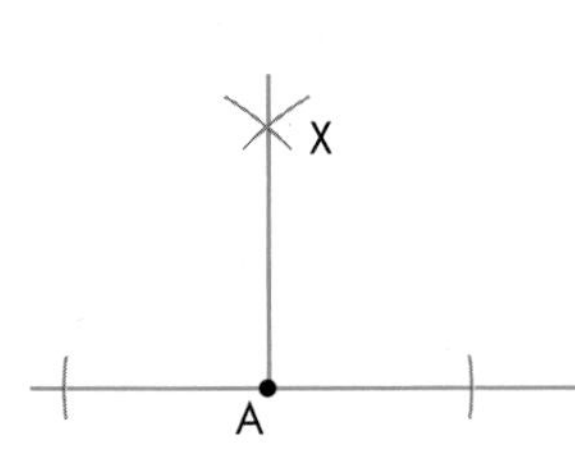

Note: If you needed to construct a 90° angle at the end of a line, you would first have to extend the line.

You could be even more accurate by also drawing two arcs underneath the line, which would give three points in line.

To construct a perpendicular from a point to a line

This construction will produce a perpendicular from a point A to a line.

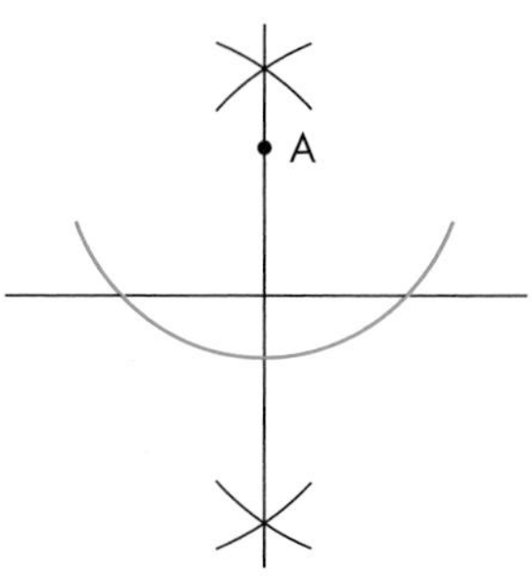

- With point A as centre, draw an arc which intersects the line at two points.
- With centres at these two points of intersection, draw two arcs to intersect each other both above and below the line.
- Join the two points at which the arcs intersect. The resulting line passes through point A and is perpendicular to the line.

Examination note: When a question says *construct*, you must *only* use compasses, not a protractor. When it says *draw*, you may use whatever you can to produce an accurate diagram. But also note, when constructing you may use your protractor to check your accuracy.

EXERCISE 10B

HINTS AND TIPS

Remember that examiners want to see your construction lines.

1 Draw a line 7 cm long and bisect it. Check your accuracy by seeing if each half is 3.5 cm.

2 Draw a circle of about 4 cm radius.

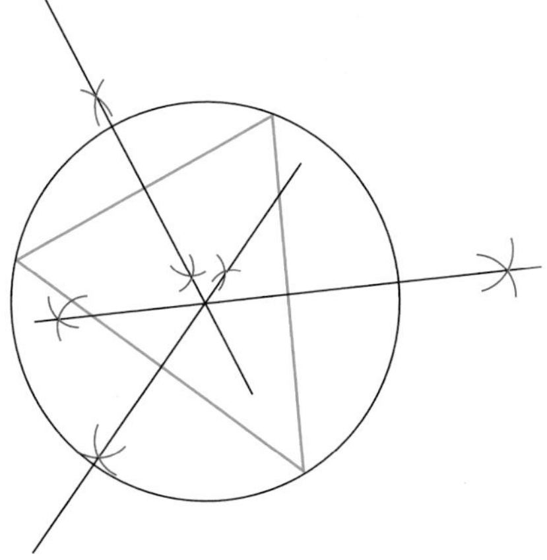

Draw a triangle inside the circle so that the corners of the triangle touch the circle.

Bisect each side of the triangle.

The bisectors should all meet at the same point, which should be the centre of the circle.

3 **a** Draw any triangle with sides that are between 5 cm and 10 cm.

b On each side construct the line bisector.

All your line bisectors should intersect at the same point.

c Using this point as the centre, draw a circle that goes through every vertex of the triangle.

C

4 Repeat question **2** with a different triangle and check that you get a similar result.

5 **a** Draw the following quadrilateral.

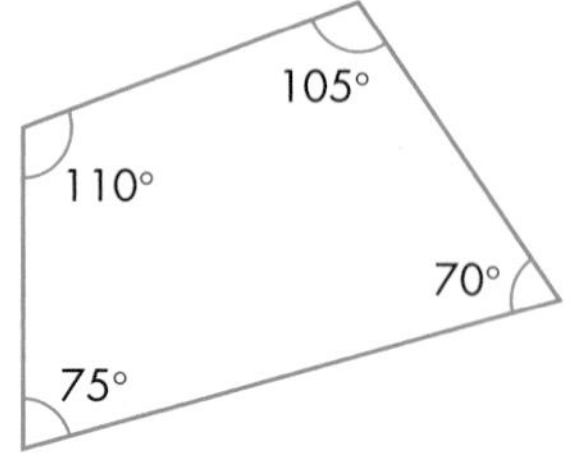

b Construct the line bisector of each side. These all should intersect at the same point.

c Use this point as the centre of a circle that goes through the quadrilateral at each vertex. Draw this circle.

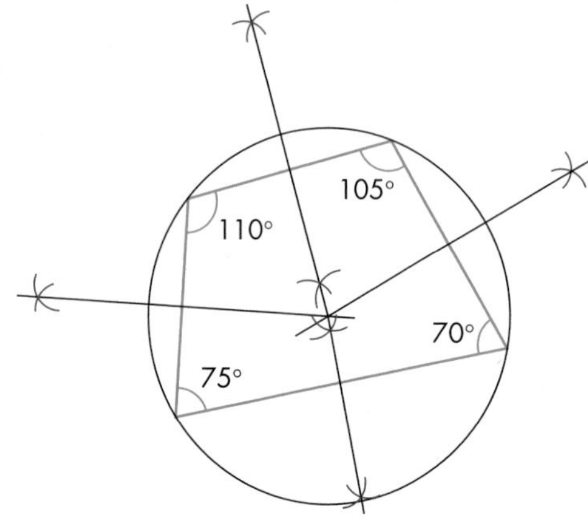

6 **a** Draw an angle of 50°.

b Construct the angle bisector.

c Check how accurate you have been by measuring each half. Both should be 25°.

7 Draw a circle with a radius of about 3 cm.

Draw a triangle so that the sides of the triangle are tangents to the circle.

Bisect each angle of the triangle.

The bisectors should all meet at the same point, which should be the centre of the circle.

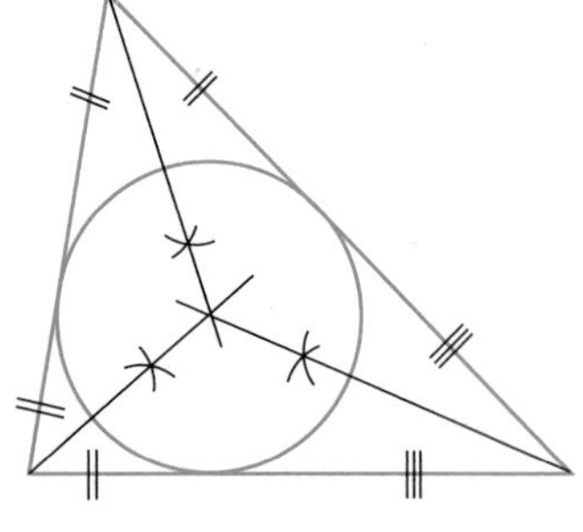

8 **a** Draw any triangle with sides that are between 5 cm and 10 cm.

b At each angle construct the angle bisector.
All three bisectors should intersect at the same point.

c Use this point as the centre of a circle that just touches the sides of the triangle.

9 Repeat question **8** with a different triangle.

FM 10 Gianni and Anna have children living in Bristol and Norwich. Gianni is about to start a new job in Birmingham. They are looking on a map of Britain for places they might move to.

Anna says, "I want to be the same distance from both children."

Gianni says, "I want to be as close to Birmingham as possible."

Find the largest city that would suit both Gianni and Anna. Use a map of the UK to help you.

C

PS **11** Draw a circle with radius about 4 cm.

Draw a quadrilateral, **not** a rectangle, inside the circle so that each vertex is on the circumference.

Construct the bisector of each side of the quadrilateral.

Where is the point where these bisectors all meet?

AU **12** Briefly outline how you would construct a triangle with angles 90°, 60° and 30°.

13 **a** Draw a line AB, 6 cm long, and construct an angle of 90° at A.

b Bisect this angle to construct an angle of 45°.

14 **a** Draw a line AB, 6 cm long, and construct an angle of 60° at A.

b Bisect this angle to construct an angle of 30°.

15 Draw a line AB, 6 cm long, and mark a point C, 4 cm above the middle of the line.

Construct the perpendicular from the point C to the line AB.

10.3 Defining a locus

This section will show you how to:
- draw a locus for a given rule

Key words
equidistant
loci
locus

A **locus** (plural **loci**) is the movement of a point according to a given rule.

EXAMPLE 4

A point P that moves so that it is always at a distance of 5 cm from a fixed point A will have a locus that is a circle of radius 5 cm.

You can express this mathematically by saying the locus of the point P is such that AP = 5 cm.

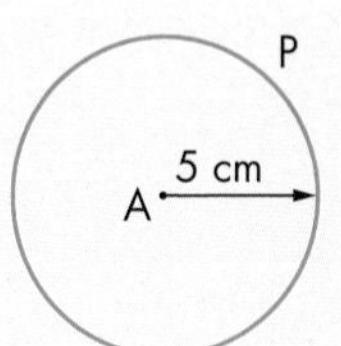

EXAMPLE 5

A point P that moves so that it is always the same distance from two fixed points A and B will have a locus that is the perpendicular bisector of the line joining A and B.

You can express this mathematically by saying:

the locus of the point P is such that $AP = BP$.

A point that is always the same distance from two points is **equidistant** from the two points.

EXAMPLE 6

A point that is always 5 m from a long, straight wall will have a locus that is a line parallel to the wall and 5 m from it.

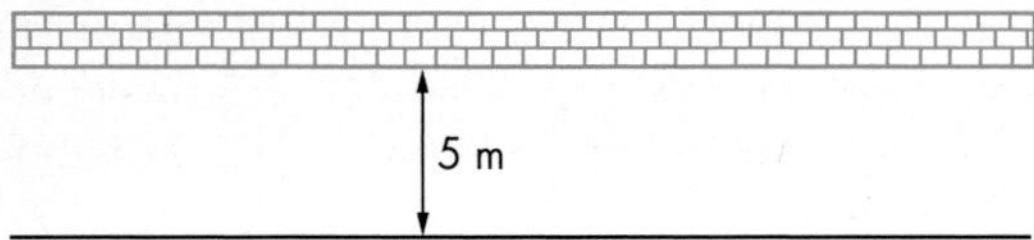

EXAMPLE 7

A point that moves so that it is always 5 cm from a line AB will have a locus that is a racetrack shape around the line.

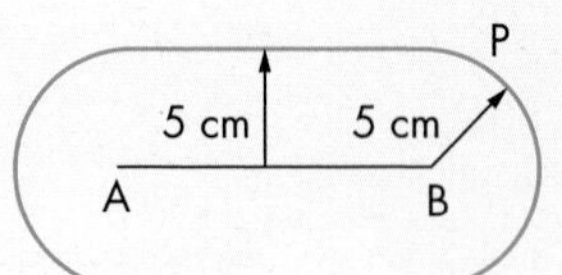

In your GCSE examination, you will usually get practical situations rather than abstract mathematical ones.

EXAMPLE 8

Imagine a grassy, flat field in which a horse is tethered to a stake by a rope that is 10 m long. What is the shape of the area that the horse can graze?

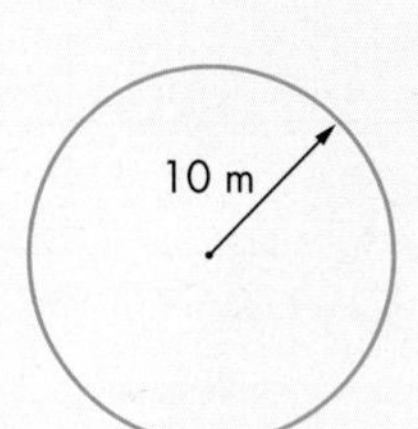

In reality, the horse may not be able to reach the full 10 m if the rope is tied round its neck but ignore fine details like that. You 'model' the situation by saying that the horse can move around in a 10 m circle and graze all the grass within that circle.

In this example, the locus is the whole of the area inside the circle.

You can express this mathematically as:

the locus of the point P is such that $AP \leqslant 10$ m.

EXERCISE 10C

> **HINTS AND TIPS**
>
> Sketch the situation before doing an accurate drawing.

1 A is a fixed point. Sketch the locus of the point P in each of these situations.

a AP = 2 cm **b** AP = 4 cm **c** AP = 5 cm

2 A and B are two fixed points 5 cm apart. Sketch the locus of the point P for each of these situations.

a AP = BP **b** AP = 4 cm and BP = 4 cm

c P is always within 2 cm of the line AB

FM 3 **a** A horse is tethered in a field on a rope 4 m long. Describe or sketch the area that the horse can graze.

b The horse is still tethered by the same rope but there is now a long, straight fence running 2 m from the stake. Sketch the area that the horse can now graze.

4 ABCD is a square of side 4 cm. In each of the following loci, the point P moves only inside the square. Sketch the locus in each case.

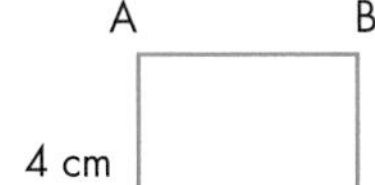
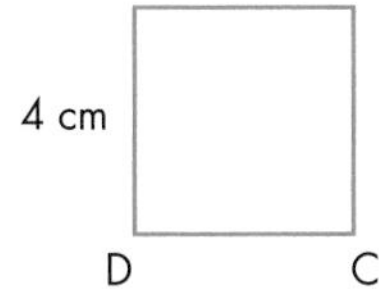

a AP = BP **b** AP < BP **c** AP = CP

d CP < 4 cm **e** CP > 2 cm **f** CP > 5 cm

5 One of the following diagrams is the locus of a point on the rim of a bicycle wheel as it moves along a flat road. Which is it?

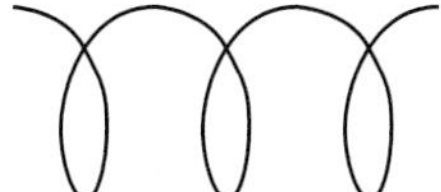
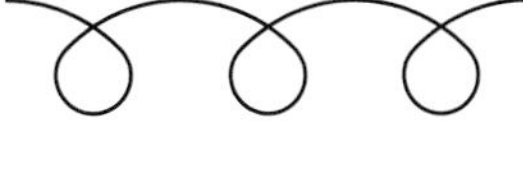

6 Draw the locus of the centre of the wheel for the bicycle in question **5**.

PS 7 ABC is a triangle.

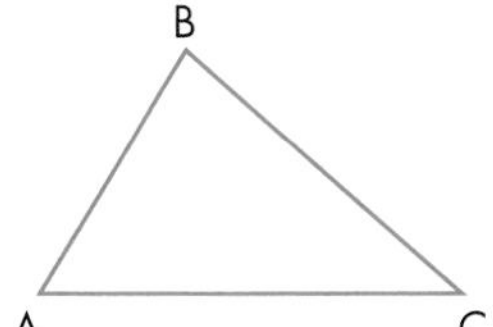

The region R is defined as the set of points inside the triangle such that:

- they are closer to the line AB than the line AC
- they are closer to the point A than the point C.

Using a ruler and compasses, construct the region R.

AU 8 ABCD is a rectangle.

Copy the diagram and draw the locus of all points that are 2 cm from the edges of the rectangle.

10.4 Loci problems

This section will show you how to:
- solve practical problems using loci

Key words
loci
scale

Most of the **loci** problems in your GCSE examination will be of a practical nature, as in the next example.

EXAMPLE 9

Imagine that a radio company wants to find a site for a transmitter. The transmitter must be the same distance from Doncaster and Leeds and within 20 miles of Sheffield.

In mathematical terms, this means they are concerned with the perpendicular bisector between Leeds and Doncaster and the area within a circle of radius 20 miles from Sheffield.

The diagram, drawn to a **scale** of 1 cm = 10 miles, illustrates the situation and shows that the transmitter can be built anywhere along the thick part of the blue line.

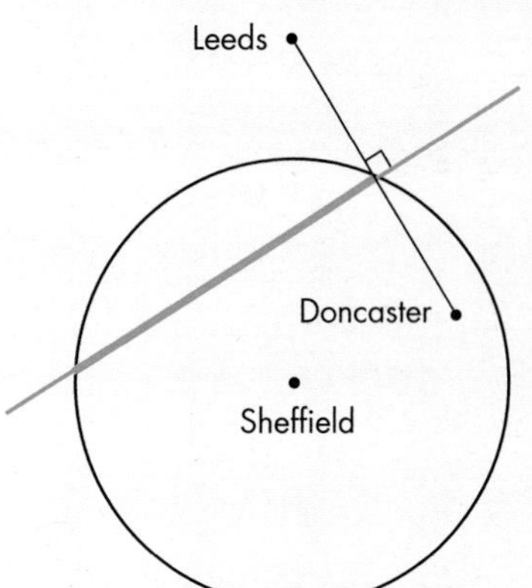

EXAMPLE 10

A radar station in Birmingham has a range of 150 km (that is, it can pick up any aircraft within a radius of 150 km). Another radar station in Norwich has a range of 100 km.

Can an aircraft be picked up by both radar stations at the same time?

The situation is represented by a circle of radius 150 km around Birmingham and another circle of radius 100 km around Norwich. The two circles overlap, so an aircraft could be picked up by both radar stations when it is in the overlap.

EXAMPLE 11

A dog is tethered by a rope, 3 m long, to the corner of a shed, 4 m by 2 m. What is the area that the dog can guard effectively?

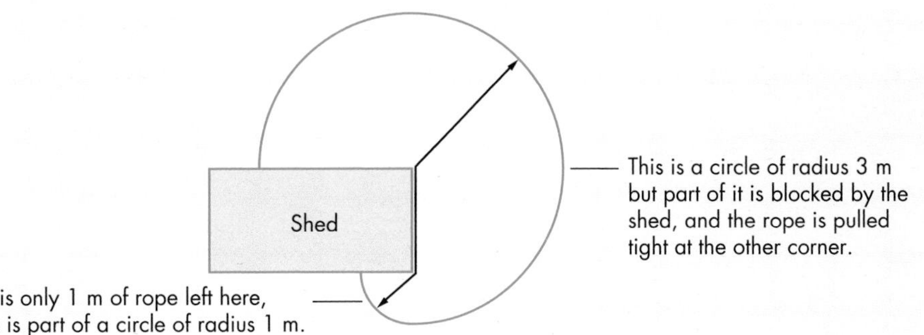

EXERCISE 10D

For questions **1** to **7**, you should start by sketching the picture given in each question on a 6 × 6 grid, each square of which is 1 cm by 1 cm. The scale for each question is given.

FM **1** A goat is tethered by a rope, 7 m long, in a corner of a field with a fence at each side. What is the locus of the area that the goat can graze? Use a scale of 1 cm ≡ 2 m.

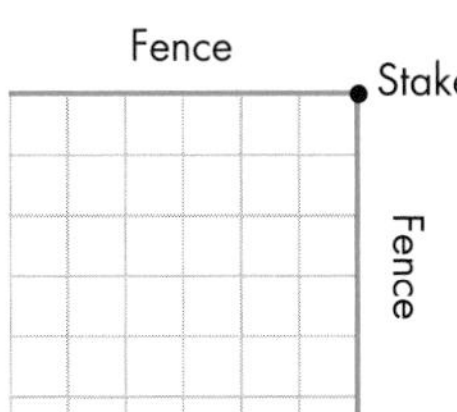

FM **2** In a field, a horse is tethered to a stake by a rope 6 m long. What is the locus of the area that the horse can graze? Use a scale of 1 cm ≡ 2 m.

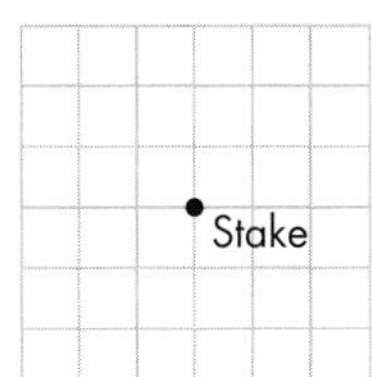

FM **3** A cow is tethered to a rail at the top of a fence 6 m long. The rope is 3 m long. Sketch the area that the cow can graze. Use a scale of 1 cm ≡ 2 m.

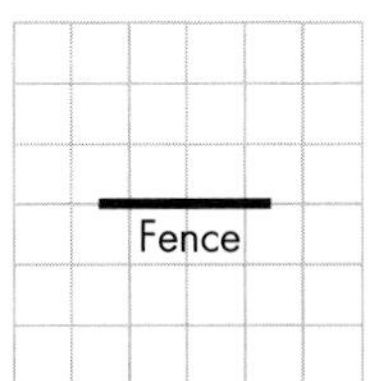

FM **4** A horse is tethered to a stake near a corner of a fenced field, at a point 4 m from each fence. The rope is 6 m long. Sketch the area that the horse can graze. Use a scale of 1 cm ≡ 2 m.

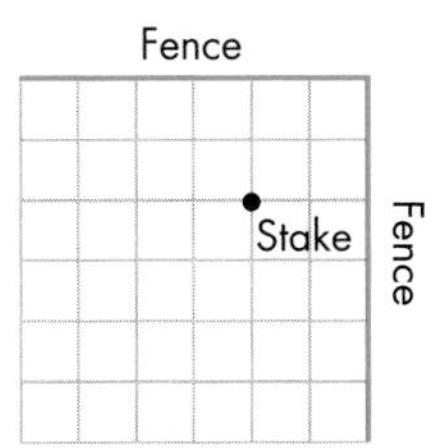

C

FM 5 A horse is tethered to a corner of a shed, 2 m by 1 m. The rope is 2 m long. Sketch the area that the horse can graze. Use a scale of 1 cm ≡ 1 m.

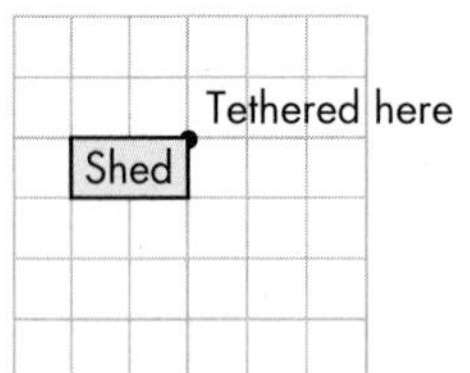

FM 6 A goat is tethered by a 4 m rope to a stake at one corner of a pen, 4 m by 3 m. Sketch the area of the pen on which the goat cannot graze. Use a scale of 1 cm ≡ 1 m.

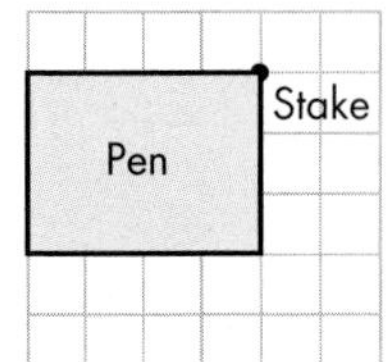

FM 7 A puppy is tethered to a stake by a rope, 1.5 m long, on a flat lawn on which are two raised brick flower beds. The stake is situated at one corner of a bed, as shown. Sketch the area that the puppy is free to roam in. Use a scale of 1 cm ≡ 1 m.

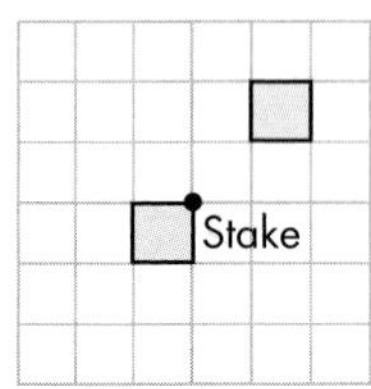

For questions **8** to **15**, you should use a copy of the map opposite. For each question, trace the map and mark on those points that are relevant to that question.

FM 8 A radio station broadcasts from London on a frequency of 1000 kHz with a range of 300 km. Another radio station broadcasts from Glasgow on the same frequency with a range of 200 km.

a Sketch the area to which each station can broadcast.

b Will they interfere with each other?

c If the Glasgow station increases its range to 400 km, will they then interfere with each other?

FM 9 The radar at Leeds airport has a range of 200 km. The radar at Exeter airport has a range of 200 km.

a Will a plane flying over Birmingham be detected by the Leeds radar?

b Sketch the area where a plane can be picked up by both radars at the same time.

FM 10 A radio transmitter is to be built according to these rules.

i It has to be the same distance from York and Birmingham.

ii It must be within 350 km of Glasgow.

iii It must be within 250 km of London.

a Sketch the line that is the same distance from York and Birmingham.

b Sketch the area that is within 350 km of Glasgow and 250 km of London.

c Show clearly the possible places at which the transmitter could be built.

Glasgow
Newcastle upon Tyne
North Sea
York
Leeds
Irish Sea
Manchester
Sheffield
Norwich
Birmingham
London
Bristol
Exeter
0
50
100
150
200
250
300
350 km

C

FM **11** A radio transmitter centred at Birmingham is designed to give good reception in an area greater than 150 km and less than 250 km from the transmitter. Sketch the area of good reception.

FM **12** Three radio stations pick up a distress call from a boat in the Irish Sea. The station at Glasgow can tell from the strength of the signal that the boat is within 300 km of the station. The station at York can tell that the boat is between 200 km and 300 km from York. The station at London can tell that it is less than 400 km from London. Sketch the area where the boat could be.

FM **13** Sketch the area that is between 200 km and 300 km from Newcastle upon Tyne, and between 150 km and 250 km from Bristol.

FM **14** An oil rig is situated in the North Sea in such a position that it is the same distance from Newcastle upon Tyne and Manchester. It is also the same distance from Sheffield and Norwich. Draw the line that shows all the points that are the same distance from Newcastle upon Tyne and Manchester. Repeat for the points that are the same distance from Sheffield and Norwich and find out where the oil rig is located.

FM **15** Whilst looking at a map, Fred notices that his house is the same distance from Glasgow, Norwich and Exeter. Where is it?

16 Wathsea Harbour is as shown in the diagram. A boat sets off from point A and steers so that it stays the same distance from the sea wall and the West Pier. Another boat sets off from B and steers so that it stays the same distance from the East Pier and the sea wall. Copy the diagram. On your diagram show accurately the path of each boat.

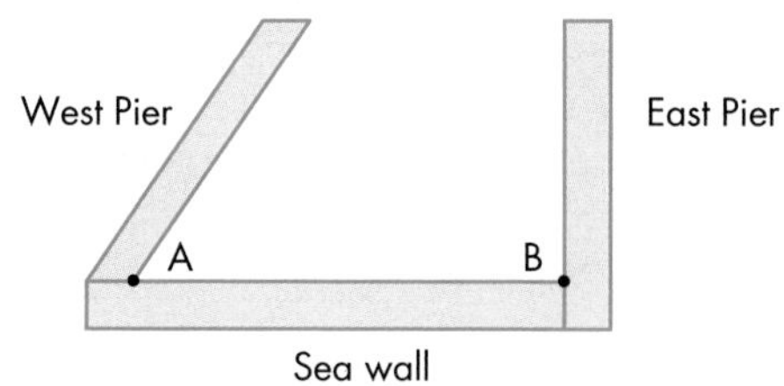

PS **17** Tariq wanted to fly himself from the Isle of Wight north, towards Scotland. He wanted to remain at the same distance from London as Bristol as far as he could.

Once he is past London and Bristol, which city should he aim toward to keep him, as accurately as possible, the same distance from London and Bristol? Use the map to help you.

AU **18** A distress call is heard by coastguards in both Newcastle and Bristol. The signal strength suggests that the call comes from a ship that is the same distance from both places. Explain how the coastguards could find the area of sea to search.

GRADE BOOSTER

- **C** You can construct line and angle bisectors
- **C** You can describe and draw the locus of a point from a given rule
- **C** You can use loci to solve problems
- **B** You can construct a perpendicular from a point on a line
- **B** You can construct a perpendicular from a point to a line
- **B** You can construct angles of 60° and 90°

What you should know now

- How to construct line and angle bisectors
- How to construct perpendiculars
- How to construct angles without using a protractor
- Understand what is meant by a locus
- How to solve problems, using loci

1 The diagram shows a sketch of a parallelogram.

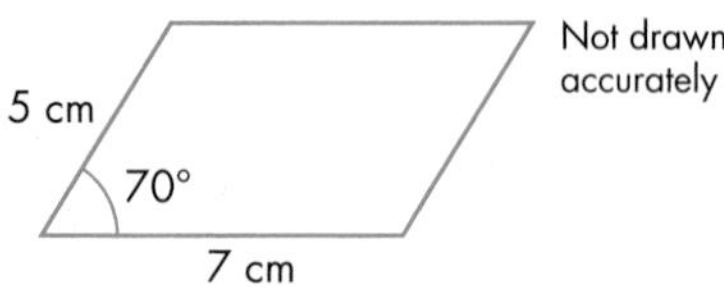

Make an accurate drawing of the parallelogram. *(3 marks)*

AQA, November 2008, Module 5, Paper 1 Higher, Question 3

2 In trapezium PQRS, the sides PQ and SR are parallel.

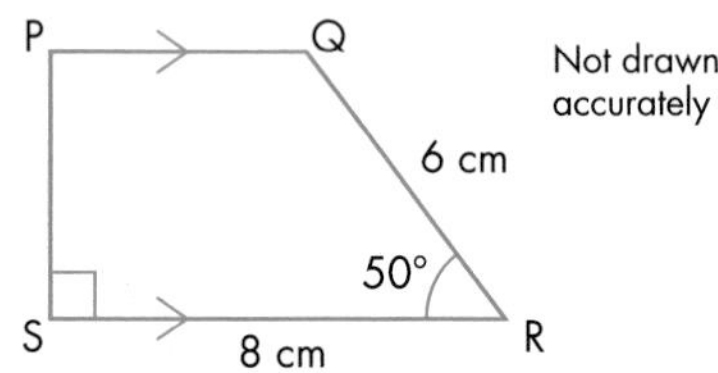

Make an accurate drawing of the trapezium. *(4 marks)*

AQA, June 2009, Module 5, Paper 2 Higher, Question 5

3 

a Draw a line PQ 10 cm long. Now, using ruler and compasses only, construct the perpendicular bisector of the line PQ. *(2 marks)*

b Complete the sentence.

The perpendicular bisector of the line PQ is the locus of points that are… *(1 mark)*

AQA, June 2008, Module 5, Paper 2 Higher, Question 9

4 ABCD is a quadrilateral.

The region R is defined as the set of points inside ABCD that are:

closer to the side AB than the side AD

and closer to the point D than the point C.

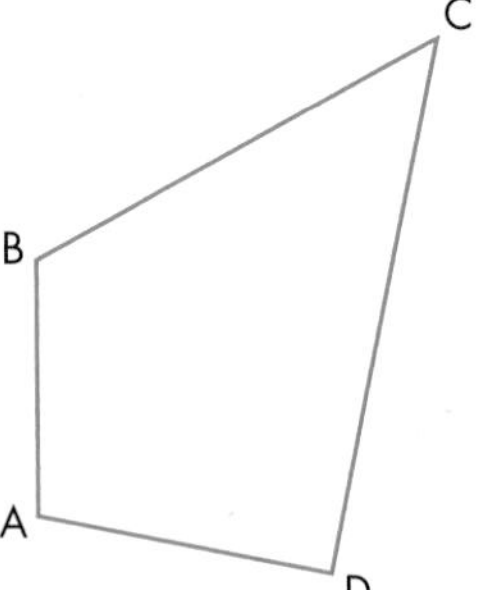

Using a ruler and compasses, copy the diagram and construct **accurately** the region R.

Label the region clearly with the letter R. *(4 marks)*

AQA, June 2009, Paper 2 Higher, Question 20

5 The diagram shows an L shape.

Copy the diagram and draw the locus of all points 2 cm from the L shape. *(3 marks)*

AQA, June 2005, Paper 1 Higher, Question 5

6 **a** Using a ruler and compasses only, construct an angle of 60°. *(2 marks)*

b Two lifeboat stations A and B receive a distress call from a boat.

The boat is within 6 kilometres of station A.

The boat is within 8 kilometres of station B.

Trace the diagram and shade the possible area where the boat could be. *(2 marks)*

Scale: 1 cm represents 1 km

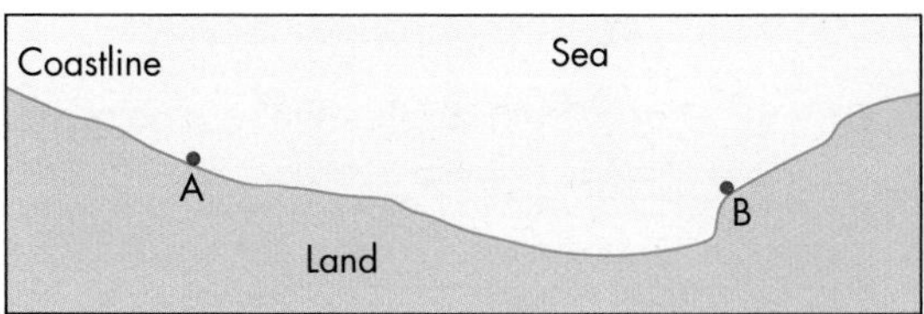

AQA, Specimen Paper 2008, Module 5, Paper 2 Higher, Question 8

7  The diagram shows a scale drawing of a straight road.

A walker is at point P.

P ×

Scale
1 cm represents 0.5 km

Road

a Copy the diagram and use a ruler and compasses to construct the perpendicular from the point P to the road. You **must** show all your construction lines and arcs. *(3 marks)*

b Find the shortest real distance from the walker to the road. *(2 marks)*

AQA, November 2007, Paper 2 Higher, Question 3

Worked Examination Questions

1 Here is a sketch of a triangle. PR = 6.4 cm, QR = 7.7 cm and angle R = 35°.

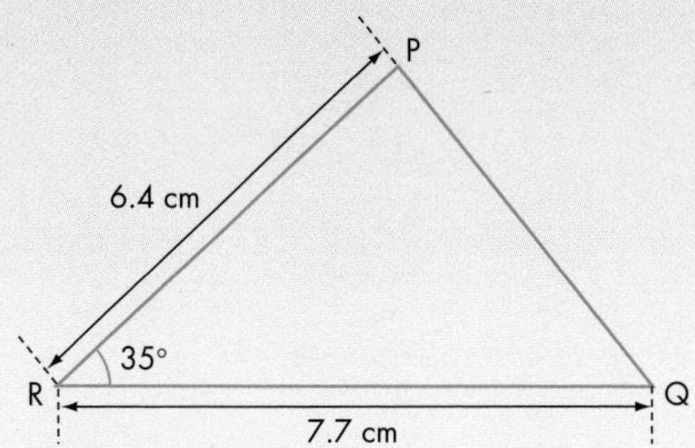

a Make an accurate drawing of the triangle.

b Measure the size of angle Q on your drawing.

1 a **Make an accurate drawing, using these steps.**

Step 1: Draw the base as a line 7.7 cm long.
You can draw this and measure it with a ruler, although using a pair of compasses is more accurate.

Step 2: Measure the angle at R as 35°.
Draw a faint line at this angle.

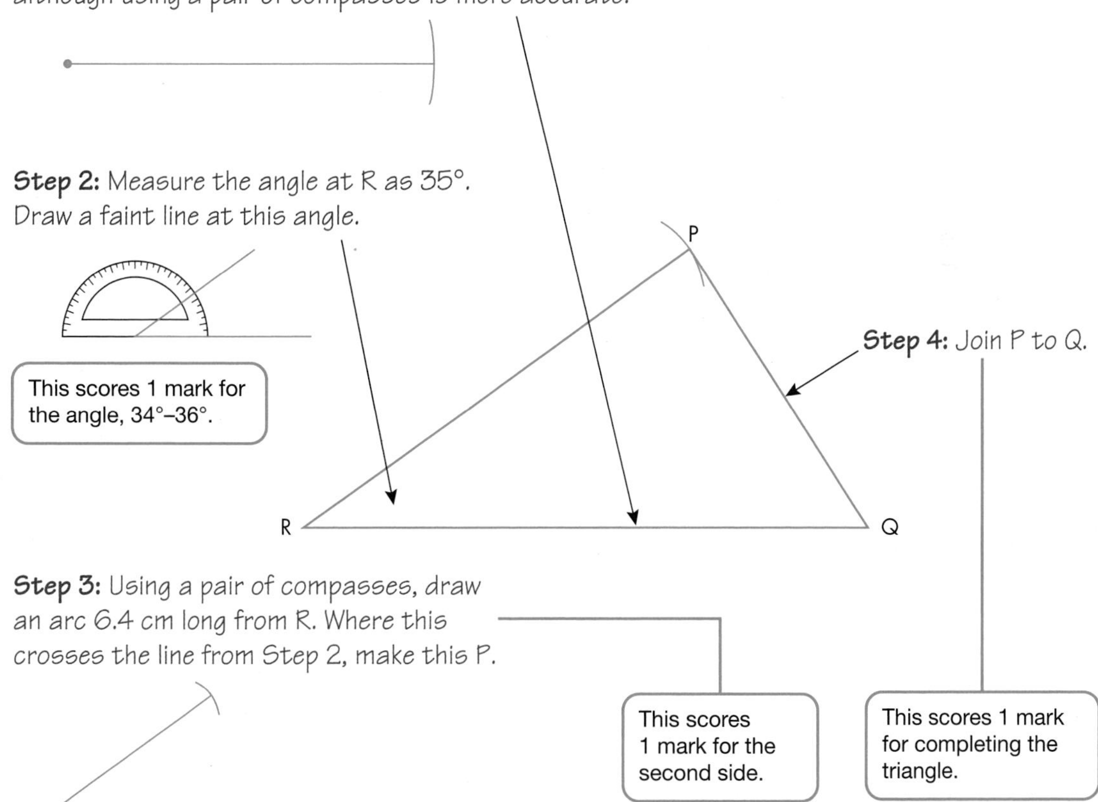

This scores 1 mark for the angle, 34°–36°.

Step 3: Using a pair of compasses, draw an arc 6.4 cm long from R. Where this crosses the line from Step 2, make this P.

This scores 1 mark for the second side.

This scores 1 mark for completing the triangle.

b **Measure the angle Q. It is 56°.**

This scores 1 mark for this angle 56° ± 2°. If your actual angle was not 56° but you measured it accurately, then you would still get the mark.

Total: 4 marks

Worked Examination Questions

2 The map shows three boats, A, B and C, on a lake. Along one edge of the lake there is a straight path.

Treasure lies at the bottom of the lake.

The treasure is:

between 150 m and 250 m from B,

nearer to A than C,

more than 100 m from the path.

Using a ruler and compasses only, shade the region in which the treasure lies.

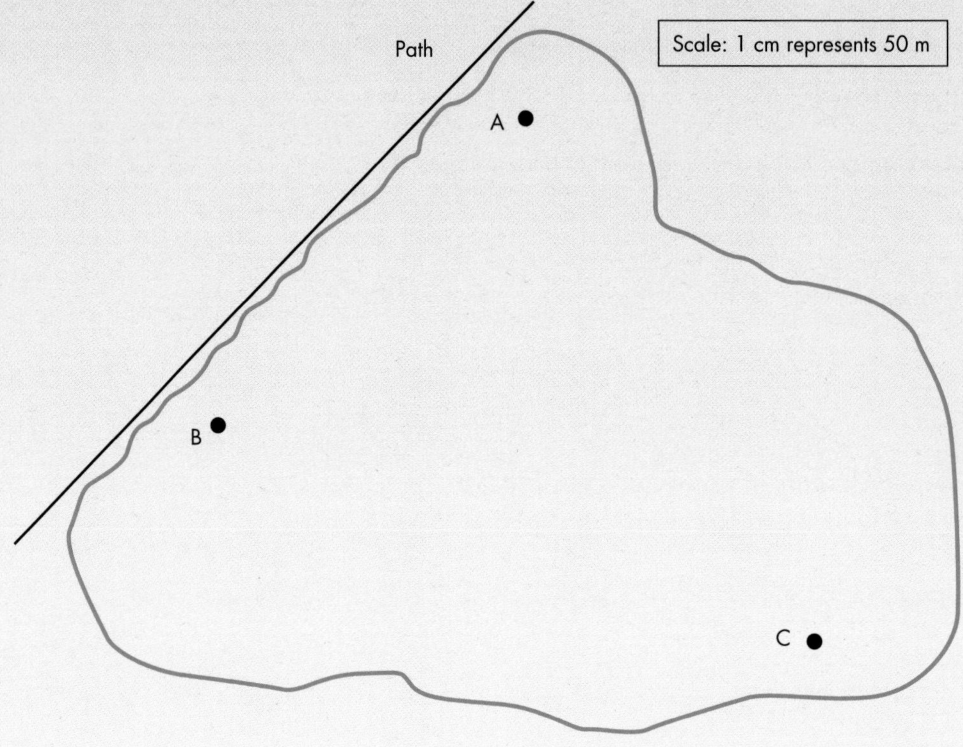

You must show clearly all your construction arcs.

Worked Examination Questions

2 Draw two circles with centre at B with radii 3 cm and 5 cm. — This scores 1 mark.

Draw the perpendicular bisector of AC. — This scores 1 mark.

Draw a parallel line 2 cm from the path. — This scores 1 mark.

The region required is shaded on the diagram. — This scores 1 mark.

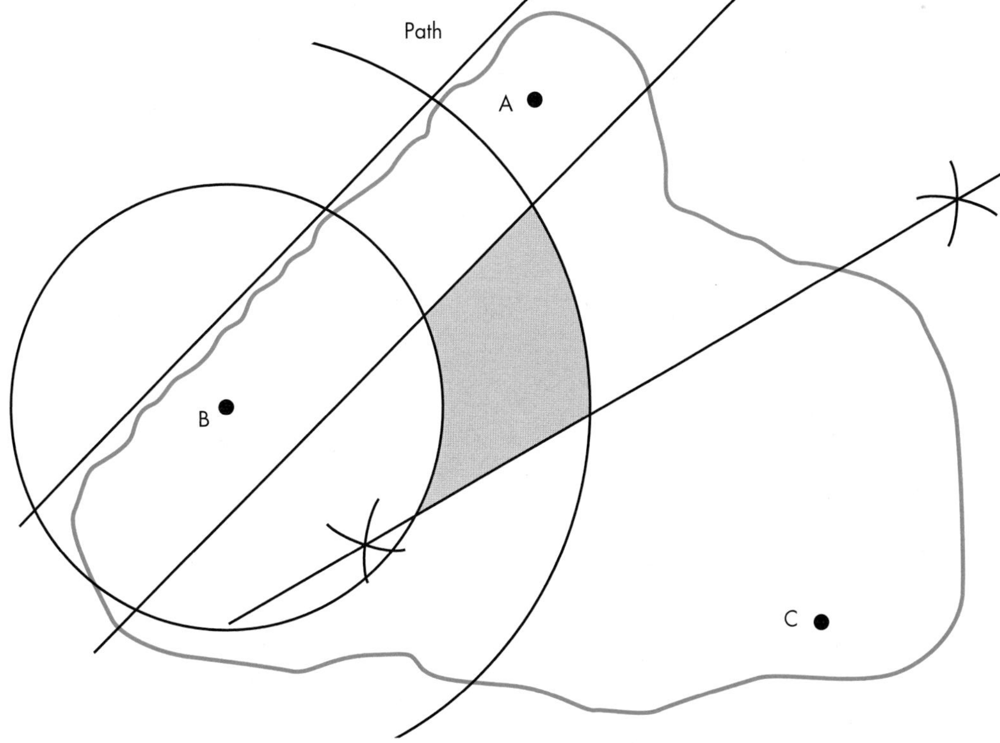

Total: 4 marks

10

Problem Solving
Planning a football pitch

You already know that architects and engineers must construct accurate diagrams, to be certain that the buildings and constructions they have designed will be built correctly. However, did you know that the same principles used by architects and engineers to construct diagrams are also used when planning sports pitches, whether it is your local playing field or a Premier League football club?

In this task you will take on the role of the grounds staff of a football pitch. You will need to negotiate many variables to ensure that the pitch is drawn up correctly and can be maintained thoroughly.

Your task

As a member of the grounds staff you have been asked to prepare the pitch ready for the new football season.

The club needs the pitch designed according to FIFA's specifications.

1. Construct a scale drawing of the football pitch, to be used in laying out the pitch on the football field. Be sure to label all the dimensions of the pitch.
2. The pitch will need to be regularly watered in order to keep it in good condition. For this, you will need to design a comprehensive sprinkler system. On a copy of your drawing of the pitch, mark up where the sprinklers should go, ensuring that the maximum possible area of the pitch is watered at any one time.

Getting started

Discuss these questions with a partner.

- What shapes do you see on sports fields? Do these shapes vary, depending on the sports that take place on these pitches?
- What angles do you see on sports fields?
- How are shapes drawn on to sports fields?

FIFA specifications

The field

The field of play must be rectangular, divided into two halves by a halfway line. The centre mark is indicated at the midpoint of the halfway line and a circle with a radius of 9.15 m (10 yards) is marked around it.

The field dimensions should be as follows.

	Minimum	Maximum
Length	90 m (approx. 100 yards)	120 m (approx. 130 yards)
Width	45 m (approx. 50 yards)	90 m (approx. 100 yards)

The goal area

The goal mouth is 7.3 metres (8 yards) wide. The goal area is 5.5 m (6 yards) wide by 18.3 m (12 yards) long.

The penalty area

The penalty area is 16.5 m (18 yards) wide by 40.3 m (44 yards) long.

Within each penalty area there is a penalty mark 11 m (12 yards) from the midpoint between the goalposts and equidistant from them. An arc with a radius of 9.15 m (10 yards) from each penalty mark is drawn outside the penalty area.

The corner arc

A flagpost is placed in each corner. A quarter-circle with a radius of 1 m (approximately 1 yard) is drawn at each corner flag post, inside the field of play.

Hint: Use the internet to research football pitches and FIFA's specifications further.

Why this chapter matters

How many sides does a strip of paper have? Two or one?

Take a strip of paper about 20 cm by 2 cm.

How many sides does it have? Easy! You can see that this has two sides, a topside and an underside. If you were to draw a line along one side of the strip, you would have one side with a line 20 cm long on it and one side blank.

Now mark the ends A and B, put a single twist in the strip of paper and tape (or glue) the two ends together, as shown.

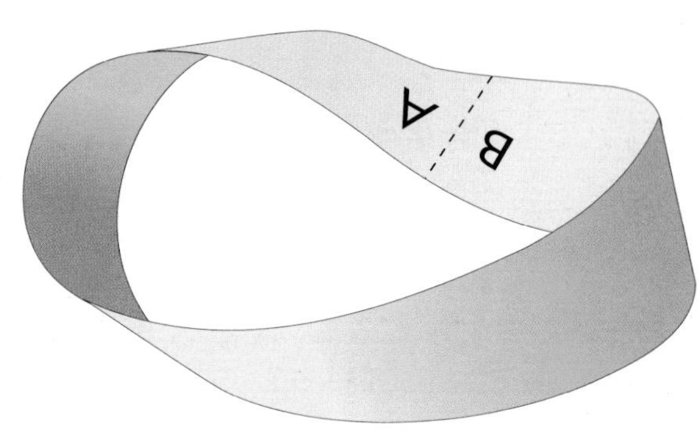

How many sides does this strip of paper have now?

Take a pen and draw a line on the paper, starting at any point you like. Continue the line along the length of the paper – you will eventually come back to your starting point. Your strip has only one side now! There is no blank side.

You have transformed a two-sided piece of paper into a one-sided piece of paper.

This curious shape is called a Möbius strip. It is named after August Ferdinand Möbius, a 19th-century German mathematician and astronomer. Möbius, along with others, caused a revolution in geometry.

Möbius strips have a number of surprising applications that exploit its remarkable property of one-sidedness, including conveyor belts in industry as well as in domestic vacuum cleaners. Have a look at the belt that turns the rotor in the vacuum cleaner at home.

The Möbius strip has become the universal symbol of recycling. The symbol was created in 1970 by Gary Anderson, who was a senior at the University of Southern California, as part of a contest sponsored by a paper company.

The Möbius strip is a form of transformation. In this chapter, you will look at some other transformations of shapes.

Chapter

Geometry: Transformation geometry

1. Congruent triangles
2. Translations
3. Reflections
4. Rotations
5. Enlargements
6. Combined transformations

This chapter will show you ...

- **D** what is meant by a transformation
- **D** to **B** how to translate, reflect, rotate and enlarge 2D shapes
- **B** to **A** how to show that two triangles are congruent

Visual overview

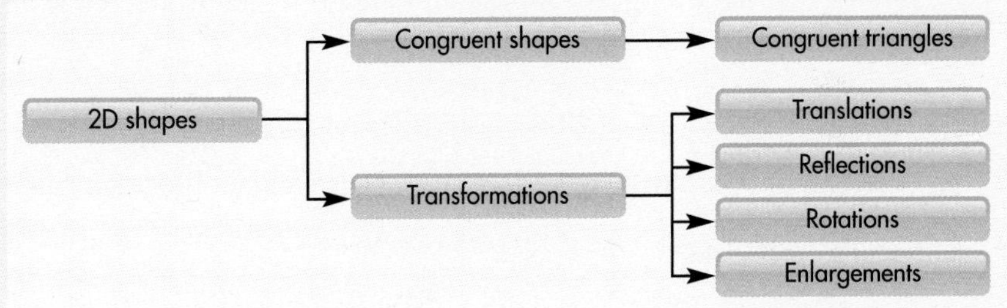

What you should already know

- How to recognise congruent shapes **(KS3 level 4, GCSE grade G)**
- How to find the lines of symmetry of a 2D shape **(KS3 level 4, GCSE grade F)**
- How to find the order of rotational symmetry of a 2D shape **(KS3 level 4, GCSE grade F)**
- How to draw the lines with equations $x = \pm a$, $y = \pm b$, $y = x$ and $y = -x$ **(KS3 level 5, GCSE grade E)**

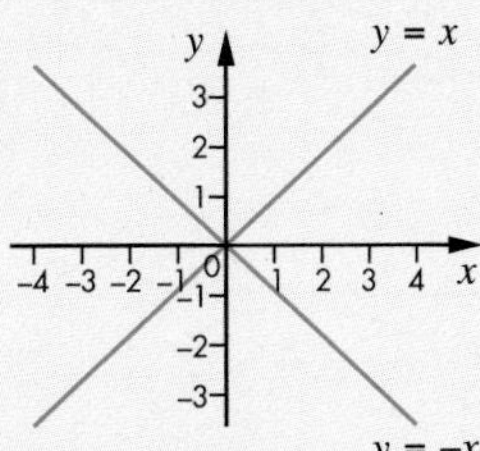

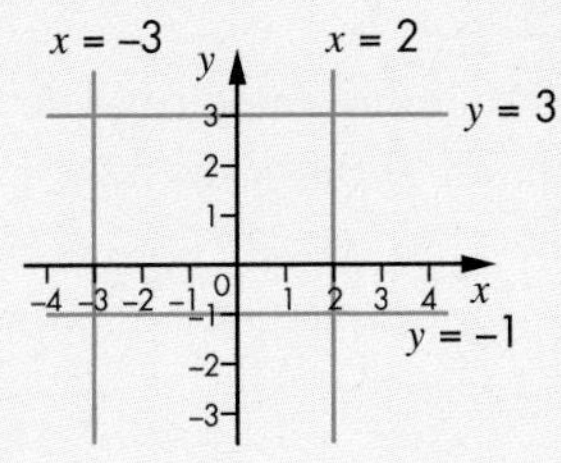

Quick check

Which of these shapes is not congruent to the others?

a 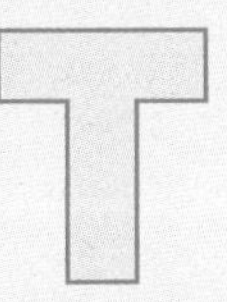

b 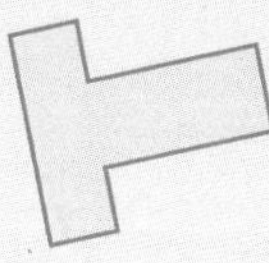

c 

d

11.1 Congruent triangles

This section will show you how to:

- show that two triangles are congruent

Key words
congruent

Two shapes are **congruent** if they are exactly the same size and shape.

For example, these triangles are all congruent.

Notice that the triangles can be differently oriented (reflected or rotated).

Conditions for congruent triangles

Any one of the following four conditions is sufficient for two triangles to be congruent.

- **Condition 1**

  All three sides of one triangle are equal to the corresponding sides of the other triangle.

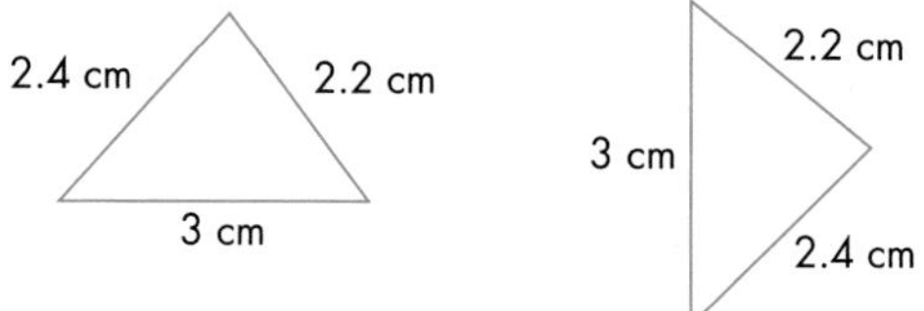

  This condition is known as SSS (side, side, side).

- **Condition 2**

  Two sides and the angle between them of one triangle are equal to the corresponding sides and angle of the other triangle.

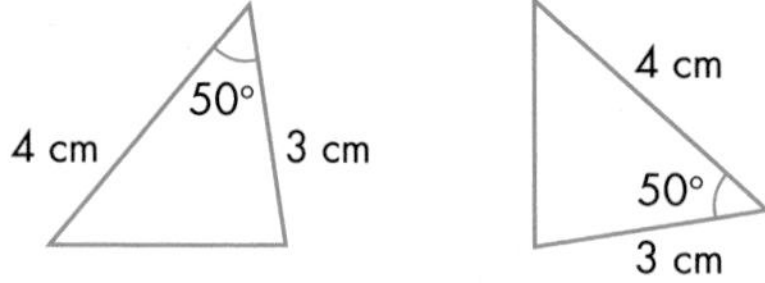

  This condition is known as SAS (side, angle, side).

- **Condition 3**

  Two angles and a side of one triangle are equal to the corresponding angles and side of the other triangle.

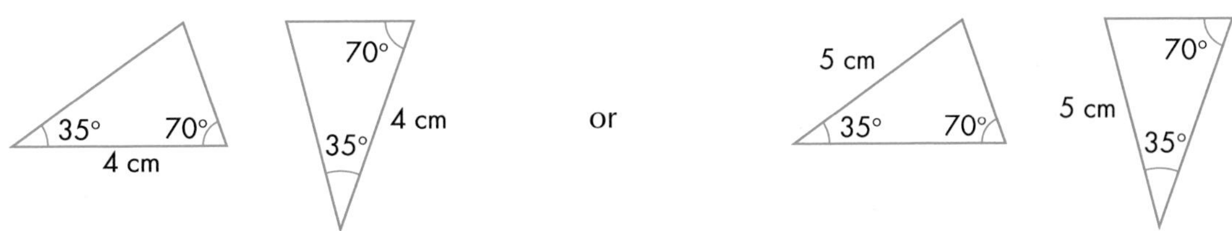

  This condition is known as ASA (angle, side, angle) or AAS (angle, angle, side).

- **Condition 4**

  Both triangles have a right angle, an equal hypotenuse and another equal side.

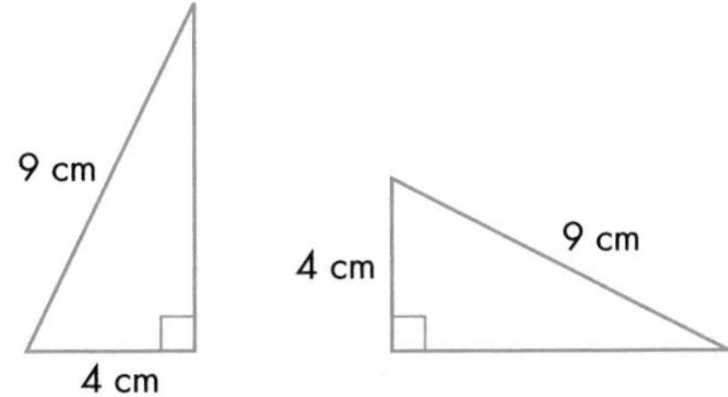

  This condition is known as RHS (right angle, hypotenuse, side).

Notation

Once you have shown that triangle ABC is congruent to triangle PQR by one of the above conditions, it means that:

∠A = ∠P AB = PQ

∠B = ∠Q BC = QR

∠C = ∠R AC = PR

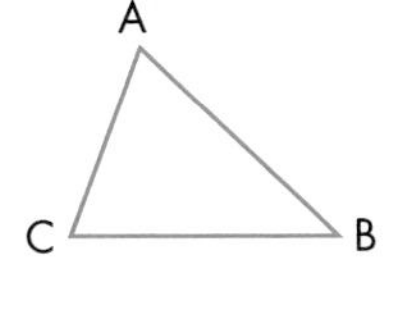

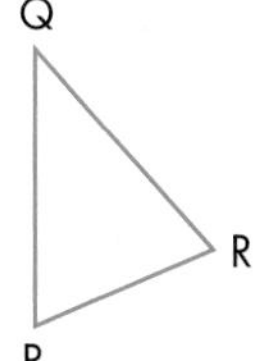

In other words, the points ABC correspond exactly to the points PQR in that order. Triangle ABC is congruent to triangle PQR can be written as ΔABC ≡ ΔPQR.

EXAMPLE 1

ABCD is a kite. Show that triangle ABC is congruent to triangle ADC.

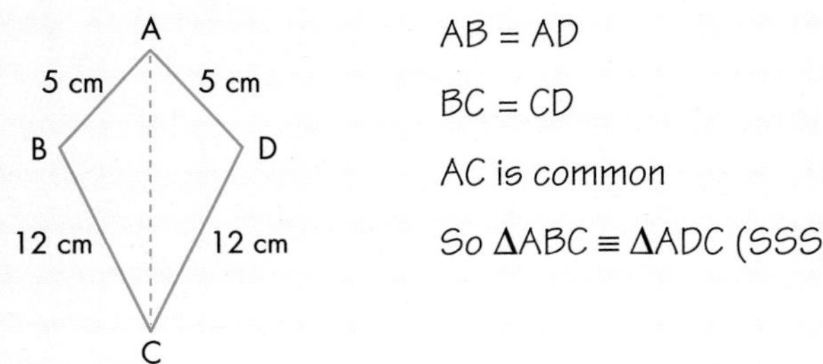

AB = AD

BC = CD

AC is common

So ΔABC ≡ ΔADC (SSS)

EXERCISE 11A

B

1 The triangles in each pair are congruent. State the condition that shows that the triangles are congruent.

a

5 cm
100°
3 cm

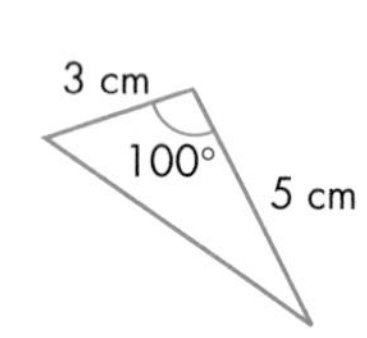

b

5 cm
4 cm
7 cm

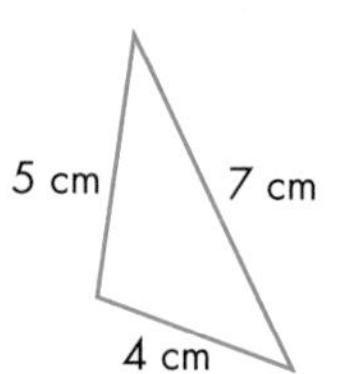

c

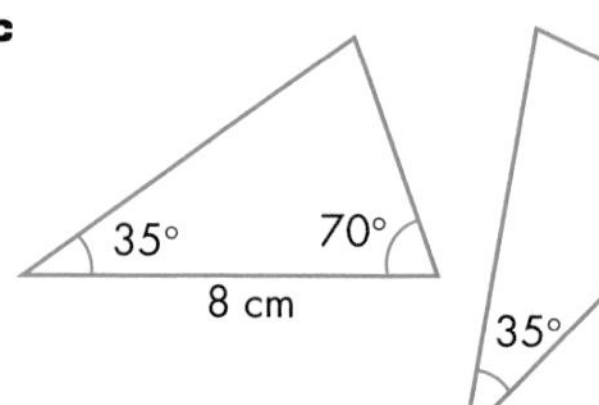

70°
8 cm
35°

d

7 cm
5 cm

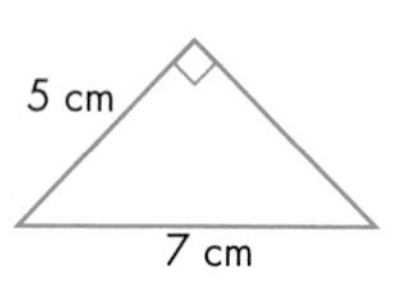

e

4 cm
6 cm
6.5 cm

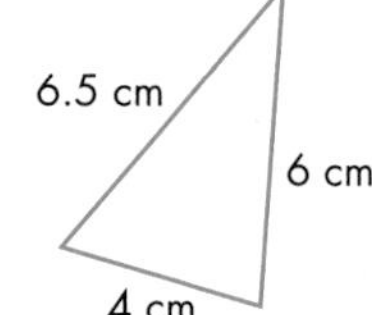

f

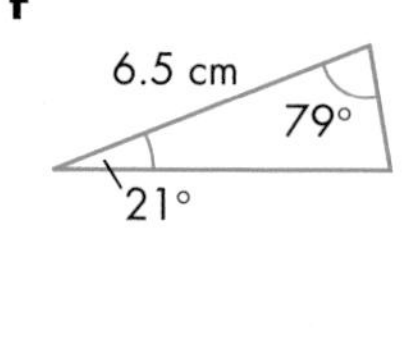

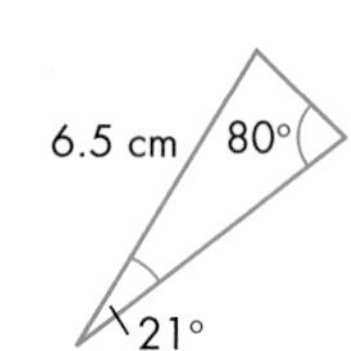

2 The triangles in each pair are congruent. State the condition that shows that the triangles are congruent and say which points correspond to which.

a ABC where AB = 8 cm, BC = 9 cm, AC = 7.4 cm
PQR where PQ = 9 cm, QR = 7.4 cm, PR = 8 cm

b ABC where AB = 5 cm, BC = 6 cm, angle B = 35°
PQR where PQ = 6 cm, QR = 50 mm, angle Q = 35°

3 Triangle ABC is congruent to triangle PQR, ∠A = 60°, ∠B = 80° and AB = 5 cm. Find these.

a ∠P **b** ∠Q **c** ∠R **d** PQ

4 ABCD is congruent to PQRS, ∠A= 110°, ∠B = 55°, ∠C = 85° and RS = 4 cm. Find these.

a ∠P **b** ∠Q **c** ∠R **d** ∠S **e** CD

A

5 Draw a rectangle EFGH. Draw in the diagonal EG. Prove that triangle EFG is congruent to triangle EHG.

6 Draw an isosceles triangle ABC where AB = AC. Draw the line from A to X, the midpoint of BC. Prove that triangle ABX is congruent to triangle ACX.

PS 7 In the diagram ABCD and DEFG are squares.
Use congruent triangles to prove that AE = CG.

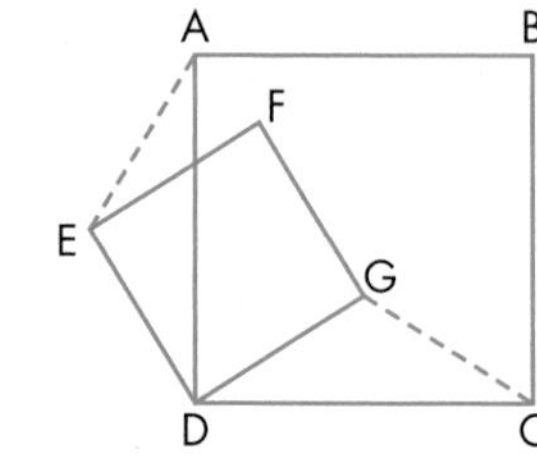

AU 8 Jez says that these two triangles are congruent because two angles and a side are the same.

Explain why he is wrong.

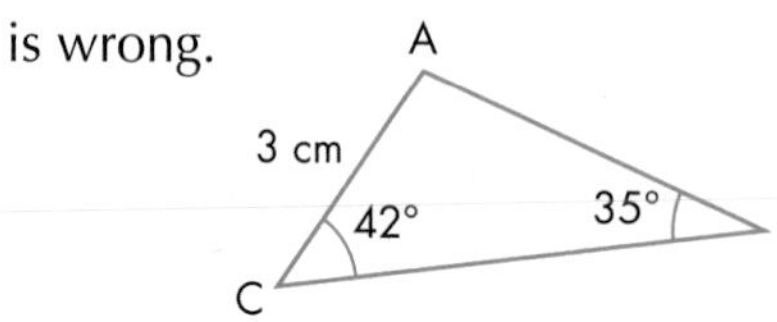

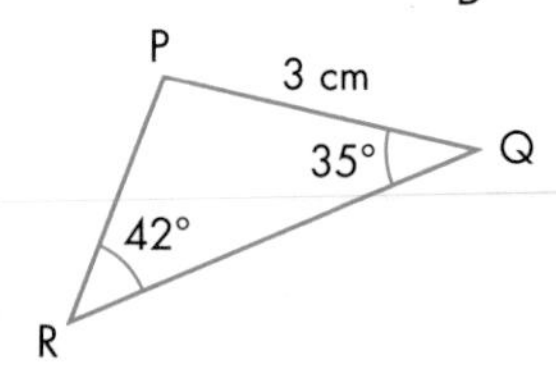

11.2 Translations

This section will show you how to:

- translate a 2D shape

Key words
transformation
translation
vector

A **transformation** changes the position or the size of a shape.

There are four basic ways of changing the position and size of 2D shapes: a **translation**, a reflection, a rotation or an enlargement. All of these transformations, except enlargement, keep shapes congruent.

A translation is the 'movement' of a shape from one place to another without reflecting it or rotating it. It is sometimes called a glide, since the shape appears to glide from one place to another. Every point in the shape moves in the same direction and through the same distance.

We describe translations by using **vectors**. A vector is represented by the combination of a horizontal shift and a vertical shift.

EXAMPLE 2

Use vectors to describe the translations of the following triangles.

a A to B

b B to C

c C to D

d D to A

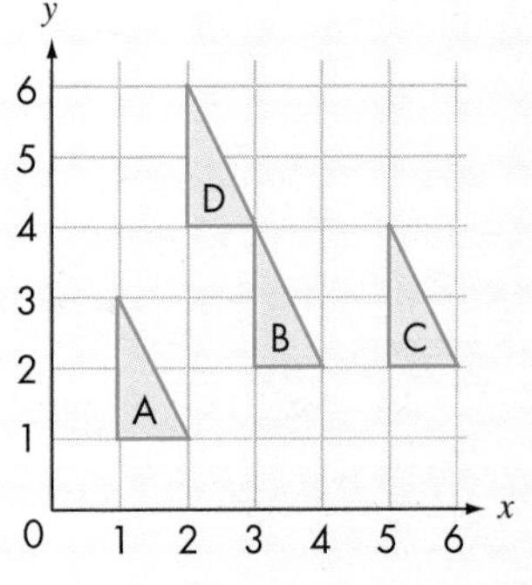

a The vector describing the translation from A to B is $\begin{pmatrix} 2 \\ 1 \end{pmatrix}$.

b The vector describing the translation from B to C is $\begin{pmatrix} 2 \\ 0 \end{pmatrix}$.

c The vector describing the translation from C to D is $\begin{pmatrix} -3 \\ 2 \end{pmatrix}$.

d The vector describing the translation from D to A is $\begin{pmatrix} -1 \\ 3 \end{pmatrix}$.

Note:

- The top number in the vector describes the horizontal movement. To the right +, to the left –.
- The bottom number in the vector describes the vertical movement. Upwards +, downwards –.
- These vectors are also called *direction vectors*.

EXERCISE 11B

C

1 Use vectors to describe the following translations of the shapes on the grid below.

a **i** A to B **ii** A to C **iii** A to D
iv A to E **v** A to F **vi** A to G

b **i** B to A **ii** B to C **iii** B to D
iv B to E **v** B to F **vi** B to G

c **i** C to A **ii** C to B **iii** C to D
iv C to E **v** C to F **vi** C to G

d **i** D to E **ii** E to B **iii** F to C
iv G to D **v** F to G **vi** G to E

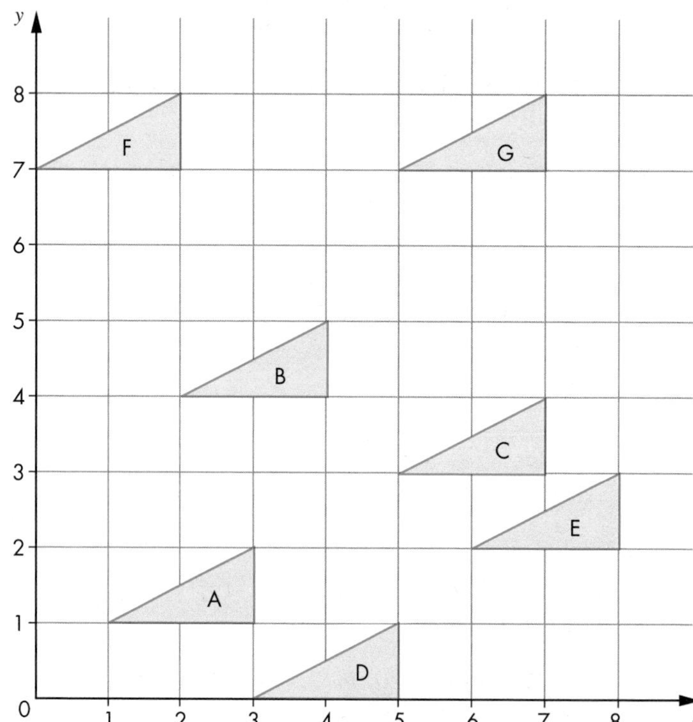

2 **a** Draw a set of coordinate axes and on it the triangle with coordinates A(1, 1), B(2, 1) and C(1, 3).

b Draw the image of ABC after a translation with vector $\begin{pmatrix} 2 \\ 3 \end{pmatrix}$. Label this triangle P.

c Draw the image of ABC after a translation with vector $\begin{pmatrix} -1 \\ 2 \end{pmatrix}$. Label this triangle Q.

d Draw the image of ABC after a translation with vector $\begin{pmatrix} 3 \\ -2 \end{pmatrix}$. Label this triangle R.

e Draw the image of ABC after a translation with vector $\begin{pmatrix} -2 \\ -4 \end{pmatrix}$. Label this triangle S.

C

3 Using your diagram from question **2**, use vectors to describe the translation that will move:

a P to Q **b** Q to R **c** R to S **d** S to P

e R to P **f** S to Q **g** R to Q **h** P to S

PS **4** Draw a 10 × 10 coordinate grid and on it the triangle A(0, 0), B(1, 0) and C(0, 1). How many different translations are there that use integer values only and will move the triangle ABC to somewhere in the grid?

PS **5** In a game of *Snakes and ladders*, each of the snakes and ladders can be described by a translation.

Use the following vectors.

Ladders $\begin{pmatrix} 1 \\ 2 \end{pmatrix}, \begin{pmatrix} 2 \\ 5 \end{pmatrix}, \begin{pmatrix} -3 \\ 4 \end{pmatrix}, \begin{pmatrix} -2 \\ 3 \end{pmatrix}, \begin{pmatrix} 3 \\ 2 \end{pmatrix}$

Snakes $\begin{pmatrix} 1 \\ -3 \end{pmatrix}, \begin{pmatrix} 3 \\ -4 \end{pmatrix}, \begin{pmatrix} -2 \\ -2 \end{pmatrix}, \begin{pmatrix} -1 \\ -3 \end{pmatrix}, \begin{pmatrix} 2 \\ -5 \end{pmatrix}$

Put all five ladders and all five snakes onto a 10 × 10 coordinate grid in order to design a *Snakes and ladders* game board.

AU **6** If a translation is given by:

$$\begin{pmatrix} x \\ y \end{pmatrix}$$

describe the translation that would take the image back to the original position.

FM **7** A plane flies between three cities A, B and C. It uses direction vectors, with distances in kilometres.

The direction vector for the flight from A to B is $\begin{pmatrix} 500 \\ 200 \end{pmatrix}$ and the direction vector for the flight from B to C is $\begin{pmatrix} -200 \\ 300 \end{pmatrix}$.

Using centimetre-squared paper, draw a diagram to show the three flights. Use a scale of 1 cm represents 100 km.

Work out the direction vector for the flight from C to A.

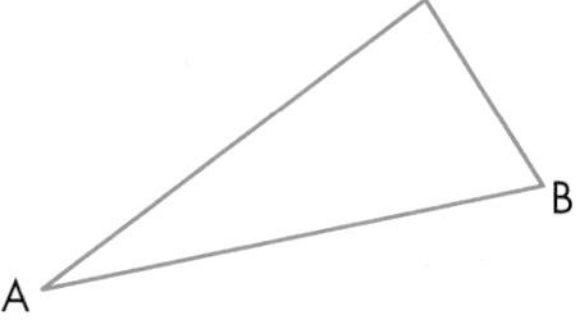

FM **8** A pleasure cruise travels between three jetties X, Y and Z on a lake. It uses direction vectors, with distance in kilometres.

The direction vector from X to Y is $\begin{pmatrix} 3 \\ -1 \end{pmatrix}$ and the direction vector from Y to Z is $\begin{pmatrix} -2 \\ -3 \end{pmatrix}$.

Using centimetre-squared paper, draw a diagram to show journeys between X, Y and Z. Use a scale of 1 cm represents 1 km. Work out the direction vector for the journey from Z to X.

11.3 Reflections

This section will show you how to:
- reflect a 2D shape in a mirror line

Key words
image
mirror line
object
reflection

A **reflection** transforms a shape so that it becomes a mirror image of itself.

EXAMPLE 3

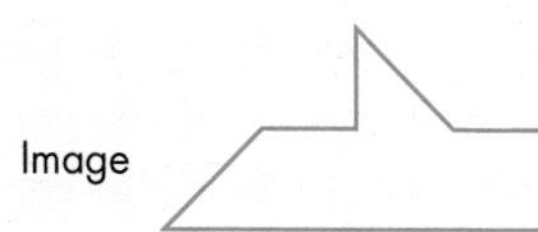

Notice the reflection of each point in the original shape, called the **object**, is perpendicular to the mirror line. So if you 'fold' the whole diagram along the **mirror line**, the object will coincide with its reflection, called its **image**.

EXERCISE 11C

D

1 Copy the diagram below and draw the reflection of the given triangle in the following lines.

a $x = 2$ **b** $x = -1$ **c** $x = 3$

d $y = 2$ **e** $y = -1$ **f** y-axis

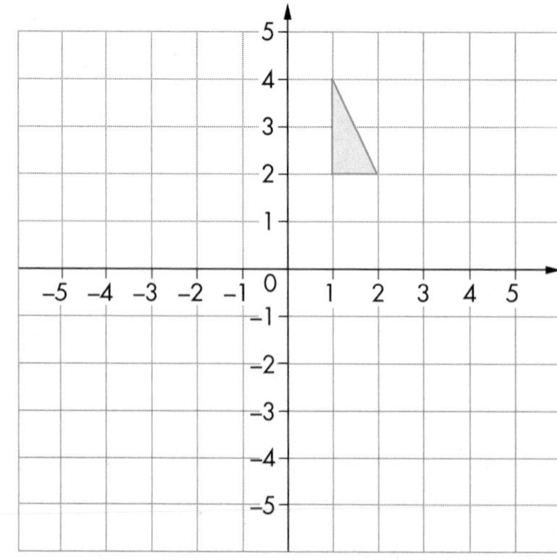

D

2 **a** Draw a pair of axes. Label the x-axis from –5 to 5 and the y-axis from –5 to 5.

b Draw the triangle with coordinates A(1, 1), B(3, 1), C(4, 5).

c Reflect the triangle ABC in the x-axis. Label the image P.

d Reflect triangle P in the y-axis. Label the image Q.

e Reflect triangle Q in the x-axis. Label the image R.

f Describe the reflection that will move triangle ABC to triangle R.

AU **3** **a** Draw a pair of axes. Label the x-axis from –5 to +5 and the y-axis from –5 to +5.

b Reflect the points A(2, 1), B(5, 0), C(–3, 3), D(3, –2) in the x-axis.

c What do you notice about the values of the coordinates of the reflected points?

d What would the coordinates of the reflected point be if the point (a, b) were reflected in the x-axis?

AU **4** **a** Draw a pair of axes. Label the x-axis from –5 to +5 and the y-axis from –5 to +5.

b Reflect the points A(2, 1), B(0, 5), C(3, –2), D(–4, –3) in the y-axis.

c What do you notice about the values of the coordinates of the reflected points?

d What would the coordinates of the reflected point be if the point (a, b) were reflected in the y-axis?

PS **5** By using the middle square as a starting square ABCD, describe how to keep reflecting the square to obtain the final shape in the diagram.

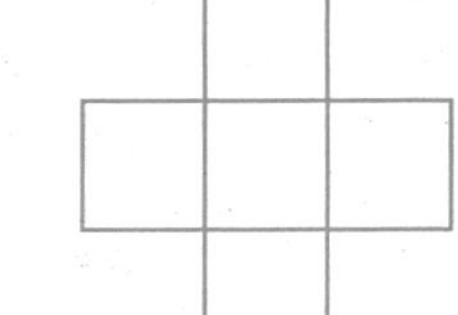

AU **6** Triangle A is drawn on a grid.

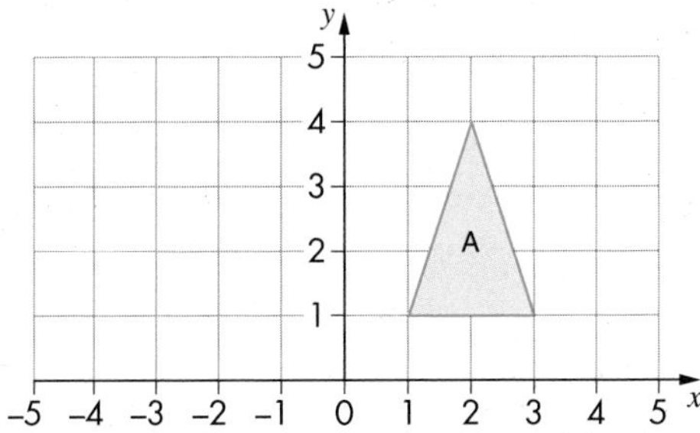

Triangle A is reflected to form a new triangle B.
The coordinates of B are (–4, 4), (–3, 1) and (–5, 1).

Work out the equation of the mirror line.

7 A designer used the following instructions to create a design.

- Start with any rectangle ABCD.
- Reflect the rectangle ABCD in the line AC.
- Reflect the rectangle ABCD in the line BD.

Draw a rectangle and use the above to create a design.

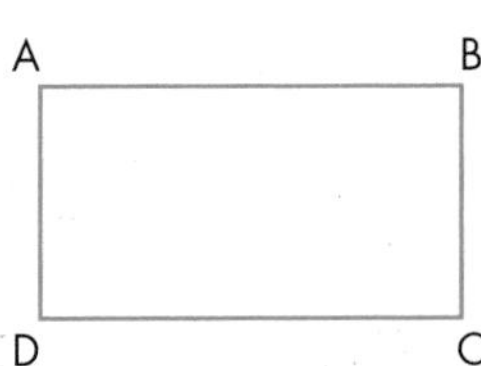

C

8 Draw each of these triangles on squared paper, leaving plenty of space on the opposite side of the given mirror line. Then draw the reflection of each triangle.

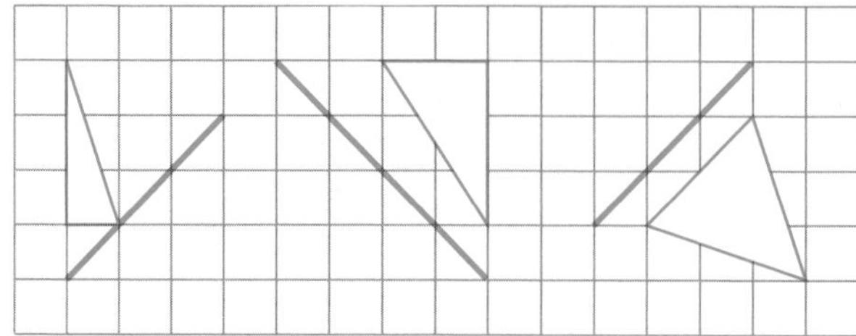

HINTS AND TIPS

Turn the page around so that the mirror lines are vertical or horizontal.

9 **a** Draw a pair of axes and the lines $y = x$ and $y = -x$, as shown.

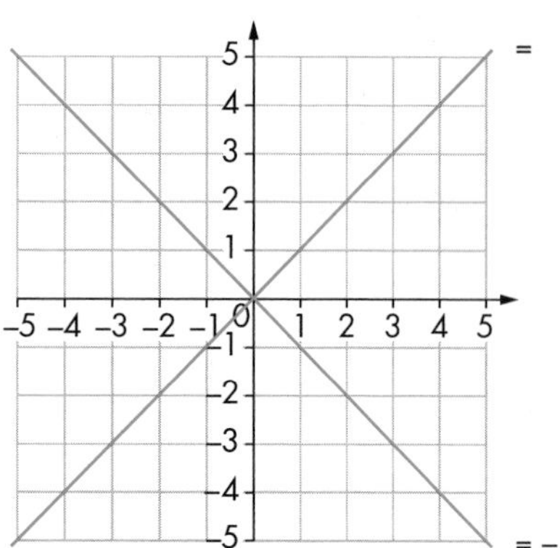

b Draw the triangle with coordinates A(2, 1), B(5, 1), C(5, 3).

c Draw the reflection of triangle ABC in the x-axis and label the image P.

d Draw the reflection of triangle P in the line $y = -x$ and label the image Q.

e Draw the reflection of triangle Q in the y-axis and label the image R.

f Draw the reflection of triangle R in the line $y = x$ and label the image S.

g Draw the reflection of triangle S in the x-axis and label the image T.

h Draw the reflection of triangle T in the line $y = -x$ and label the image U.

i Draw the reflection of triangle U in the y-axis and label the image W.

j What single reflection will move triangle W to triangle ABC?

10 Copy the diagram and reflect the triangle in these lines.

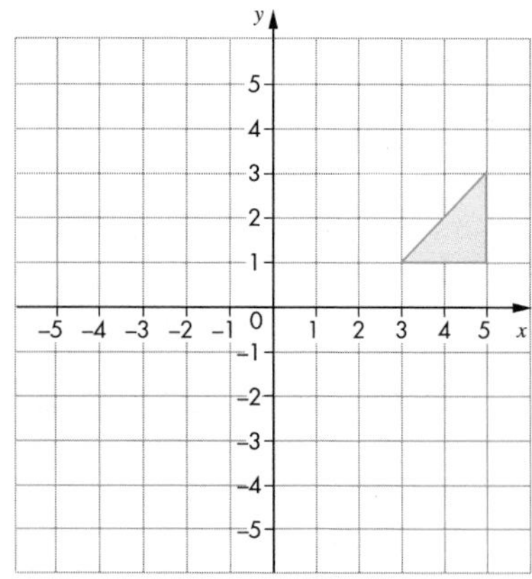

a $y = x$ **b** $x = 1$

c $y = -x$ **d** $y = -1$

PS **11** **a** Draw a pair of axes. Label the x-axis from –5 to +5 and the y-axis from –5 to +5.

b Draw the line $y = x$.

c Reflect the points A(2, 1), B(5, 0), C(–3, 2), D(–2, –4) in the line $y = x$.

d What do you notice about the values of the coordinates of the reflected points?

e What would the coordinates of the reflected point be if the point (a, b) were reflected in the line $y = x$?

12 **a** Draw a pair of axes. Label the x-axis from –5 to +5 and the y-axis from –5 to +5.

b Draw the line $y = -x$.

c Reflect the points A(2, 1), B(0, 5), C(3, –2), D(–4, –3) in the line $y = -x$.

d What do you notice about the values of the coordinates of the reflected points?

e What would the coordinates of the reflected point be if the point (a, b) were reflected in the line $y = -x$?

C

11.4 Rotations

This section will show you how to:

- rotate a 2D shape about a point

Key words

angle of rotation
anticlockwise
centre of rotation
clockwise
rotation

A **rotation** transforms a shape to a new position by turning it about a fixed point called the **centre of rotation**.

EXAMPLE 4

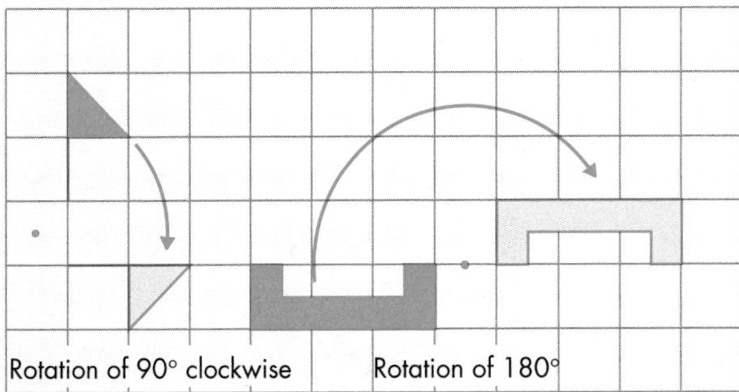

Note:

- The direction of turn or the **angle of rotation** is expressed as **clockwise** or **anticlockwise**.
- The position of the centre of rotation is always specified.
- The rotations 180° clockwise and 180° anticlockwise are the same.

The rotations that most often appear in examination questions are 90° and 180°.

EXERCISE 11D

D

1 On squared paper, draw each of these shapes and its centre of rotation, leaving plenty of space all round the shape.

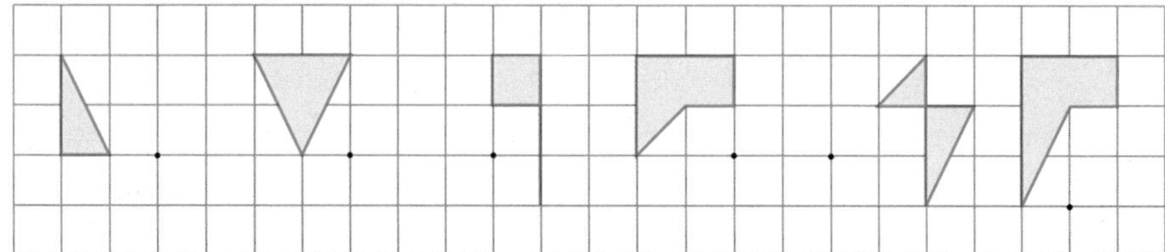

a Rotate each shape about its centre of rotation:

i first by 90° clockwise (call the image A)

ii then by 90° anticlockwise (call the image B).

b Describe, in each case, the rotation that would take:

i A back to its original position **ii** A to B.

2 A graphics designer came up with the following routine for creating a design.

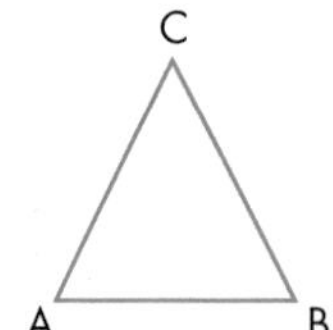

- Start with a triangle ABC.
- Reflect the triangle in the line AB.
- Rotate the whole shape about point C clockwise 90°, then a further clockwise 90°, then a further clockwise 90°.

From any triangle of your choice, create a design using the above routine.

PS **3** By using the middle square as a starting square ABCD, describe how to keep rotating the square to obtain the final shape in the diagram.

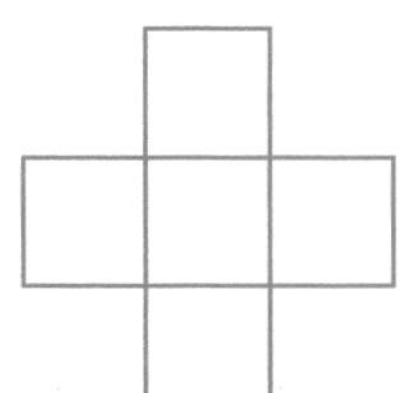

C

4 Copy the diagram and rotate the given triangle by the following.

a 90° clockwise about (0, 0)

b 180° about (3, 3)

c 90° anticlockwise about (0, 2)

d 180° about (–1, 0)

e 90° clockwise about (–1, –1)

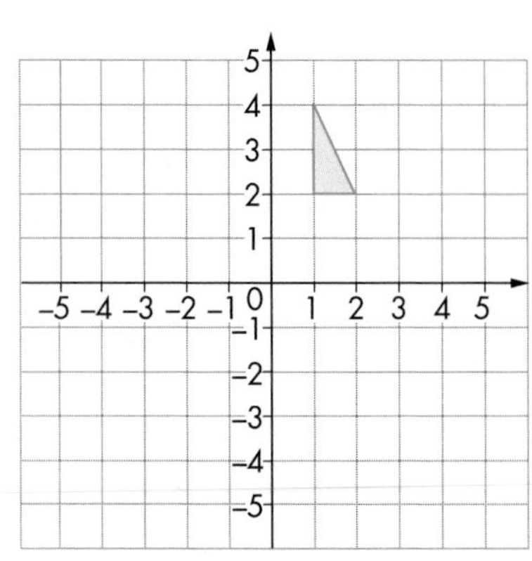

5 What other rotations are equivalent to these rotations?

a 270° clockwise **b** 90° clockwise

c 60° anticlockwise **d** 100° anticlockwise

6 **a** Draw a pair of axes where both the x-values and y-values are from –5 to 5.

b Draw the triangle ABC, where A = (1, 2), B = (2, 4) and C = (4, 1).

c **i** Rotate triangle ABC 90° clockwise about the origin (0, 0) and label the image A′, B′, C′, where A′ is the image of A, etc.

ii Write down the coordinates of A′, B′, C′.

iii What connection is there between A, B, C and A′, B′, C′?

iv Will this connection always be so for a 90° clockwise rotation about the origin?

7 Repeat question **6**, but rotate triangle ABC through 180°.

8 Repeat question **6**, but rotate triangle ABC 90° anticlockwise.

PS **9** Show that a reflection in the x-axis followed by a reflection in the y-axis is equivalent to a rotation of 180° about the origin.

PS **10** Show that a reflection in the line $y = x$ followed by a reflection in the line $y = -x$ is equivalent to a rotation of 180° about the origin.

11 **a** Draw a regular hexagon ABCDEF with centre O.

b Using O as the centre of rotation, describe a transformation that will result in the following movements.

i Triangle AOB to triangle BOC **ii** Triangle AOB to triangle COD

iii Triangle AOB to triangle DOE **iv** Triangle AOB to triangle EOF

c Describe the transformations that will move the rhombus ABCO to these positions.

i Rhombus BCDO **ii** Rhombus DEFO

AU **12** Triangle A, as shown on the grid, is rotated to form a new triangle B.

The coordinates of the vertices of B are (0, –2), (–3, –2) and (–3, –4).

Describe fully the rotation that maps triangle A onto triangle B.

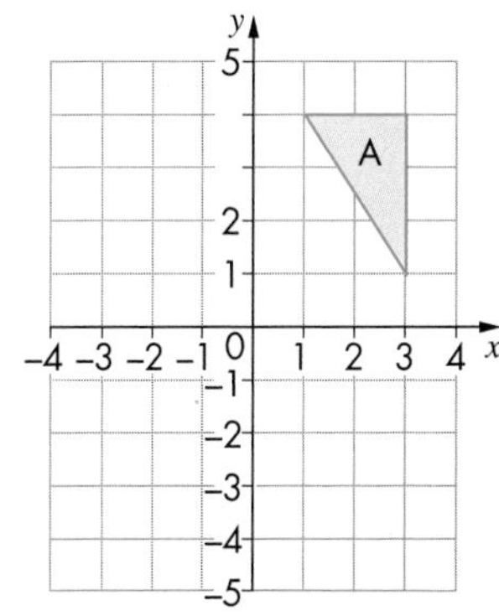

11.5 Enlargements

This section will show you how to:

- enlarge a 2D shape by a scale factor

Key words
centre of enlargement
enlargement
scale factor

An **enlargement** changes the size of a shape to give a similar image. It always has a **centre of enlargement** and a **scale factor**. Every length of the enlarged shape will be:

original length × scale factor

The distance of each image point on the enlargement from the centre of enlargement will be:

distance of original point from centre of enlargement × scale factor

EXAMPLE 5

The diagram shows the enlargement of triangle ABC by scale factor 3 about the centre of enlargement X.

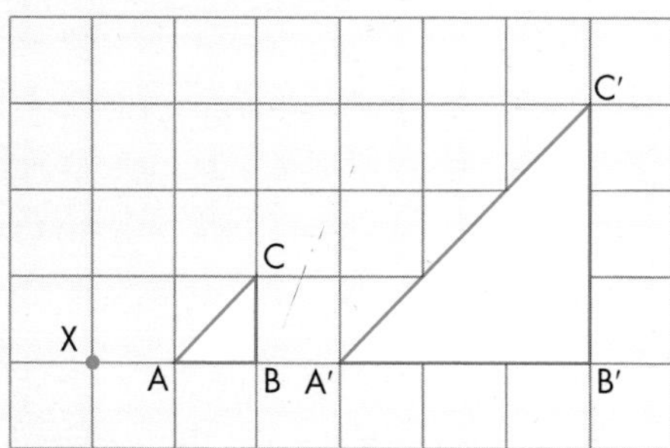

Note:

- Each length on the enlargement A′B′C′ is three times the corresponding length on the original shape.

  This means that the corresponding sides are in the same ratio:

  AB : A′B′ = AC : A′C′ = BC : B′C′ = 1 : 3
- The distance of any point on the enlargement from the centre of enlargement is three times the distance from the corresponding point on the original shape to the centre of enlargement.

There are two distinct ways to enlarge a shape: the ray method and the coordinate, or counting squares, method.

Ray method

This is the *only* way to construct an enlargement when the diagram is not on a grid.

EXAMPLE 6

Enlarge triangle ABC by scale factor 3 about the centre of enlargement X.

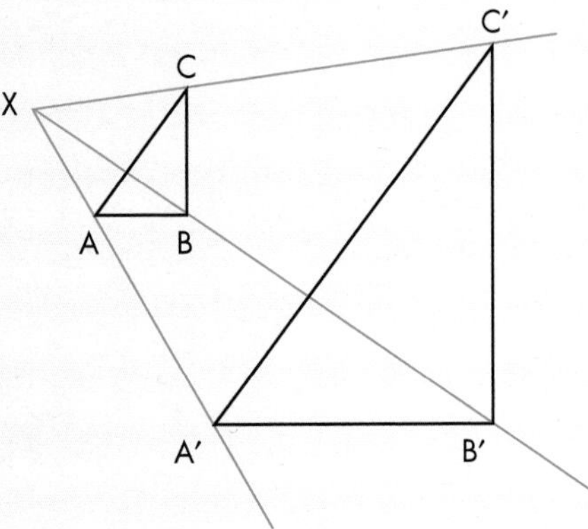

Notice that the rays have been drawn from the centre of enlargement to each vertex and beyond.

The distance from X to each vertex on triangle ABC is measured and multiplied by 3 to give the distance from X to each vertex A′, B′ and C′ for the enlarged triangle A′B′C′.

Once each image vertex has been found, the whole enlarged shape can then be drawn.

Check the measurements and see for yourself how the calculations have been done.

Notice again that the length of each side on the enlarged triangle is three times the length of the corresponding side on the original triangle.

Counting squares method

In this method, you use the coordinates of the vertices to 'count squares'.

EXAMPLE 7

Enlarge the triangle ABC by scale factor 3 from the centre of enlargement (1, 2).

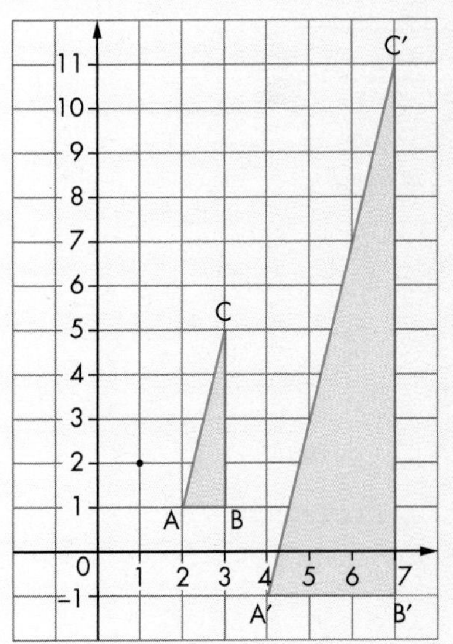

To find the coordinates of each image vertex, first work out the horizontal and vertical distances from each original vertex to the centre of enlargement.

Then multiply each of these distances by 3 to find the position of each image vertex.

For example, to find the coordinates of C′ work out the distance from the centre of enlargement (1, 2) to the point C(3, 5).

horizontal distance = 2

vertical distance = 3

Make these 3 times longer to give:

new horizontal distance = 6

new vertical distance = 9

So the coordinates of C′ are:

$(1 + 6, 2 + 9) = (7, 11)$

Notice again that the length of each side is three times as long in the enlargement.

Negative enlargement

A negative enlargement produces an image shape on the opposite side of the centre of enlargement to the original shape.

EXAMPLE 8

Triangle ABC has been enlarged by scale factor −2, with the centre of enlargement at (1, 0).

You can enlarge triangle ABC to give triangle A′B′C′ by either the ray method or the coordinate method. You calculate the new lengths on the opposite side of the centre of enlargement to the original shape.

Notice how a negative scale factor also inverts the original shape.

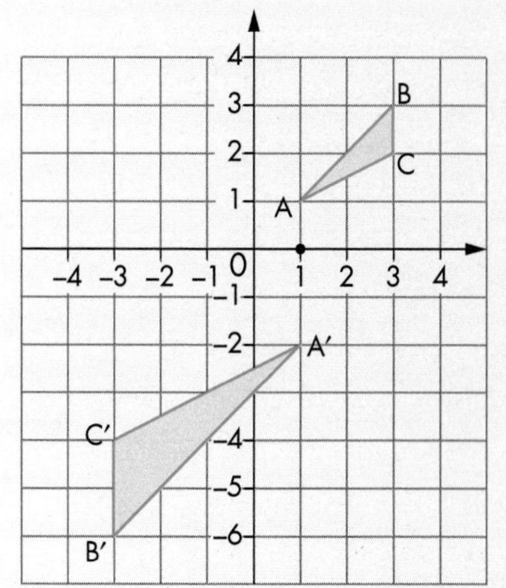

Fractional enlargement

Strange but true – you can have an enlargement in mathematics that is actually smaller than the original shape!

EXAMPLE 9

Triangle ABC has been enlarged by a scale factor of $\frac{1}{2}$ about the centre of enlargement O to give triangle A′B′C′.

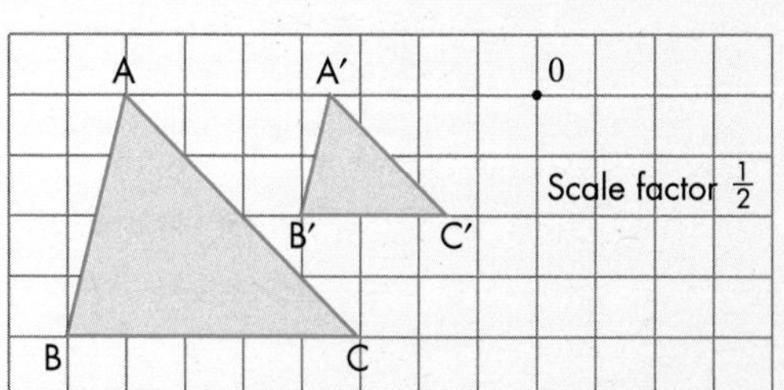

EXERCISE 11E

D

1 Copy each of these figures with its centre of enlargement. Then enlarge it by the given scale factor, using the ray method.

a

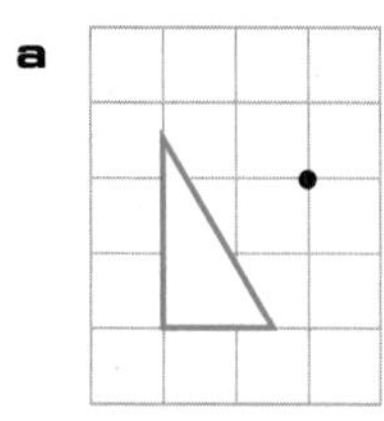

Scale factor 2

b

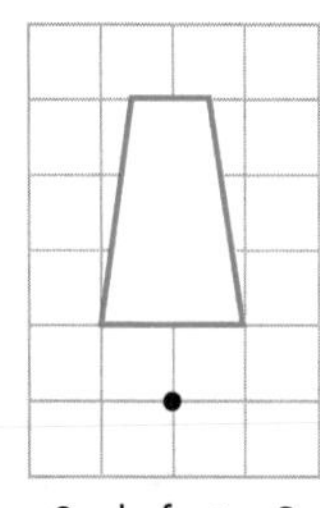

Scale factor 3

c

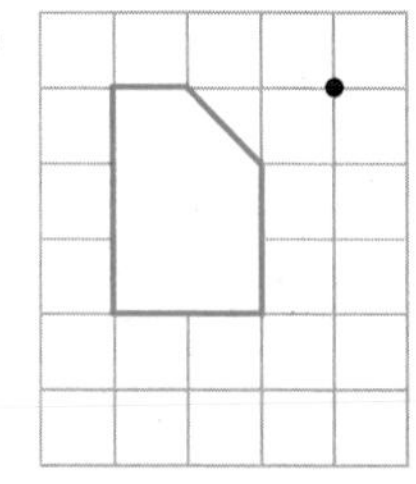

Scale factor 2

d

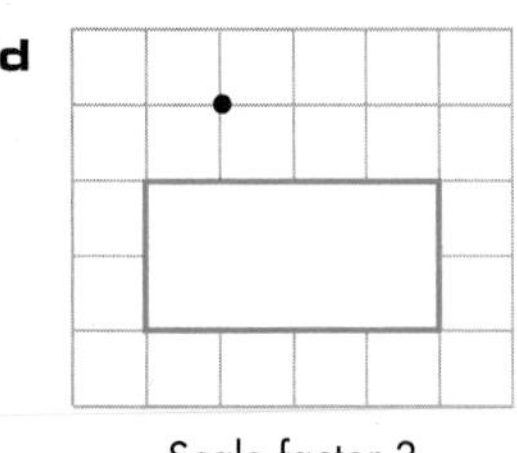

Scale factor 3

D

2 Copy each of these diagrams onto squared paper and enlarge it by scale factor 2, using the origin as the centre of enlargement.

a

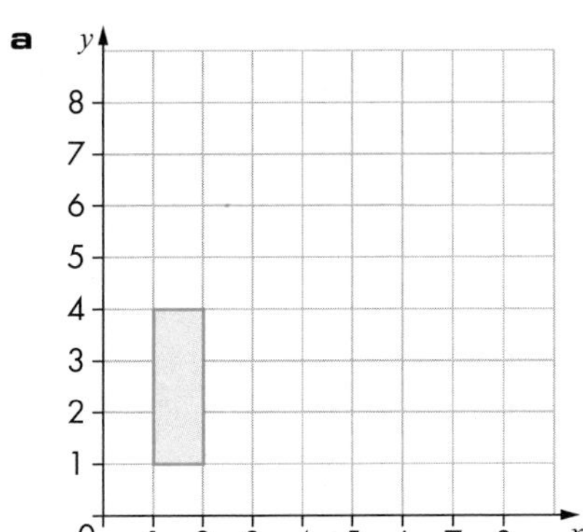

b

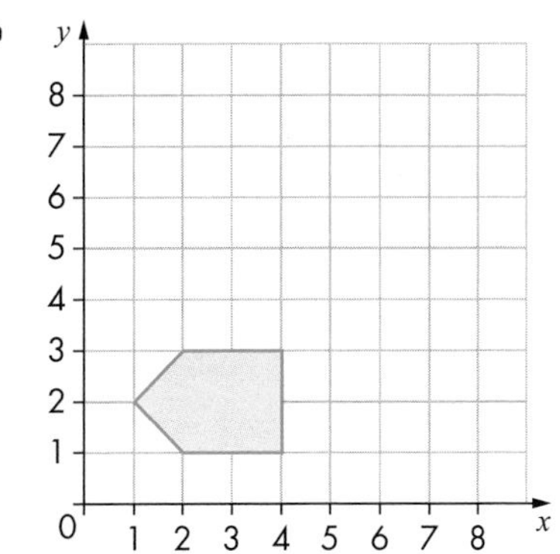

c

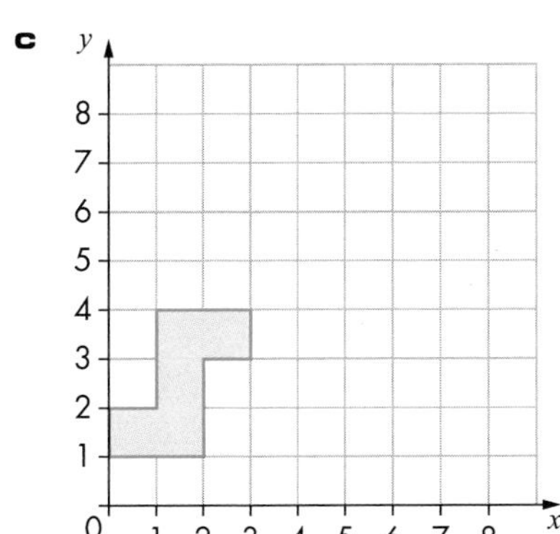

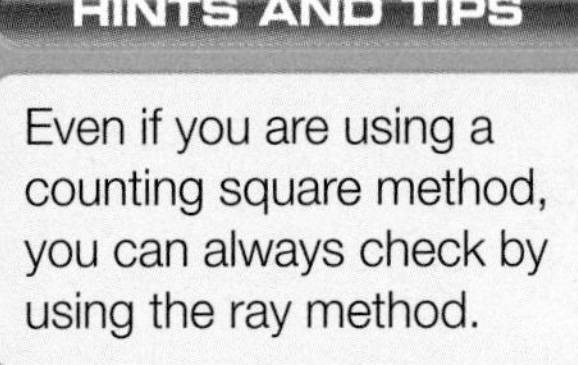

3 Copy each of these diagrams onto squared paper and enlarge it by scale factor 2, using the given centre of enlargement.

a

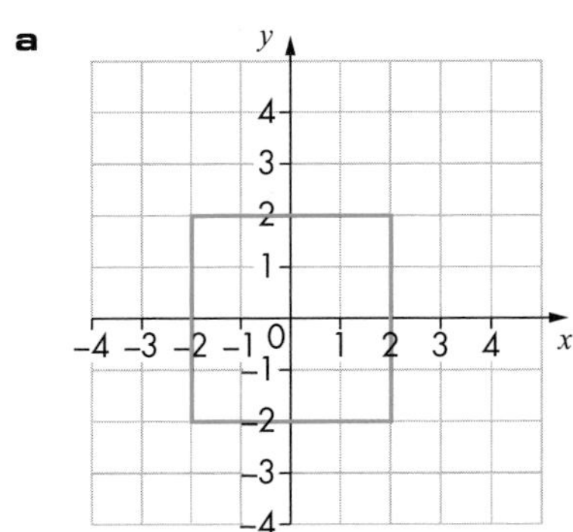

Centre of enlargement (–1,1)

b

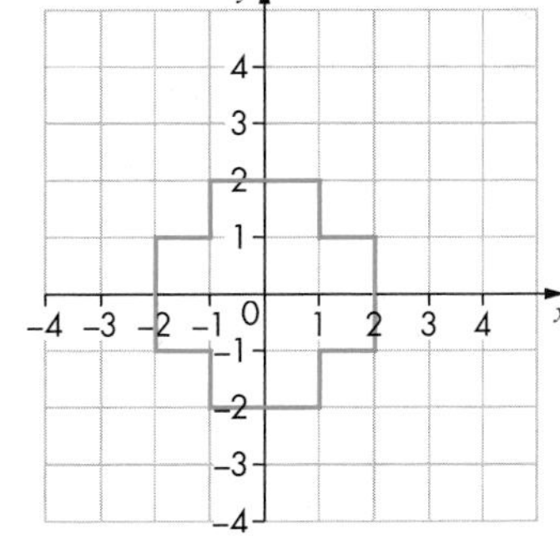

Centre of enlargement (–2,–3)

4 A designer is told to use the following routine.

- Start with a rectangle ABCD.
- Reflect ABCD in the line AC.
- Rotate the whole new shape about C through 180°.
- Enlarge the whole shape scale factor 2, centre of enlargement point A.

Start with any rectangle of your choice and create the design above.

C

5 Enlarge each of these shapes by a scale factor of $\frac{1}{2}$ about the given centre of enlargement.

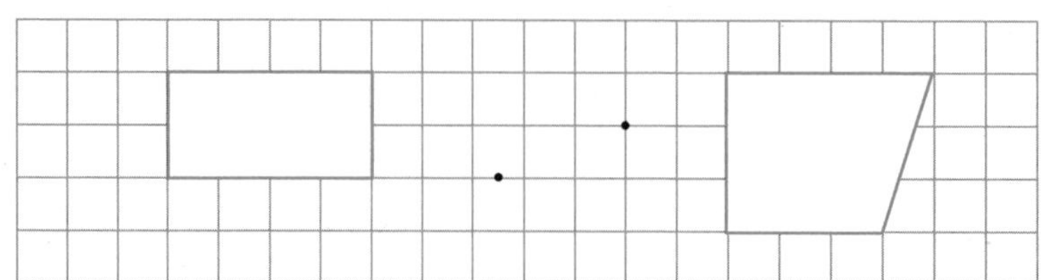

6 Copy this diagram onto squared paper.

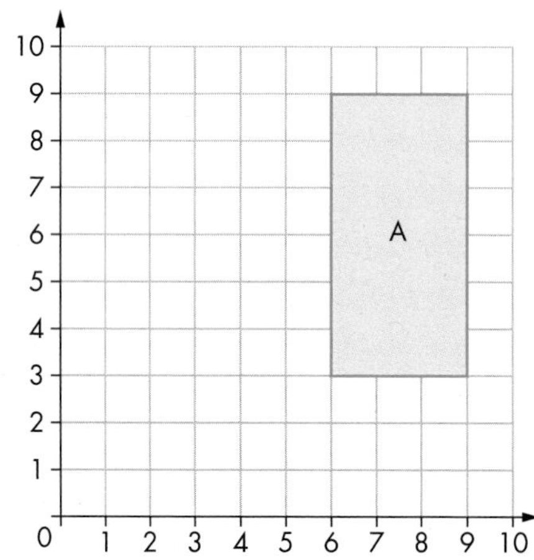

a Enlarge the rectangle A by scale factor $\frac{1}{3}$ about the origin. Label the image B.

b Write down the ratio of the lengths of the sides of rectangle A to the lengths of the sides of rectangle B.

c Work out the ratio of the perimeter of rectangle A to the perimeter of rectangle B.

d Work out the ratio of the area of rectangle A to the area of rectangle B.

B

AU 7 Copy this diagram onto squared paper.

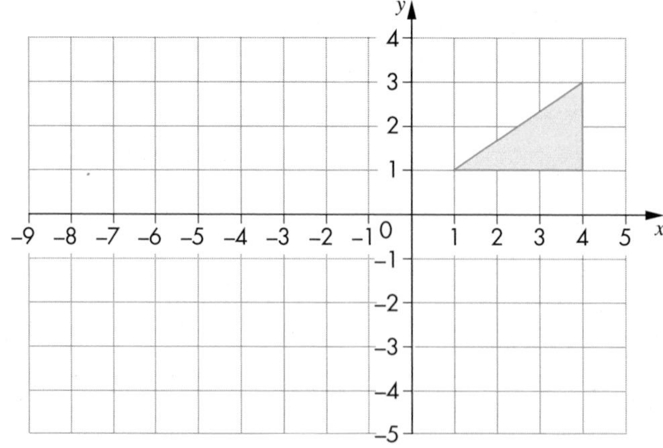

Enlarge the triangle by scale factor –2 about the origin.

B

8 Copy the diagram onto squared paper.

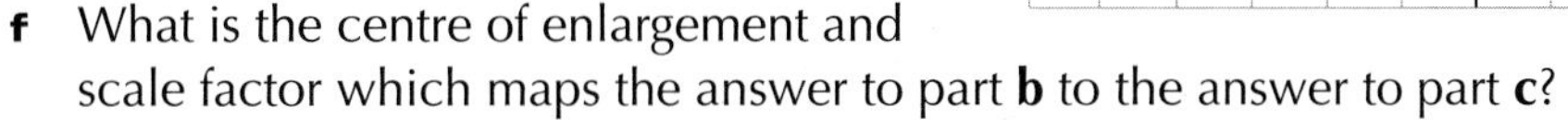

a Enlarge A by a scale factor of 3 about a centre (4, 5).

b Enlarge B by a scale factor $\frac{1}{2}$ about a centre (–1, –3).

c Enlarge B by scale factor $-\frac{1}{2}$ about a centre (–3, –1).

d What is the centre of enlargement and scale factor which maps B onto A?

e What is the centre of enlargement and scale factor which maps A onto B?

f What is the centre of enlargement and scale factor which maps the answer to part **b** to the answer to part **c**?

g What is the centre of enlargement and scale factor which maps the answer to part **c** to the answer to part **b**?

h What is the connection between the scale factors and the centres of enlargement in parts **d** and **e**, and in parts **f** and **g**?

PS **9** Triangle A has vertices with coordinates (2, 1), (4, 1) and (4, 4).
Triangle B has vertices with coordinates (–5, 1), (–5, 7) and (–1, 7).

Describe fully the single transformation that maps triangle A onto triangle B.

11.6 Combined transformations

This section will show you how to:

- combine transformations

Key words
enlargement
reflection
rotation
transformation
translation

Examination questions often require you to use more than one **transformation**. In this exercise, you will revise the transformations you have met so far.

Remember, to describe:

- a **translation** fully, you need to use a vector
- a **reflection** fully, you need to use a mirror line
- a **rotation** fully, you need a centre of rotation, an angle of rotation and the direction of turn
- an **enlargement** fully, you need a centre of enlargement and a scale factor

EXERCISE 11F

D

1 The point P(3, 4) is **reflected** in the x-axis, then rotated by 90° clockwise about the origin. What are the coordinates of the image of P?

2 A point Q(5, 2) is rotated by 180°, then reflected in the x-axis.

a What are the coordinates of the image point of Q?

b What single transformation would have taken point Q directly to the image point?

C

3 Describe fully the transformations that will map the shaded triangle onto each of the triangles A–F.

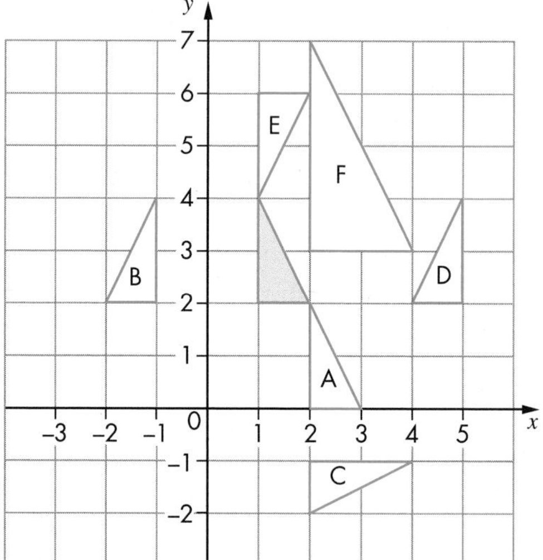

4 Describe fully the transformations that will result in the following movements.

a T_1 to T_2

b T_1 to T_6

c T_2 to T_3

d T_6 to T_2

e T_6 to T_5

f T_5 to T_4

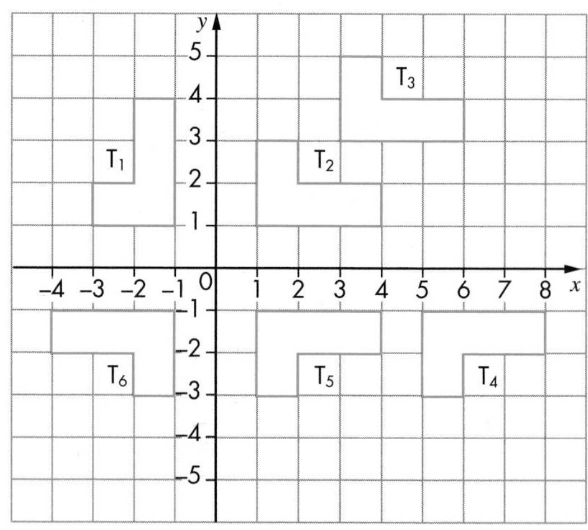

5 **a** Plot a triangle T with vertices (1, 1), (2, 1), (1, 3).

b Reflect triangle T in the y-axis and label the image T_b.

c Rotate triangle T_b 90° anticlockwise about the origin and label the image T_c.

d Reflect triangle T_c in the y-axis and label the image T_d.

e Describe fully the transformation that will move triangle T_d back to triangle T.

6 Find the coordinates of the image of the point (3, 5) after a clockwise rotation of 90° about the point (1, 3).

PS 7 Describe fully at least three different transformations that could move the square labelled S to the square labelled T.

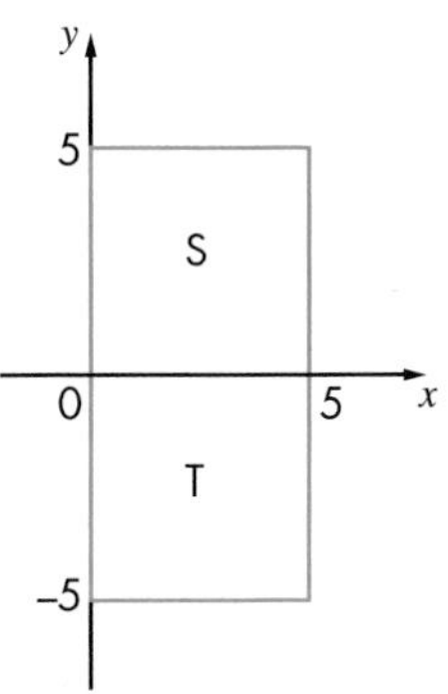

PS 8 The point A(4, 4) has been transformed to the point A′(4, –4). Describe as many different transformations as you can that could transform point A to point A′.

AU 9 Copy the diagram onto squared paper.

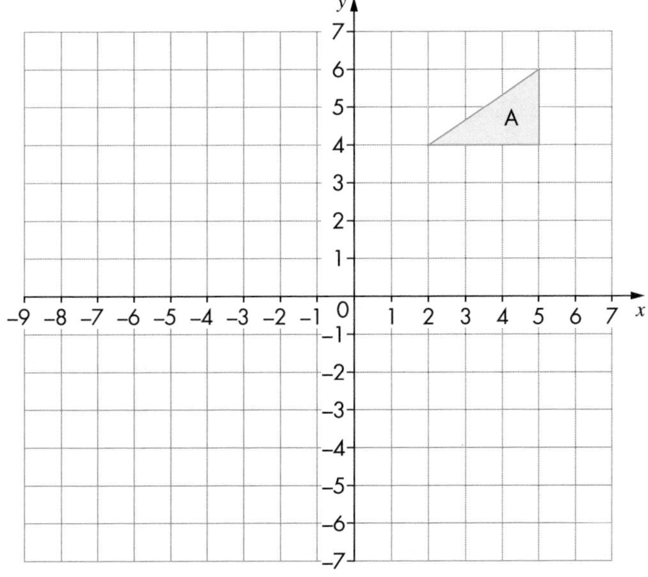

a Triangle A is translated by the vector $\begin{pmatrix} -1.5 \\ -3 \end{pmatrix}$ to give triangle B.

Triangle B is then enlarged by a scale factor –2 about the origin to give triangle C.

Draw triangles B and C on the diagram.

b Describe fully the single transformation that maps triangle C onto triangle A.

GRADE BOOSTER

- **D** You can reflect a 2D shape in a line $x = a$ or $y = b$
- **D** You can rotate a 2D shape about the origin
- **D** You can enlarge a 2D shape by a whole number scale factor
- **C** You can translate a 2D shape by a vector
- **C** You can reflect a 2D shape in the line $y = x$ or $y = -x$
- **C** You can rotate a 2D shape about any point
- **C** You can enlarge a 2D shape by a fractional scale factor
- **C** You can enlarge a 2D shape about any point
- **B** You know the conditions to show two triangles are congruent
- **B** You can enlarge a 2D shape by a negative scale factor
- **A** You can prove two triangles are congruent

What you should know now

- How to translate a 2D shape by a vector
- How to reflect a 2D shape in any line
- How to rotate a 2D shape about any point and through any angle
- How to enlarge a 2D shape about any point using a positive, fractional or negative scale factor
- How to show that two triangles are congruent

1 Enlarge the shape by scale factor 2, using the origin as the centre of enlargement. *(3 marks)*

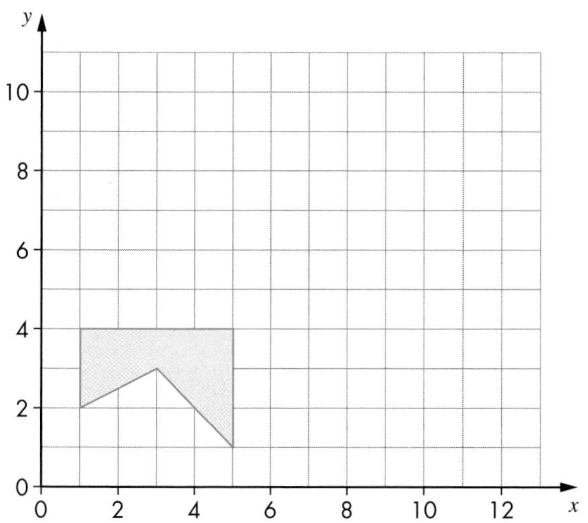

AQA, June 2009, Module 5, Paper 2 Higher, Question 1

2 Triangles A, B and C are shown on the grid.

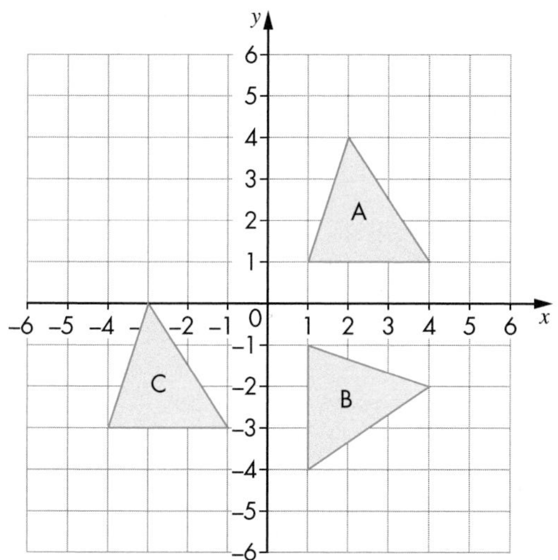

a Describe fully the **single** transformation that maps triangle A onto triangle B. *(3 marks)*

b Write down the vector which describes the translation of triangle A onto triangle C. *(1 mark)*

AQA, May 2009, Paper 1, Question 10(a)(b)

3 The diagram shows two rectangles A and B.

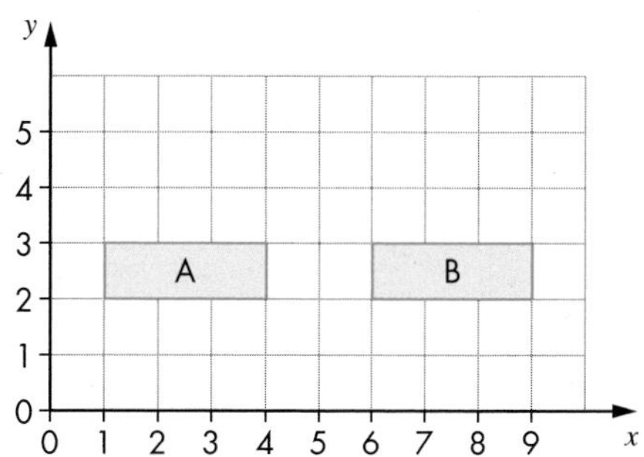

Complete the sentences.

a Rectangle B is a reflection of rectangle A in the line … *(1 mark)*

b Rectangle B is a translation of rectangle A by the vector… *(1 mark)*

c Rectangle B is a rotation of rectangle A through … degrees about the point … *(2 marks)*

AQA, June 2007, Module 5, Paper 1 Higher, Question 1(a)(b)(c)

4

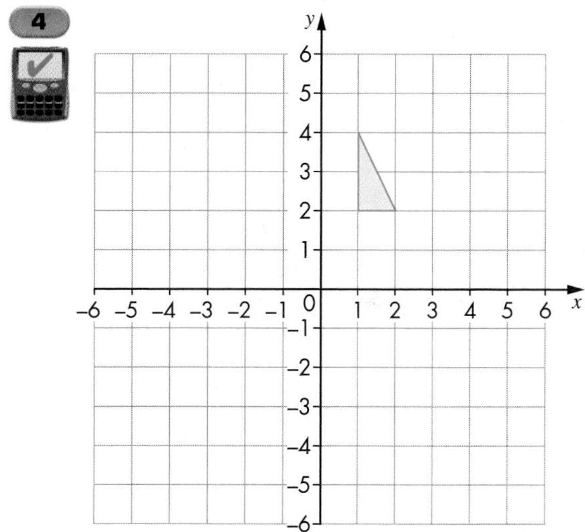

a Reflect the shaded triangle in the line $y = -x$.

Label this new triangle with the letter A
. *(2 marks)*

b Rotate the original shaded triangle by a quarter-turn anticlockwise about (0, 1).

Label this new triangle with the letter B. *(2 marks)*

AQA, June 2005, Paper 2 Higher, Question 3(a)(b)

5 The diagram shows two triangles, A and B.

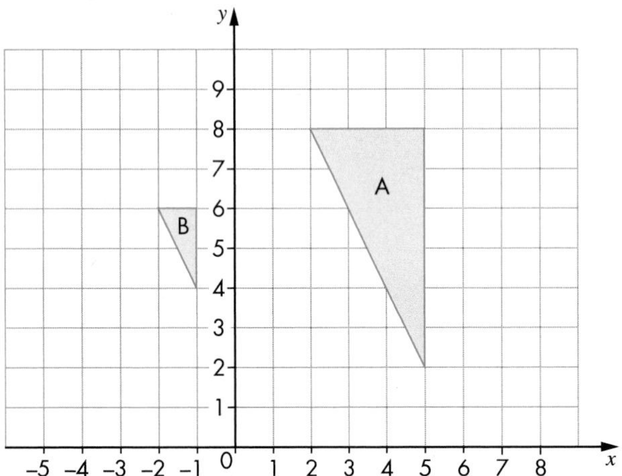

a Describe fully the single transformation that maps triangle A onto triangle B. *(3 marks)*

b On the diagram, draw the image of triangle A after it has been reflected in the line $y = x$. Label your image C. *(2 marks)*

AQA, June 2007, Paper 1 Higher, Question 6(a)(b)

6 The diagram shows four shapes, A, B, C and D.

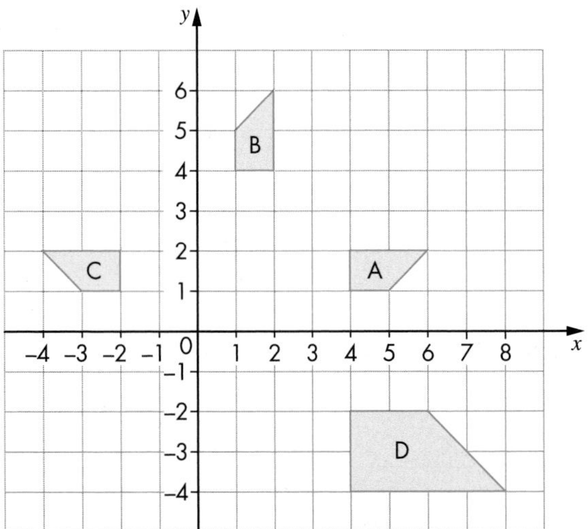

a Describe fully the single transformation that takes shape A onto shape B. *(2 marks)*

b Describe fully the single transformation that takes shape B onto shape C. *(3 marks)*

c Describe fully the single transformation that takes shape C onto shape D. *(3 marks)*

AQA, June 2006, Paper 1 Higher, Question 4(a)(b)(c)

7 ABC is an isosceles triangle.

M is the midpoint of AC.

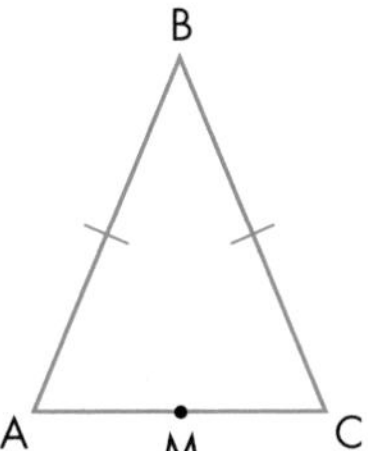

Prove that triangles ABM and CBM are congruent. *(4 marks)*

AQA, November 2007, Paper 2 Higher, Question 15

8 XYZ is an isosceles triangle in which XZ = XY.

M and N are points on XZ and XY such that angle MYZ = angle NZY.

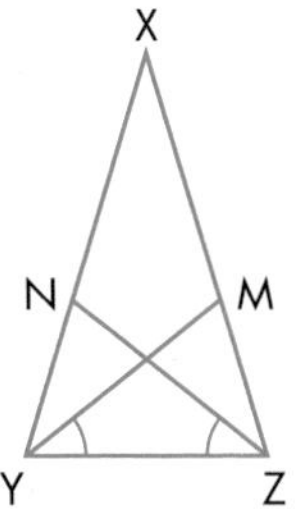

Prove that triangles YMZ and ZNY are congruent. *(4 marks)*

9 *ABCD* is a parallelogram.
E and *F* are points on *AB* and *CD* such that *BE* = *DF*

Not drawn accurately

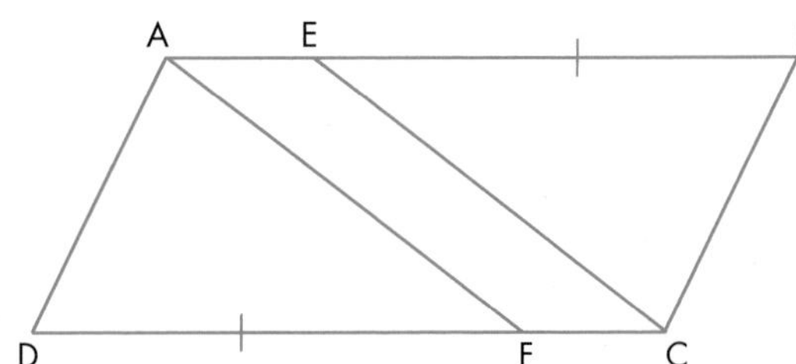

a Prove that triangles BCE and DAF are congruent. *(4 marks)*

b Deduce that *EC* is parallel to *AF*.

AQA, June 2011, H1, Question 25

Worked Examination Questions

AU **1** The grid shows several transformations of the shaded triangle.

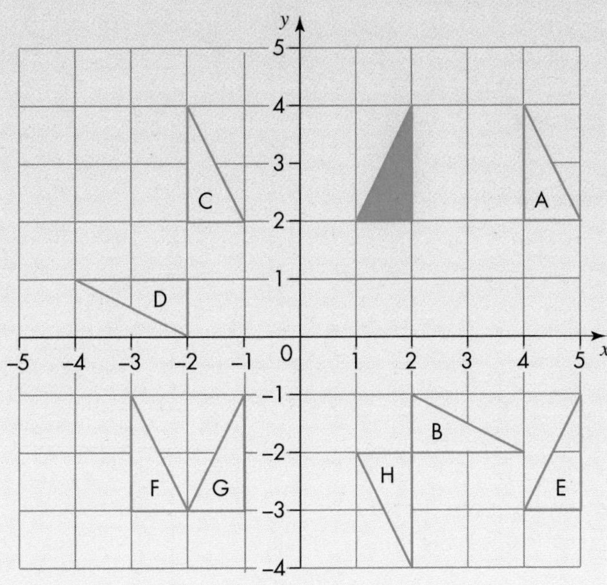

a Write down the letter of the shape:

i after the shaded triangle is reflected in the line $x = 3$

ii after the shaded triangle is translated by the vector $\begin{pmatrix} 3 \\ -5 \end{pmatrix}$

iii after the shaded triangle is rotated 90° clockwise about 0.

b Describe fully the single transformation that takes triangle F onto triangle G.

1 a i A — $x = 3$ is the vertical line passing through $x = 3$ on the x-axis. This scores 1 mark.

ii E — Move the triangle 3 squares to the right and 5 squares down. This scores 1 mark.

iii B — Use tracing paper to help you. Trace the shaded triangle, pivot the paper on 0 with your pencil point and rotate the paper through 90° clockwise. This scores 1 mark.

b A reflection in the line $x = -2$. — The vertical mirror line passes through $x = -2$ on the x-axis. This scores 1 mark for method of identifying reflection and 1 mark for accuracy of mirror line.

Total: 5 marks

Worked Examination Questions

2 Triangle ABC has vertices A(6, 0), B(6, 9), C(9, 3).

a Rotate triangle ABC through 180° about the point (2, 4). Label the image triangle R.

b Enlarge triangle ABC by scale factor $\frac{1}{3}$ from the centre of enlargement (3, 0). Label the image triangle E.

c Describe fully the single transformation which maps triangle E to triangle R.

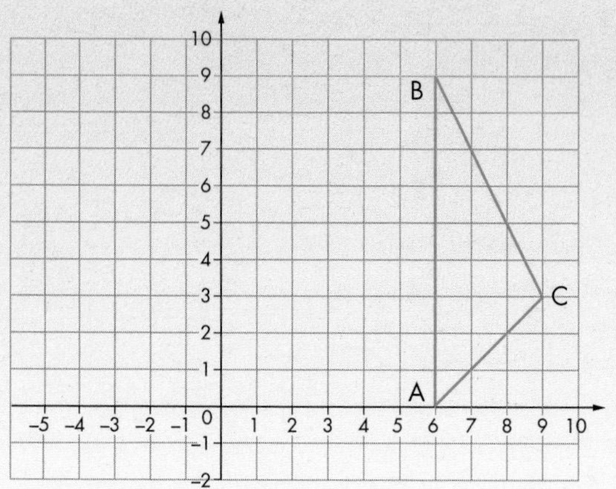

2 a *Join each vertex to (2, 4) and rotate each line through 180° or use tracing paper.*

This scores 1 method mark for rotating the triangle 180° about any point.

This scores 1 accuracy mark for the correct rotation.

b *Use the ray method or the counting squares method. Remember, a fractional scale factor makes the image smaller.*

This scores 1 method mark for enlarging the triangle by scale factor $\frac{1}{3}$ about any point.

This scores 1 accuracy mark for the correct enlargement.

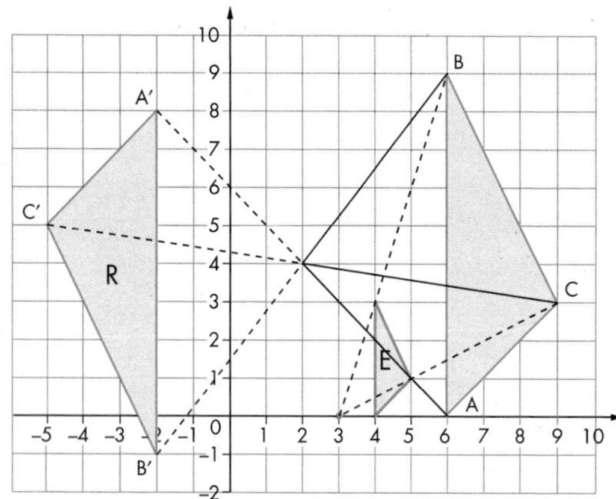

c

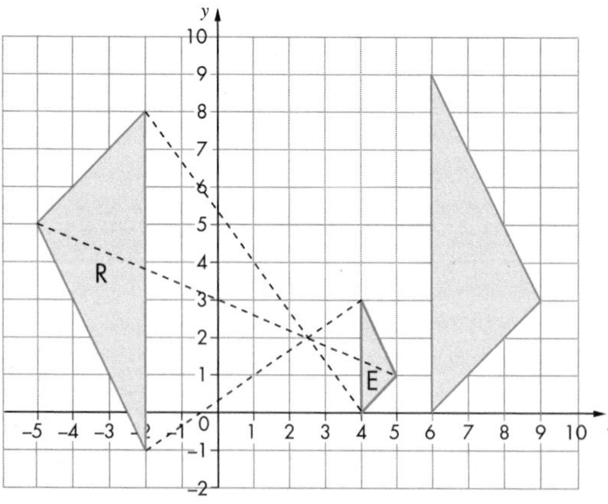

An enlargement of scale factor –3 about the point ($2\frac{1}{2}$, 2). Draw in the ray lines to find the centre of enlargement.

This scores 1 mark for enlargement of scale factor –3.

This scores 1 mark for ($2\frac{1}{2}$, 2).

Total: 6 marks

Worked Examination Questions

3 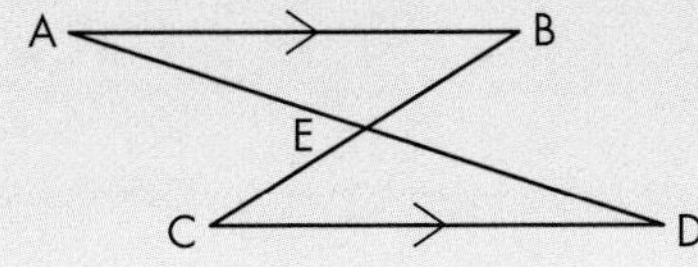

In the diagram AB and CD are parallel.

E is the midpoint of AD.

Prove triangle ABE is congruent to triangle CDE.

3 You will be expected to give reasons for each statement you make.

AE = DE (E is midpoint of AD)

$\angle$BAE = $\angle$CDE (alternate angles)

$\angle$AEB = $\angle$CED (opposite angles)

So ΔABE $\equiv$ ΔCDE (ASA)

Total: 3 marks

This scores 1 mark for method.

This scores 1 mark for method.

Or you could use $\angle$ABE = $\angle$DCE (alternate angles).

This scores 1 accuracy mark for third statement with correct conclusion.

You would lose a mark if you missed out any of the reasons.

4 ABCD and CPQR are squares. Prove that triangles ACR and ACP are congruent.

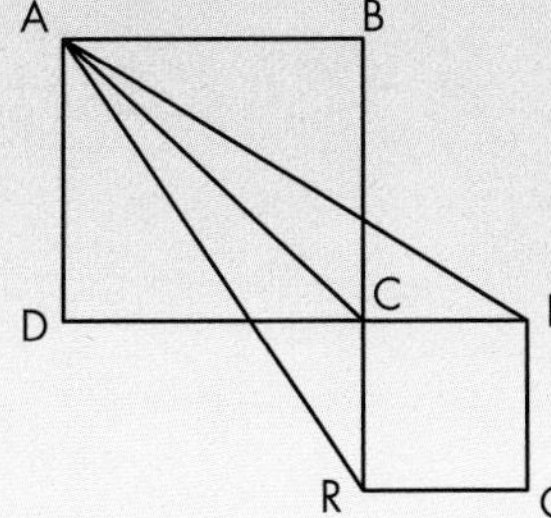

4 AC is a common side

CP = CR (sides of a square)

Angle ACP = Angle ACP = 45 + 90 + 135

Hence triangles ACR and ACP are congruent because of SAS.

Total: 4 marks

If two triangles have a common side then it must be of equal length in both triangles. This scores 1 mark.

You must give a reason in a proof. Sides of a square must be the same length. This scores 1 mark.

As AC bisects the square it must make an angle of 45° with each side. Between the two squares is a right angle. This scores 1 mark.

You need to give the reason why the triangles are congruent. This scores 1 mark.

11

Problem Solving
Developing photographs

Enlargement is used in many aspects of life. How many can you think of?

Discuss these with the person sitting next to you or as a whole class.

Getting started

- What two things do you need to describe an enlargement?
- When an object is enlarged, what stays the same?
- When an object is enlarged, what changes?
- If an object is enlarged by a scale factor of 3, what is the ratio of the lengths in the image to the corresponding lengths in the object?
- If an object is enlarged by a scale factor of 2, what is the ratio of the area of the image to the area of the object?

Getting started (continued)

Which set of photographs is the odd one out? Explain why.

Look at these images. How can you tell that they are all based on the same original photograph?

Which do you think is the original? Which are enlargements? Why?

Your task

Use a digital camera to take at least two photographs. Use them to make a display or poster to explain enlargement. To complete the task successfully, you should answer at least two of the following questions about enlargement.

1. What happens to the image when you move the centre of enlargement?
2. Compare what happens to the image when you have a scale factor that is:
   - a whole number
   - a number between 0 and 1
   - a negative number.
3. What is the relationship between the scale factor and the properties of the object and image? For example, consider how the scale factor affects the relationships between lengths or areas in the object and image.

Note

If you do not have a digital camera available, use a suitable computer program to produce and enlarge a simple image, logo or other shape for your display or poster.

Why this chapter matters

In Britain today, taking the pulse of the nation is an involved affair. Does the teacher in London have the same views as the banker in Sheffield? Will a milkman in Cornwall agree with an IT specialist in Scotland? And if not, how can the opinions of all the people be sought?

Opinion polls

Opinion polls have been used by politicians for centuries.

The Greeks and Romans used opinion polls years ago. At the time of Plato, Athens only had a population of around 30 000 people who could vote (all men!) and if an Ancient Greek wanted to know the opinions of his townsfolk he would simply ask a question in a public forum.

The politicians didn't see rival groups as an important section of society. These groups were small, for example, small groups of farmers, of old men, of wine makers.

Plato

Straw polls

Polling, as we know it, began in Pennsylvania, USA. In 1824, the *Harrisburg Pennsylvanian* newspaper surveyed about 500 potential voters and found that Andrew Jackson held a commanding lead over John Quincy Adams (70% vs. 23%). Jackson went on to win the election by 38 000, so the poll was accurate.

This type of poll is called a 'straw poll' – it involves no analysis of statistical trends but is just a vote.

Gallup polls

For the next 100 years, straw polling was used but it was not always very accurate. Eventually a change was made to the way polls were taken. In 1936, the USA Presidential election between Franklin Roosevelt and Alf Landon was about to take place. A popular newspaper mailed over 10 million questionnaires. They received about two million responses and, a few days before election day, they predicted an overwhelming win for Alf Landon. The results of the straw poll were:

Landon	1 293 669 (57%)
Roosevelt	972 897 (43%)

At this time, George Gallup had created a new type of poll—one that used a much smaller sample but involved some statistical analysis. Gallup's polls showed a victory for Roosevelt. Roosevelt won a huge victory, gaining 63% of the vote.

George Gallup

The American public asked how the newspaper could be so wrong. The questionnaires were sent to their own subscribers, so the poll was biased from the start. Then only those who were interested replied, so another bias. The Gallup polls did not have this same bias as they used a representative, stratified sampling.

George Gallup went from strength to strength and, to this day, his firm are responsible for finding out opinions for a whole range of people from politicians to farmers, from trade union leaders to football clubs.

Sampling is used frequently today. We are regularly being asked our opinions in the street. However, they don't ask everyone; they select particular people so that their sample is truly representative.

Statistics: Data handling

This chapter will show you ...

- D to C how to calculate and use the mode, median, mean and range from frequency tables of discrete data
- D to C how to decide which is the best average for different types of data
- D to C how to recognise the modal class and calculate an estimate of the mean from frequency tables of grouped data
- D to C how to design questions for questionnaires and surveys
- C to B how to use the data-handling cycle
- A to A* how to draw frequency polygons and histograms

Visual overview

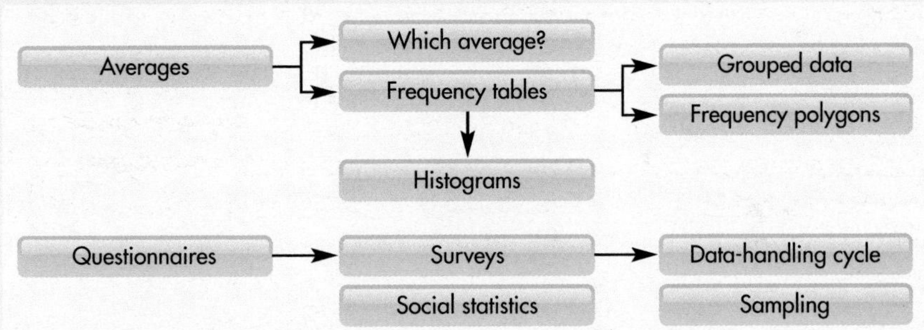

What you should already know

- How to work out the mean, mode, median and range of small sets of discrete data **(KS3 level 4, GCSE grade F)**
- How to extract information from tables and diagrams **(KS3 level 4, GCSE grade F)**

Quick check

1 The marks for 15 students in a maths test are as follows

2, 3, 4, 5, 5, 6, 6, 6, 7, 7, 7, 7, 7, 8, 10

a What is the modal mark?

b What is the median mark?

c What is the range of the marks?

d What is the mean mark?

12.1 Averages

This section will show you how to:

- use averages
- solve more complex problems using averages
- identify the advantages and disadvantages of each type of average and learn which one to use in different situations

Key words

mean
measure of location
median
mode

Average is a term you will often use when describing or comparing sets of data. The average is also known as a **measure of location**. For example, you may refer to the average rainfall in Britain, the average score of a batsman, an average weekly wage, the average mark in an examination. In each of these examples, you are representing the whole set of many values by just one single, typical value, called the average.

The idea of an average is extremely useful, because it enables you to compare one set of data with another set by comparing just two values—their averages.

There are several ways of expressing an average, but the most commonly used averages are the **mode**, the **median** and the **mean**.

An average must be truly representative of a set of data. So, when you have to find an average, it is crucial to choose the *correct type of average* for this particular set of data. If you use the wrong average, your results will be distorted and give misleading information.

This table, which compares the advantages and disadvantages of each type of average, will help you to make the correct decision.

	Mode	Median	Mean
Advantages	Very easy to find Not affected by extreme values Can be used for non-numerical data	Easy to find for ungrouped data Not affected by extreme values	Easy to find Uses all the values The total for a given number of values can be calculated from it
Disadvantages	Doesn't use all the values May not exist	Doesn't use all the values Often not understood	Extreme values can distort it Has to be calculated
Used for	Non-numerical data For finding the most likely value	Data with extreme values	Data with values that are spread in a balanced way

EXAMPLE 1

The ages of 20 people attending a conference are as follows.

23, 25, 26, 28, 28, 34, 34, 34, 37, 45, 47, 48, 52, 53, 56, 63, 67, 70, 73, 77

a Find **i** the mode **ii** the median **iii** the mean of the data.

b Which average best represents the age of the people at the conference?

a **i** The mode is 34 **ii** the median is 46 **iii** the mean is 920 ÷ 20 = 46

b The mean is distorted because of the few very old people at the conference. The median is also distorted by the larger values, so in this case the mode would be the most representative average.

EXERCISE 12A

D

FM 1 Shopkeepers always want to keep the most popular items in stock. Which average do you think is often known as the shopkeeper's average?

2 A list contains seven even numbers. The largest number is 24. The smallest number is half the largest. The mode is 14 and the median is 16. Two of the numbers add up to 42. What are the seven numbers?

3 The marks of 25 students in an English examination are as follows.

55, 63, 24, 47, 60, 45, 50, 89, 39, 47, 38, 42, 69, 73, 38, 47, 53, 64, 58, 71, 41, 48, 68, 64, 75

Find the median.

AU 4 Decide which average you would use for each of the following. Give a reason for your answer.

a The average mark in an examination.

b The average pocket money for a group of 16-year-old students.

c The average shoe size for all the girls in Year 10.

d The average height for all the artistes on tour with a circus.

e The average hair colour for students in your school.

f The average weight of all newborn babies in a hospital's maternity ward.

AU 5 A pack of matches consists of 12 boxes. The contents of each box are as follows.

34 31 29 35 33 30 31 28 29 35 32 31

On the box it states that the average contents is 32 matches. Is this correct?

Explain your answer.

D

FM 6 This table shows the annual salaries for a firm's employees.

a What is the **i** modal salary **ii** median salary and **iii** mean salary?

b The management has suggested a pay rise for all of 6%. The shopfloor workers want a pay rise for all of £1500. What difference to the mean salary would each suggestion make?

Chairman	£83 000
Managing director	£65 000
Floor manager	£34 000
Skilled worker 1	£28 000
Skilled worker 2	£28 000
Machinist	£20 000
Computer engineer	£20 000
Secretary	£20 000
Office junior	£8 000

AU 7 Mr Brennan, a caring maths teacher, told each student their individual test mark and only gave the test statistics to the whole class. He gave the class the modal mark, the median mark and the mean mark.

a Which average would tell a student whether he/she was in the top half or the bottom half of the class?

b Which average tells the students nothing really?

c Which average allows a student to gauge how well he/she has done compared with everyone else?

FM 8 Three players were hoping to be chosen for the basketball team.

The table shows their scores for the last few games they played.

The teacher said they would be selected by their best average score.

Tom	16, 10, 12, 10, 13, 8, 10
David	16, 8, 15, 25, 8
Mohammed	15, 2, 15, 3, 5

By which average would each boy choose to be selected?

C

9 Here are three expressions, where x represents a number.

$x + 5$ $\quad 3x$ $\quad 2x - 2$

a Work out the median when $x = 4$.

b Work out the mean when $x = 3$.

10 These expressions represent five numbers.

$3x$ $\quad 4x$ $\quad 4x$ $\quad 5x$ $\quad 9x$

a Write down the range in terms of x.

b Work out the value of x when the mean = 40.

PS 11 Here are three cards with algebraic expressions on two of them.
The mean of the expression on all three cards is $3x - 2$

$2x + 1$		$5x - 3$

What expression is missing from the third card?

12 These expressions represent four numbers.

$x + 1$ $\quad x + 5$ $\quad 2x - 1$ $\quad 4x + 7$

Work out the mean in terms of x.

Give your answer in its simplest form.

13 A list of nine numbers has a mean of 7.6. What number must be added to the list to give a new mean of 8?

14 A dance group of 17 teenagers had a mean weight of 44.5 kg. To enter a competition, there needs to be 18 people in the group with an average weight of 44.4 kg or less. What is the maximum weight that the 18th person could be?

15 The mean age of a group of eight walkers is 42. Joanne joins the group and the mean age changes to 40. How old is Joanne?

PS **16** **a** Find five numbers that have **both** the properties below.

i A range of 5 **ii** A mean of 5

b Find five numbers that have **all** the properties below.

i A range of 5 **ii** A median of 5 **iii** A mean of 5

FM AU **17** What is the average pay at a factory with 10 employees?

The boss said, "£43 295." A worker said, "£18 210."

They were both correct. Explain how this can be.

PS **18** Five numbers written in order are 3, 5, x, y, 9

The mean and median are the same.

Work out an expression for the value of y in terms of x.

PS **19** The rule for this sequence is that each term is the mean of the two previous terms.

$\boldsymbol{a}$	x	y	$\frac{x+y}{2}$	$\boldsymbol{b}$

a Find an expression for $\boldsymbol{a}$ in terms of x and y.

b Find an expression for $\boldsymbol{b}$ in terms of x and y.

Simplify your answer.

PS **20** Five people, Arnie, Beth, Chas, Dave and Ed weigh themselves.

You are given the following information about their weights.

The mean weight of Arnie, Beth and Chas is 70 kg.
The mean weight of Beth, Chas and Dave is 76kg.
The mean weight of Chas, Dave and Ed is 77kg.
Chas and Ed weigh the same.
Arnie weighs 75kg.

Find the weights of the other four people.

C B A

12.2 Frequency tables

This section will show you how to:
- calculate the mode and median from a frequency table
- calculate the mean from a frequency table

Key words
frequency table

When a lot of information has been gathered, it is often convenient to put it together in a **frequency table**. From this table, you can then find the values of the mode, median, mean and range of the data.

EXAMPLE 2

A survey was done on the number of people in each car leaving the Meadowhall Shopping Centre, in Sheffield. The results are summarised in the table.

Calculate **a** the mode, **b** the median, **c** the mean number of people in a car.

Number of people in each car	1	2	3	4	5	6
Frequency	45	198	121	76	52	13

a The modal number of people in a car is easy to spot. It is the number with the largest frequency (198). Hence, the modal number of people in a car is 2.

b The median number of people in a car is found by working out where the middle of the set of numbers is located. First, add up frequencies to get the total number of cars surveyed, which comes to 505. Next, calculate the middle position.

$(505 + 1) \div 2 = 253$

You now need to add the frequencies across the table to find which group contains the 253rd item. The 243rd item is the end of the group with 2 in a car. Therefore, the 253rd item must be in the group with 3 in a car. Hence, the median number of people in a car is 3.

c The mean number of people in a car is found by calculating the total number of people, and then dividing this total by the number of cars surveyed.

Number in car	Frequency	Number in these cars
1	45	$1 \times 45 = 45$
2	198	$2 \times 198 = 396$
3	121	$3 \times 121 = 363$
4	76	$4 \times 76 = 304$
5	52	$5 \times 52 = 260$
6	13	$6 \times 13 = 78$
Totals	505	1446

Hence, the mean number of people in a car is $1446 \div 505 = 2.9$ (2 significant figures)

Using your calculator

The previous example can also be done by using the statistical mode which is available on some calculators. However, not all calculators are the same, so you will either have to read your instruction manual or experiment with the statistical keys on your calculator.

You may find one labelled

DATA or M+ or Σ+ or $\bar{x}$ where $\bar{x}$ is printed in blue.

Try the following key strokes.

1 Find **i** the mode, **ii** the median and **iii** the mean from each frequency table below.

a A survey of the shoe sizes of all the Year 10 boys in a school gave these results.

Shoe size	4	5	6	7	8	9	10
No. of students	12	30	34	35	23	8	3

b This is a record of the number of babies born each week over one year in a small maternity unit.

No. of babies	0	1	2	3	4	5	6	7	8	9	10	11	12	13	14
Frequency	1	1	1	2	2	2	3	5	9	8	6	4	5	2	1

2 A survey of the number of children in each family of a school's intake gave these results.

No. of children	1	2	3	4	5
Frequency	214	328	97	26	3

a Assuming each child at the school is shown in the data, how many children are at the school?

b Calculate the mean number of children in a family.

c How many families have this mean number of children?

FM **d** How many families would consider themselves average from this survey?

3 A dentist kept records of how many teeth he extracted from his patients.

In 1989 he extracted 598 teeth from 271 patients.

In 1999 he extracted 332 teeth from 196 patients.

In 2009 he extracted 374 teeth from 288 patients.

a Calculate the average number of teeth taken from each patient in each year.

AU **b** Explain why you think the average number of teeth extracted falls each year.

D

4 The teachers in a school were asked to indicate the average number of hours they spent each day marking. The table summarises their replies.

No. of hours spent marking	1	2	3	4	5	6
No. of teachers	10	13	12	8	6	1

a How many teachers are at the school?

b What is the modal number of hours spent marking?

c What is the mean number of hours spent marking?

5 Two friends often played golf together. They recorded their scores for each hole over five games to determine who was more consistent and who was the better player. The results are summarised in the table.

No. of shots to hole ball	1	2	3	4	5	6	7	8	9
Roger	0	0	0	14	37	27	12	0	0
Brian	5	12	15	18	14	8	8	8	2

a What is the modal score for each player?

b What is the range of scores for each player?

c What is the median score for each player?

d What is the mean score for each player?

e Which player is the more consistent? Explain why.

AU **f** Who would you say is the better player. State why.

C

6 The number of league goals scored by a football team over a season is given in the table.

No. of goals scored	0	1	2	3	4	5	6	7
No. of matches	3	8	10	11	4	2	1	1

a How many games were played that season?

b What is the range of goals scored?

c What is the modal number of goals scored?

d What is the median number of goals scored?

e What is the mean number of goals scored?

FM **f** Which average do you think the team's supporters would say is the average number of goals scored by the team that season?

g If the team also scored 20 goals in 10 cup matches that season, what was the mean number of goals the team scored throughout the whole season?

7 A survey of the number of children in each family of a school's intake gave these results.

No. of children	1	2	3	4	5
Frequency	214	328	97	26	3

a State the median number of children per family.

b Calculate the mean number of children in a family.

FM **c** What percentage of families could consider themselves average from this survey?

FM PS **8** The number of sweets in some tubes is shown in the table below, but a coffee stain has deleted one of the figures.

No. of sweets	Frequency
32	4
33	
34	9
35	1
36	1

The mean number of sweets in a tube is known to be 33.5.

Find out what the missing number is in the table.

FM AU **9** I have been given a frequency table by Corrin. She says, "I can calculate the mean to be an integer but not the median. Why is that?"

Can you give a possible explanation?

10 The table shows the number of passengers in each of 100 taxis leaving London Heathrow Airport one day.

No. of passengers in a taxi	1	2	3	4
No. of taxis	x	40	y	26

a Find the value of $x + y$.

b If the mean number of passengers per taxi is 2.66, show that $x + 3y = 82$.

c Find the values of x and y by solving appropriate equations.

d State the median of the number of passengers per taxi.

B

FM **11** Sam, a farmer, thinks that the amount of rainfall is decreasing each year.

He records the amount of rain that falls each month for a year. The table shows the results (in mm).

Months	Jan	Feb	Mar	Apr	May	Jun	Jul	Aug	Sep	Oct	Nov	Dec
Rainfall (mm)	27	43	30	37	54	20	16	21	32	36	41	39

From the internet Sam finds that the mean rainfall for 2008 is 36 mm with a range of 34. Investigate the hypothesis:

'The amount of rainfall is decreasing each year.'

12 The table shows the number of times that Aston fills up his car each week for 15 weeks.

Number of times	**Frequency**
0	1
1	4
2	x
y	3

a Work out x.

b The mean number of times is 2.

Work out y.

c Over the next 5 weeks Aston fills the car up once each week.

Work out the mean for the 20 weeks.

A*

PS **13** In this table a and b are integers.

x	f
2	5
3	6
6	a
7	b
10	2

The mean of x is 5.

Find a possible pair of values for a and b.

12.3 Grouped data

This section will show you how to:
- identify the modal group
- calculate and estimate the mean from a grouped table

Key words
continuous data
discrete data
estimated mean
group
modal group

Sometimes the information you are given is grouped in some way, as in the table in Example 3, which shows the range of weekly pocket money given to Year 10 students in a particular class.

EXAMPLE 3

From the data in the table:

a write down the **modal group**

b calculate an estimate of the mean weekly pocket money.

Pocket money, p, (£)	$0 < p \leqslant 1$	$1 < p \leqslant 2$	$2 < p \leqslant 3$	$3 < p \leqslant 4$	$4 < p \leqslant 5$
No. of students	2	5	5	9	15

a The modal group is still easy to pick out, since it is simply the one with the largest frequency. Here the modal group is £4 to £5.

b The mean can only be estimated, since you do not have all the information. To estimate the mean, you simply assume that each person in each **group** has the midway amount, then you can proceed to build up the table as before.

To find the midway value. Add the two end values and divide the total by two.

Pocket money, p, (£)	Frequency (f)	Midway (m)	$f \times m$
$0 < p \leqslant 1$	2	0.50	1.00
$1 < p \leqslant 2$	5	1.50	7.50
$2 < p \leqslant 3$	5	2.50	12.50
$3 < p \leqslant 4$	9	3.50	31.50
$4 < p \leqslant 5$	15	4.50	67.50
Totals	36		120

The **estimated mean** is £120 ÷ 36 = £3.33 (rounded)

Note the notation used for the groups.

$0 < p \leqslant 1$ means any amount above 0p up to and including £1.

$1 < p \leqslant 2$ means any amount above £1 up to and including £2.

If you had written 0.01 – 1.00, 1.01 – 2.00, … for the groups, the midway values would have been 0.505, 1.505, … Although technically correct, this makes the calculation of the mean harder and does not have a significant effect on the final answer, which is an estimate anyway.

This issue only arises because money is **discrete data**, which is data that consists of separate numbers, such as goals scored, marks in a test, number of children and shoe sizes. Normally, grouped tables use **continuous data**, which is data that can have an infinite number of different values, such as height, weight, time, area and capacity. It is always rounded information.

Whatever the type of data, remember to find the midway value by adding the two end values of the group and dividing by 2.

EXERCISE 12C 

C

1 For each table of values, find the following.

i The modal group **ii** An estimate for the mean

a

x	$0 < x \leqslant 10$	$10 < x \leqslant 20$	$20 < x \leqslant 30$	$30 < x \leqslant 40$	$40 < x \leqslant 50$
Frequency	4	6	11	17	9

b

y	$0 < y \leqslant 100$	$100 < y \leqslant 200$	$200 < y \leqslant 300$	$300 < y \leqslant 400$	$400 < y \leqslant 500$	$500 < y \leqslant 600$
Frequency	95	56	32	21	9	3

c

z	$0 < z \leqslant 5$	$5 < z \leqslant 10$	$10 < z \leqslant 15$	$15 < z \leqslant 20$
Frequency	16	27	19	13

d

Weeks	1–3	4–6	7–9	10–12	13–15
Frequency	5	8	14	10	7

HINTS AND TIPS

When you copy the tables, draw them vertically, as in Example 3.

2 Jason brought 100 pebbles back from the beach and weighed them all, to the nearest gram. His results are summarised in this table.

Weight, w (grams)	$40 < w \leqslant 60$	$60 < w \leqslant 80$	$80 < w \leqslant 100$	$100 < w \leqslant 120$	$120 < w \leqslant 140$	$140 < w \leqslant 160$
Frequency	5	9	22	27	26	11

Find the following.

a The modal weight of the pebbles **b** An estimate of the total weight of all the pebbles

c An estimate of the mean weight of the pebbles

3 One hundred light bulbs were tested by their manufacturer to see whether the average life span of the bulbs was over 200 hours. The table summarises the results.

Life span, h (hours)	$150 < h \leqslant 175$	$175 < h \leqslant 200$	$200 < h \leqslant 225$	$225 < h \leqslant 250$	$250 < h \leqslant 275$
Frequency	24	45	18	10	3

a What is the modal length of time a bulb lasts?

b What percentage of bulbs last longer than 200 hours?

FM **c** Estimate the mean life span of the light bulbs.

FM **d** Do you think the test shows that the average life span is over 200 hours? Explain your answer fully.

FM **4** The table shows the distances run by an athlete who is training for a marathon.

Distance, *d*, (miles)	$0 < d \leqslant 5$	$5 < d \leqslant 10$	$10 < d \leqslant 15$	$15 < d \leqslant 20$	$20 < d \leqslant 25$
Frequency	3	8	13	5	2

a It is recommended that an athlete's daily average mileage should be at least one-third of the distance of the race being trained for. A marathon is 26.2 miles. Is this athlete doing enough training?

b The athlete records the times of some runs and calculates that her average pace for all runs is $6\frac{1}{2}$ minutes for a mile. Explain why she is wrong to expect a finishing time of $26.2 \times 6\frac{1}{2}$ minutes ≈ 170 minutes for the marathon.

c The athlete claims that the difference between her shortest and longest run is 21 miles. Could this be correct? Explain your answer.

FM **5** The owners of a boutique did a survey to find the average age of people using the boutique. The table summarises the results.

Age (years)	14–18	19–20	21–26	27–35	36–50
Frequency	26	24	19	16	11

What do you think is the average age of the people using the boutique?

FM AU **6** Three supermarkets each claimed to have the lowest average price increase over the year. The table summarises their average price increases.

Price increase (p)	1–5	6–10	11–15	16–20	21–25	26–30	31–35
Soundbuy	4	10	14	23	19	8	2
Springfields	5	11	12	19	25	9	6
Setco	3	8	15	31	21	7	3

Using their average price increases, make a comparison of the supermarkets and write a report on which supermarket, in your opinion, has the lowest price increases over the year. Don't forget to justify your answers.

7 A survey was conducted to see how quickly the AOne attended calls that were not on a motorway.

The following table summarises the results.

Time (min)	1–15	16–30	31–45	46–60	61–75	76–90	91–105
Frequency	2	23	48	31	27	18	11

a How many calls were used in the survey?

b Estimate the mean time taken per call.

FM **c** Which average would the AOne use for the average call-out time?

d What percentage of calls do the AOne get to within the hour?

C

PS **8** In a cricket competition, the table shows the runs scored by all the batsmen.

Runs	0–9	10–19	20–29	30–39	40–49
Frequency	8	5	10	5	2

Helen noticed that two numbers were in the wrong part of the table and that this made a difference of 1.7 to the arithmetic mean.

Which two numbers were the wrong way round?

AU **9** The profit made each week by a charity shop is shown in the table below.

Profit (£)	0–500	501–1000	1001–1500	1501–2000
Frequency	15	26	8	3

Explain how you would estimate the mean profit made each week.

AU **10** The table shows the number of members of 100 football clubs.

Weight	20–29	30–39	40–49	50–59	60–69
Frequency	16	34	27	18	5

a Roger claims that the median number of members is 39.5.

Is he correct? Explain your answer

b He also says that the range of the number of members is 34.

Could he be correct? Explain your answer.

12.4 Frequency diagrams

This section will show you how to:

- draw frequency polygons for discrete and continuous data
- draw histograms for continuous data with equal intervals
- draw pie charts

Key words

continuous data
discrete data
frequency density
grouped data
frequency polygon
histogram
pie chart
sector

Pie charts

Bar charts and line graphs are easy to draw but they can be difficult to interpret when there is a big difference between the frequencies or there are only a few categories. In these cases, it is often more convenient to illustrate the data on a **pie chart.**

In a pie chart, the whole of the data is represented by a circle (the 'pie') and each category is represented by a **sector** of the circle (a 'slice of the pie'). The angle of each sector is proportional to the frequency of the category it represents.

A pie chart cannot show individual frequencies, like a bar chart can, for example. It can only show proportions.

Sometimes the pie chart will be marked off in equal sections rather than angles. In these cases, the numbers are always easy to work with.

EXAMPLE 4

In a survey on holidays, 120 people were asked to state which type of transport they used on their last holiday. This table shows the results of the survey. Draw a pie chart to illustrate the data.

Type of transport	Train	Coach	Car	Ship	Plane
Frequency	24	12	59	11	14

You need to find the angle for the fraction of 360° that represents each type of transport. This is usually done in a table, as shown below.

Types of transport	Frequency	Calculation	Angle
Train	24	$\frac{24}{120} \times 360° = 72°$	72°
Coach	12	$\frac{12}{120} \times 360° = 36°$	36°
Car	59	$\frac{59}{120} \times 360° = 177°$	177°
Ship	11	$\frac{11}{120} \times 360° = 33°$	33°
Plane	14	$\frac{14}{120} \times 360° = 42°$	42°
Totals	120		360°

Draw the pie chart, using the calculated angle for each sector.

Note:

- Use the frequency total (120 in this case) to calculate each fraction.
- Check that the sum of all the angles is 360°.
- Label each sector.
- The angles or frequencies do not have to be shown on the pie chart.

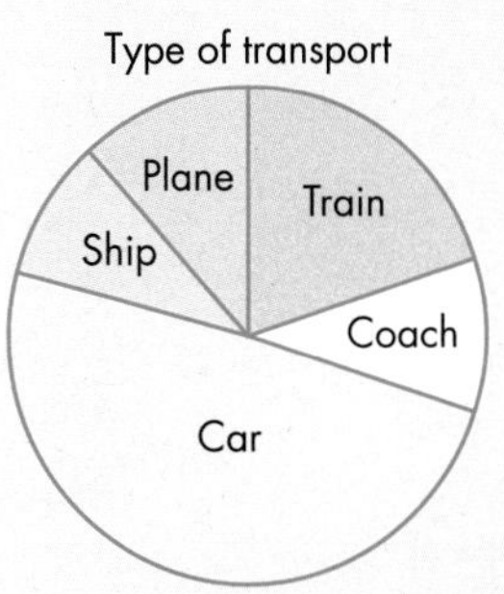

EXERCISE 12D 

D

1 Andy wrote down the number of lessons he had per week in each subject on his school timetable.

Mathematics 5	English 5	Science 8	Languages 6
Humanities 6	Arts 4	Games 2	

a How many lessons did Andy have on his timetable?

b Draw a pie chart to show the data.

c Draw a bar chart to show the data.

d Which diagram better illustrates the data? Give a reason for your answer.

2 In the run up to an election, 720 people were asked in a poll which political party they would vote for. The results are given in the table.

Conservative	248
Labour	264
Liberal-Democrat	152
Green Party	56

a Draw a pie chart to illustrate the data.

b Why do you think pie charts are used to show this sort of information during elections?

3 This pie chart shows the proportions of the different shoe sizes worn by 144 students in Year 11 in a London school.

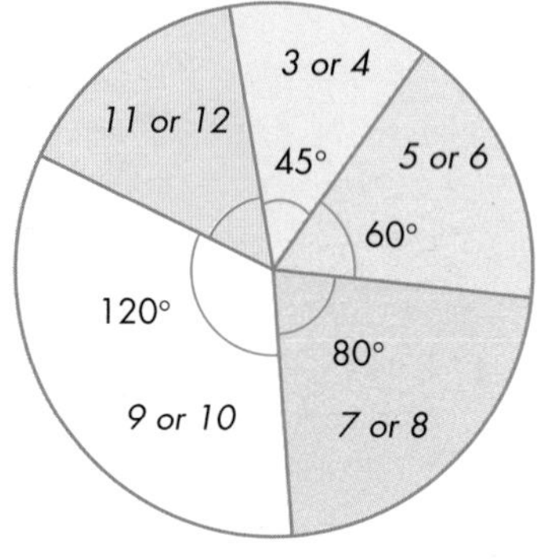

a What is the angle of the sector representing shoe sizes 11 and 12?

b How many students had a shoe size of 11 or 12?

c What percentage of students wore the modal size?

FM 4 The table below shows the numbers of candidates, at each grade, taking music examinations in Strings and Brass.

	Grades					Total no. of candidates
	3	4	5	6	7	
Strings	200	980	1050	600	70	3000
Brass	250	360	300	120	70	1100

a Draw a pie chart to represent each of the two examinations.

b Compare the pie charts to decide which group of candidates, Strings or Brass, did better overall. Give reasons to justify your answer.

PS 5 In a survey, a rail company asked passengers whether their service had improved.

What is the probability that a person picked at random from this survey answered "Don't know"?

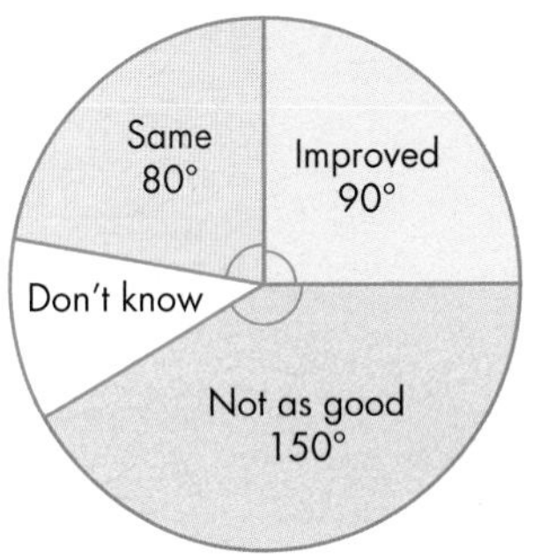

AU 6 You have been asked to draw a pie chart representing the different ways in which students come to school one morning.

What data would you collect, to do this?

Frequency polygons

To help people understand it, statistical information is often presented in pictorial or diagrammatic form. For example, you should have seen pie charts, bar charts and stem-and-leaf diagrams. Another method of showing data is by **frequency polygons**.

Frequency polygons can be used to represent both ungrouped data and **grouped data**, as shown in Example 5 and Example 6 respectively. They are useful to show the shapes of distributions, and can be used to compare distributions.

EXAMPLE 5

No. of children	0	1	2	3	4	5
Frequency	12	23	36	28	16	11

This is the frequency polygon for the ungrouped data in the table.

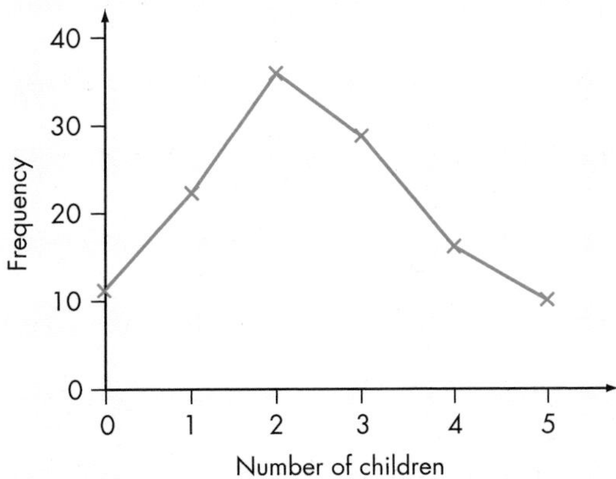

- You simply plot the coordinates from each ordered pair in the table.
- You complete the polygon by joining up the plotted points with straight lines.

EXAMPLE 6

Weight, w (kilograms)	$0 < w \leqslant 5$	$5 < w \leqslant 10$	$10 < w \leqslant 15$	$15 < w \leqslant 20$	$20 < w \leqslant 25$	$25 < w \leqslant 30$
Frequency	4	13	25	32	17	9

This is the frequency polygon for the grouped data in the table.

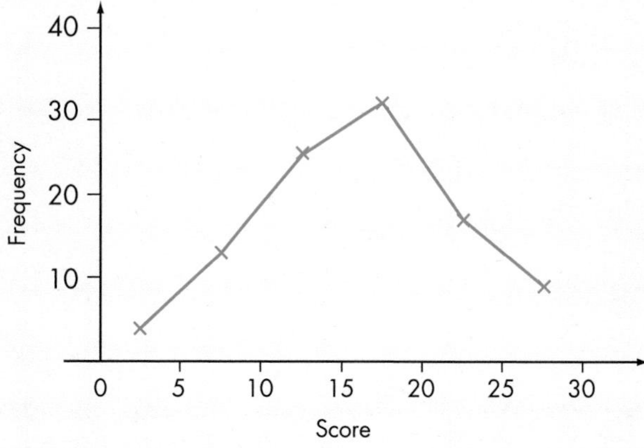

- You use the midway value of each group, just as in estimating the mean.
- You plot the ordered pairs of midway values with frequency, namely,
  (2.5, 4), (7.5, 13), (12.5, 25), (17.5, 32), (22.5, 17), (27.5, 9)
- You do not know what happens above and below the groups in the table, so do not draw lines before (2.5, 4) or after (27.5, 9). The diagram shows the shape of the distribution.

Bar charts and histograms

You should already be familiar with the bar chart in which the vertical axis represents frequency, and the horizontal axis represents the type of data. (Sometimes it is more convenient to have the axes the other way.)

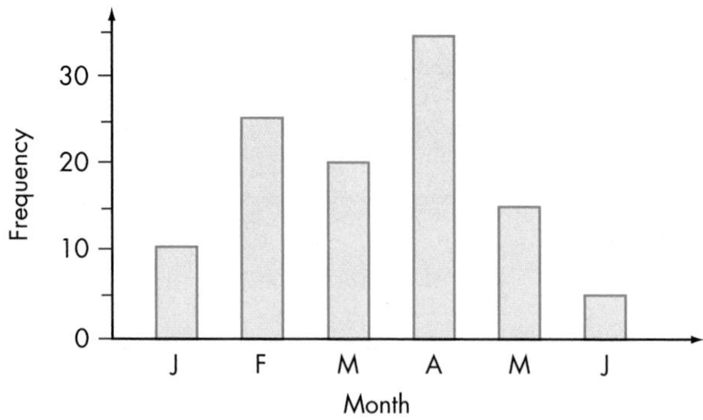

You should also recognise a composite bar chart, which can be used to compare two sets of related data.

EXAMPLE 7

This dual bar chart shows the average daily maximum temperatures for England and Turkey over a five-month period.

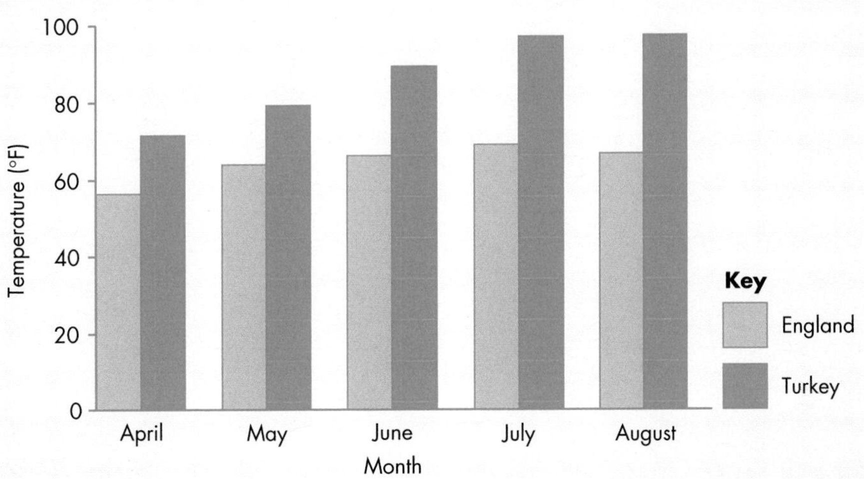

In which month was the difference between temperatures in England and Turkey the greatest?

The largest difference can be seen in August.

A **histogram** looks similar to a bar chart, but there are four fundamental differences.

- There are no gaps between the bars.
- The horizontal axis has a continuous scale since it represents **continuous data**, such as time, weight or length.
- The area of each bar represents the class or group frequency of the bar.
- The vertical axis is labelled '**frequency density**', where

$$\text{frequency density} = \frac{\text{frequency of class interval}}{\text{width of class interval}}$$

When the data is not continuous, a simple bar chart is used. For example, you would use a bar chart to represent the runs scored in a test match or the goals scored by a hockey team since these are integer values and are **discrete data**.

The histogram below has been drawn from this table of times it takes people to walk to work.

Time, t (min)	$0 < t \leqslant 4$	$4 < t \leqslant 8$	$8 < t \leqslant 12$	$12 < t \leqslant 16$
Frequency	8	12	10	7
Frequency density	2	3	2.5	1.75

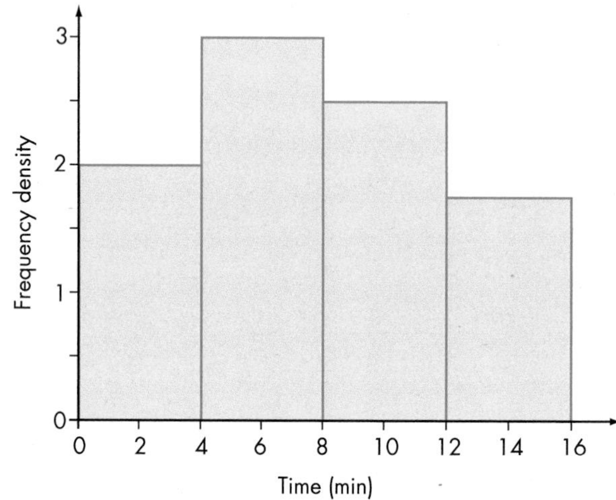

Notice that each histogram bar starts at the *least possible* time and finishes at the *greatest possible* time for its group.

Using your calculator

Histograms can also be drawn on graphics calculators or by using computer software packages. If you have access to either of these, try to use them.

EXERCISE 12E

D

1 The table shows how many students were absent from one particular class throughout the year.

Students absent	1	2	3	4	5
Frequency	48	32	12	3	1

a Draw a frequency polygon to illustrate the data.

b Calculate the mean number of absences each lesson.

2 The table shows the number of goals scored by a hockey team in one season.

Goals	1	2	3	4	5
Frequency	3	9	7	5	2

a Draw the frequency polygon for this data.

b Calculate the mean number of goals scored per game in the season.

C

3 The frequency polygon shows the amount of money spent in a corner shop by the first 40 customers on one day.

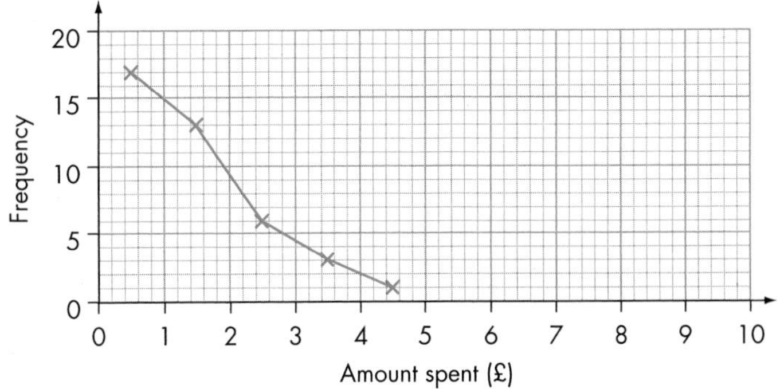

a **i** Use the frequency polygon to complete the table for the amounts spent by the first 40 customers.

Amount spent, *m* (£)	$0 < m \leqslant 1$	$1 < m \leqslant 2$	$2 < m \leqslant 3$	$3 < m \leqslant 4$	$4 < m \leqslant 5$
Frequency					

ii Work out the mean amount of money spent by these 40 customers.

b Mid-morning the shopkeeper records the amount spent by another 40 customers. The table below shows the data.

Amount spent, *m* (£)	$0 < m \leqslant 2$	$2 < m \leqslant 4$	$4 < m \leqslant 6$	$6 < m \leqslant 8$	$8 < m \leqslant 10$
Frequency	3	5	18	10	4

i On a copy of the graph above, draw the frequency polygon to show this data.

ii Calculate the mean amount spent by the 40 mid-morning customers.

FM **c** Comment on the differences between the frequency polygons and the average amounts spent by the different sets of customers.

4 The table shows the range of heights of the girls in Year 11 at a London school.

Height, *h* (cm)	$120 < h \leqslant 130$	$130 < h \leqslant 140$	$140 < h \leqslant 150$	$150 < h \leqslant 160$	$160 < h \leqslant 170$
Frequency	15	37	25	13	5

a Draw a frequency polygon for this data. **b** Draw a histogram for this data.

c Estimate the mean height of the girls.

5 A doctor was concerned at the length of time her patients had to wait to see her when they came to the morning surgery. The survey she did gave her these results.

Time, *m* (minutes)	$0 < m \leqslant 10$	$10 < m \leqslant 20$	$20 < m \leqslant 30$	$30 < m \leqslant 40$	$40 < m \leqslant 50$	$50 < m \leqslant 60$
Monday	5	8	17	9	7	4
Tuesday	9	8	16	3	2	1
Wednesday	7	6	18	2	1	1

a On the same pair of axes draw a frequency polygon for each day.

b What is the average amount of time spent waiting each day?

FM **c** Why might the average time for each day be different?

C

6 After a spelling test, all the results were collated for girls and boys separately as below.

Number correct, *N*	$1 \leqslant N \leqslant 4$	$5 \leqslant N \leqslant 8$	$9 \leqslant N \leqslant 12$	$13 \leqslant N \leqslant 16$	$17 \leqslant N \leqslant 20$
Boys	3	7	21	26	15
Girls	4	8	17	23	20

a Draw frequency polygons to illustrate the differences between the boys' scores and the girls' scores.

FM **b** Estimate the mean scores for boys and girls separately, then comment on your results.

FM PS **7** The frequency polygon shows the lengths of time that students spent on homework one weekend.

Calculate an estimate of the mean time spent on homework by the students.

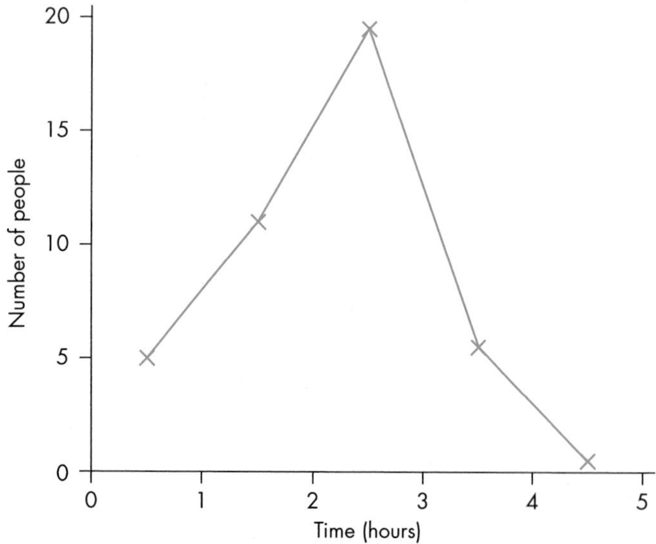

FM AU **8** The frequency polygon shows the times that a number of people waited at a Post Office before being served one morning.

Julie said, "Most people spent 30 seconds waiting."

Explain why this might be wrong.

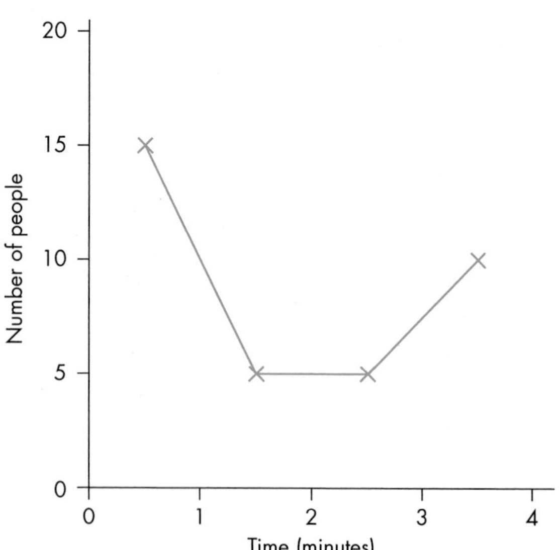

12.5 Histograms with bars of unequal width

This section will show you how to:

- draw and read histograms where the bars are of unequal width
- find the median, quartiles and interquartile range from a histogram

Key words

class interval
interquartile range
lower quartile
median
upper quartile

Sometimes the data in a frequency distribution are grouped into classes with intervals that are different. In this case, the resulting histogram has bars of unequal width.

The key fact that you should always remember is that the area of a bar in a histogram represents the class frequency of the bar. So, in the case of an unequal-width histogram, you find the height to draw each bar by dividing its class frequency by its **class interval** width (bar width), which is the difference between the lower and upper bounds for each interval. Conversely, given a histogram, you can find any of its class frequencies by multiplying the height of the corresponding bar by its width.

It is for this reason that the scale on the vertical axes of histograms is nearly always labelled 'frequency density', where

$$\text{frequency density} = \frac{\text{frequency of class interval}}{\text{width of class interval}}$$

EXAMPLE 8

The heights of a group of girls were measured. The results were classified as shown in the table.

Height, h (cm)	$151 \leqslant h < 153$	$153 \leqslant h < 154$	$154 \leqslant h < 155$	$155 \leqslant h < 159$	$159 \leqslant h < 160$
Frequency	64	43	47	96	12

It is convenient to write the table vertically and add two columns, class width and frequency density.

The class width is found by subtracting the lower class boundary from the upper class boundary. The frequency density is found by dividing the frequency by the class width.

Height, h (cm)	Frequency	Class width	Frequency density
$151 \leqslant h < 153$	64	2	32
$153 \leqslant h < 154$	43	1	43
$154 \leqslant h < 155$	47	1	47
$155 \leqslant h < 159$	96	4	24
$159 \leqslant h < 160$	12	1	12

The histogram can now be drawn. The horizontal scale should be marked off as normal, from a value below the lowest value in the table to a value above the largest value in the table. In this case, mark the scale from 150 cm to 160 cm. The vertical scale is always frequency density and is marked up to at least the largest frequency density in the table. In this case, 50 is a sensible value.

continued

EXAMPLE 8 continued

Each bar is drawn between the lower class interval and the upper class interval horizontally, and up to the frequency density vertically.

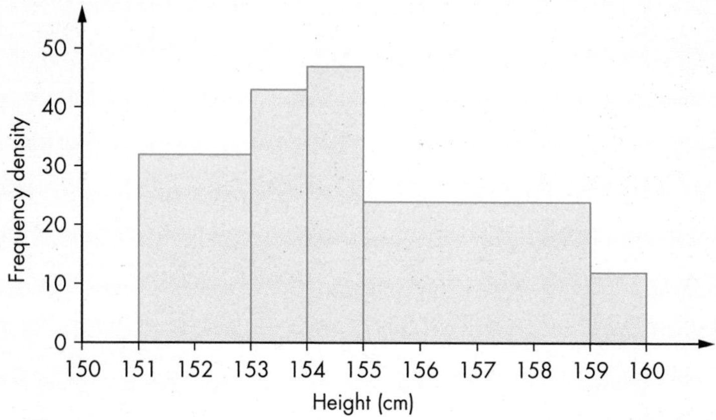

EXAMPLE 9

This histogram shows the distribution of heights of daffodils in a greenhouse.

a Complete a frequency table for the heights of the daffodils, and show the cumulative frequency.

b Find the **median** height.

c Find the **interquartile range** of the heights.

d Estimate the mean of the distribution.

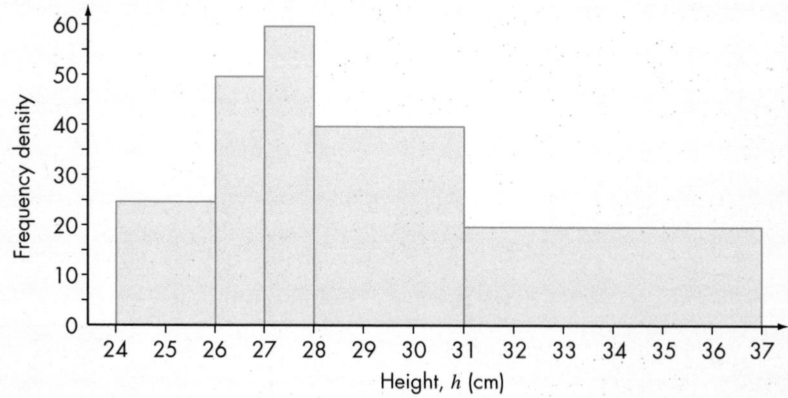

a The frequency table will have groups of $24 \leqslant h < 26$, $26 \leqslant h < 27$, etc. These are read from the height axis. The frequencies will be found by multiplying the width of each bar by the frequency density. Remember that the value on the vertical axis is not the frequency.

Height, h (cm)	$24 \leqslant h < 26$	$26 \leqslant h < 27$	$27 \leqslant h < 28$	$28 \leqslant h < 31$	$31 \leqslant h < 37$
Frequency	50	50	60	120	120
Cumulative frequency	50	100	160	280	400

b There are 400 values so the median will be the 200th value. Counting up the frequencies from the beginning you reach the third row of the table above.

The median occurs in the $28 \leq h < 31$ group. There are 160 values before this group and 120 in it. To get to the 200th value you need to go 40 more values into this group. 40 out of 120 is one-third. One-third of the way through this group is the value 29 cm. Hence the median is 29 cm.

c The interquartile range is the difference between the **upper quartile** and the **lower quartile,** the quarter and three-quarter values respectively. In this case, the lower quartile is the 100th value (found by dividing 400, the total number of values, by 4) and the upper quartile is the 300th value. So, in the same way that you found the median, you can find the lower (100th value) and upper (300th value) quartiles. The 100th value is at 27 cm and the 300th value is at 32 cm. The interquartile range is 32 cm – 27 cm = 5 cm.

d To estimate the mean, use the table to get the midway values of the groups and multiply these by the frequencies. The sum of these divided by 400 will give the estimated mean.

So, the mean is:

$(25 \times 50 + 26.5 \times 50 + 27.5 \times 60 + 29.5 \times 120 + 34 \times 120) \div 400$
$= 11\,845 \div 400 = 29.6$ cm (3 significant figures)

EXERCISE 12F

1 Draw histograms for these grouped frequency distributions.

a

Temperature, t (°C)	$8 \leq t < 10$	$10 \leq t < 12$	$12 \leq t < 15$	$15 \leq t < 17$	$17 \leq t < 20$	$20 \leq t < 24$
Frequency	5	13	18	4	3	6

b

Wage, w (£1000)	$6 \leq w < 10$	$10 \leq w < 12$	$12 \leq w < 16$	$16 \leq w < 24$
Frequency	16	54	60	24

c

Age, a (nearest year)	$11 \leq a < 14$	$14 \leq a < 16$	$16 \leq a < 17$	$17 \leq a < 20$
Frequency	51	36	12	20

d

Pressure, p (mm)	$745 \leq p < 755$	$755 \leq p < 760$	$760 \leq p < 765$	$765 \leq p < 775$
Frequency	4	6	14	10

e

Time, t (min)	$0 \leq t < 8$	$8 \leq t < 12$	$12 \leq t < 16$	$16 \leq t < 20$
Frequency	72	84	54	36

A

A

FM **2** The following information was gathered about the weekly pocket money given to 14-year-olds.

Pocket money, *p* (£)	$0 \leqslant p < 2$	$2 \leqslant p < 4$	$4 \leqslant p < 5$	$5 \leqslant p < 8$	$8 \leqslant p < 10$
Girls	8	15	22	12	4
Boys	6	11	25	15	6

a Represent the information about the boys on a histogram.

b Represent both sets of data with a frequency polygon, using the same pair of axes.

c What is the mean amount of pocket money given to each sex? Comment on your answer.

3 The sales of the *Star* newspaper over 70 years are recorded in this table.

Years	1940–60	1961–80	1981–90	1991–2000	2001–05	2006–2010
Copies	62 000	68 000	71 000	75 000	63 000	52 000

Illustrate this information on a histogram. Take the class boundaries as 1940, 1960, 1980, 1990, 2000, 2005, 2010.

4 The London trains were always late, so one month a survey was undertaken to find how many trains were late, and by how many minutes (to the nearest minute). The results are illustrated by this histogram.

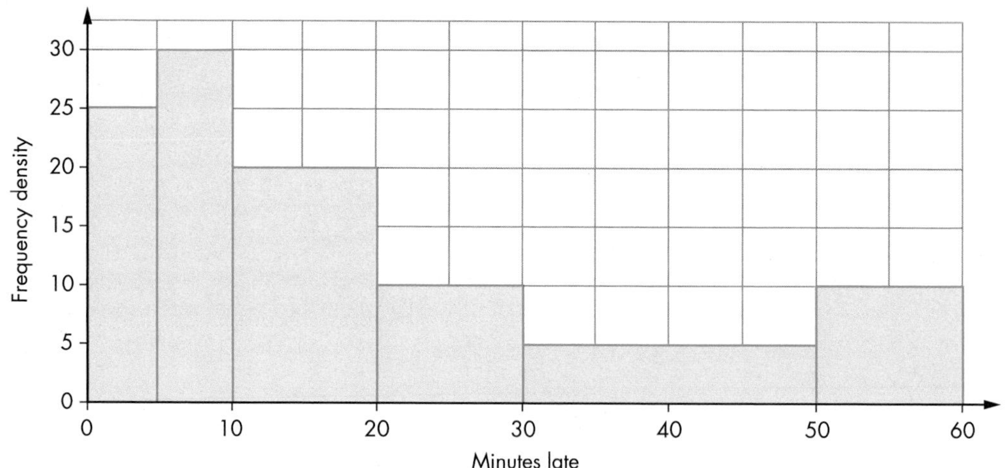

a How many trains were in the survey?

b How many trains were delayed for longer than 15 minutes?

AU **5** Hannah was asked to create a histogram.

Explain how Hannah will find the height of each bar on the frequency density scale.

6 For each of the frequency distributions illustrated in the histograms:

i write down the grouped frequency table

ii state the modal group

iii estimate the median

iv find the lower and upper quartiles and the interquartile range

v estimate the mean of the distribution.

a

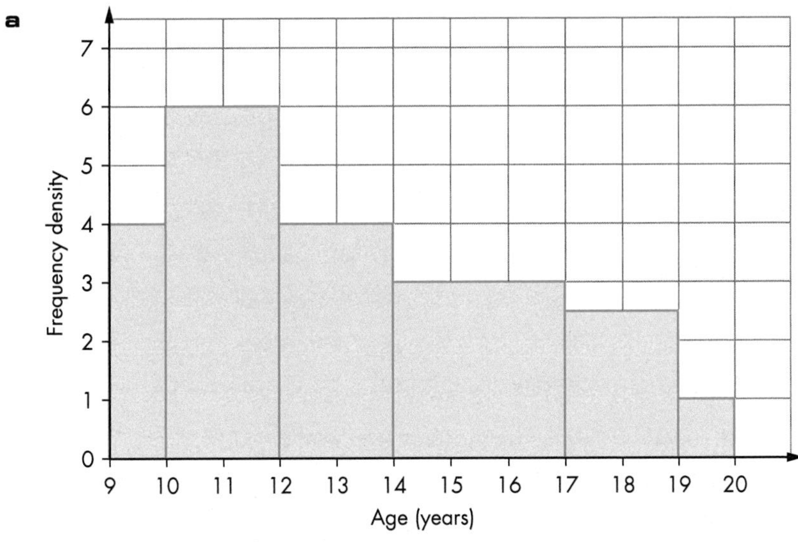

b

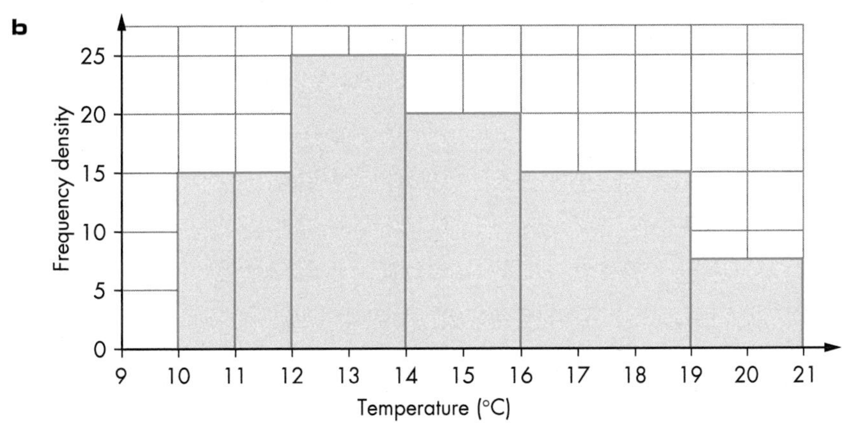

c

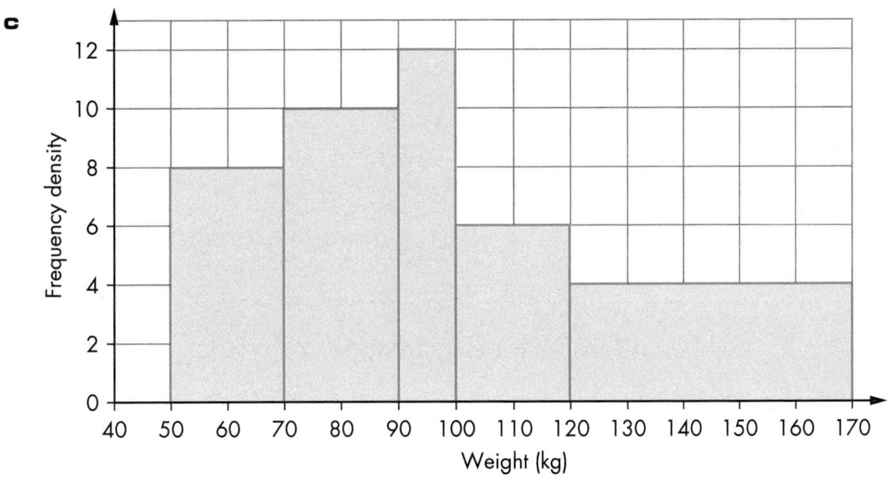

7 All the patients in a hospital were asked how long it was since they last saw a doctor. The results are shown in the table.

Hours, h	$0 \leqslant h < 2$	$2 \leqslant h < 4$	$4 \leqslant h < 6$	$6 \leqslant h < 10$	$10 \leqslant h < 16$	$16 \leqslant h < 24$
Frequency	8	12	20	30	20	10

a Find the median time since a patient last saw a doctor.

b Estimate the mean time since a patient last saw a doctor.

c Find the interquartile range of the times.

8 One summer, Albert monitored the weight of the tomatoes grown on each of his plants. His results are summarised in this table.

Weight, w (kg)	$6 \leqslant w < 10$	$10 \leqslant w < 12$	$12 \leqslant w < 16$	$16 \leqslant w < 20$	$20 \leqslant w < 25$
Frequency	8	15	28	16	10

a Draw a histogram for this distribution.

b Estimate the median weight of tomatoes the plants produced.

c Estimate the mean weight of tomatoes the plants produced.

d How many plants produced more than 15 kg?

9 A survey was carried out to find the speeds of cars passing a particular point on the M1. The histogram illustrates the results of the survey.

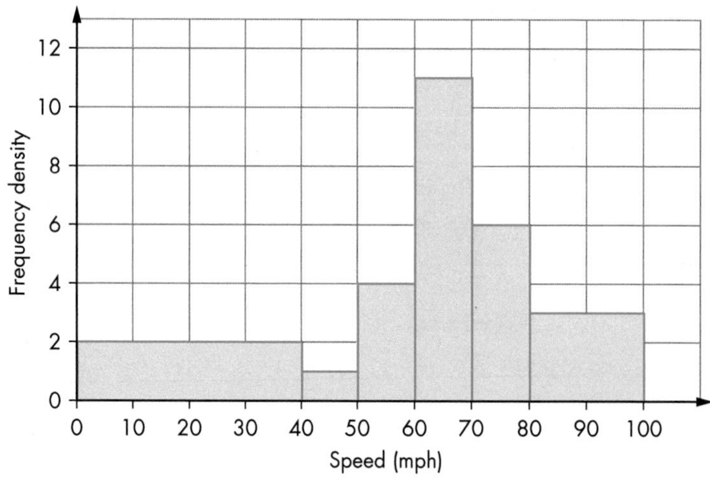

a Copy and complete this table.

Speed, v (mph)	$0 < v \leqslant 40$	$40 < v \leqslant 50$	$50 < v \leqslant 60$	$60 < v \leqslant 70$	$70 < v \leqslant 80$	$80 < v \leqslant 100$
Frequency		10	40	110		

b Find the number of cars included in the survey.

c Work out an estimate of the median speed of the cars on this part of the M1.

d Work out an estimate of the mean speed of the cars on this part of the M1.

A*

10 The histogram shows the test scores for 320 students in a school.

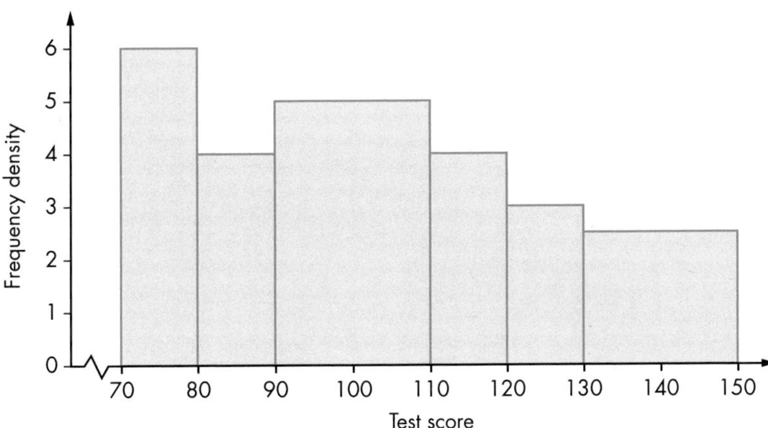

a Find the median score.

b Find the interquartile range of the scores.

c Find an estimate for the mean score.

FM **d** What was the pass score if 90% of the students passed this test?

FM PS **11** The distances employees of a company travel to work are shown in the histogram below.

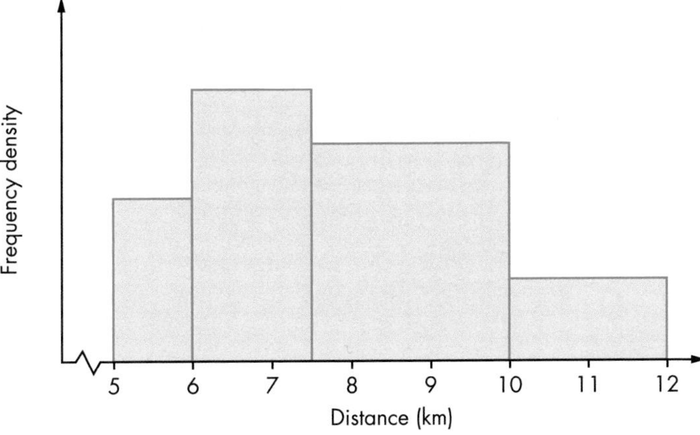

It is known that 18 workers travel between 10 and 12 km to work. What is the probability of choosing a worker at random who travels less than 7.5 km to work?

12.6 Surveys

This section will show you how to:
- conduct surveys

Key words
data collection sheet
hypothesis
survey

A **survey** is an organised way of asking a lot of people a few, well-constructed questions, or of making a lot of observations in an experiment, in order to reach a conclusion about something.

Surveys are used to test out people's opinions or to test a **hypothesis**.

Simple data collection sheet

If you just need to collect some data to analyse, you will have to design a simple **data collection sheet**. This section will show you how to design a clear, easy-to-fill-in data collection sheet.

For example, if you want to find out Year 10 students' preferences for the end-of-term trip from four options you could ask:

Where do you want to go for the Year 10 trip at the end of term — Blackpool, Alton Towers, The Great Western Show or London?

You would put this question, on the same day, to a lot of Year 10 students, and enter their answers straight onto a data collection sheet, as below.

Place	Tally	Frequency
Blackpool	𝍸 𝍸 𝍸 𝍸 |||	23
Alton Towers	𝍸 𝍸 𝍸 𝍸 𝍸 𝍸 𝍸 𝍸 𝍸 |	46
The Great Western Show	𝍸 𝍸 ||||	14
London	𝍸 𝍸 𝍸 𝍸 ||	22

Notice how plenty of space is available for the tally marks, and how the tallies are gated in groups of five to make counting easier when the survey is complete.

This is a good, simple data collection sheet because:

- only one question *(Where do you want to go?)* has to be asked
- all the four possible venues are listed
- the answer from each interviewee can be easily and quickly tallied, then on to the next interviewee.

Notice, too, that since the question listed specific places, they must appear on the data collection sheet.

You would lose many marks in an examination if you just asked the open question: *Where do you want to go?*

Data sometimes needs to be collected to obtain responses for two different categories. The data collection sheet is then in the form of a two-way table.

EXAMPLE 10

The head of a school carries out a survey to find out how much time students in different year groups spend on their homework during a particular week. He asks a sample of 60 students and fills in a two-way table with headings as follows.

	0–5 hours	0–10 hours	10–20 hours	More than 20 hours
Year 7				

This is not a good table as the headings overlap. A student who does 10 hours' work a week could tick either of two columns. Response sections should not overlap, so that there is only one possible place to put a tick.

A better table would be like this.

	0 up to 5 hours	More than 5 and up to 10 hours	More than 10 and up to 15 hours	More than 15 hours
Year 7	𝍸 ||	𝍸		
Year 8	𝍸	𝍸 ||		
Year 9	|||	𝍸 ||	||	
Year 10	|||	𝍸	|||	|
Year 11	||	||||	||||	||

This gives a clearer picture of the amount of homework done in each year group.

EXERCISE 12G

D

FM 1 'People like the supermarket to open on Sundays.'

a To see whether this statement is true, design a data collection sheet that will allow you to capture data while standing outside a supermarket.

b Does it matter on which day you collect data outside the supermarket?

FM 2 The school tuck shop wants to know which types of chocolate it should get in to sell — plain, milk, fruit and nut, wholenut or white chocolate.

a Design a data collection sheet that you could use to ask students in your school which of these chocolate types are their favourite.

b Invent the first 30 entries on the chart.

HINTS AND TIPS

Include space for tallies.

D

3 When you throw two dice together, what number are you most likely to get?

a Design a data collection sheet on which you can record the data from an experiment in which two dice are thrown together and note the sum of the two numbers shown on the dice.

b Carry out this experiment for at least 100 throws.

c Which sums are most likely to occur?

d Illustrate your results on a frequency polygon.

FM **4** Who uses the buses the most in the mornings? Is it pensioners, mums, schoolchildren, the unemployed or some other group? Design a data collection sheet to be used in a survey of bus passengers.

> **HINTS AND TIPS**
>
> Make sure all possible responses are covered.

C

FM **5** Design two-way tables to show:

a how students in different year groups travel to school in the morning

b the types of programme that different age groups prefer to watch on TV

c the favourite sports of boys and girls

d the amount of time students in different year groups spend on the computer in the evening.

Invent about 40 entries for each one.

FM **6** Hassan wanted to find out who eats healthy food.

He decided to investigate the hypothesis:

'Boys are less likely to eat healthy food than girls are.'

a Design a data collection sheet that Hassan could use to help him do this.

b Hassan records information from a sample of 40 boys and 25 girls. He finds that 17 boys and 15 girls eat healthy food.

Based on this sample, is the hypothesis correct? Explain your answer.

7 What kind of tariffs do your classmates use on their mobile phones?

Design a data collection sheet to help you find this out.

FM AU **8** You are asked to find out what shops the parents of the students at your school like to use.

When creating a data collection sheet for this information what two things must you include on the collection sheet?

12.7 Questionnaires

This section will show you how to:

- ask good questions in order to collect reliable and valid data

Key words
data collection sheet
hypothesis
leading question
questionnaire
survey

This section will show you how to put together a clear, easy-to-use **questionnaire**.

When you are putting together a questionnaire for a **survey**, you must think very carefully about the sorts of question you are going to ask. Here are five rules that you should always follow.

- Never ask a **leading question** designed to get a particular response.
- Never ask a personal, irrelevant question.
- Keep each question as simple as possible.
- Include questions that will get a response from whomever is asked.
- Make sure the responses do not overlap and keep the number of choices to a reasonable number (six at the most).

The following questions are badly constructed and should *never* appear in any questionnaire.

What is your age? This is personal. Many people will not want to answer. It is always better to give a range of ages.

☐ Under 15 ☐ 16–20 ☐ 21–30 ☐ 31–40 ☐ Over 40

Slaughtering animals for food is cruel to the poor defenceless animals. Don't you agree? This is a leading question, designed to get a 'yes' response. It is better to ask an impersonal question.

Are you a vegetarian? ☐ Yes ☐ No

Do you go to discos when abroad? This can be answered only by people who have been abroad. It is better to ask a starter question, with a follow-up question.

Have you been abroad for a holiday? ☐ Yes ☐ No

If yes, did you go to a disco whilst you were away? ☐ Yes ☐ No

When you first get up in a morning and decide to have some sort of breakfast that might be made by somebody else, do you feel obliged to eat it all or not? This question is too complicated. It is better to ask a series of shorter questions.

What time do you get up for school? ☐ Before 7 ☐ Between 7 and 8 ☐ After 8

Do you have breakfast every day? ☐ Yes ☐ No

If No, on how many school days do you have breakfast? ☐ 0 ☐ 1 ☐ 2 ☐ 3 ☐ 4 ☐ 5

A questionnaire is usually a specialised **data collection sheet**, put together to test a **hypothesis** or a statement. For example, a questionnaire might be constructed to test this statement.

People buy cheaper milk from the supermarket as they don't mind not getting it on their doorstep. They'd rather go out to buy it.

A questionnaire designed to test whether this statement is true or not should include these questions.

Do you have milk delivered to your doorstep?

Do you buy cheaper milk from the supermarket?

Would you buy your milk only from the supermarket?

Once these questions have been answered, the responses can be checked to see whether the majority of people hold views that agree with the statement.

EXERCISE 12H

D

1 These are questions from a questionnaire on healthy eating.

a

Fast food is bad for you. Don't you agree?		
☐ Strongly agree	☐ Agree	☐ Don't know

Give two criticisms of the question.

b

Do you eat fast food?	☐ Yes	☐ No	
If yes, how many times on average do you eat fast food a week?			
☐ Once or less	☐ 2 or 3 times	☐ 4 or 5 times	☐ More than 5 times

Give two reasons why these are good questions.

2 This is a question from a survey on pocket money.

How much pocket money do you get each week?			
☐ £0–£2	☐ £0–£5	☐ £5–£10	☐ £10 or more

a Give a reason why this is not a good question.

b Rewrite the question to make it a good question.

C

3 Design a questionnaire to test this statement.

'People under 16 do not know what is meant by all the jargon used in the business news on TV, but the over-twenties do.'

C

FM PS **4** Design a questionnaire to test this statement.

'The under-twenties feel quite at ease with computers, while the over-forties would rather not bother with them. The 20–40s are all able to use computers effectively.'

FM PS **5** Design a questionnaire to test this hypothesis.

'The older you get, the less sleep you need.'

FM PS **6** A headteacher wants to find out if her students think they have too much, too little or just the right amount of homework. She also wants to know the parents' views about homework.

Design a questionnaire that could be used to find the data that the headteacher needs to look at.

7 Anja and Andrew are doing a survey on the type of music people buy.

a This is one question from Anja's survey.

> Folk music is just for country people.
>
> Don't you agree?
>
> ☐ Strongly agree ☐ Agree ☐ Don't know

Give two criticisms of Anja's question.

b This is a question from Andrew's survey.

> How many CDs do you buy each month?
>
> ☐ 2 or fewer ☐ 3 or 4 ☐ more than 4

Give two reasons why this is a good question.

c Make up another good question with responses that could be added to this survey.

PS **8** Design a questionnaire to test this hypothesis.

'People with back problems do not sit properly.'

AU **9** As each customer left a store, an assistant gave them a questionnaire containing the following question.

> Question: How much do you normally spend in this shop?
>
> Response: ☐ Less than £15 ☐ More than £25
>
> ☐ Less than £25 ☐ More than £50

Explain why the response section of this questionnaire is poor.

The data-handling cycle

This section will show you how to:

- use the data-handling cycle

Key words
hypothesis
primary data
secondary data

The data-handling cycle

Testing out a **hypothesis** involves a cycle of planning, collecting data, evaluating the significance of the data and then interpreting the results, which may or may not show the hypothesis to be true. This cycle often leads to a refinement of the problem, which starts the cycle all over again.

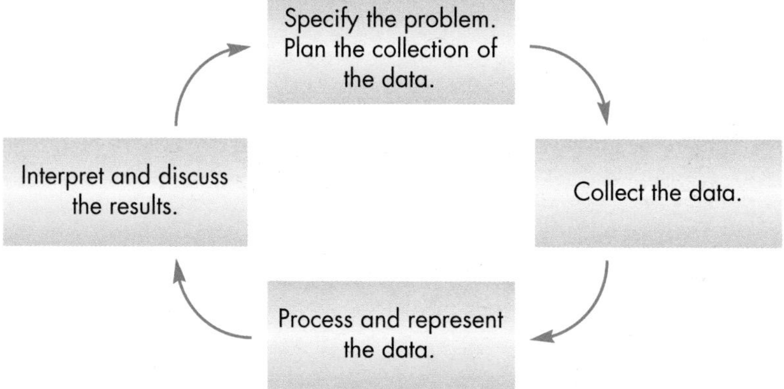

There are four parts to the data-handling cycle.

1. State the hypothesis, which is the idea being tested, outlining the problem and planning what needs to be done.
2. Plan the data collection and collect the data.
3. Choose the best way to process and represent the data. This will normally mean calculating averages (mean, median, mode) and measures of spread, then representing data in suitable diagrams.
4. Interpret the data and make conclusions.

Then the hypothesis can be refined or changes can be made in respect to the data, whether in the type to be studied or in method of collection, and the process becomes a cycle of improving reliability.

EXAMPLE 11

A gardener grows tomatoes, both in a greenhouse and outside.

He wants to investigate whether tomato plants grown in the greenhouse produce more tomatoes than those grown outside.

Describe the data-handling cycle that may be applied to this problem.

State the hypothesis: 'Tomato plants grown in the greenhouse produce more tomatoes than those grown outside.'

Plan the data collection and collect the data. Consider 10 tomato plants grown in the greenhouse, and 10 plants grown outside. Count the tomatoes on each plant. Record the numbers of tomatoes collected from the plants between June and September. Only count those that are 'fit for purpose'.

Choose the best way to process and represent the data. Calculate the mean number collected per plant as well as the range.

Interpret the data and make conclusions. Look at the statistics. *What do they show? Is there a clear conclusion or do you need to alter the hypothesis in any way?* Discuss results, refine the method and continue the cycle.

As you see, *in describing the data-handling cycle, you must refer to each of the four parts.*

Data collection

Data that you collect yourself is called **primary data**. You control it, in terms of accuracy and amount.

Data collected by someone else is called **secondary data**. Generally, there is a lot of this type of data available on the internet or in newspapers. This provides a huge volume of data but you have to rely on the sources being reliable, for accuracy.

EXERCISE 12I

C

1 Use the data-handling cycle to describe how you would test each of the following statements or scenarios. Write your own hypothesis to suit the situation. In each case state whether you would use primary or secondary data.

- **a** Oliver is investigating which month of the year is the hottest.
- **b** Andrew wants to compare how good boys and girls are at estimating distances.
- **c** Joy thinks that more men than women go to football matches.
- **d** Sheehab wants to know if tennis is watched by more women than men.
- **e** A headteacher said that the more revision you do, the better your examination results.
- **f** A newspaper suggested that the older you are, the more likely you are to shop at Marks and Spencers.

FM **2** You are asked to compare the number of news programmes on different TV channels.

- **a** Write down a suitable hypothesis that you could test.
- **b** Design a suitable observation sheet to record the data.
- **c** Show how the data-handling cycle would be used in this investigation.

AU **3** Kath thinks that girls are better at mental arithmetic than boys.

Explain how Kath could test this. Use the stages of the data-handling cycle in your answer.

12.9 Other uses of statistics

This section will show you how to:
- apply statistics in everyday situations

Key words
margin of error
national census
polls
Retail Price Index
social statistics
time series

Many situations occur in daily life where statistical techniques are used to produce data. The results of surveys appear in newspapers every day. There are many on-line **polls** and phone-ins to vote in reality TV shows, for example.

Results for these polls are usually given as a percentage with a **margin of error**, which is a measure of how accurate the information is.

Here are some common **social statistics** in daily use.

General Index of Retail Prices

This is also know as the **Retail Price Index** (RPI). It measures how much the daily cost of living increases (or decreases). One year is chosen as the base year and given an index number, usually 100. The costs of subsequent years are compared to this and given a number proportional to the base year, say 103, etc.

Note the numbers do not represent actual values but just compare current prices to the base year.

Time series

Like the RPI, a **time series** measures changes in a quantity over time. Unlike the RPI, the actual values of the quantity are used. This might measure how the exchange rate between the pound and the dollar changes over time.

National census

A **national census** is a survey of all people and households in a country. Data about age, gender, religion, employment status, etc. is collected to enable governments to plan where to allocate resources in the future. In Britain, a national census is taken every 10 years. The last census was in 2001.

EXERCISE 12J 

D

FM **1** In 2000, the cost of a litre of petrol was 78p. Using 2000 as a base year, the price index of petrol for the next five years is shown in this table.

Year	2000	2001	2002	2003	2004	2005
Index	100	103	108	109	112	120
Price	78p					

Work out the price of petrol in each subsequent year. Give your answers to 1 decimal place.

FM **2** The graph shows the exchange rate for the dollar against the pound for each month in one year.

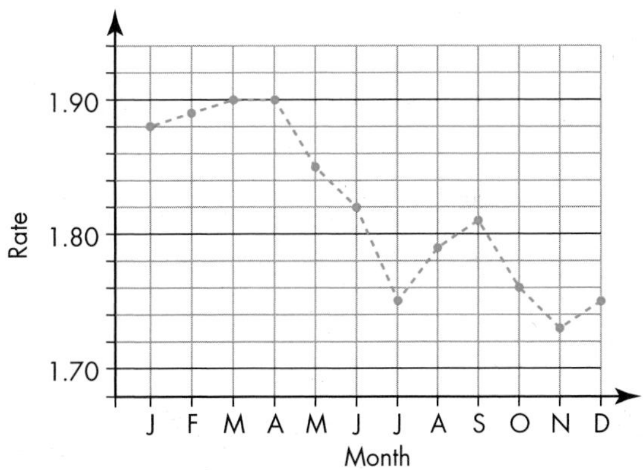

a What was the exchange rate in January?

b Between which two months did the exchange rate fall the most?

c Explain why you could not use the graph to predict the exchange rate in January of the next year.

FM **3** The following is taken from the UK government statistics website.

> In mid-2004 the UK was home to 59.8 million people, of which 50.1 million lived in England. The average age was 38.6 years, an increase on 1971 when it was 34.1 years. In mid-2004 approximately one in five people in the UK were aged under 16 and one in six people were aged 65 or over.

Use this extract to answer these questions.

a How many of the population of the UK do *not* live in England?

b By how much has the average age increased since 1971?

c Approximately how many of the population are under 16?

d Approximately how many of the population are over 65?

FM **4** The General Index of Retail Prices started in January 1987 when it was given a base number of 100. In January 2006 the index number was 194.1.

If the 'standard weekly shopping basket' cost £38.50 in January 1987, how much would it be in January 2006?

C

FM AU **5** The Retail Price Index measures how much the daily cost of living increases or decreases. If 2008 is given a base index number of 100, then 2009 is given 98. What does this mean?

B

FM 6 This time series shows car production in Britain from November 2004 to November 2005.

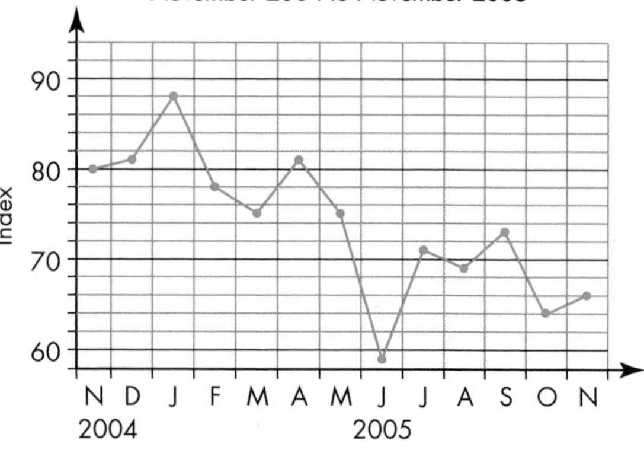

a Why was there a sharp drop in production in July?

b The average production over the first three months shown was 172 thousand cars.

i Work out an approximate number for the average production over the last three months shown.

ii The base month for the index is January 2000 when the index was 100. What was the approximate production in January 2000?

12.10 Sampling

This section will show you how to:

- understand different methods of sampling
- collect unbiased reliable data

Key words
population
random
sample
stratified
unbiased

Statisticians often have to carry out surveys to collect information and test hypotheses about the **population** for a wide variety of purposes. (In statistics, population does not only mean a group of people, it also means a group of objects or events.)

It is seldom possible to survey a whole population, mainly because such a survey would cost too much and take a long time. Also, there are populations for which it would be physically impossible to survey every member. For example, if you wanted to find the average length of eels in the North Sea, it would be impossible to find and measure every eel. So a statistician chooses a small part of the population to survey and assumes that the results for this **sample** are representative of the whole population.

Therefore, to ensure the accuracy of a survey, you must consider two questions.

- Will the sample be representative of the whole population and thereby eliminate bias?
- How large should the sample be to give results that are valid for the whole population?

Sampling methods

There are two main types of sample: **random** and **stratified**.

In a random sample, every member of the population has an equal chance of being chosen. For example, it may be the first 100 people met in a survey, or 100 names picked from a hat, or 100 names taken at random from the electoral register or a telephone directory.

In a stratified sample, the population is first divided into categories and the number of members in each category determined. The sample is then made up of members from these categories in the same proportions as they are in the population. The required sample in each category is chosen by random sampling.

EXAMPLE 12

A school's student numbers are given in the table. The headteacher wants to take a stratified sample of 100 students for a survey.

a Calculate the number of boys and girls in each year that should be interviewed.

b Explain how the students could then be chosen to give a random sample.

School year	Boys	Girls	Total
7	52	68	120
8	46	51	97
9	62	59	121
10	47	61	108
11	39	55	94
Total number in school			540

a To get the correct number in each category, say, boys in Year 7, the calculation is done as follows.

$\frac{52}{540} \times 100 = 9.6$ (1 decimal place)

After all calculations are done, you should get the values in this table.

School year	Boys	Girls
7	9.6	12.6
8	8.5	9.4
9	11.5	10.9
10	8.7	11.3
11	7.2	10.2

Obviously you cannot have a decimal fraction of a student, so round all values and make sure that the total is 100. This gives the final table.

School year	Boys	Girls	Total
7	10	13	23
8	8	9	17
9	12	11	23
10	9	11	20
11	7	10	17

b Within each category, choose students to survey at random. For example, all the Year 7 girls could have their names put into a hat and 13 names drawn out or they could be listed alphabetically and a random number generator used to pick out 13 names from 68.

Sample size

Before the sampling of a population can begin, it is necessary to determine how much data needs to be collected to ensure that the sample is representative of the population. This is called the sample size.

Two factors determine sample size:

- the desired precision with which the sample represents the population
- the amount of money available to meet the cost of collecting the sample data.

The greater the precision desired, the larger the sample size needs to be. But the larger the sample size, the higher the cost will be. Therefore, the benefit of achieving high accuracy in a sample will always have to be set against the cost of achieving it.

There are statistical procedures for determining the most suitable sample size, but these are beyond the scope of the GCSE syllabus.

The next example addresses some of the problems associated with obtaining an **unbiased** sample.

EXAMPLE 13

You are going to conduct a survey among an audience of 30 000 people at a rock concert. How would you choose the sample?

1 You would not want to question all of them, so you might settle for a sample size of 2%, which is 600 people.

2 Assuming that there will be as many men at the concert as women, you would need the sample to contain the same proportion of each, namely, 300 men and 300 women.

3 Assuming that about 20% of the audience will be aged under 20, you would also need the sample to contain 120 people aged under 20 (20% of 600) and 480 people aged 20 and over (600 – 120 or 80% of 600).

4 You would also need to select people from different parts of the auditorium in equal proportions so as to get a balanced view. Say this breaks down into three equal groups of people, taken respectively from the front, the back and the middle of the auditorium. So, you would further need the sample to consist of 200 people at the front, 200 at the back and 200 in the middle.

5 If you now assume that one researcher can survey 40 concert-goers, you would arrive at this sampling strategy

$600 \div 40 = 15$ researchers to conduct the survey

$15 \div 3 = 5$ researchers in each part of the auditorium

Each researcher would need to question four men aged under 20, 16 men aged 20 and over, four women aged under 20 and 16 women aged 20 and over.

EXERCISE 12K

D

FM **1** Comment on the reliability of the following ways of finding a sample.

a Find out about smoking by asking 50 people in a non-smoking part of a restaurant.

b Find out how many homes have video recorders by asking 100 people outside a video hire shop.

c Find the most popular make of car by counting 100 cars in a city car park.

d Find a year representative on a school's council by picking a name out of a hat.

e Decide whether the potatoes have cooked properly by testing one with a fork.

C

FM **2** Comment on the way the following samples have been taken. For those that are not satisfactory, suggest a better way to find a more reliable sample.

a Joseph had a discussion with his dad about pocket money. To get some information, he asked 15 of his friends how much pocket money they each received.

b Douglas wanted to find out what proportion of his school went abroad for holidays, so he asked the first 20 people he came across in the school yard.

c A teacher wanted to know which lesson his students enjoyed most. So he asked them all.

d It has been suggested that more females than males go to church. So Ruth did a survey in her church that Sunday and counted the number of females there.

e A group of local people asked for a crossing on a busy road. The council conducted a survey by asking 100 randomly-selected people in the neighbourhood.

FM **3** For a school project you have been asked to do a presentation of the social activities of the students in your school. You decide to interview a sample of students. Explain how you will choose the students you wish to interview if you want your results to be:

a reliable **b** unbiased **c** representative **d** random.

FM **4** A fast-food pizza chain attempted to estimate the number of people in a certain town who eat pizzas. One evening they telephoned 50 people living in the town and asked: "Have you eaten a pizza in the last month?" Eleven people said "Yes." The pizza chain stated that 22% of the town's population eat pizzas. Give three criticisms of this method of estimation.

5 Mr Charlton, the deputy head at High Storrs School, wanted to find out how often the upper-school students in his school visited a fast-food outlet. The numbers of students in each upper-school year are given in the table.

	Boys	Girls
Y9	119	85
Y10	107	118
Y11	104	110

a Create a questionnaire that Mr Charlton could use to sample the school.

b Mr Charlton wanted to do a stratified sample using 60 students. To how many of each group should he give the questionnaire?

C

PS 6 Naysha belonged to a school of 1850 students. She was in a class of 30 students. She noticed one day that the school was doing a survey over the whole school. Four boys and five girls in her class were involved in the survey.

Estimate how many students in the whole school were involved in the sample.

AU 7 You are asked to conduct a survey at a football match where the attendance is approximately 20 000.

Explain how you could create a stratified sample of the crowd.

FM 8 **a** Adam is writing a questionnaire for a survey about the Meadowhall shopping centre in Cambridge. He is told that fewer local people than people from further away visit Meadowhall. He is also told that the local people spend less money per visit. Write two questions that would help him to test these ideas. Each question should include at least three options for a response. People are asked to choose one of these options.

b For another survey, Adam investigates how much is spent at the chocolate machines by students at his school. The numbers of students in each year group are shown in the table. Explain, with calculations, how Adam should obtain a stratified random sample of 100 students for his survey.

Year group	7	8	9	10	11
No. of students	143	132	156	131	108

9 Claire made a survey of students in her school. She wanted to find out their opinions on the eating facilities in the school. The size of each year group in the school is shown in the table.

Year group	Boys	Girls	Total
8	96	78	174
9	84	86	170
10	84	91	175
11	82	85	167
6th form	83	117	200
			886

Claire took a sample of 90 students.

a Explain why she should not have sampled equal numbers of boys and girls in the sixth form.

b Calculate the number of students she should have sampled in the sixth form.

ACTIVITY

Using the internet

Through the internet you have access to a vast amount of data on many topics, which you can use to carry out statistical investigations. This data will enable you to draw statistical diagrams, answer a variety of questions and test all manner of hypotheses.

Here are some examples of hypotheses you can test.

- Football teams are most likely to win when they are playing at home.
- Boys do better than girls at GCSE mathematics.
- The number 3 gets drawn more often than the number 49 in the National Lottery.
- The literacy rate in a country is linked to that country's average income.
- People in the north of England have larger families than people who live in the south.

The following websites are a useful source of data for some of the above.

www.statistics.gov.uk

www.national-lottery.co.uk

www.cia.gov/cia/publications/the-world-factbook/index.html

GRADE BOOSTER

- **D** You can draw and interpret pie charts
- **D** You can find the mean from a frequency table of discrete data and also draw a frequency polygon for such data
- **C** You can find an estimate of the mean from a grouped table of continuous data and draw a frequency polygon for continuous data
- **C** You can design questionnaires and surveys
- **C** You can use the data-handling cycle
- **A** You can draw histograms from frequency tables with unequal class intervals
- **A** You can calculate the numbers to be surveyed for a stratified sample
- **A*** You can find the median, quartiles and the interquartile range from a histogram

What you should know now

- Which average to use in different situations
- How to find the modal class and an estimated mean for continuous data
- How to draw and interpet pie charts
- How to draw frequency polygons and histograms for discrete and continuous data
- How to design questionnaires and surveys
- How to use the data-handling cycle to test and refine a hypothesis
- How to work out the numbers to survey within each group for a stratified sample.

1 **a** The two-way table shows the number of televisions and radios in 50 households.

		Televisions			
		0	1	2	3
Radios	0	0	3	0	0
	1	2	5	8	3
	2	1	6	12	5
	3	0	0	3	2

i How many households have two televisions and one radio? *(1 mark)*

ii How many households have two televisions? *(2 marks)*

iii How many households have the same number of televisions as radios? *(2 marks)*

b Louise wanted to find out the number of hours of television watched last Sunday.

This is her question and response section.

Question: How many hours of television did you watch last Sunday?
Response: Tick a box

☐ 1 to 3 hours ☐ 4 to 6 hours ☐ 6 to 8 hours ☐ more than 8 hours

Write down two criticisms of the response section. *(2 marks)*

AQA, June 2008, Module 1 Higher, Question 6

2 A factory manager surveys the owners of the cars parked in the car park.
One of the questions is:
When you drive to work how many people, including yourself, are in your car?
The responses are summarised in the table below.

Number of people	Number of cars
1	42
2	26
3	12
4	2
5	0
6	1
Total	83

a Calculate the mean number of people per car. *(3 marks)*

b The manager wants to find out if car owners are prepared to car-share.

Write a suitable question with a response section to find out which days from Monday to Friday the owners will car-share. *(2 marks)*

AQA, March 2008, Module 1 Higher, Question 2

3 A student recorded the time, in minutes, that 50 people spent in the library.

Time, t (minutes)	Frequency
$0 < t \leqslant 10$	2
$10 < t \leqslant 20$	8
$20 < t \leqslant 30$	20
$30 < t \leqslant 40$	12
$40 < t \leqslant 50$	8

Calculate an estimate of the mean number of minutes spent in the library. *(4 marks)*

AQA, March 2007, Module 1 Higher, Question 2

4 **a** The table shows the frequency of the variable, x, for various values.

x	Frequency
25	16
35	38
45	26
55	14
65	6
Total	100

Show that the mean of x is 40.6 *(3 marks)*

b The table shows the heights, h (in centimetres), of 100 girls in year 10.

Height, h (cm)	Frequency
$120 < h \leqslant 130$	16
$130 < h \leqslant 140$	38
$140 < h \leqslant 150$	26
$150 < h \leqslant 160$	14
$160 < h \leqslant 170$	6
Total	100

i What is the midpoint of the group $120 < h \leqslant 130$? *(1 mark)*

ii Using the mean value of x from part **a**, write down an estimate for the mean height of the 100 girls. *(1 mark)*

c The frequency diagram shows the distribution of the heights of 100 boys in Y10.

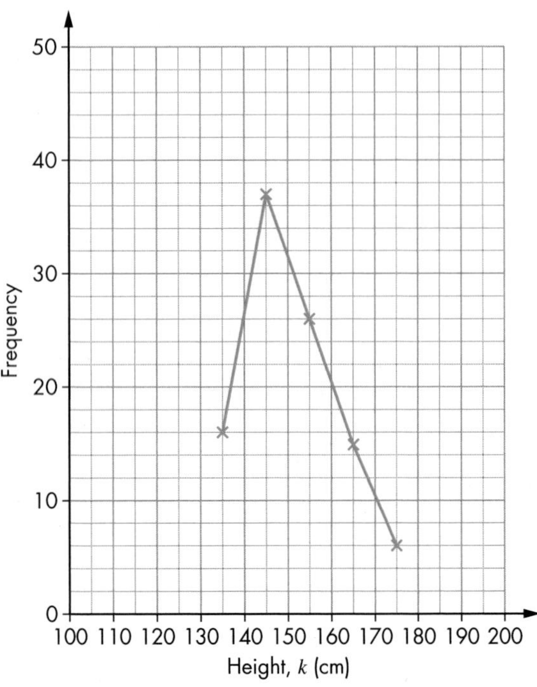

i On a copy of the same grid, draw a frequency diagram for the heights of the girls in Y10. *(2 marks)*

ii Make two comments to compare the heights of the boys and the girls. *(2 marks)*

AQA, June 2008, Paper 2 Higher, Question 13

5 The headteacher of a school sends a questionnaire to each head of department.

One of the questions is:

How many hours do you think you are working each week?

The results are shown in the table.

Hours worked each week, t	Number of heads of department
$25 \leqslant t < 30$	0
$30 \leqslant t < 35$	2
$35 \leqslant t < 40$	3
$40 \leqslant t < 45$	8
$45 \leqslant t < 50$	6
$50 \leqslant t < 55$	1

a Draw a frequency diagram to represent this data. *(3 marks)*

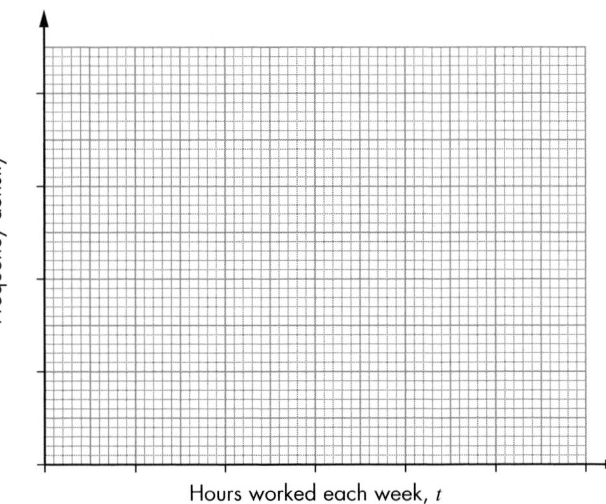

b There are 80 teachers in the school.

Use the figures in the table to calculate an estimate of how many teachers in the school are working 40 or more hours each week. *(3 marks)*

c Explain why the answer to part **b** is likely to be unrealistic. *(1 mark)*

AQA, March 2008, Module 1 Higher, Question 5

6 In Year 9 there are 30 students who study both German and French. Their National Curriculum levels in these subjects are shown in the two-way table.

		Level in French					
		1	2	3	4	5	6
Level in German	1	0	0	0	0	0	0
	2	1	0	0	0	0	0
	3	2	1	1	0	0	0
	4	0	3	4	1	0	0
	5	0	1	2	3	2	0
	6	0	0	3	3	2	1

a What is the median level for German?

b What is the mean level for French?

c The teacher claims that the students are better at German than at French. How can you tell from the table that this is true?

7 The weekly pocket money of one class is shown in the table below.

Pocket money	£0–£5	£5–£10	£10–£15	£15–£20
Frequency	4	6	12	8

Sean says that he has estimated the mean amount of pocket money as £9.50.

Explain how you can tell Sean must be wrong without having to calculate the estimated mean.

8 The table shows information about the times taken by a group of 50 girls to complete a challenge.

Time taken, t (minutes)	Frequency		
$0 < t \leqslant 10$	6		
$10 < t \leqslant 20$	10		
$20 < t \leqslant 30$	20		
$30 < t \leqslant 40$	8		
$40 < t \leqslant 50$	6		

Calculate an estimate of the mean time taken for the girls to complete the challenge. *(4 marks)*

AQA, November 2011, H1, Question 10

9 The frequency polygon shows the times taken by a group of 50 boys to complete the same challenge.

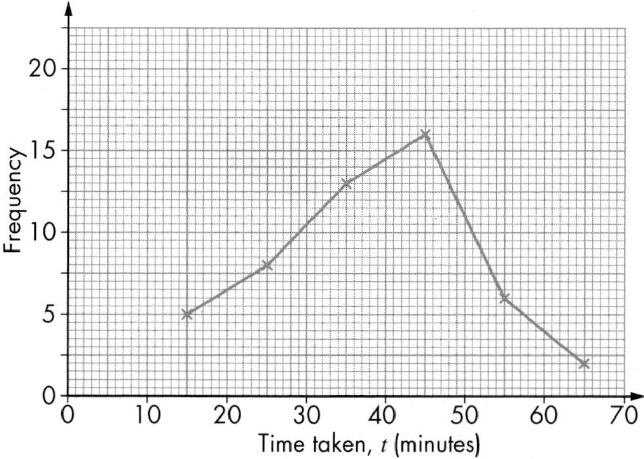

a On the same grid, draw a frequency polygon to show the times taken by the girls to complete the challenge. *(2 marks)*

b Make two comments to compare the times taken by the boys and the times taken by the girls to complete the challenge. *(2 marks)*

AQA, November 2011, H1, Question 10

10 The histogram shows the distribution of ages of 100 members of a chess club.

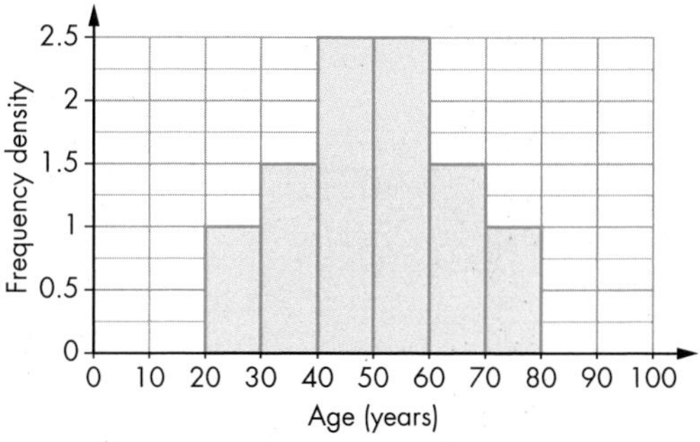

a How many members of the club were less than 40 years old? *(1 mark)*

b How many members of the club are between 40 and 60 years old? *(1 mark)*

c Work out the interquartile range of the ages. *(2 marks)*

AQA, November 2008, Paper 2 Higher, Question 16

11 The histogram represents the times that a number of runners took to complete a cross-country race.

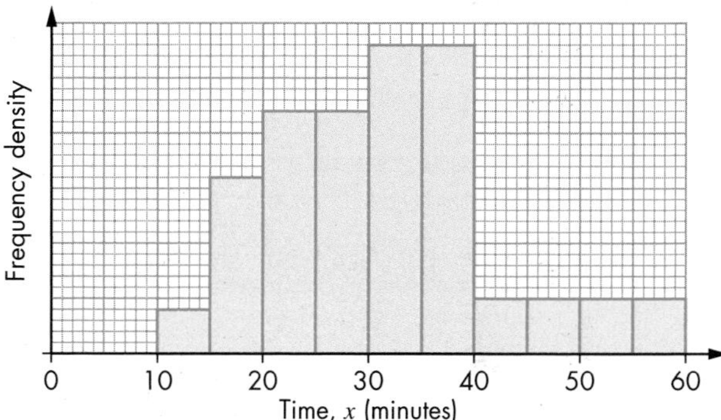

Ten runners completed the race in under 20 minutes.

How many runners completed the race? *(3 marks)*

AQA, June 2008, Module 1 Higher, Question 9

12 A mobile speed camera recorded the speed of some vehicles on a motorway.

The table shows the results.

Speed, s (mph)	Frequency
$0 < s \leqslant 30$	42
$30 < s \leqslant 50$	54
$50 < s \leqslant 60$	82
$60 < s \leqslant 70$	116
$70 < s \leqslant 80$	70
$80 < s \leqslant 120$	36
Total	400

a Draw a histogram to illustrate the data. *(3 marks)*

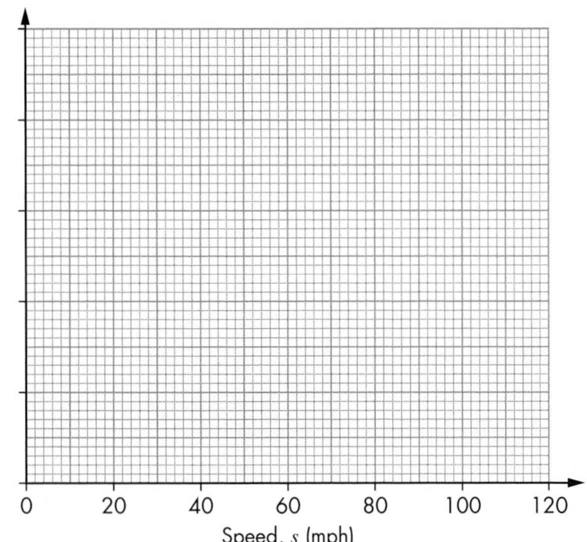

b Drivers of vehicles doing more than 77 miles per hour were given a speeding ticket.

Estimate the number of drivers who receive a ticket. *(1 mark)*

AQA, June 2008, Paper 2 Higher, Question 22

13 The examination scores of a group of students are summarised in the table.

Examination score, s	Frequency
$0 \leqslant s < 20$	10
$20 \leqslant s < 40$	18
$40 \leqslant s < 50$	25
$50 \leqslant s < 60$	20
$60 \leqslant s < 80$	16
$80 \leqslant s < 100$	2

a Draw a histogram for this data. *(3 marks)*

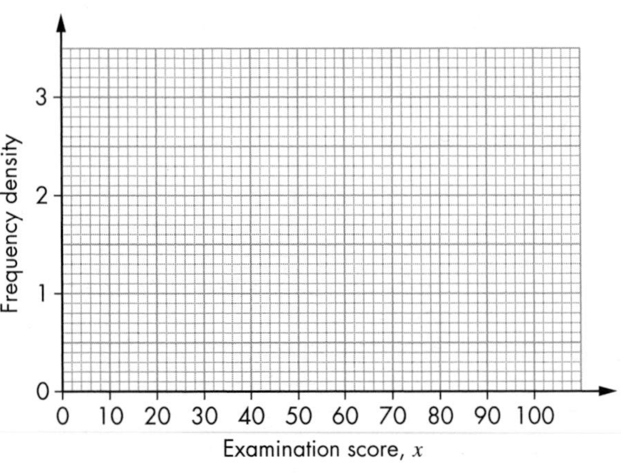

b A Merit is awarded for a mark between 48 and 75.

Calculate an estimate of the number of students awarded a Merit. *(2 marks)*

AQA, November 2008, Module 1 Higher, Question 8

14 A company wants to obtain a stratified sample of total size 2000 from the members of three teaching unions.

The table shows the number of members, in thousands, of the three unions.

Union	No. of members in thousands
NUT	260
ATL	170
NATFHE	70

Calculate the number of members of each union selected for the stratified sample.

15 The table shows the number of students in each year group of a junior school.

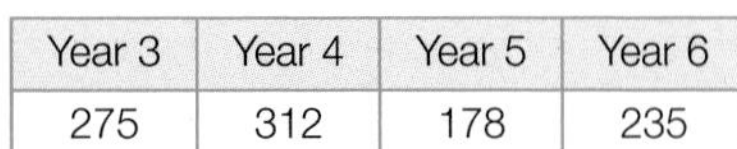

Year 3	Year 4	Year 5	Year 6
275	312	178	235

There are 1000 students in the school.

The headteacher wants to take a stratified sample of 50 children to carry out a survey.

How many more year 6 students than year 5 students would be chosen? *(3 marks)*

AQA, June 2008, Module 1 Higher, Question 4

16 The table shows the bookings at a hotel for one month.

Single person	Couple	Family
27	70	103

The hotel manager wants to send questionnaires to a stratified sample of 30 of these bookings.

Calculate the number of each type of booking he should include. *(3 marks)*

AQA, March 2009, Module 1 Higher, Question 4

Worked Examination Questions

1 The distances travelled by 100 cars using 10 litres of petrol is shown in the histogram and table.

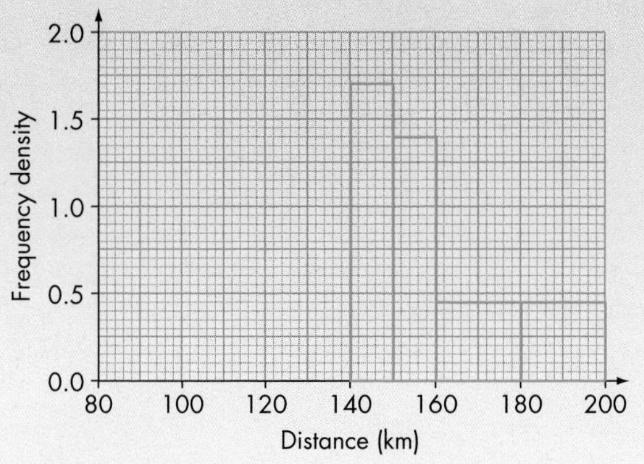

a Complete the histogram and the table.

b Estimate the number of cars that travel between 155 km and 185 km using 10 litres of petrol.

Distance (km)	80–110	110–130	130–140	140–150	150–160	160–200
Frequency	9	22	20			

1 a Set up the table with columns for class width and frequency density and fill in the given information, reading frequency densities from the graph (be careful with scales).

Now fill in the rest of the information using f.d. $= \dfrac{\text{frequency}}{\text{class width}}$ and

frequency = f.d. × class width

These values are shown in red.

Distance (km)	Frequency	Class width	Frequency density
80–110	9	30	0.3
110–130	22	20	1.1
130–140	20	10	2
140–150	17	10	1.7
150–160	14	10	1.4
160–200	18	40	0.45

This scores 1 mark for accuracy in completing the frequency column from the information in the histogram. This scores 1 mark for method in creating the frequency density column. This scores 2 marks for accuracy of completing the frequency density column – 1 mark lost for each error.

Complete the graph.

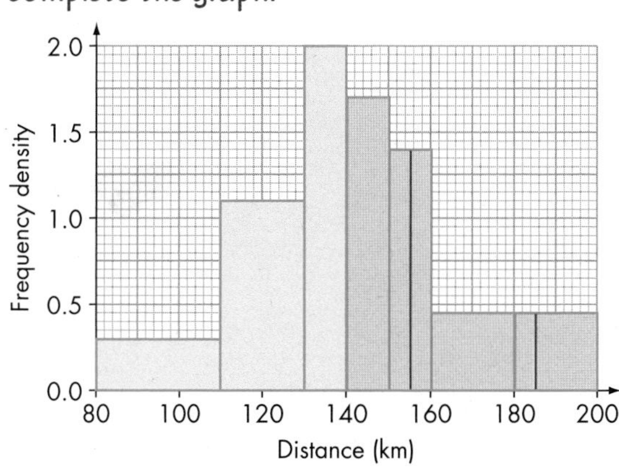

This scores 3 marks for accurately drawing the three columns.

b Draw lines at 155 and 185. The number of cars is represented by the area between these lines.
In the 150–160 bar the area is $\frac{1}{2}$ of the total.
In the 160–200 bar the area is $\frac{5}{8}$ of the total.

This scores 1 mark for method of attempting to find area between the two lines.

Number of cars $= \frac{1}{2} \times 14 + \frac{5}{8} \times 18 = 18.25 \approx 18$ cars

This scores 1 mark for method and 1 mark for accuracy.

Total: 10 marks

Worked Examination Questions

FM **2** A survey was taken to find how many days people had to wait to get a dental appointment. One hundred people were surveyed in three different cities with the following results.

Days, d	$0 < d \leqslant 2$	$2 < d \leqslant 5$	$5 < d \leqslant 10$	$10 < d \leqslant 15$	$15 < d \leqslant 20$
London	25	47	20	7	1
Sheffield	13	44	35	6	2
York	8	37	41	10	4

a What is the average number of days spent waiting for a dental appointment in each city?

b Give two reasons why the average number of days for each city might be so different.

2 a London

[(1 × 25) + (3.5 × 47) + (7.5 × 20) + (12.5 × 7) + (17.5)] ÷ 100

= 444.5 ÷ 100 = 4.445

London = 4.4 days

Sheffield

[(1 × 13) + (3.5 × 44) + (7.5 × 35) + (12.5 × 6) + (17.5 × 2)] ÷ 100

= 539.5 ÷ 100 = 5.395

Sheffield = 5.4 days

York

[(1 × 8) + (3.5 × 37) + (7.5 × 41) + (12.5 × 10) + (17.5 × 4] ÷ 100

= 640 ÷ 100 = 6.4

York = 6.4 days

> This scores 1 mark for method of using midpoints in at least one city.
>
> This scores 1 mark for method of multiplying the midpoints by each frequency in at least one city.
>
> This scores 3 marks for accuracy for the mean in each city.

b There might be more dentists in London, or there might be more people with problem teeth in York.

> This scores 1 mark for each valid reason, to a maximum of two marks.

Total: 7 marks

PS **3** The mean speed of each member of a cycling club over a long-distance race was recorded and a frequency polygon was drawn.

From the frequency polygon, estimate the mean speed.

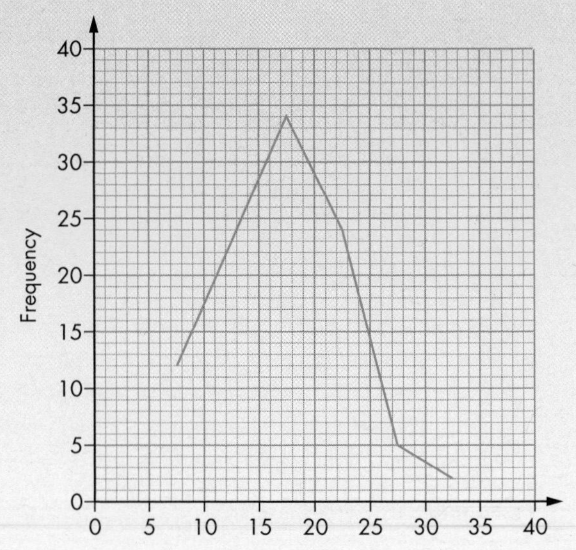

Worked Examination Questions

3 Create a grouped frequency table.

Speed, s (mph)	Frequency, f	Midpoint, m	$f \times m$
5–10	12	7.5	90
10–15	23	12.5	287.5
15–20	34	17.5	595
20–25	24	22.5	540
25–30	5	27.5	137.5
30–35	2	32.5	65
	100		1715

The table must include the midpoint values, the frequency and $f \times m$.

This gets 1 method mark for giving the frequencies and showing a total.

This gets 1 method mark for working out the midpoints.

This gets 1 method mark for attempting to work out $f \times m$ and showing a total.

An estimate of the mean is 1715 ÷ 100 = 17.15 mph

This gets 1 method mark for dividing the total $f \times m$ by the total f.

This gets 1 mark for accuracy.

Total: 5 marks

4 An inspector visits a large company to check their vehicles. The company has large-load vehicles, light vans and cars. The inspector decides to sample 100 vehicles. Each type of vehicle is to be represented in the sample.

a What is this kind of sampling procedure called?

b How will the inspector decide how many of each type of vehicle he should inspect?

4 a Stratified sampling

This gets 1 mark.

b Find the total number of vehicles and the numbers of each type.

This gets 1 mark for stating he needs to find the numbers of each type of vehicle.

Work out the proportion of each type of vehicle in the total number.

Multiply each proportion by 100.

This gets 1 mark.

This will give the number of each type of vehicle to inspect.

This gets 1 mark for stating he needs to multiply the proportions by 100.

Total: 4 marks

12 Functional Maths
Fishing competition on the Avon

Averages are used to compare data collected through surveys and investigations. They are used every day for a wide variety of purposes, from describing the weather to analysing the economy. In this activity surveys and averages will be looked at in the context of a sporting event.

Your task

Every summer, Kath's family runs a fishing competition on their land by the river Avon. Kath plans to use this year's fishing competition to learn more about competitive fishing and the 'average' angler.

- Show how Kath could collect information on competitive fishing.
- Kath collects this data during the first four weeks in July:

	Week 1	Week 2	Week 3	Week 4	Week 5
Mean number of fish caught	12.1	12.3	7.2	11.8	
Mean time spent fishing (h)	6.1	5.6	4.5	5.4	
Mean weight of fish caught (g)	1576.0	1728.0	1635.0	1437.0	
Mean length of longest fish caught (cm)	21.7	17.6	21.6	19.2	

In the fifth week of July she collects this information (given on the right) from each angler.

- What conclusions can Kath draw from this data about fishing and the 'average' angler? Is it possible for Kath to apply her conclusions to competitive fishing in general?

Support your ideas with figures and diagrams.

Number of fish caught per angler
Up to 5: 6
Between 6 and 10: 11
Between 11 and 15: 8
Between 16 and 20: 5

Weight of largest fish caught (g)
Up to 500: 1
Between 501 and 1000: 8
Between 1001 and 1500: 18
Between 1501 and 2000: 3
Over 2000: 0

Time spent fishing per angler (h/min)
Up to 4: 2
Between 4 and 4.59: 14
Between 5 and 5.59: 6
Between 6 and 6.59: 8
7 and over: 0

Length of longest fish caught (cm)
Up to 10: 2
Between 10.1 and 15: 6
Between 15.1 and 20: 12
Between 20.1 and 25: 10

Getting started

Use the questions below to get you started in thinking about averages and the organisation of data.

- Start by thinking about how averages are calculated.
  - What do the mean, median, mode and range show when applied to sets of data? How do they apply to frequency tables?
  - What are the mean, median, mode and range for this set of data?
    3, 4, 8, 3, 2, 4, 5, 3, 4, 6, 8, 4, 2, 9, 1
  - Which averages might you need in your presentation? How could you represent these averages graphically?
- Think about how data can be ordered.
  - Write down a number smaller than five. Are you sure that you have written down the smallest number possible?
  - Write down the largest number that is smaller than five. Are you sure that you've written down the largest possible number?
  - Write this set of numbers using inequalities.

Handy hints

Inequalities can be used to accurately represent ranges in data.

$>$ means 'greater than' (for example, $5 > 2$)

$<$ means 'less than' (for example, $2 < 5$)

$\geqslant$ means 'greater than or equal to'

$\leqslant$ means 'less than or equal to'

Try to make use of these signs when representing your data.

Why this chapter matters

Graphs are to be found in a host of different media, including newspapers and the textbooks of most of the subjects that you learn in school.

Graphs are used to show the relationship between two variables. Very often, one of these variables is time. Then the graph shows how the other variable changes over time.

For example, this graph shows how the exchange rate between the dollar and the pound changed over five months in 2009.

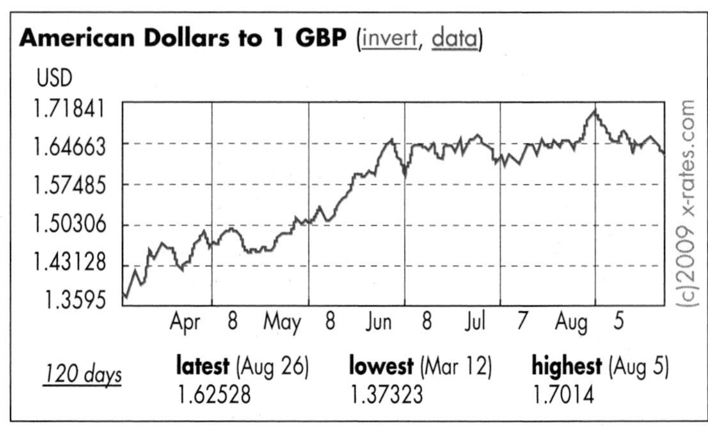

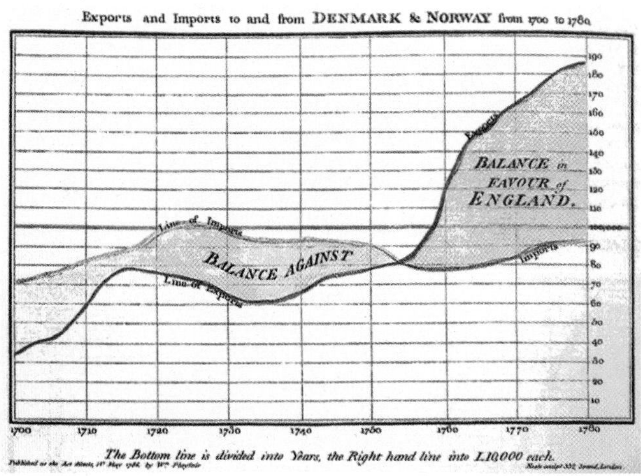

The earliest line graphs, such as the one shown here, appeared in the book *A Commercial and Political Atlas*, written in 1786 by Henry Playfair, who also used bar charts and pie charts for the first time. Playfair argued that charts and graphs communicated information to an audience better than tables of data.

Do you think that this is true?

This graph shows all the data from a racing car going around a circuit. Engineers can use this to fine-tune parts of the car to give the best performance. It illustrates that graphs give a visual representation of how variables change and can be used to compare data in a way that looking at lists of data cannot.

Think about instances in school and everyday life where a line graph would help you to communicate information more effectively.

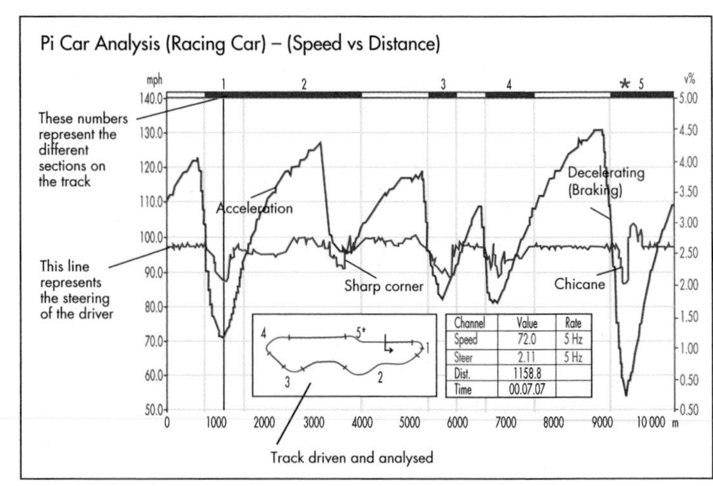

Chapter

Algebra: Real-life graphs

1 Straight-line distance–time graphs

2 Other types of graphs

This chapter will show you ...

D how to interpret distance–time graphs

C to A how to interpret other types of graphs associated with real-life situations

Visual overview

Distance–time graphs

Other types of graphs

What you should already know

- How to plot coordinate points **(KS3 level 4, GCSE grade F)**
- How to read scales **(KS3 level 4, GCSE grade F)**

Quick check

1 Give the coordinates of points A, B and C.

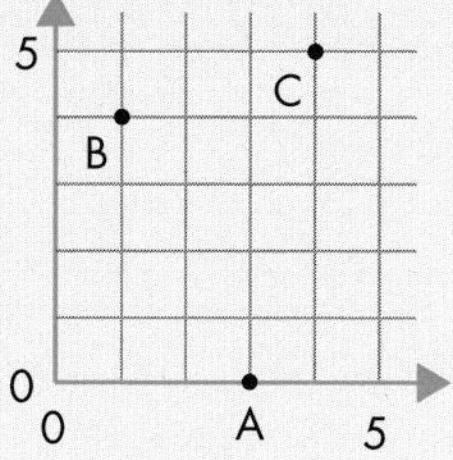

2 What are the values shown on the following scales?

a

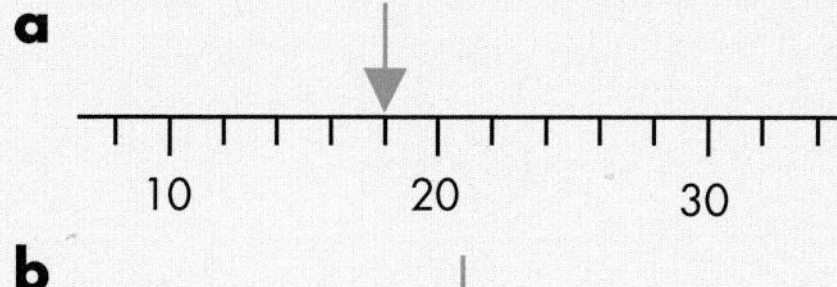

b

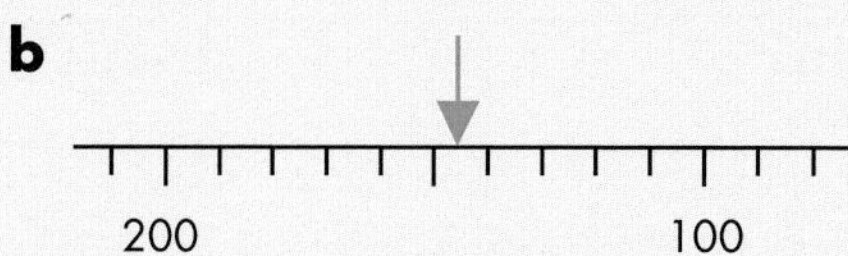

13.1 Straight-line distance–time graphs

This section will show you how to:
- interpret distance–time graphs

Key words
average speed, distance, distance–time graph, gradient, speed, time

As the name suggests, a **distance–time graph** gives information about how far someone or something has travelled over a given **time** period.

A travel graph is read in a similar way to the conversion graphs you have just done. But you can also find the **average speed** from a distance–time graph, using the formula:

$$\text{average speed} = \frac{\text{total distance travelled}}{\text{total time taken}}$$

EXAMPLE 1

The distance–time graph below represents a car journey from Barnsley to Nottingham, a distance of 50 km, and back again.

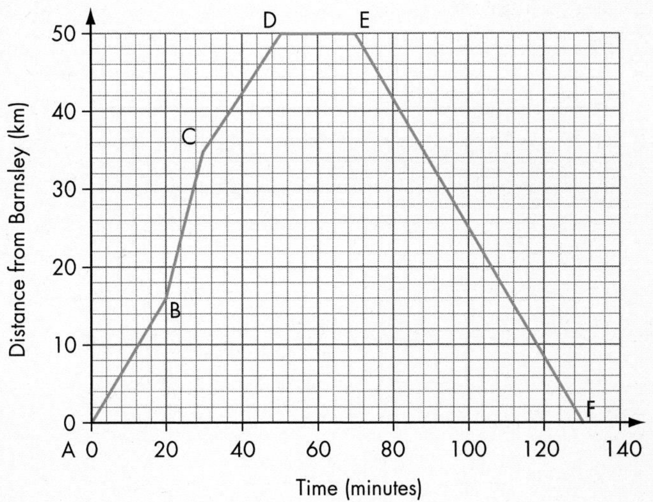

a What can you say about points B, C and D?

b What can you say about the journey from D to F?

c Work out the average speed for each of the five stages of the journey.

From the graph:

a B: After 20 minutes the car was 16 km away from Barnsley.
C: After 30 minutes the car was 35 km away from Barnsley.
D: After 50 minutes the car was 50 km away from Barnsley, so at Nottingham.

b D–F: The car stayed at Nottingham for 20 minutes, and then took 60 minutes for the return journey.

c The average speeds over the five stages of the journey are worked out as follows.

A to B represents 16 km in 20 minutes.

20 minutes is $\frac{1}{3}$ of an hour, so we need to multiply by 3 to give distance/hour.

Multiplying both numbers by 3 gives 48 km in 60 minutes, which is 48 km/h.

B to C represents 19 km in 10 minutes.

Multiplying both numbers by 6 gives 114 km in 60 minutes, which is 114 km/h.

C to D represents 15 km in 20 minutes.

Multiplying both numbers by 3 gives 45 km in 60 minutes, which is 45 km/h.

D to E represents a stop: no further distance travelled.

E to F represents the return journey of 50 km in 60 minutes, which is 50 km/h.

So, the return journey was at an average speed of 50 km/h.

EXERCISE 13A

D

FM 1 Paul was travelling in his car to a meeting. This distance–time graph illustrates his journey.

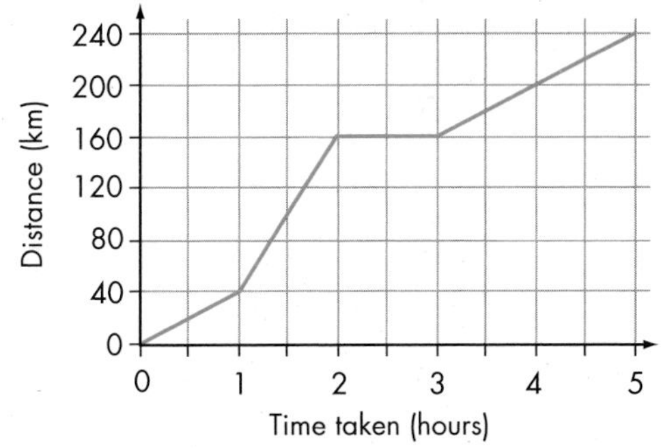

a How long after he set off did he:

i stop for his break

ii set off after his break

iii get to his meeting place?

b At what average speed was he travelling:

i over the first hour

ii over the second hour

iii for the last part of his journey?

c The meeting was scheduled to start at 10.30 am. What is the latest time he should have left home?

HINTS AND TIPS

If a part of a journey takes 30 minutes, just double the distance to get the average speed.

FM 2 James was travelling to Cornwall on his holiday. This distance–time graph illustrates his journey.

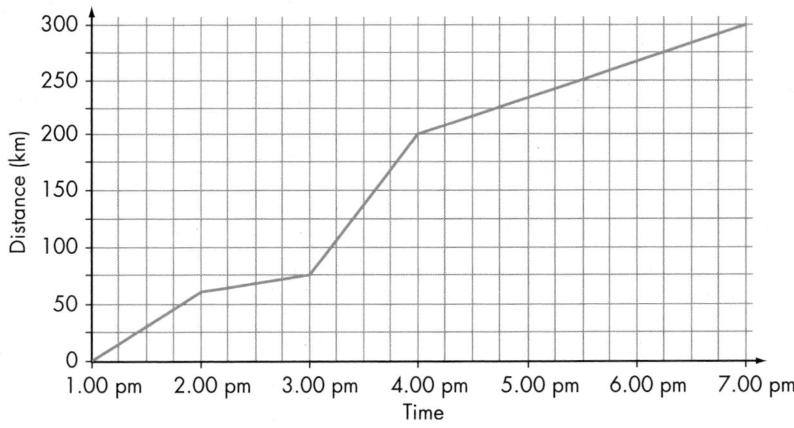

a His greatest speed was on the motorway.

i How far did he travel on the motorway?

ii What was his average speed on the motorway?

b **i** When did he travel the most slowly? **ii** What was his lowest average speed?

D

FM **3** A small bus set off from Leeds to pick up Mike and his family. It then went on to pick up Mike's parents and grandparents. It then travelled further, dropping them all off at a hotel. The bus then went on a further 10 km to pick up another party and it took them back to Leeds. This distance–time graph illustrates the journey.

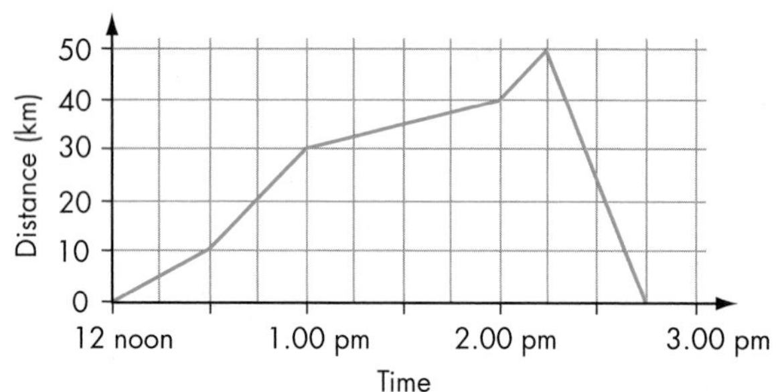

a How far from Leeds did Mike's parents and grandparents live?

b How far from Leeds is the hotel at which they all stayed?

c What was the average speed of the bus on its way back to Leeds?

PS **4** Richard and Paul took part in a 5000-m race. It is illustrated in this graph.

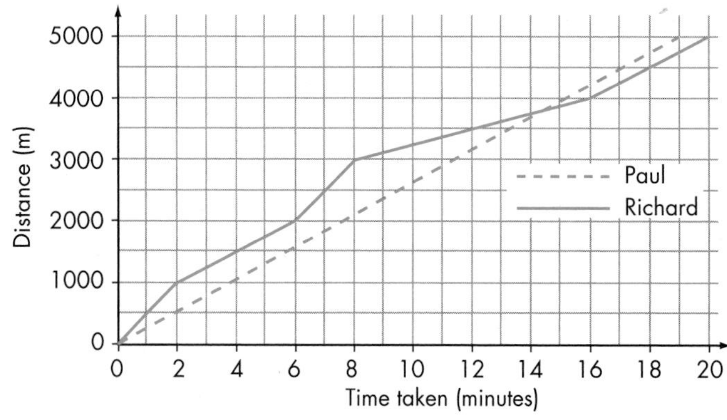

a Paul ran a steady race. What is his average speed in:

i metres per minute

ii kilometres per hour?

b Richard ran in spurts. What was his highest average speed?

c Who won the race and by how much?

FM PS **5** Three friends, Patrick, Araf and Sean, ran a 1000-m race. The race is illustrated on the distance–time graph shown here.

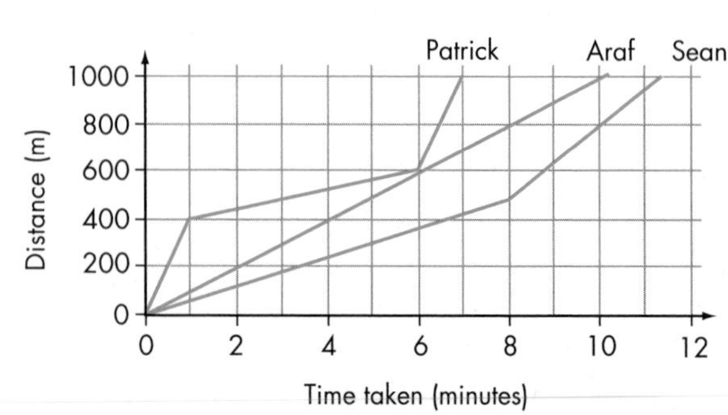

a Describe how each of them completed the race.

b i What is Araf's average speed in m/s?

ii What is this speed in km/h?

D

PS 6 A walker sets off at 9.00 am from point P to walk along a trail at a steady pace of 6 km per hour.

90 minutes later, a cyclist sets off from P on the same trail at a steady pace of 15 km per hour.

At what time did the cyclist overtake the walker?

You may use a graph to help you solve this question.

> **HINTS AND TIPS**
>
> This question can be done by many methods, but drawing a distance–time graph is the easiest. Mark a grid with a horizontal axis as time from 9 am to 1 pm, and the vertical axis as distance from 0 to 24. Draw lines for both walker and cyclist. Remember that the cyclist doesn't start until 10.30.

C

AU 7 Three school friends set off from school at the same time, 3.45 pm. They all lived 12 km away from the school. The distance–time graph illustrates their journeys.

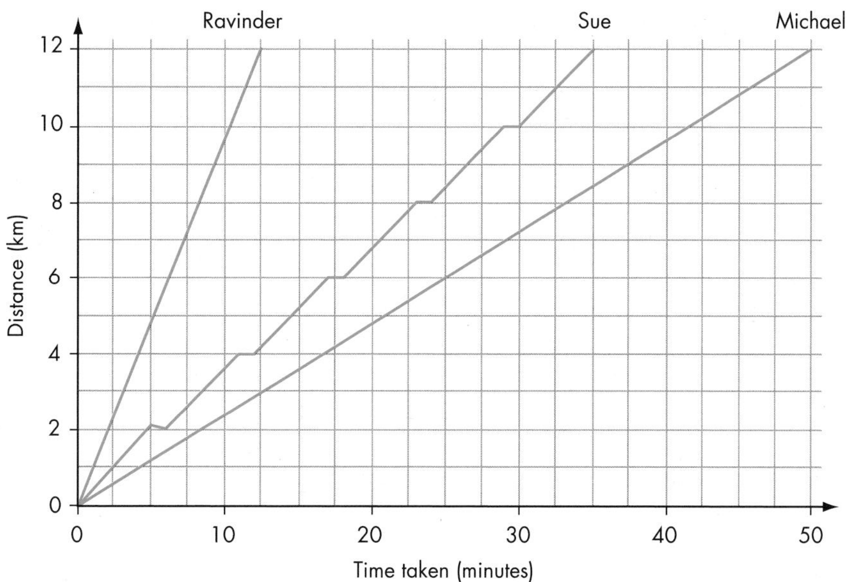

One of them went by bus, one cycled and one was taken by car.

a **i** Explain how you know that Sue used the bus.

ii Who went by car?

b At what times did each of them get home?

c **i** When the bus was moving, it covered 2 km in 5 minutes. What is this speed in kilometres per hour?

ii Overall, the bus covered 12 km in 35 minutes. What is this speed, in kilometres per hour?

iii How many stops did the bus make before Sue got off?

Gradient of straight-line distance–time graphs

The **gradient** of a straight line is a measure of its slope.

You can find the gradient of this line by constructing a right-angled triangle of which the hypotenuse (sloping side) is on the line. Then:

$$\text{average speed} = \frac{\text{distance measured vertically}}{\text{distance measured horizontally}} = \frac{6}{4} = 1.5$$

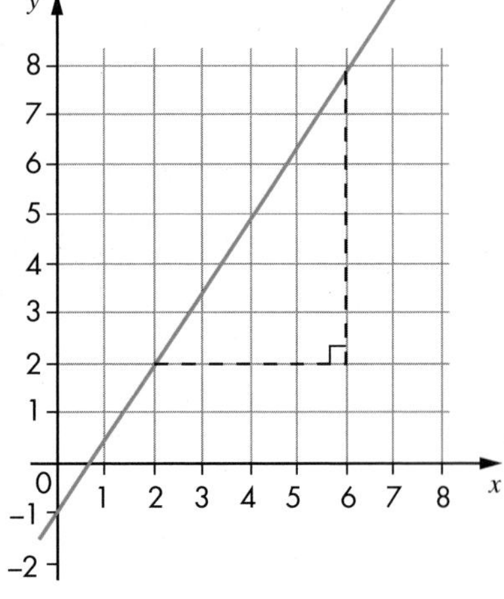

Look at the following examples of straight lines and their gradients.

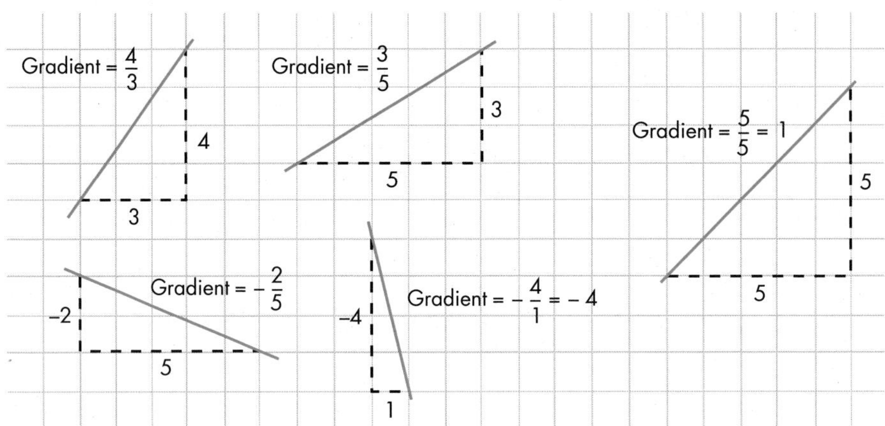

Note: Lines which slope downwards from left to right have *negative gradients.*

You can only count squares to find the gradient if the scale on both axes is one square to one unit. For a straight-line graph that compares two quantities, you must find the gradient by using the *scales* on its axes, *not* the actual number of grid squares. The gradient usually represents a third quantity, the value of which you want to know. For example, look at the next graph.

The gradient on this distance–time graph represents average speed.

$$\text{Gradient} = \frac{500 \text{ km}}{2 \text{ h}} = 250 \text{ km/h}$$

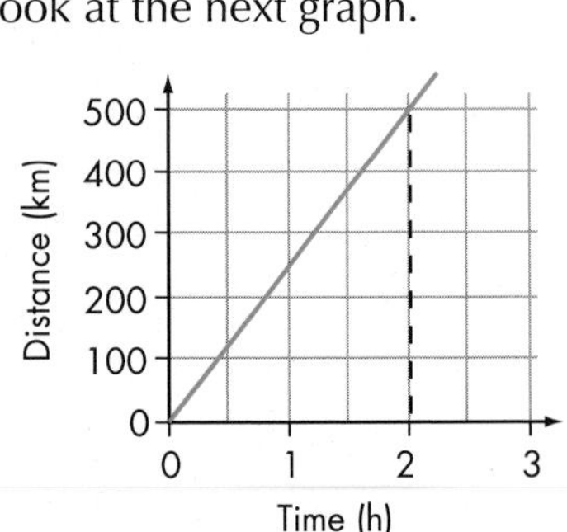

EXERCISE 13B

D

FM **1** Ravi was ill in hospital.

This is his temperature chart for the two weeks he was in hospital.

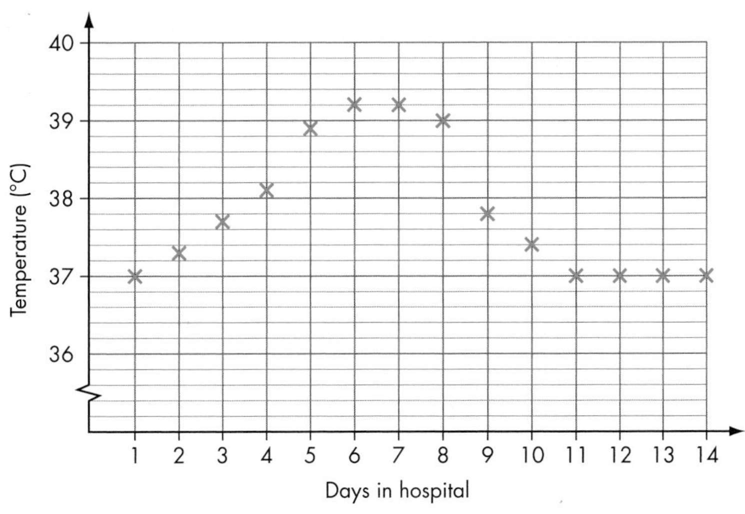

a What was Ravi's highest temperature?

b Between which days did Ravi's temperature increase the fastest? Explain how you can tell.

c Between which days did Ravi's temperature fall the fastest? Explain how you can tell.

d When Ravi's temperature went over 38.5 °C he was put on an antibiotic drip.

i On what day did Ravi go on the drip?

ii How many days did it take for the antibiotics to work before Ravi's temperature was below 38.5 again?

e Once Ravi's temperature had returned to normal and was stable for four days, he was allowed home. What is the normal body temperature?

C

2 Calculate the gradient of each line, using the scales on the axes.

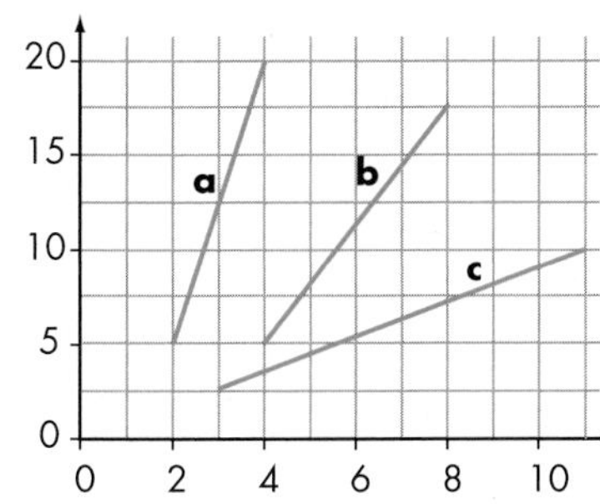

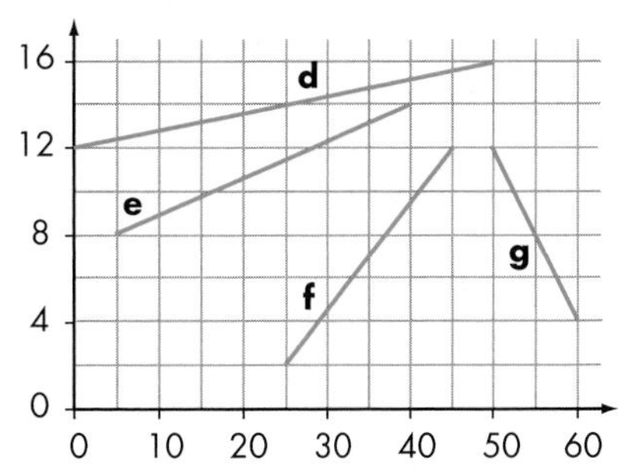

C

3 Calculate the average speed of the journey represented by the line in each of the following diagrams.

a

Distance (km)
0 10 20 30
Time (hours)
0 1 2 3 4

b

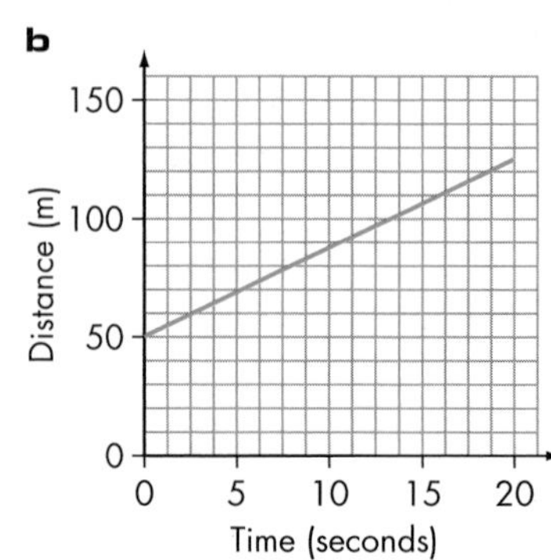

c

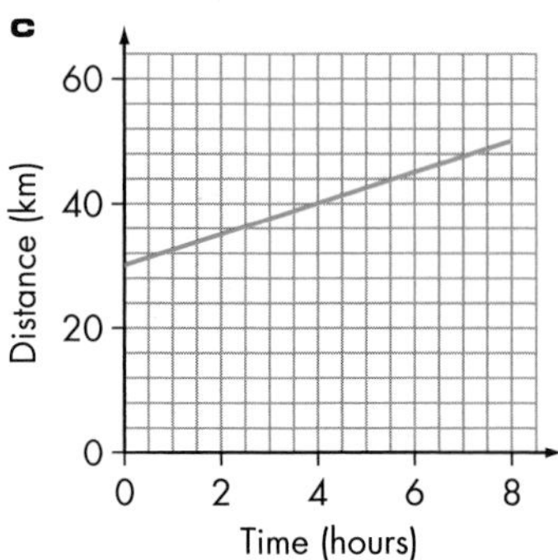

4 From the diagrams below, calculate the speed for each stage of each journey.

a

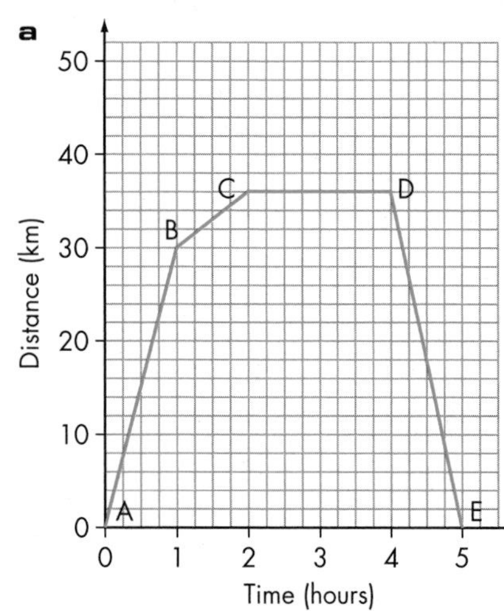

b

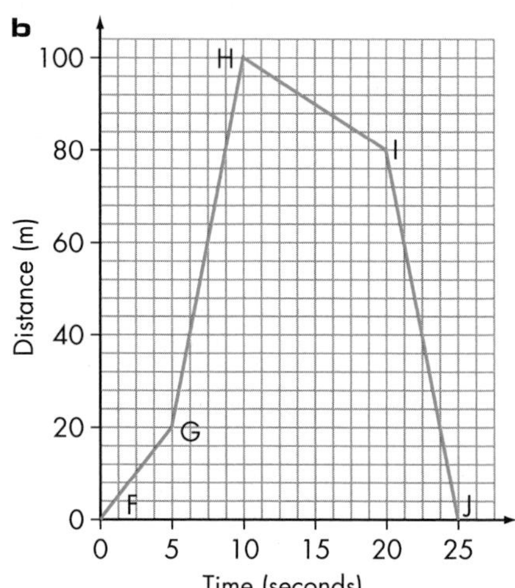

AU 5 Students in a class were asked to predict the y-value for an x-value of 10 for this line.

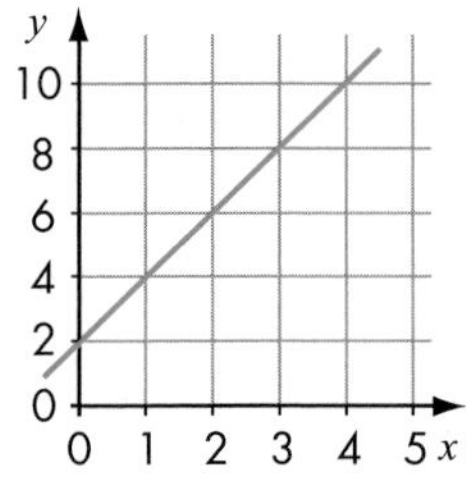

Rob says, "The gradient is 1, so the line is $y = x + 2$. When $x = 10$, $y = 12$."

Rob is wrong.

Explain why and work out the correct y-value.

FM PS 6 The Health and Safety regulations for vent pipes from gas appliances state that the minimum height depends on the pitch (gradient) of the roof.

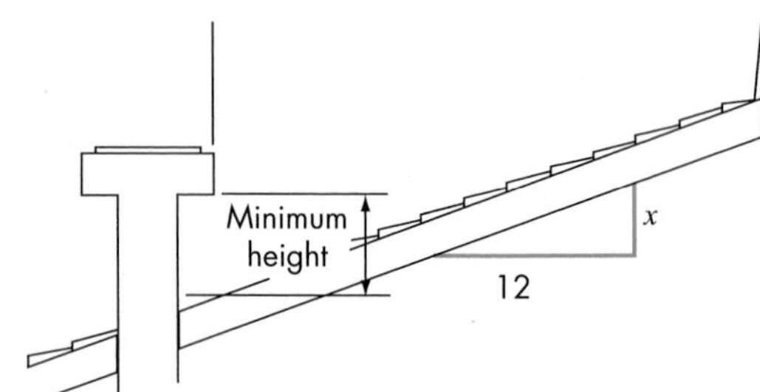

This is the rule:

minimum height = 1 m or twice the pitch in metres, whichever is greater

a What is the minimum height for a roof with a pitch of 2?

b What is the minimum height for a roof with a pitch of 0.4?

c What is the minimum height for these two roofs?

i

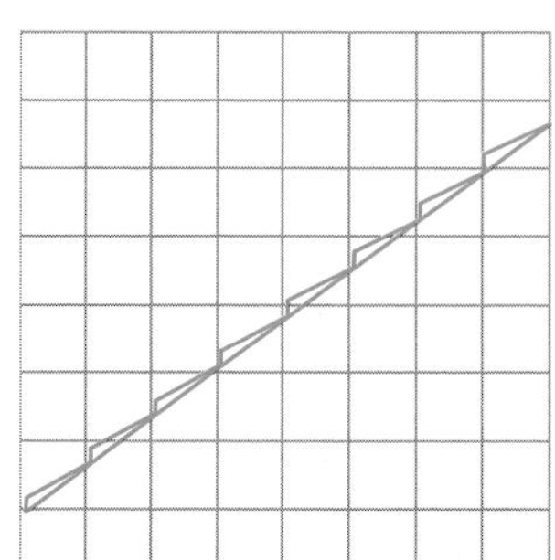

ii

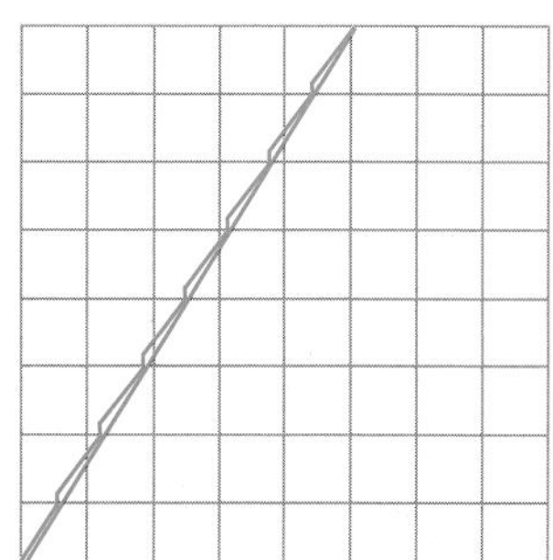

13.2 Other types of graphs

This section will show you how to:

- identify and draw some of the more unusual types of real-life graphs

Some situations can lead to unusual graphs. For example, this graph represents the cost of postage of a first-class letter against its weight.

This next graph shows the change in the depth of water in a flat-bottomed flask, as it is filled from a tap delivering water at a steady rate. The graph shows that at first the depth of water increases quickly then slows down as the flask gets wider. As the flask gets narrower, the depth increases at a faster rate. Finally, when the water reaches the neck, which has a constant cross-section, the depth increases at a constant rate to the top of the neck.

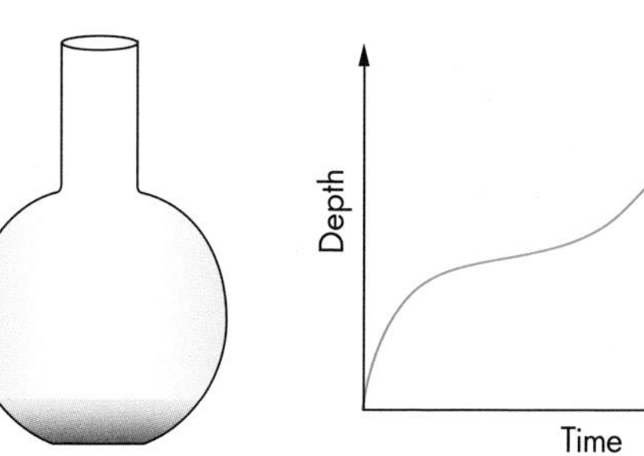

The application of graphs to describe the rate of change of depth as a container is filled with water is also covered in question **1** (Exercise 13C).

Another example of the use of graphs is set out in question **3** of Exercise 13C, in the calculation of personal income tax.

EXERCISE 13C

AU 1 **a** Liquid is poured at a steady rate into the bottle shown in the diagram.

As the bottle is filled, the depth, d, of the liquid in the bottle changes.

Which of the four graphs below shows the change in depth?

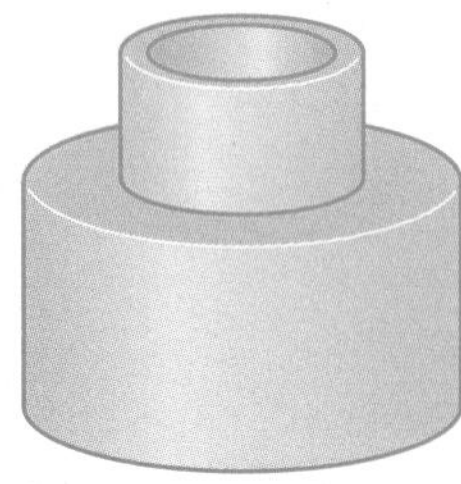

A: d, Time

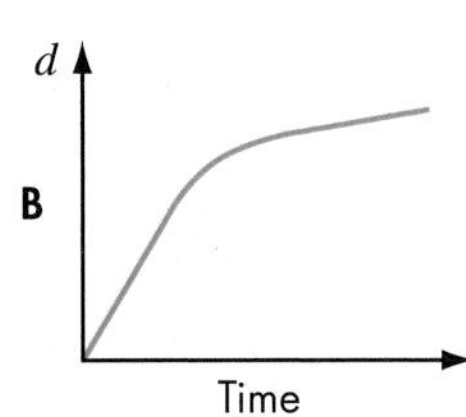

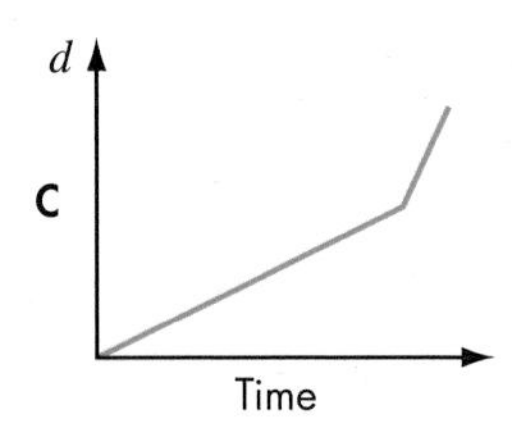

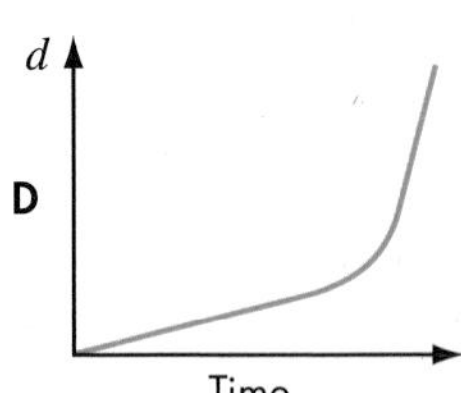

b Liquid is poured at a steady rate into another container.

The graph shows how the depth, d, changes.

Sketch a picture of this container.

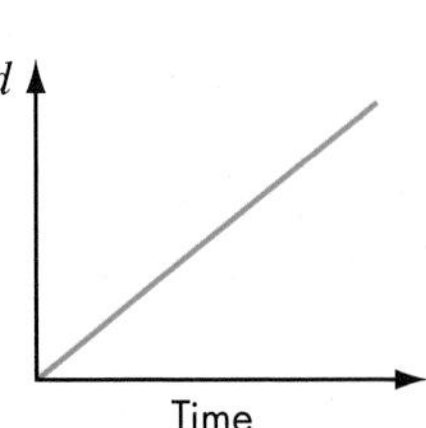

PS 2 Draw a graph of the depth of water in each of these containers as it is filled steadily.

a

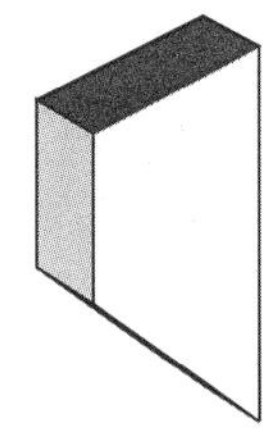

b

c

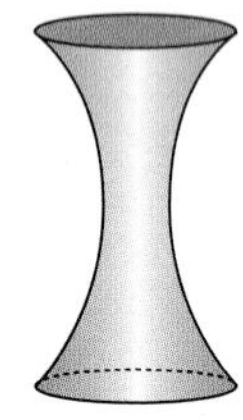

d

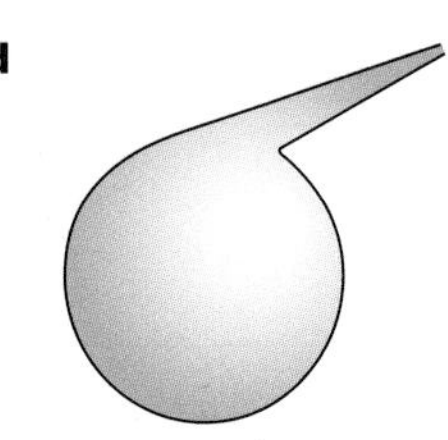

e

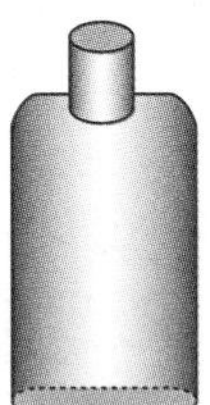

f

FM 3 The following is a simplified model of how income tax is calculated for an individual.

The first £5000 earned is tax free. Any income over £5000 up to £35 000 is taxed at 20%.

For example, a person who earns £20 000 per year would pay 20% of £15 000 i.e. £3000.

Any income over £35 000 is taxed at 40%.

Draw a graph to show the amount of tax paid by people who earn up to £50 000 per year.

Take the horizontal axis as 'income' from £0 to £50 000. Take the vertical axis as 'tax paid' from £0 to £12 000.

GRADE BOOSTER

D You can draw and read information from a distance–time graph

C You can calculate the gradient of a straight line and use this to find speed from a distance–time graph

B You can interpret real-life graphs

A You can interpret and draw more complex real-life graphs

What you should know now

- How to find the speed from a distance–time graph
- How to interpret real-life graphs

1 Mr Smith leaves home at 10 am to go to the shopping mall. He walks to the station where he catches a train. He gets off at the mall. The travel graph shows his journey.

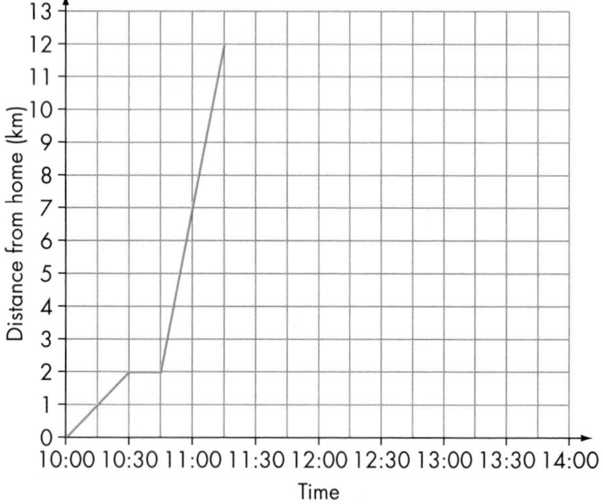

After shopping Mr Smith goes home by taxi. The taxi leaves the mall at 1 pm and arrives at his home at 1:45 pm.

a Complete the travel graph. *(2 marks)*

b Calculate the average speed of the taxi. *(2 marks)*

AQA, June 2005, Paper 2 (2-tier trial), Question 1

2 Hannah cycles from home to her friend's house. The graph shows her journey.

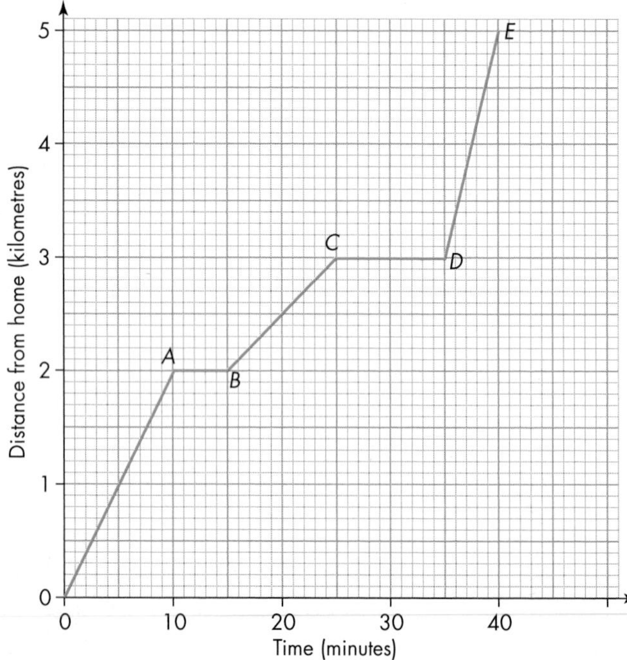

a What is the total length of time she stops on her journey? *(1 mark)*

b How far does she travel between *B* and *E*? *(1 mark)*

c On which part of the journey is her average speed the fastest? Give a reason for your answer. *(1 mark)*

AQA, June 2011, F1, Question 21

3 Kevin drove from Leeds to Luton.

The distance–time graph shows his journey.

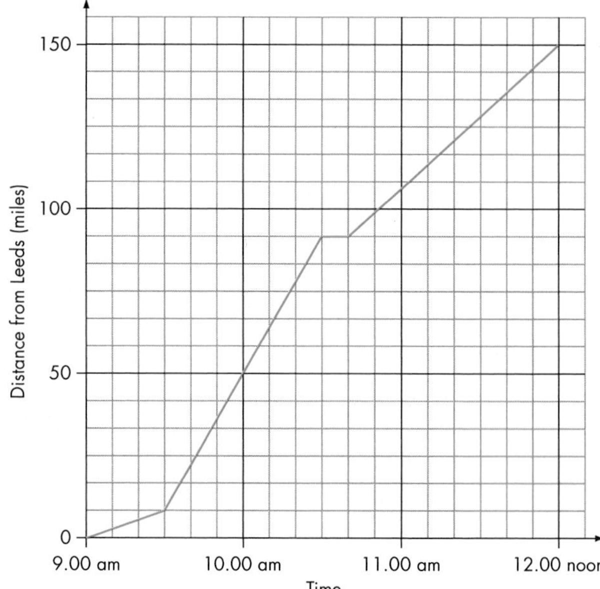

a How far is it from Leeds to Luton? *(1 mark)*

b Kevin stopped at a service station for petrol.

How long did he stop for? *(1 mark)*

c What was Kevin's average speed for the whole journey? *(2 marks)*

AQA, June 2008, Paper 2 Foundation, Question 20

4 A train travels from Glasgow to London in $4\frac{3}{4}$ hours. The distance travelled is 323 miles. Find the average speed of the train in miles per hour.

5 A motorbike drives from Sheffield to Plymouth. The journey is 468 kilometres in total. 372 kilometres are on motorway and 96 kilometres on normal roads. On normal roads the bike does 15 kilometres to a litre of petrol. In total the bike uses 25 litres of petrol on the journey. How many kilometres per litre does the bike do on average on motorways?

6 Grant and Mark race each other over two lengths of a 50 metre swimming pool.

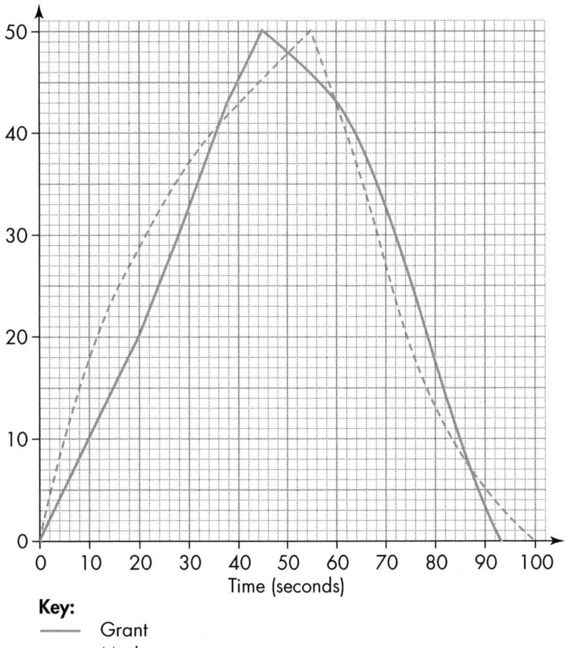

a Who won the race? *(1 mark)*

b What was the winning time? *(1 mark)*

c What was Grant's average speed during the first 30 seconds of the race? *(2 marks)*

d **i** Who was swimming faster at 60 seconds? *(1 mark)*

ii How can you tell from the graph? *(1 mark)*

AQA, November 2007, Paper 2 Intermediate, Question 10

7 A cyclist sets off on a ride over the moors at 9 am. The climb up to the highest point is 25 km. It takes the cyclist 1 hour 30 minutes to do this. She rests at the highest point for 15 minutes then sets off back. She arrives home at 11:45 am.

a Show this information on a travel graph with a horizontal axis showing time from 9 am to 12 pm and a vertical axis showing distance from home from 0 to 40 km. *(2 marks)*

b Calculate the average speed of the return journey in km per hour. *(2 marks)*

AQA, June 2005, Paper 2 Higher, Question 8

8 The waist-to-hip ratio has been found to be an important predictor of health problems.

The ratio is expressed as $1 : n$

where $n = \dfrac{\text{waist circumference}}{\text{hip circumference}}$

The table shows the health risk associated with different ratios.

Risk	Men	Women
High Risk	$n > 1.2$	$n > 1$
Moderate Risk	$1 \leq n \leq 1.2$	$0.8 \leq n \leq 1$
Low Risk	$n < 1$	$n < 0.8$

This graph shows the health risk for men for various waist and hip circumferences.

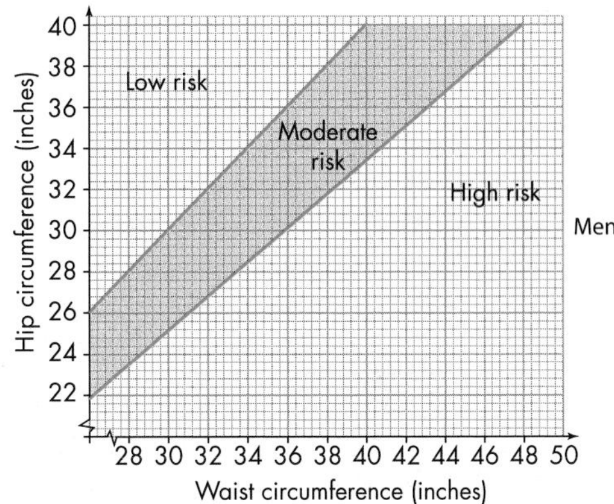

a Alf has a waist circumference of 38 inches and a hip circumference of 30 inches. Is Alf at high, moderate or low health risk? *(1 mark)*

b Marlene has a 24 inch waist circumference. What would her hip circumference be if $n = 0.8$? *(1 mark)*

c On the graph below the boundary line between low and moderate health risk is shown for women.

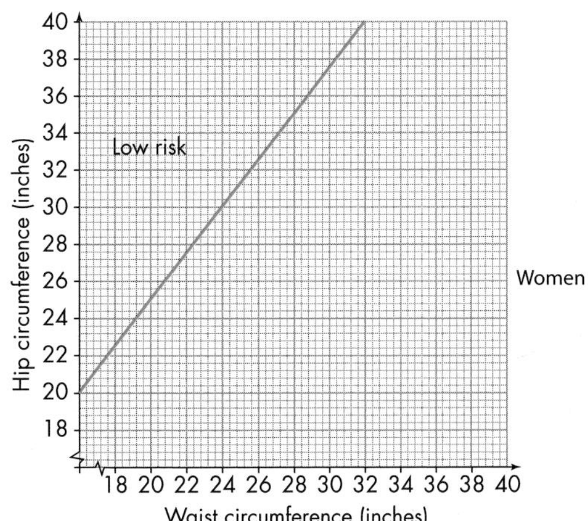

Complete the graph to show the health risk factors for women. *(2 marks)*

AQA, November 2010, H2, Question 13

Worked Examination Questions

1 The distance–time graph shows the journey of a train between two stations. The stations are 6 km apart.

a During the journey the train stopped at a signal. For how long was the train stopped?

b What was the average speed of the train for the whole journey? Give your answer in kilometres per hour.

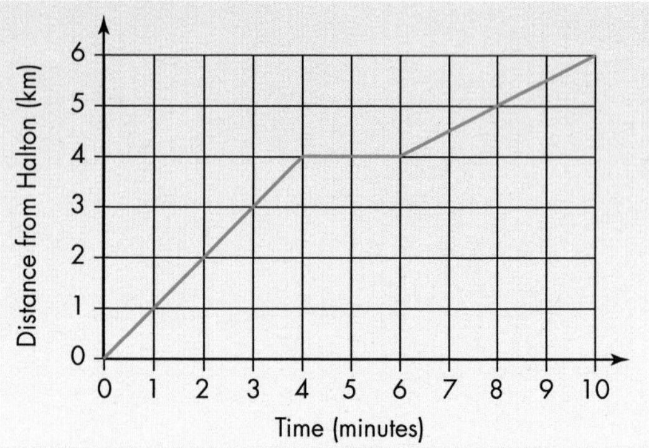

1 a *The train stopped for 2 minutes (where the line is horizontal).*

This gets 1 mark.

b *The train travels 6 km in 10 minutes. This is 36 km in 60 minutes. So the average speed is 36 km/h.*

Multiply both numbers by 6. This gets 1 mark for writing down any distance and an equivalent time, and 1 mark for the answer.

Total: 3 marks

FM **2** The graph shows the increase in rail fares as a percentage of the fares in 1997 since the railways were privatised in 1997.

a In what year did first-class fares double in price from 1997?

b Approximately how much would a regulated standard-class fare that cost £20 in 1997 cost in 2002?

c Approximately how much would a first-class fare that cost £100 in 1997 cost in 2010?

d During which period did first-class fares rise the most? How can you tell?

e An unregulated standard-class fare cost £35 in 2008. Approximately how much would this fare have been in 1997?

2 a *2007*

Read when the fares were 200%. This gets 1 mark for method.

b *£22.50*

Read the percentage increase in 2002 and multiply the fare by this figure (about 12.5%). This gets 1 mark for method and 1 mark for accuracy.

c *£245*

Read the percentage increase in 2010. This gets 1 mark for method.

d *2005–2010*

This is the steepest part of the graph. This gets 1 mark.

e *£20*

Unregulated fares have gone up by 75%. 75% of £20 is £15. This gets 1 mark for accuracy.

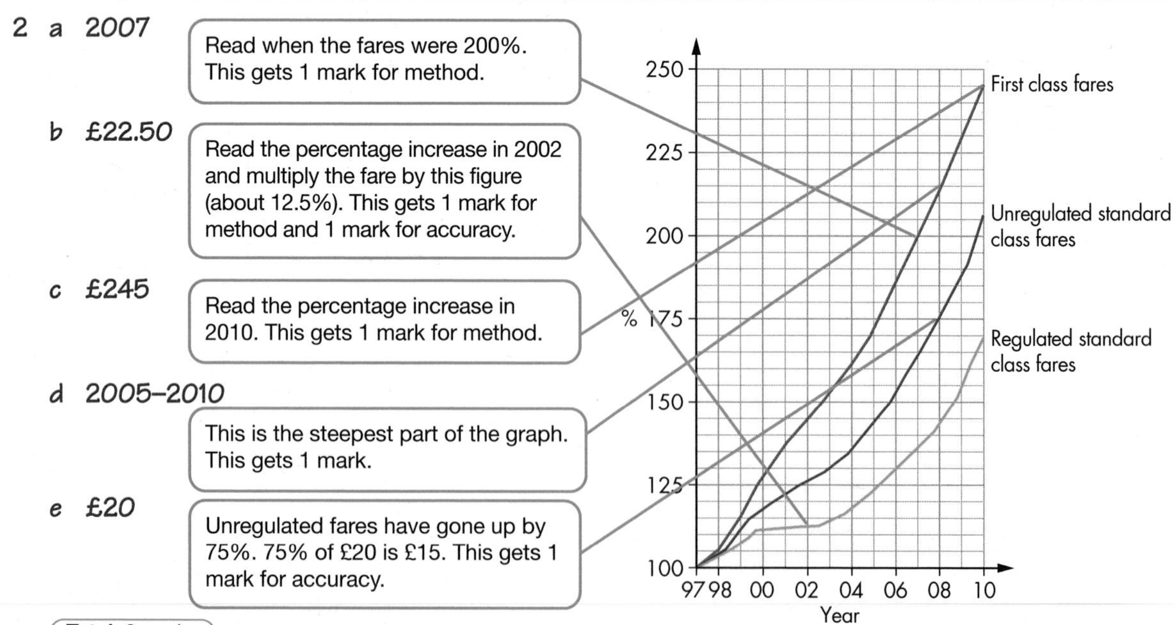

Total: 6 marks

Worked Examination Questions

PS **3** A cyclist sets off at midday from a point P and cycles at a steady speed of 20 km/h for 90 minutes. She then rests for 30 minutes and continues at a steady speed of 15 km/h.

A car sets off from P on the same road. The car drives at a steady speed of 45 km/h and overtakes the cyclist at 4 pm.

What time did the car set off from P?

You may use a copy of the axes shown here to help you with your answer.

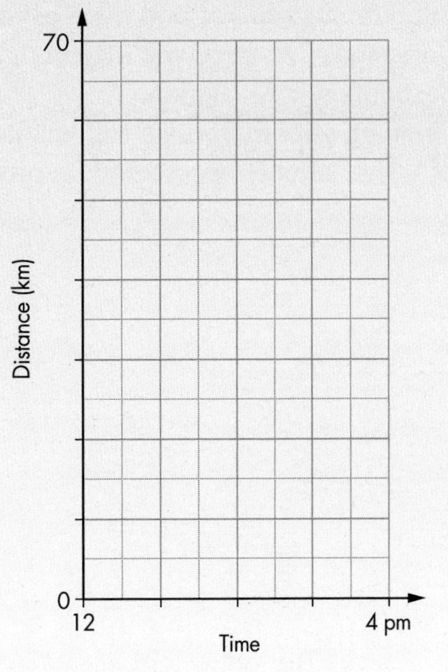

3 Line from (12, 0) to (1.30, 30)

Line from (1.30, 30) to (2, 30)

Line from (2, 30) to (4, 60)

> Draw a distance–time graph of the cyclist's journey. This gets 1 mark for method and 1 mark for accuracy.

60 km ÷ 45 km/hour = 1 h 20 m

Line from (4, 60) with a gradient of 45 km/h.

> There are two methods to calculate the time it takes the car to overtake the cyclist. Either is OK. This gets 1 mark for method.

4pm – 1 h 20 m = 2.40 pm

> Work out the time the motorist set off. This gets 1 mark for accuracy.

Total: 4 marks

13 Functional Maths
Planning a motorbike trip to France

A group of friends are going on a holiday to France. They have decided to go on motorbikes and have asked you to join them as a pillion passenger. The ferry will take your group to Boulogne and the destination is Perpignan.
Planning your motorbike trip will involve a range of mathematics, much of which can be represented on graphs.

Your task

The following task will require you to work in groups of 2–3.

Using all the information that you gather from these pages and your own knowledge, investigate the key mathematical elements of your motorbike trip.

Remember, sometimes events outside our control can change our travel plans. Take the following scenarios into account when planning your motorbike trip.

There is a problem at the ferry port and you must use the Channel Tunnel instead.

Just prior to travelling, the British pound drops against the euro, giving a new exchange rate of 1 euro to 98p.

One friend would like to extend the trip to Barcelona.

You must draw at least three graphs and use as many different mathematical methods as possible when drawing up your final travel plans and your contingency recommendations.

Getting started

Start by thinking about the mathematics that you use when you go on holiday. Here are a few questions to get you going.

- What information do you need before you travel?
- How many euros (€) are there in one British pound (£)?
- What differences might you find when you travel abroad?
- What are the differences between metric and imperial units of measure? (It may help to list the conversion facts that you know.)
- After approximately how long would you need to stop to rest when travelling?

Handy hints

There are a number of measures and units in France that will need converting when you get there. Two of the most noticeable are:

- **Currency:** in France the currency is in euros.
- **Distances:** in France, distances are measured in kilometres, so speeds are in kilometres per hour.

1 euro ≈ £0.90

1 gallon ≈ 4 litres

1 mile ≈ 1.6 km

A motorbike fuel tank holds 3 gallons and travels about 45 miles per gallon.

The cost of petrol in France is €1.2 per litre.

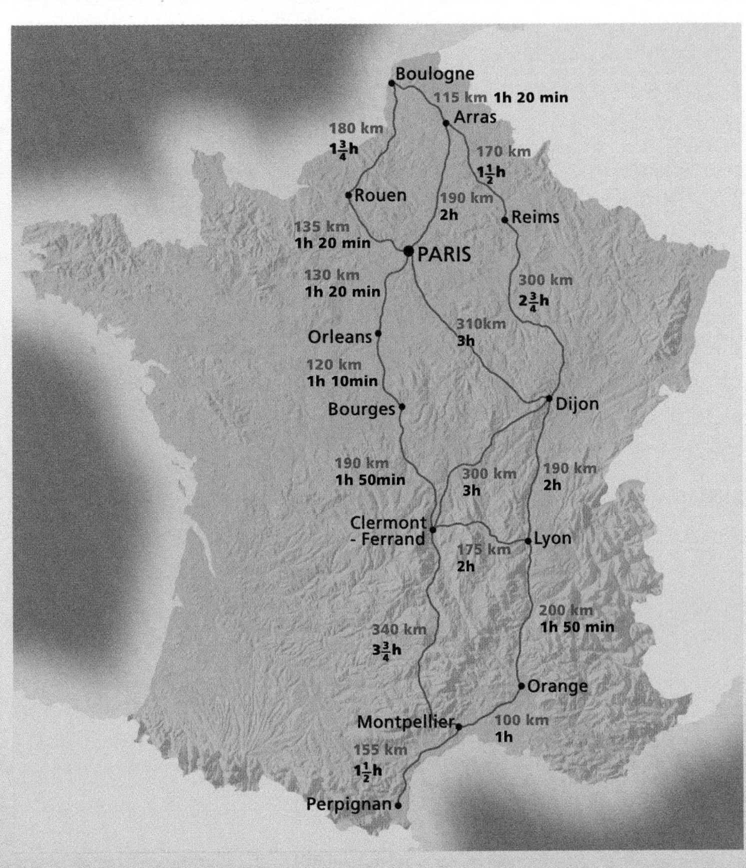

Why this chapter matters

Statistical distributions can help us to understand our society and the world we live in. There are many different sorts of distribution, two of which are described here.

Population distribution

Where on Earth do people live?

Population distribution is the number of people living in a square mile (or kilometre). Calculating this over the whole planet shows that it is far from uniform.

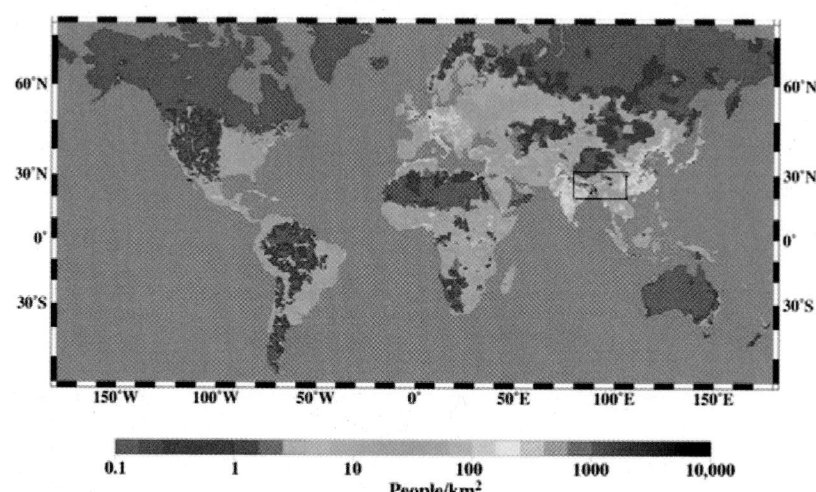

The map shows areas of the world that are densely populated and areas where very few people live. In some areas it is easy to understand the distribution of the population. For example, central regions of Australia are so hot and so thickly covered in scrubland that very few people would want to live there, hence they are thinly populated. However, coastal regions that enjoy good climate are much more thickly populated.

The south-east part of the UK is far more densely populated than the rest of the UK and much of Europe. In fact, the southern part of England has a very similar population density to India and Pakistan.

The normal distribution

This can be a confusing description of a distribution, as it often tells us little about what is normal. For example, the bar chart shows the normal distribution of IQ scores amongst the UK population.

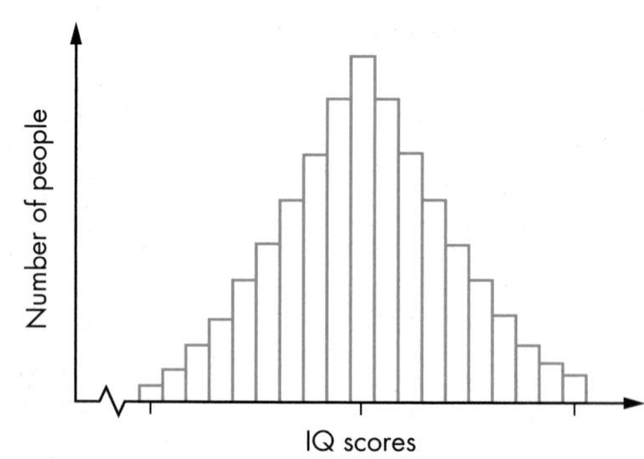

Although this shows there are more people with the 'average' score of 100 than any other score, there is no such thing as a normal IQ since there are many more people who score higher than 100 and less than 100.

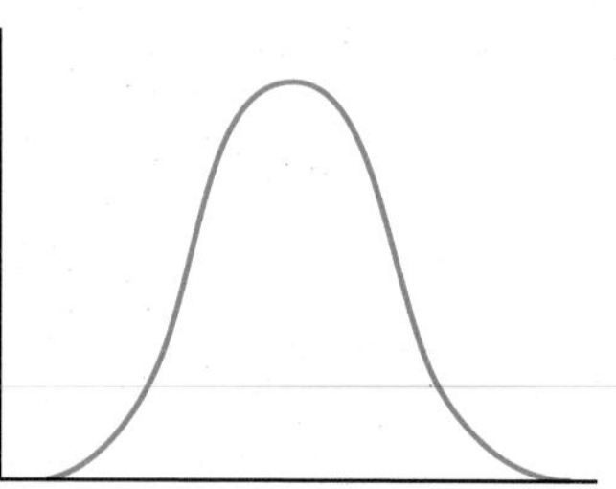

So what is normal?

What is normal is that none of us is normal! It is normal for some of us to be clever and others not to be so clever, but no particular IQ can be considered normal. And it is normal that this causes confusion, so it is not covered in your GCSE (you'll be pleased to know).

In this chapter we examine ways of looking at distributions of sets of people or data in order to compare them with each other.

Statistics: Statistical representation

1. Line graphs
2. Stem-and-leaf diagrams
3. Scatter diagrams

This chapter will show you ...

- **D** how to interpret and draw line graphs and stem-and-leaf diagrams
- **D to C** how to draw scatter diagrams and lines of best fit
- **D to C** how to interpret scatter diagrams and the different types of correlation

Visual overview

Averages → Stem-and-leaf diagrams

Scatter diagrams

What you should already know

- How to plot coordinate points **(KS3 level 4, GCSE grade F)**
- How to read information from charts and tables **(KS3 level 4, GCSE grade F)**
- How to calculate the mean of a set of data from a frequency table **(KS3 level 5, GCSE grade E)**
- How to recognise a positive or negative gradient **(KS3 level 6, GCSE grade D)**

Quick check

The table shows the numbers of children in 10 classes in a primary school.

Calculate the mean number of children in each class.

Number of children	27	28	29	30	31
Frequency	1	2	4	2	1

14.1 Line graphs

This section will show you how to:
- draw a line graph to show trends in data

Key words
line graphs
trends

Line graphs are usually used in statistics to show how data changes over a period of time. One use is to indicate **trends**: for example, line graphs can be used to show whether the Earth's temperature is increasing as the concentration of carbon dioxide builds up in the atmosphere, or whether a firm's profit margin is falling year on year.

Line graphs are best drawn on graph paper.

EXAMPLE 1

This line graph shows the outside temperature one day in November.

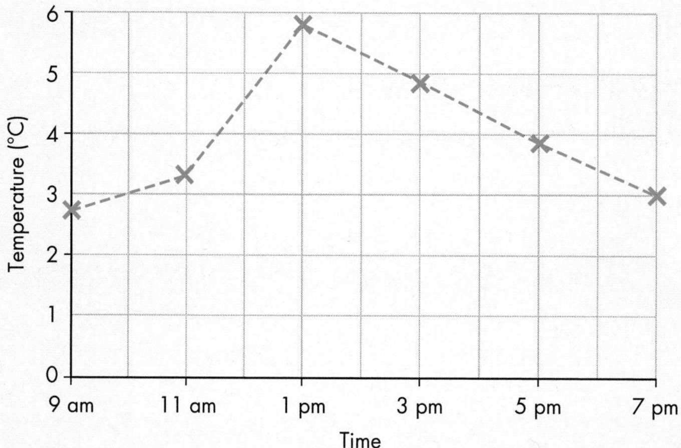

For this graph, the values between the plotted points have no true meaning because only the temperatures at the plotted points are known. However, by joining the points with a dashed line, as shown, you can estimate the temperatures at points in between. Although the graph shows the temperature falling in the early evening, it would not be sensible to try to predict what will happen after 7 pm that night.

EXERCISE 14A

D

1 The table shows the estimated numbers of tourists worldwide.

Year	1975	1980	1985	1990	1995	2000	2005	2010
No. of tourists (millions)	100	150	220	280	290	320	340	345

a Draw a line graph for the data.

b Use your graph to estimate the number of tourists in 2002.

c In which five-year period did tourism increase the most?

FM **d** Explain the trend in tourism. What reasons can you give to explain this trend?

2 The table shows the maximum and minimum daily temperatures for London over a week.

Day	Sunday	Monday	Tuesday	Wednesday	Thursday	Friday	Saturday
Maximum (°C)	12	14	16	15	16	14	10
Minimum (°C)	4	5	7	8	7	4	3

a Draw line graphs on the *same axes* to show the maximum and minimum temperatures.

b Find the smallest and greatest differences between the maximum and minimum temperatures.

3 Maria opened a coffee shop. She was interested in how trade was picking up over the first few weeks. The table shows the number of coffees sold in these weeks.

Week	1	2	3	4	5
Coffees sold	46	71	89	103	113

a Draw a line graph for this data.

b From your graph, estimate the number of coffees Maria hopes to sell in week 6.

FM **c** Give a possible reason for the way in which the number of coffees sold increased.

PS **4** A puppy is weighed at the end of each week.

Week	1	2	3	4	5
Weight (g)	850	920	940	980	1000

Estimate how much the puppy would weigh after eight weeks.

AU **5** When plotting a graph to show the summer midday temperatures in Spain, Abbass decided to start his graph at the temperature 20 °C.

Explain why he might have done that.

14.2 Stem-and-leaf diagrams

This section will show you how to:
- draw and read information from an ordered stem-and-leaf diagram

Key words
discrete data
ordered
raw data
unordered

Raw data

If you are recording the ages of the first 20 people who line up at a bus stop in the morning, the **raw data** might look like this.

23, 13, 34, 44, 26, 12, 41, 31, 20, 18, 19, 31, 48, 32, 45, 14, 12, 27, 31, 19

This data is **unordered** and is difficult to read and analyse. When the data is **ordered**, it looks like this.

12, 12, 13, 14, 18, 19, 19, 20, 23, 26, 27, 31, 31, 31, 32, 34, 41, 44, 45, 48

This is easier to read and analyse.

Another method for displaying **discrete data**, such as this, is a stem-and-leaf diagram. The tens values will be the 'stem' and the units values will be the 'leaves'.

Key: 1 | 2 represents 12

1	2	2	3	4	8	9	9
2	0	3	6	7			
3	1	1	1	2	4		
4	1	4	5	8			

This is called an ordered stem-and-leaf diagram. It gives a better idea of how the data is distributed.

A stem-and-leaf diagram should always have a key.

EXAMPLE 2

Put the following data into an ordered stem-and-leaf diagram.

45, 62, 58, 58, 61, 49, 61, 47, 52, 58, 48, 56, 65, 46, 54

a What is the modal value?

b What is the median value?

c What is the range of the values?

First, decide on the stem and leaf.

In this case, the tens digit will be the stem and the units digit will be the leaf.

Key: 4 | 5 represents 45

4	5	6	7	8	9	
5	2	4	6	8	8	8
6	1	1	2	5		

a The modal value is the most common, which is 58.

b There are 15 values, so the median will be the value that is $(15 + 1) \div 2$, or the 8th value. Counting from either the top or the bottom, the median is 56.

c The range is the difference between the largest and the smallest value, which is $65 - 45 = 20$

EXERCISE 14B

1 The heights of 15 tulips are measured.

43 cm, 39 cm, 41 cm, 29 cm, 36 cm,
34 cm, 43 cm, 48 cm, 38 cm, 35 cm,
41 cm, 38 cm, 43 cm, 28 cm, 48 cm

a Show the results in an ordered stem-and-leaf diagram, using this key.

Key: 4 | 3 represents 43 cm

b What is the modal height?

c What is the median height?

d What is the range of the heights?

2 A student records the number of text messages she receives each day for two weeks.

12, 18, 21, 9, 17, 23, 8, 2, 20, 13, 17, 22, 9, 9

a Show the results in an ordered stem-and-leaf diagram, using this key.

Key: 1 | 2 represents 12 messages

b What was the modal number of text messages received in a day?

c What was the median number of text messages received in a day?

3 Zachia wanted to know how many people attended a daily youth club each day for a month. She recorded the data.

13, 19, 20, 9, 18, 24, 7, 8, 19, 14, 18, 23, 9, 10, 15, 31, 28, 26, 12, 24

a Show these results in an ordered stem-and-leaf diagram.

b What is the median number of people at the youth club?

c What is the range of the numbers of people who attended the youth club?

D

D

PS 4 This stem-and-leaf diagram shows some heights of boys and girls in the same form.

6	15	3 5 5 7 7 9
2 5 5 6 8 9 9	16	1 1 5 6 6 7 8 9
3 4 5 5 7	17	4
2 3 3	18	

Explain what the diagram is telling you about the students in the form.

AU 5 The matches in a set of boxes were counted and the following numbers were recorded.

50, 52, 51, 53, 52, 51, 51, 53, 54, 55, 54, 52, 52, 51, 50, 53

Explain why a stem-and-leaf diagram is *not* a good way to represent this information.

14.3 Scatter diagrams

This section will show you how to:

- draw, interpret and use scatter diagrams

Key words
line of best fit
negative correlation
no correlation
positive correlation
scatter diagram
variable

A **scatter diagram** (also called a scattergraph or scattergram) is a method of comparing two **variables** by plotting on a graph their corresponding values (usually taken from a table). In other words, the variables are treated just like a set of (x, y) coordinates.

In this scatter diagram, the marks scored by students in an English test are plotted against the marks they scored in a mathematics test. This graph shows **positive correlation**. This means that the students who got high marks in the mathematics test also tended to get high marks in the English test.

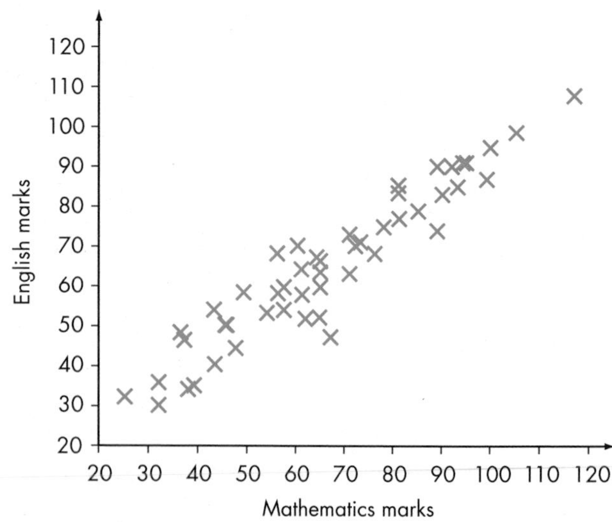

Correlation

This section will explain the different types of correlation.

Here are three statements that may or may not be true.

The taller people are, the wider their arm span is likely to be.
The older a car is, the lower its value will be.
The distance you live from your place of work will affect how much you earn.

These relationships could be tested by collecting data and plotting the data on a scatter diagram.

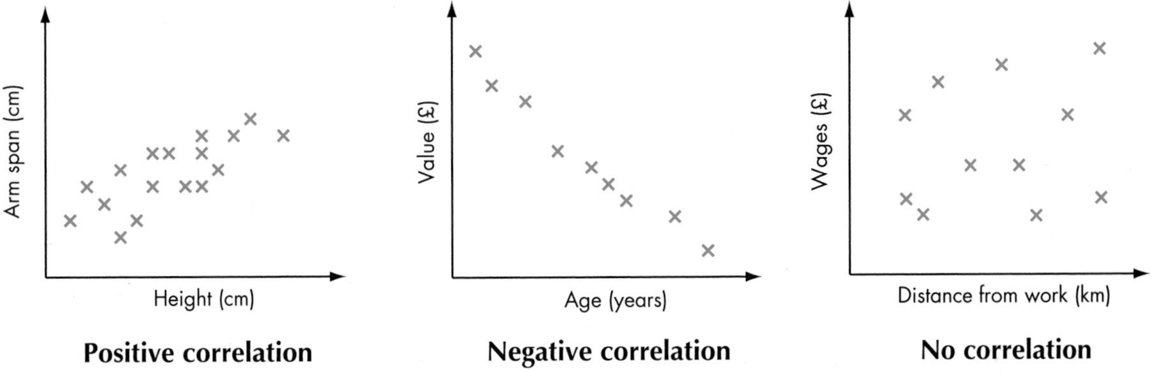

Positive correlation **Negative correlation** **No correlation**

For example, the first statement may give a scatter diagram like the first one above. This diagram has positive correlation because as one quantity increases so does the other. From such a scatter diagram you could say that the taller someone is, the wider the arm span.

Testing the second statement may give a scatter diagram like the second one. This diagram has **negative correlation** because as one quantity increases, the other quantity decreases. From such a scatter diagram you could say that as a car gets older, its value decreases.

Testing the third statement may give a scatter diagram like the third one. This scatter diagram has **no correlation**. There is no relationship between the distance a person lives from work and how much that person earns.

EXAMPLE 3

The graphs show the relationship between the temperature and the amount of ice cream sold, and that between the age of people and the amount of ice cream they eat.

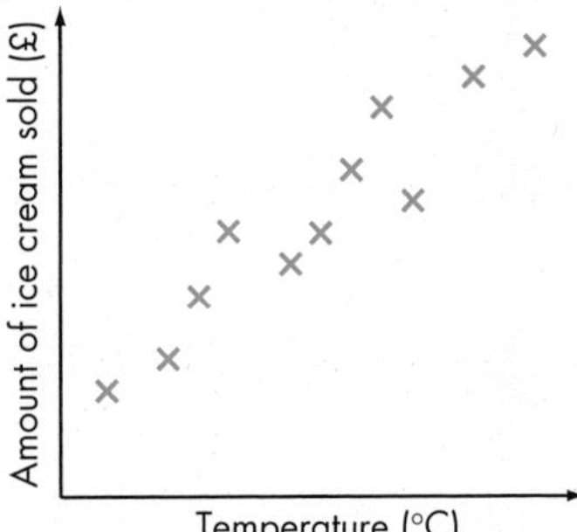

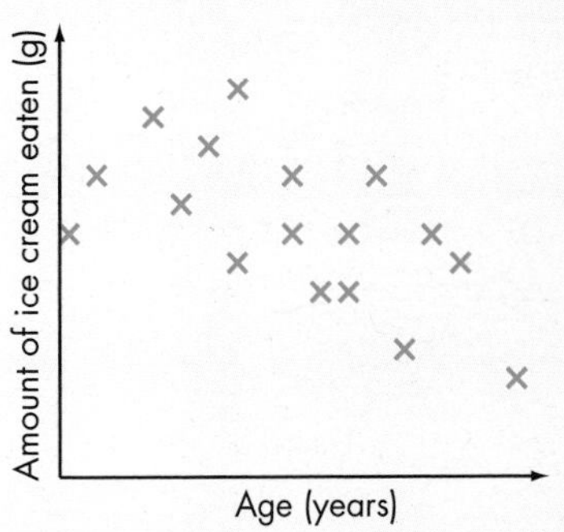

a Comment on the correlation of each graph. **b** What does each graph tell you?

The first graph has positive correlation and tells us that as the temperature increases, the amount of ice cream sold increases.

The second graph has negative correlation and tells us that as people get older, they eat less ice cream.

Line of best fit

This section will explain how to draw and use a **line of best fit**.

The line of best fit is a straight line that goes between all the points on a scatter diagram, passing as close as possible to all of them. You should try to have the same number of points on both sides of the line. Because you are drawing this line by eye, examiners make a generous allowance for the correct answer. The line of best fit for the scatter diagram on page 352 is shown below, left.

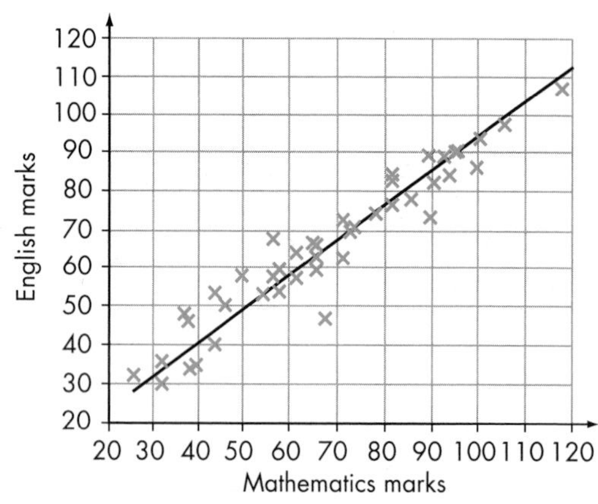

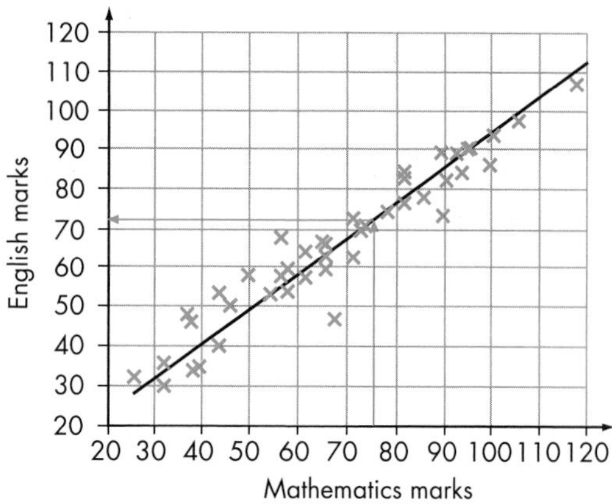

The line of best fit can be used to answer the following type of question: A girl took the mathematics test and scored 75 marks, but was ill for the English test. How many marks was she likely to have scored?

The answer is found by drawing a line up from 75 on the mathematics axis to the line of best fit and then drawing a line across to the English axis (above right). This gives 73, which is the mark she is likely to have scored in the English test.

EXERCISE 14C

D

1 Describe the correlation of each of these four graphs and write in words what each graph tells you.

a

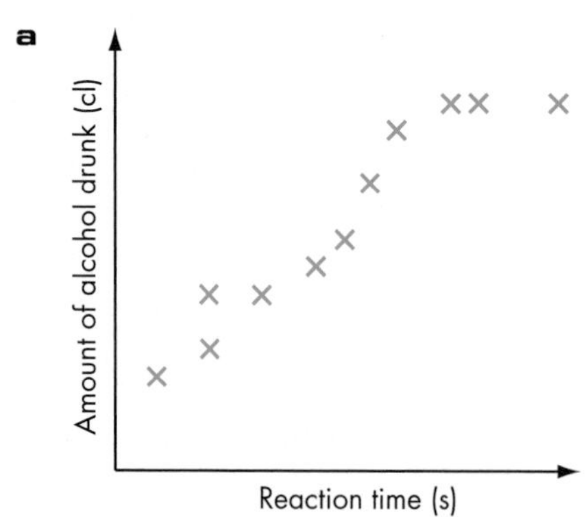

b

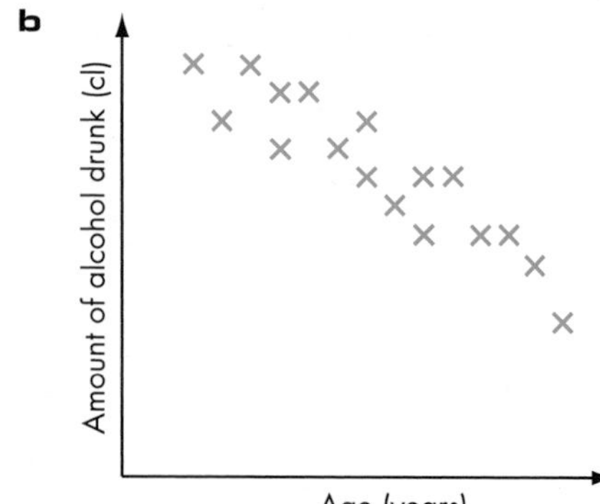

c

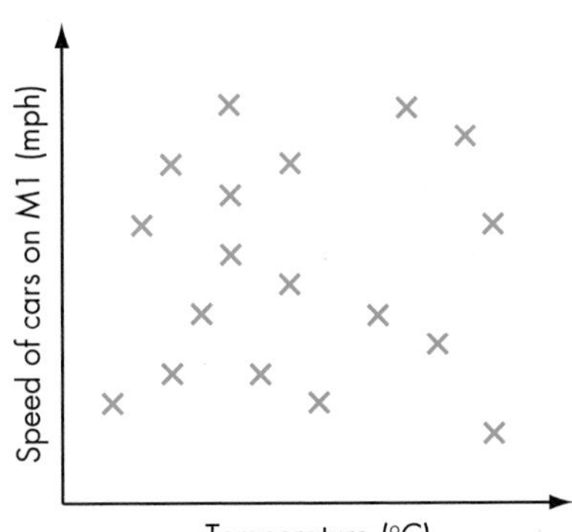

d

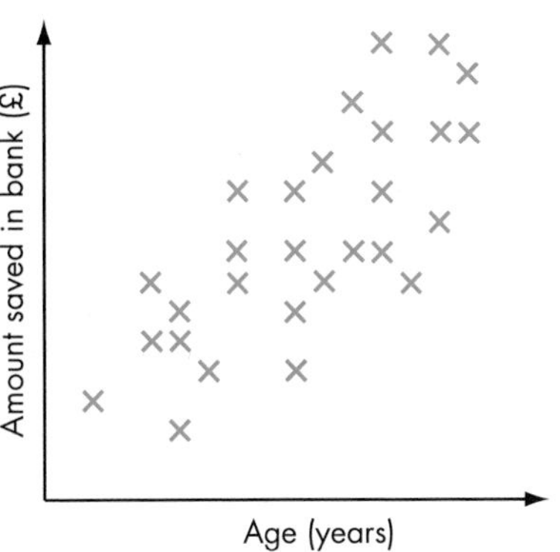

D

C

2 The table shows the results of a science experiment in which a ball is rolled along a desk top. The speed of the ball is measured at various points.

Distance from start (cm)	10	20	30	40	50	60	70	80
Speed (cm/s)	18	16	13	10	7	5	3	0

a Plot the data on a scatter diagram.

b Draw the line of best fit.

c If the ball's speed had been measured at 5 cm from the start, what is it likely to have been?

FM **d** Estimate how far from the start the ball was when its speed was 12 cm/s.

HINTS AND TIPS

Often in exams axes are given and most, if not all, of the points are plotted.

3 The table shows the marks for 10 students in their mathematics and geography examinations.

Student	Anna	Beryl	Cath	Dema	Ethel	Fatima	Greta	Hannah	Imogen	Joan
Maths	57	65	34	87	42	35	59	61	25	35
Geog	45	61	30	78	41	36	35	57	23	34

a Plot the data on a scatter diagram. Take the x-axis for the mathematics scores and mark it from 20 to 100. Take the y-axis for the geography scores and mark it from 20 to 100.

b Draw the line of best fit.

c One of the students was ill when she took the geography examination. Which student was it most likely to be?

FM **d** If another student, Kate, was absent for the geography examination but scored 75 in mathematics, what mark would you expect her to have got in geography?

e If another student, Lynne, was absent for the mathematics examination but scored 65 in geography, what mark would you expect her to have got in mathematics?

C

FM 4 The heights, in centimetres, of 20 mothers and their 15-year-old daughters were measured. These are the results.

Mother	153	162	147	183	174	169	152	164	186	178
Daughter	145	155	142	167	167	151	145	152	163	168

Mother	175	173	158	168	181	173	166	162	180	156
Daughter	172	167	160	154	170	164	156	150	160	152

a Plot these results on a scatter diagram. Take the x-axis for the mothers' heights from 140 to 200. Take the y-axis for the daughters' heights from 140 to 200.

b Is it true that the tall mothers have tall daughters?

FM 5 A teacher carried out a survey of his class. He asked students to say how many hours per week they spent playing sport and how many hours per week they spent watching TV. This table shows the results of the survey.

Student	1	2	3	4	5	6	7	8	9	10
Hours playing sport	12	3	5	15	11	0	9	7	6	12
Hours watching TV	18	26	24	16	19	27	12	13	17	14

Student	11	12	13	14	15	16	17	18	19	20
Hours playing sport	12	10	7	6	7	3	1	2	0	12
Hours watching TV	22	16	18	22	12	28	18	20	25	13

a Plot these results on a scatter diagram. Take the x-axis as the number of hours playing sport and the y-axis as the number of hours watching TV.

b If you knew that another student from the class watched 8 hours of TV a week, would you be able to predict how long she or he spent playing sport? Explain why.

FM 6 The table shows the time taken and distance travelled by a taxi driver for 10 journeys one day.

Distance (km)	1.6	8.3	5.2	6.6	4.8	7.2	3.9	5.8	8.8	5.4
Time (min)	3	17	11	13	9	15	8	11	16	10

a Draw a scatter diagram with time on the horizontal axis.

b Draw a line of best fit on your diagram.

c A taxi journey takes 5 minutes. How many kilometres would you expect the journey to have been?

d How long would you expect a journey of 4 km to take?

PS 7 Oliver records the time taken, in hours, and the average speed, in mph, for several different journeys.

Time (h)	0.5	0.8	1.1	1.3	1.6	1.75	2	2.4	2.6
Speed (mph)	42	38	27	30	22	23	21	9	8

Estimate the average speed for a journey of 90 minutes.

AU 8 Describe what you would expect the scatter graph to look like if someone said that it showed negative correlation.

GRADE BOOSTER

D You can draw an ordered stem-and-leaf diagram

D You can recognise the different types of correlation

C You can draw a line of best fit on a scatter diagram

C You can interpret a scatter diagram

What you should know now

- How to read information from statistical diagrams, including stem-and-leaf diagrams
- How to plot scatter diagrams, recognise correlation, draw lines of best fit and use them to predict values

1 A team of 12 run a half-marathon.

Their times, to the nearest minute, are:

72	87	65	85	91	76
67	70	80	84	70	82

Copy and complete an ordered stem-and-leaf diagram to represent this data.

Remember to complete the key.

Key … | … represents … minutes

6	
7	
8	
9	

(3 marks)

AQA, May 2008, Paper 1 Higher, Question 9

2 The scatter diagram shows the time that seven students spent practising for a typing test and the number of errors they made in the test.

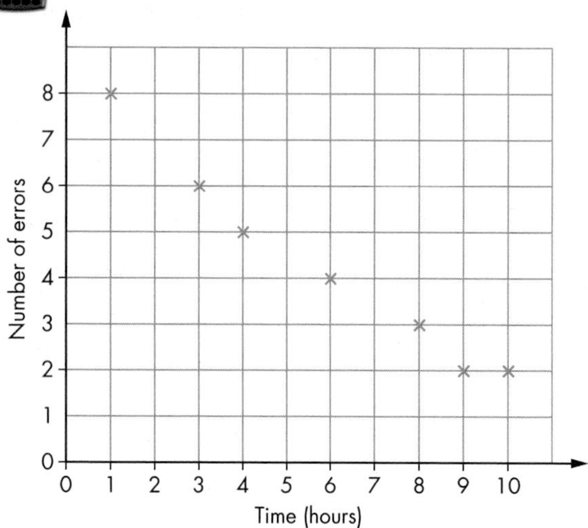

a Describe the relationship shown by the scatter graph. *(1 mark)*

b **i** Draw a line of best fit on the diagram. *(1 mark)*

ii Use your line of best fit to estimate the number of errors made if a student practised for two hours. *(1 mark)*

c Asif practised for two hours and only made one error. Give a possible reason why his number of errors was so low. *(1 mark)*

d Leanne says, "If I practise for 15 hours I will definitely not make any errors."

Is she correct?

Give a reason for your answer. *(1 mark)*

AQA, June 2008, Module 1 Higher, Question 1

3 The length and wingspan, in centimetres, of seven common garden birds is shown in the table.

Bird	Length (cm)	Wingspan (cm)
Starling	21	40
Blackbird	25	36
Blue tit	11	19
Greenfinch	15	26
Dove	32	51
Sparrow	15	23
Great tit	14	24

a Plot the data as a scatter graph on a grid as shown below.

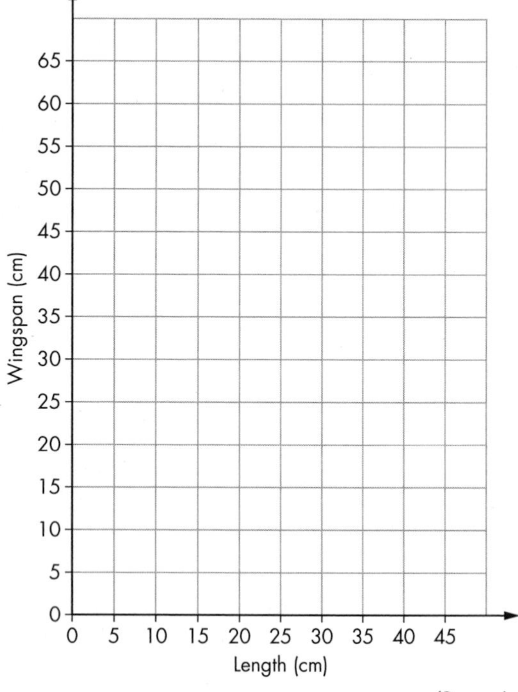

(2 marks)

b Describe the strength and type of correlation. *(1 mark)*

c Draw a line of best fit on your scatter graph. *(1 mark)*

d Use your line of best fit to estimate the wingspan of a thrush whose length is 20 cm. *(1 mark)*

e It is not sensible to use your line of best fit to estimate the wingspan of a pigeon whose length is 41 cm.

Explain why. *(1 mark)*

AQA, March 2008, Module 1 Higher, Question 1

4 The table shows the number of pages in some paperback books and their prices.

Pages	350	390	450	500	590	610	620	700	750	760
Cost (£)	6.00	5.50	6.80	7.40	6.50	8.20	7.50	8.25	8.80	7.80

a Draw a scatter diagram with prices on the horizontal axis.

b Draw a line of best fit on your diagram.

c A book has 600 pages. How much might you expect to have to pay for this book?

d How many pages would you expect in a book that cost £9?

5 A swimming club has 15 members. The stem-and-leaf diagram shows their ages, in years.

1	6 8 9
2	1 4 5 7 8
3	0 1 4 6 6
4	1 2

Key 2 | 4 represents 24 years

a How many members are aged under 21? *(1 mark)*

b What is the median age of the members? *(1 mark)*

c A new member joins the club. The range of the ages of the members is doubled.

How old is the new member? *(3 marks)*

AQA, November 2011, H1, Question 13

6 The heights of players in a school basketball squad, in centimetres, are

187 179 184 186 190
186 181 194 188 177

a Complete an ordered stem-and-leaf diagram to represent this data. Remember to include a key. *(3 marks)*

b Write down the median height. *(1 mark)*

c Another player of height 188 cm joins the squad. Explain why the median height will not change. *(1 mark)*

AQA, June 2011, H1, Question 4

7 The birth rate and the life expectancy for seven countries are shown in the table.

Country	Birth rate (number of births per 1000 people)	Life expectancy (years)
Chile	15	77
Egypt	22	72
Gambia	39	59
India	22	69
Japan	8	82
Nepal	30	64
United Kingdom	11	79

a Plot the data as a scatter graph. *(2 marks)*

b Describe the strength and type of correlation. *(2 marks)*

c Draw a line of best fit on your scatter graph. *(1 mark)*

d Use the line of best fit to estimate the life expectancy for Turkey whose birth rate is 16 births per 1000 people. *(1 mark)*

e Why might it not be reliable to use the line of best fit to estimate the life expectancy for Niger whose birth rate is 50 births per 1000 people? *(1 mark)*

AQA, November 2010, H1, Question 3

8 The amount of money that a group of 16 teenagers earn each week is shown in the stem-and-leaf diagram.

0	8 8 9
1	2 6 8
2	5 5
3	0 2 3 7
4	6
5	0 6 9

Key 1 | 8 represents £18

Calculate the mean amount of money earned by the group. *(3 marks)*

AQA, November 2009, H2, Question 3

Worked Examination Questions

FM **1** A form teacher recorded how many merits her class had received in the school year.
She recorded the data in a stem and leaf diagram.
Key 2 | 4 represents 24 merits

0	1	1	2	6	8	8	9
1	0	2	2	4	4		
2	2	4	4	7			
3	0	0	1				

11 students did not receive any merits.
Calculate the mean number of merits for the **whole** form.

1 $1 + 1 + 2 + 6 + 8 + 8 + 9 = 35$
$10 + 12 + 12 + 14 + 14 = 62$
$22 + 24 + 24 + 27 = 97$
$30 + 30 + 31 = 91$

An attempt to add up the values would get 1 method mark. The easiest way to do this is to add each row.

Total $= 35 + 62 + 97 + 91 = 285$

Getting the correct total of 285 gets 1 accuracy mark.

Mean $= 285 \div 30$

Showing that your total divided by 30 is how to calculate the mean gets 1 method mark. It is important to include the 11 students who got no merits.

Mean $= 9.5$

The correct mean gets 1 accuracy mark.

Total: 4 marks

Worked Examination Questions

PS **2**
- The older you are, the higher you score on the Speed test.
- The higher you score on the Speed test, the less TV you watch.
- The more TV you watch, the more hours you will sleep.

Suppose the above were all true. Sketch a scatter diagram to illustrate the relationship between age and hours slept.

2 Sketch a scatter diagram for each bullet.

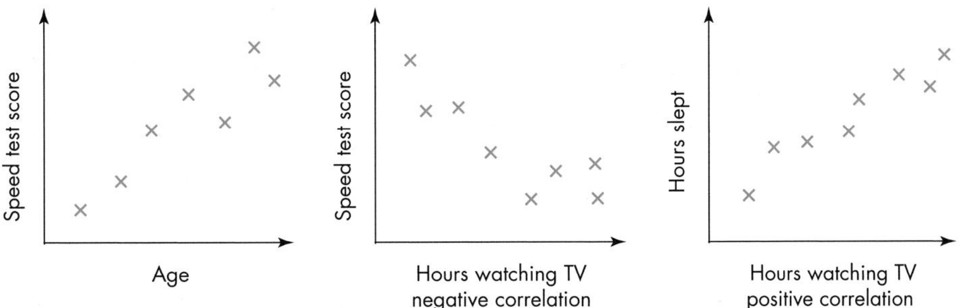

This gets 1 for showing some graphs similar to these.

Three diagrams will help to show the relationships.

Imagine the extremes of age, starting with the first diagram.
Older person scores high on Speed test.
High score on Speed test means the person watches TV for a short time.
Watching TV for a short time gives short sleep time.

This gets 2 for showing linkage for extremes like this.

Hence, 'High age relates to small hours sleep'.
Similarly, 'Low age relates to long sleep'.

This gets 1 mark for each extreme.

So the scatter diagram will show negative correlation and can be sketched as:

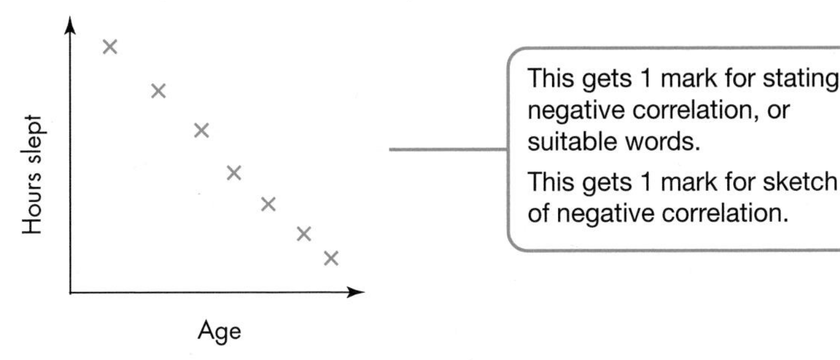

This gets 1 mark for stating negative correlation, or suitable words.

This gets 1 mark for sketch of negative correlation.

Total: 6 marks

14

Functional Maths
Reporting the weather

The weather often appears in the news headlines and a weather report is given regularly on the television. In this activity you are required to look at the data supplied in a weather report and interpret its meaning.

Your task

The type of information given on the map and in the table is found in most newspapers in the UK each day. The map gives a forecast of the weather for the day in several large towns, while the table summarises the weather on the previous day.

It is your task to use appropriate statistical diagrams and measures to summarise the data given in the map and table.

Then, you must write a report to describe fully the weather in the UK on Friday 21st April 2010. You should use your statistical analysis to support your descriptions.

Getting started

Look at the data provided in the table.

- Which cities had the most sun, which had the most rain and which were the warmest on this particular day?
- Is there any other information you can add?

Extension

Research weather in a different country. Draw together all your research, using diagrams, and compare your findings with the analysis that you have already done of the weather in the UK in April.

Friday 21st April 2010

	Sun (hrs)	Rain (mm)	Max (°C)	Min (°C)	Daytime weather
Aberdeen	5	3	10	5	rain
Barnstaple	5	0	12	8	sunny
Belfast	5	0	10	4	cloudy
Birmingham	5	0	11	5	sunny
Bournemouth	5	0	12	6	sunny
Bradford	5	1	10	5	rain
Brighton	5	0	11	5	mixed
Bristol	5	2	11	6	rain
Cardiff	5	2	11	6	rain
Carlisle	5	3	10	5	rain
Chester	5	3	10	4	rain
Eastbourne	5	0	9	5	windy
Edinburgh	5	0	9	5	cloudy
Falmouth	5	0	12	8	sunny
Glasgow	5	0	9	4	cloudy
Grimsby	5	1	10	4	mixed
Holyhead	5	0	8	3	windy
Ipswich	5	1	11	6	mixed
Isle of Man	5	0	9	3	windy
Isle of Wight	5	0	13	6	sunny

	Sun (hrs)	Rain (mm)	Max (°C)	Min (°C)	Daytime weather
Leeds	5	0	11	6	cloudy
Lincoln	5	3	10	4	rain
Liverpool	5	1	12	6	mixed
London	5	2	13	6	rain
Manchester	5	0	13	7	sunny
Middlesbrough	5	1	11	6	rain
Newcastle	5	0	11	6	cloudy
Newquay	5	0	12	8	sunny
Nottingham	5	0	12	6	cloudy
Oxford	5	2	12	5	rain
Plymouth	6	0	12	9	sunny
Rhyl	5	1	9	4	mixed
Scunthorpe	5	1	10	4	mixed
Sheffield	5	0	12	7	sunny
Shrewsbury	5	0	9	4	windy
Southampton	5	0	12	6	sunny
Swindon	5	1	11	5	mixed
Weymouth	5	0	12	6	sunny
Windermere	5	2	10	4	mixed
York	5	0	11	5	cloudy

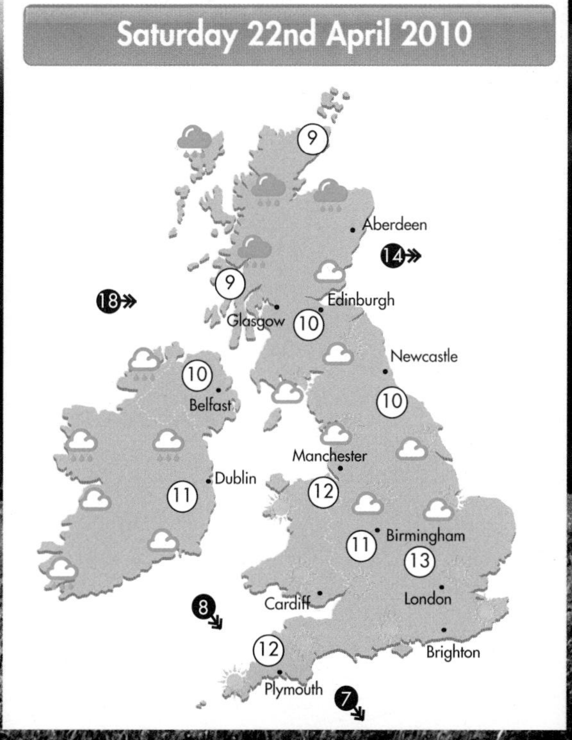

Why this chapter matters

Chance is a part of our everyday lives. Judgements are frequently made based on probability – take the weather forecast, for example. Every day we hear something like:

There is a forty per cent chance of rain today.

How do they know that?

- Records of data that predict possibility of rainfall go back as far as 1854, when meteorologists regarded the presence of nimbus clouds as an indication of a good chance of rain.
- A barometer was used to predict chances of rainfall. A sign of falling pressure on the barometer was taken as an indication of a good chance of rain.
- Finally, the direction of wind was used to determine the chances of rainfall. If the wind blew from a rainy part of the country, the chance of rain would be high.

The occurrence of all these three indicators would almost certainly mean that rain would come.

Probability originated from the study of games of chance, such as tossing a dice or spinning a roulette wheel. Mathematicians in the 16th and 17th centuries started to think about the mathematics of chance in games. Probability theory, as a branch of mathematics, came about in the 17th century when French gamblers asked mathematicians Blaise Pascal and Pierre de Fermat for help in their gambling.

Now, in the 21st century, probability is used to control the flow of traffic through road systems, the running of telephone exchanges, to look at patterns of the spread of infections and so on. These are just some of the everyday applications.

Chapter 15

Probability: Calculating probabilities

1. Experimental probability
2. Mutually exclusive and exhaustive events
3. Expectation
4. Two-way tables
5. Addition rule for events
6. Combined events

This chapter will show you ...

- D to C: how to work out the probability of events, using either theoretical models or experimental models
- D to A: how to predict outcomes, using theoretical models, and compare experimental and theoretical data
- C to A*: how to calculate probabilities for combined events

Visual overview

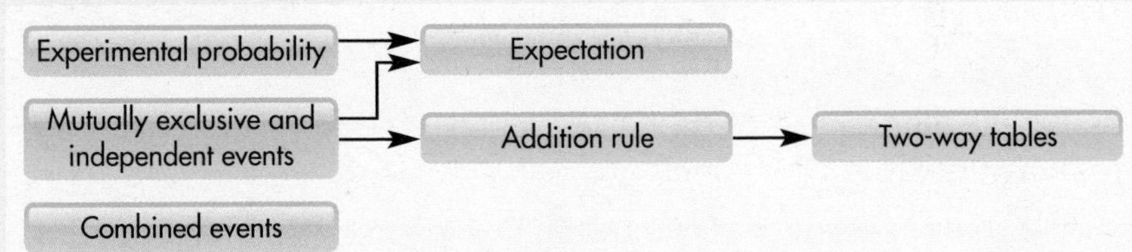

What you should already know

- That the probability scale goes from 0 to 1 **(KS3 level 4, GCSE grade F)**
- How to use the probability scale and to assess the likelihood of events, depending on their position on the scale **(KS3 level 4, GCSE grade F)**
- How to cancel, add and subtract fractions **(KS3 level 5, GCSE grade E)**

Quick check

Draw a probability scale and put an arrow to show the approximate probability of each of the following events happening.

a The next TV programme you watch will have been made in Britain.

b A person in your class will have been born in April.

c It will snow in July in Spain.

d In the next Olympic Games, a man will run the 100 m race in less than 20 seconds.

e During this week, you will drink some water or pop.

Terminology

The topic of probability has its own special terminology, which will be explained as it arises. For example, a **trial** is one go at performing something, such as throwing a dice or tossing a coin. So, if we throw a dice 10 times, we perform 10 trials.

Two other probability terms are **event** and **outcome**. An event is anything the probability of which we want to measure. An outcome is a result of the event.

Another probability term is **at random**. This means 'without looking' or 'not knowing what the outcome is in advance'.

Note: 'Dice' is used in this book in preference to 'die' for the singular form of the noun, as well as for the plural. This is in keeping with growing common usage, including in examination papers.

Probability facts

The probability of a *certain* outcome or event is 1 and the probability of an *impossible* outcome or event is 0.

Probability is *never greater than 1 or less than 0.*

Many probability examples involve coins, dice and packs of cards. Here is a reminder of their outcomes.

- Tossing a coin has two possible outcomes: head or tail.
- Throwing an ordinary six-sided dice has six possible outcomes: 1, 2, 3, 4, 5, 6.
- A pack of cards consists of 52 cards divided into four suits: Hearts (red), Spades (black), Diamonds (red) and Clubs (black). Each suit consists of 13 cards bearing the following values: 2, 3, 4, 5, 6, 7, 8, 9, 10, Jack, Queen, King and Ace. The Jack, Queen and King are called 'picture cards'. (The Ace is sometimes also called a picture card.) So the total number of outcomes is 52.

Probability is defined as:

$$\text{P(event)} = \frac{\text{number of ways the event can happen}}{\text{total number of all possible outcomes}}$$

This definition always leads to a fraction, which should be cancelled to its simplest form. Make sure that you know how to cancel fractions, with or without a calculator. It is acceptable to give a probability as a decimal or a percentage but a fraction is better.

This definition can be used to work out the probability of outcomes or events, as the following example shows.

EXAMPLE 1

A card is drawn from a normal pack of cards. What is the probability that it is one of the following?

a A red card
b A Spade
c A seven
d A picture card
e A number less than 5
f A red King

a There are 26 red cards, so P(red card) = $\frac{26}{52} = \frac{1}{2}$

b There are 13 Spades, so P(Spade) = $\frac{13}{52} = \frac{1}{4}$

c There are four sevens, so P(seven) = $\frac{4}{52} = \frac{1}{13}$

d There are 12 picture cards, so P(picture card) = $\frac{12}{52} = \frac{3}{13}$

e If you count the value of an Ace as 1, there are 16 cards with a value less than 5. So, P(number less than 5) = $\frac{16}{52} = \frac{4}{13}$

f There are two red Kings, so P(red King) = $\frac{2}{52} = \frac{1}{26}$

15.1 Experimental probability

This section will show you how to:

- calculate experimental probabilities and relative frequencies
- estimate probabilities from experiments
- use different methods to estimate probabilities

Key words

experimental probability
relative frequency
trials

The value of number of heads ÷ number of tosses is called an **experimental probability**. As the number of **trials** or experiments increases, the value of the experimental probability gets closer to the true or theoretical probability.

Experimental probability is also known as the **relative frequency** of an event. The relative frequency of an event is an estimate for the theoretical probability. It is given by:

$$\text{relative frequency of an outcome or event} = \frac{\text{frequency of the outcome or event}}{\text{total number of trials}}$$

EXAMPLE 2

The frequency table shows the speeds of 160 vehicles that pass a radar speed check on a dual carriageway.

Speed (mph)	20–29	30–39	40–49	50–59	60–69	70+
Frequency	14	23	28	35	52	8

a What is the experimental probability that a vehicle is travelling faster than 70 mph?

b If 500 vehicles pass the speed check, estimate how many will be travelling faster than 70 mph.

a The experimental probability is the relative frequency, which is $\frac{8}{160} = \frac{1}{20}$

b The number of vehicles travelling faster than 70 mph will be $\frac{1}{20}$ of 500.
That is, $500 \div 20 = 25$ vehicles

Finding probabilities

There are three ways in which the probability of an event can be found.

- If you can work out the theoretical probability of an outcome or event — for example, drawing a King from a pack of cards — this is called using *equally likely outcomes.*
- Some events, such as people buying a certain brand of dog food, cannot be calculated using equally likely outcomes. To find the probability of such an event, you can perform an experiment or conduct a survey. This is called collecting *experimental data.* The more data you collect, the better the estimate is.
- The probability of some events, such as an earthquake occurring in Japan, cannot be found by either of the above methods. One of the things you can do is to look at data collected over a long period of time and make an estimate (sometimes called a *best guess*) at the chance of the event happening. This is called looking at *historical data.*

EXAMPLE 3

Which method (A, B or C) would you use to estimate the probabilities of the events **a** to **e**?

A: Use equally likely outcomes

B: Conduct a survey/collect data

C: Look at historical data

a Someone in your class will go abroad for a holiday this year.

b You will win the National Lottery.

c Your bus home will be late.

d It will snow on Christmas Day.

e You will pick a red seven from a pack of cards.

a You would have to ask all the members of your class what they intended to do for their holidays this year. You would therefore conduct a survey, method B.

b The odds on winning are about 14 million to 1. This is an equally likely outcome, method A.

c If you catch the bus every day, you can collect data over several weeks. This would be method C.

d If you check whether it snowed on Christmas Day for the last few years you would be able to make a good estimate of the probability. This would be method C.

e There are two red sevens out of 52 cards, so the probability of picking one can be calculated: P(red seven) = $\frac{2}{52} = \frac{1}{26}$

This is method A.

EXERCISE 15A

1 Naseer throws a fair, six-sided dice and records the number of sixes that he gets after various numbers of throws. The table shows his results.

No. of throws	10	50	100	200	500	1000	2000
No. of sixes	2	4	10	21	74	163	329

a Calculate the experimental probability of a six at each stage that Naseer recorded his results.

b How many ways can a normal dice land?

c How many of these ways give a six?

d What is the theoretical probability of throwing a six with a dice?

e If Naseer threw the dice a total of 6000 times, how many sixes would you expect him to get?

2 Marie made a five-sided spinner, like the one shown in the diagram. She used it to play a board game with her friend Sarah. The girls thought that the spinner wasn't very fair as it seemed to land on some numbers more than others. They spun the spinner 200 times and recorded the results. The results are shown in the table.

Side spinner lands on	1	2	3	4	5
No. of times	19	27	32	53	69

a Work out the experimental probability of each number.

b How many times would you expect each number to occur if the spinner is fair?

c Do you think that the spinner is fair? Give a reason for your answer.

3 Sarah thought she could make a much more accurate spinner. After she had made it, she tested it and recorded how many times she scored a 5. Her results are shown in the table.

No. of spins	10	50	100	500
No. of fives	3	12	32	107

a Sarah made a mistake in recording the number of fives. Which number in the second row above is wrong? Give a reason for your answer.

b These are the full results for 500 spins.

Side spinner lands on	1	2	3	4	5
No. of times	96	112	87	98	107

Do you think the spinner is fair? Give a reason for your answer.

4 A sampling bottle is a sealed bottle with a clear plastic tube at one end. into which one of the balls can be tipped. Kenny's sampling bottle contains 20 balls, which are either black or white. Kenny conducts an experiment to see how many black balls are in the bottle. He takes various numbers of samples and records how many of them showed a black ball. The results are shown in the table.

No. of samples	No. of black balls	Experimental probability
10	2	
100	25	
200	76	
500	210	
1000	385	
5000	1987	

a Copy the table and calculate the experimental probability of getting a black ball at each stage.

b Using this information, how many black balls do you think are in the bottle?

5 Another sampling bottle contains red, white and blue balls. It is known that there are 20 balls in the bottle altogether. Carrie performs an experiment to see how many of each colour are in the bottle. She starts off putting down a tally each time a colour shows in the clear plastic tube.

Red	White	Blue
卌 卌 卌 卌 ||	卌 卌 卌 |||	卌 卌 ||

However, she forgets to count how many times she performs the experiment, so every now and again she counts up the tallies and records them in a table, like this.

Red	White	Blue	Total
22	18	12	52
48	31	16	95
65	37	24	126
107	61	32	200
152	93	62	307
206	128	84	418

The relative frequency of the red balls is calculated by dividing the frequency of red by the total number of trials, so at each stage these are:

0.423 0.505 0.516 0.535 0.495 0.493

These answers are rounded to 3 significant figures.

a Calculate the relative frequencies of the white balls at each stage, to 3 significant figures.

b Calculate the relative frequencies of the blue balls at each stage, to 3 significant figures.

c Round the final relative frequencies for Carrie's 418 trials, to 1 decimal place.

d What is the total of the answers in part **c**?

e How many balls of each colour do you think are in the bottle? Explain your answer.

6 Using card and a cocktail stick, make a six-sided spinner, as shown below.

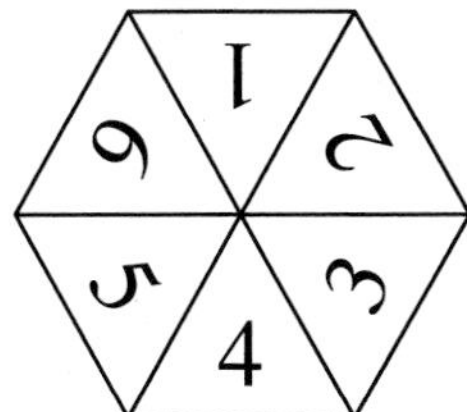

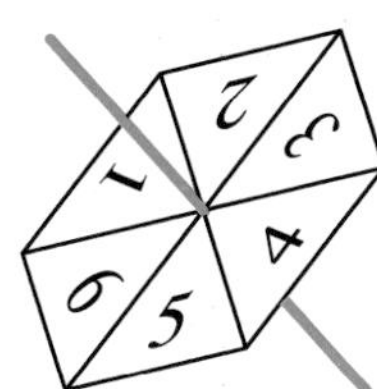

When you have made the spinner, spin it 120 times and record your results in a table like the one below.

Number	Tally	Total
1	~~IIII~~ II	
2	IIII	

a Which number occurred most frequently?

b How many times would you expect to get each number?

c Is your spinner fair?

d Explain your answer to part **c**.

7 Use a set of number cards from 1 to 10 (or make your own set) and work with a partner. Take turns to choose a card and keep a record each time of what card you get. Shuffle the cards each time and repeat the experiment 60 times. Put your results in a copy of this table.

Score	1	2	3	4	5	6	7	8	9	10
Total										

a How many times would you expect to get each number?

b Do you think you and your partner conducted this experiment fairly?

c Explain your answer to part **b**.

C

8 A four-sided dice has faces numbered 1, 2, 3 and 4. The score is the face on which it lands. Five students throw the dice to see if it is biased. They each throw it a different number of times. Their results are shown in the table.

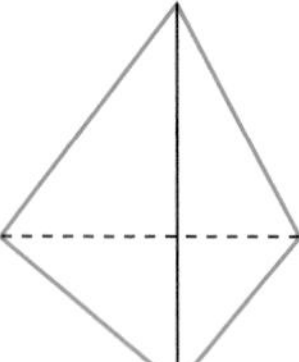

Student	Total no. of throws	Score			
		1	2	3	4
Alfred	20	7	6	3	4
Brian	50	19	16	8	7
Caryl	250	102	76	42	30
Deema	80	25	25	12	18
Emma	150	61	46	26	17

a Which student will have the most reliable set of results? Why?

b Add up all the score columns and work out the relative frequency of each score. Give your answers to 2 decimal places.

c Is the dice biased? Explain your answer.

9 Which of these methods would you use to estimate or state the probability of each of the events **a** to **h**?

Method A: Equally likely outcomes

Method B: Survey or experiment

Method C: Look at historical data

a How people will vote in the next election.

b A drawing pin dropped on a desk will land point up.

c A Premiership football team will win the FA Cup.

d You will win a school raffle.

e The next car to drive down the road will be red.

f You will throw a 'double six' with two dice.

g Someone in your class likes classical music.

h A person picked at random from your school will be a vegetarian.

10 If you were about to choose a card from a pack of yellow cards numbered from 1 to 10, what would be the chance of each of the events **a** to **i** occurring? Copy and complete each of these statements with a word or phrase chosen from 'impossible', 'not likely', '50–50 chance', 'quite likely' or 'certain'.

a The likelihood that the next card chosen will be a four is …

b The likelihood that the next card chosen will be pink is …

c The likelihood that the next card chosen will be a seven is …

d The likelihood that the next card chosen will be a number less than 11 is …

e The likelihood that the next card chosen will be a number bigger than 11 is …

f The likelihood that the next card chosen will be an even number is …

g The likelihood that the next card chosen will be a number more than five is …

h The likelihood that the next card chosen will be a multiple of 1 is …

i The likelihood that the next card chosen will be a prime number is …

AU 11 At a computer factory, tests were carried out to see how many faulty computer chips were produced in one week.

	Monday	Tuesday	Wednesday	Thursday	Friday
Sample	850	630	1055	896	450
No. faulty	10	7	12	11	4

On which day was it most likely that the highest number of faulty computer chips were produced?

PS 12 Andrew made an eight-sided spinner.

He tested it out to see if it was fair.

He spun the spinner and recorded the results.

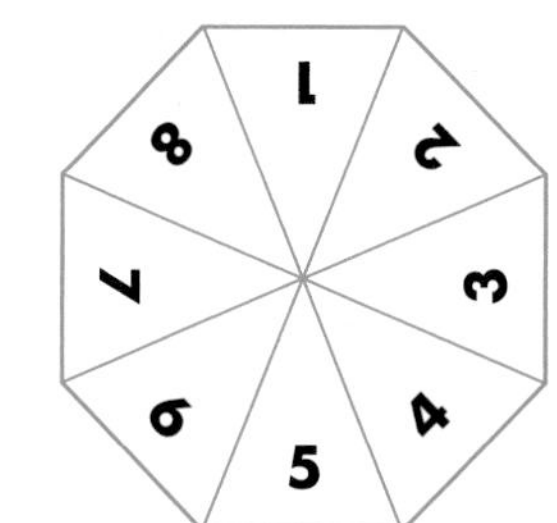

Unfortunately, his little sister spilt something over his results table, so he could not see the middle part.

No. spinner lands on	1	2	3			6	7	8
Frequency	18	19	22			19	20	22

Assuming the spinner was a fair one, complete the missing parts of the table for Andrew.

AU 13 Steve tossed a coin 1000 times to see how many heads he got.

He said, "If this is a fair coin, I should get 500 heads."

Explain why he is wrong.

AU 14 Roxy tests the eight-sided spinner shown by spinning it 100 times and recording the results.

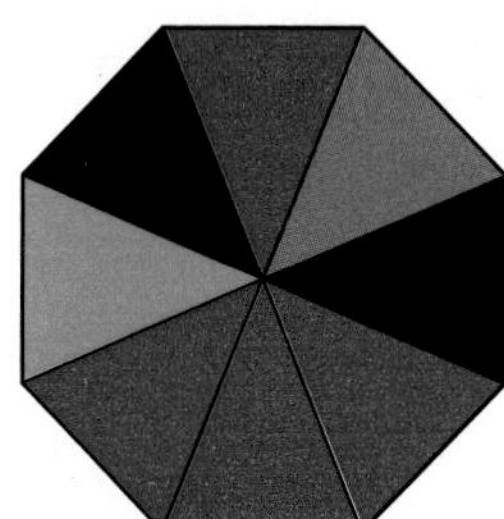

These are the results.

Colour	Red	Blue	Black	Green
Frequency	48	13	28	11

Roxy says the spinner is fair as the frequencies are around what is expected.

Sam says the spinner is unfair as there are far more reds than any other colour.

Who is correct? Back up your answer with some figures.

15.2 Mutually exclusive and exhaustive events

This section will show you how to:

- recognise mutually exclusive, complementary and exhaustive events

Key words
complementary
exhaustive events
mutually exclusive

If a bag contains three black, two yellow and five white balls and only one ball is allowed to be taken at random from the bag, then by the basic definition of probability:

$$\text{P(black ball)} = \frac{3}{10}$$

$$\text{P(yellow ball)} = \frac{2}{10} = \frac{1}{5}$$

$$\text{P(white ball)} = \frac{5}{10} = \frac{1}{2}$$

Also the probability of choosing a black ball or a yellow ball is $= \frac{5}{10} = \frac{1}{2}$

The events 'picking a yellow ball' and 'picking a black ball' can never happen at the same time when only one ball is taken out: that is, a ball can be either black or yellow. Such events are called **mutually exclusive**. Other examples of mutually exclusive events are tossing a head or a tail with a coin, drawing a King or an Ace from a pack of cards and throwing an even or an odd number with a dice.

An example of events that are not mutually exclusive would be drawing a red card and a King from a pack of cards. There are two red Kings, so these events could be true at the same time.

EXAMPLE 4

An ordinary dice is thrown.

a What is the probability of throwing:

i an even number **ii** an odd number?

b What is the total of the answers to part **a**?

c Is it possible to get a score on a dice that is both odd and even?

a **i** $\text{P(even)} = \frac{1}{2}$ **ii** $\text{P(odd)} = \frac{1}{2}$

b $\frac{1}{2} + \frac{1}{2} = 1$

c No

Events such as those in Example 4 are mutually exclusive because they can never happen at the same time. Because there are no other possibilities, they are also called **exhaustive events**. The probabilities of exhaustive events add up to 1.

EXAMPLE 5

A bag contains only black and white balls. The probability of picking at random a black ball from the bag is $\frac{7}{10}$.

a What is the probability of picking a white ball from the bag?

b Can you say how many black and white balls are in the bag?

a As 'picking a white ball' and 'picking a black ball' are mutually exclusive and exhaustive then:

$$P(\text{white}) = 1 - P(\text{black}) = 1 - \frac{7}{10} = \frac{3}{10}$$

b You cannot say precisely what the number of balls is although you can say that there could be seven black and three white, fourteen black and six white, or any combination of black and white balls in the ratio 7 : 3.

Complementary event

If there is an event A, the **complementary** event of A is:

Event A *not* happening

Any event is mutually exclusive and exhaustive to its complementary event. That is:

P(event A not happening) = 1 – P(event A happening)

which can be stated as:

P(event) + P(complementary event) = 1

For example, the probability of getting a King from a pack of cards is $\frac{4}{52} = \frac{1}{13}$, so the probability of *not* getting a King is:

$$1 - \frac{1}{13} = \frac{12}{13}$$

EXERCISE 15B

1 Say whether these pairs of events are mutually exclusive or not.

a Tossing a head with a coin/tossing a tail with a coin

b Throwing a number less than 3 with a dice/throwing a number greater than 3 with a dice

c Drawing a Spade from a pack of cards/drawing an Ace from a pack of cards

d Drawing a Spade from a pack of cards/drawing a red card from a pack of cards

e If two people are to be chosen from three girls and two boys: choosing two girls/choosing two boys

f Drawing a red card from a pack of cards/drawing a black card from a pack of cards

C

C

2 Which of the pairs of mutually exclusive events in question **1** are also exhaustive?

3 Each morning I run to work or get a lift. The probability that I run to work is $\frac{2}{5}$. What is the probability that I get a lift?

4 A letter is to be chosen at random from this set of letter-cards.

S T A T I S T I C S

a What is the probability that the letter is:

i an S **ii** a T **iii** a vowel?

b Which of these pairs of events are mutually exclusive?

i Picking an S/picking a T

ii Picking an S/picking a vowel

iii Picking an S/picking another consonant

iv Picking a vowel/picking a consonant

c Which pair of mutually exclusive events in part **b** is also exhaustive?

5 Two people are to be chosen for a job from this set of five people.

Jane Dave Anne Jack John

a List all of the possible pairs (there are 10 altogether).

b What is the probability that the pair of people chosen will:

i both be female **ii** both be male

iii both have the same initial **iv** have different initials?

c Which of these pairs of events are mutually exclusive?

i Picking two women/picking two men

ii Picking two people of the same sex/picking two people of opposite sex

iii Picking two people with the same initial/picking two men

iv Picking two people with the same initial/picking two women

d Which pair of mutually exclusive events in part **c** is also exhaustive?

6 A spinner consists of an outer ring of coloured sectors and an inner circle of numbered sectors, as shown.

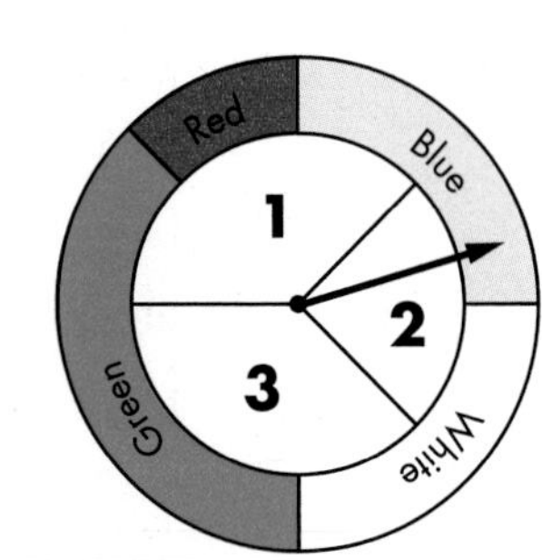

a The probability of getting 2 is $\frac{1}{4}$. The probabilities of getting 1 or 3 are equal. What is the probability of getting 3?

b The probability of getting blue is $\frac{1}{4}$. The probability of getting white is $\frac{1}{4}$. The probability of getting green is $\frac{3}{8}$. What is the probability of getting red?

c Which of these pairs of events are mutually exclusive?

i Getting 3/getting 2 **ii** Getting 3/getting green

iii Getting 3/getting blue **iv** Getting blue/getting red

d Explain why it is not possible to get a colour that is mutually exclusive to the event 'getting an odd number'.

7 At the morning break, I have the choice of coffee, tea or hot chocolate. If the probability I choose coffee is $\frac{3}{5}$, the probability I choose tea is $\frac{1}{4}$, what is the probability I choose hot chocolate?

PS **8** Four friends, Kath, Ann, Sandra and Padmini, regularly ran races against each other in the park.

The chances of:

Kath winning the race is 0.7
Ann winning the race is $\frac{1}{6}$
Sandra winning the race is 12%.

What is the chance of Padmini winning the race?

9 Assemblies at school are always taken by the head, the deputy head or the senior teacher. If the head takes the assembly, the probability that she goes over time is $\frac{1}{2}$. If the deputy takes the assembly, the probability that he goes over time is $\frac{1}{4}$. Explain why it is not necessarily true to say that the probability that the senior teacher goes over time is $\frac{1}{4}$.

FM **10** A hotelier conducted a survey of guests staying at her hotel. The table shows some of the results of her survey.

Type of guest	Probability
Man	0.7
Woman	0.3
American man	0.2
American woman	0.05
Vegetarian	0.3
Married	0.6

a A guest was chosen at random. From the table, work out these probabilities.

i The guest was American.

ii The guest was single.

iii The guest was not a vegetarian.

b Explain why it is not possible to work out from the table the probability of a guest being a married vegetarian.

c From the table, give two examples of pairs of types of guest that would form a mutually exclusive pair.

d From the table, give one example of a pair of types of guest that would form an exhaustive pair.

C B

B

FM **11** In a restaurant, the head waiter has worked out the probability of customers choosing certain dishes.

A starter	0.7
A pudding	0.4
Beef	0.3
Pork	0.2
Chicken	0.45
Vegetarian	0.08
Vegetables	0.8
Red wine	0.4
White wine	0.5

a What is the probability that the first person entering the restaurant:

i chooses a meat dish

ii chooses wine

iii does not have a starter?

b Explain why it is not possible to work out from the table the probability of someone having a starter or a pudding.

c Give an example of a choice from the table that would form a mutually exclusive pair.

AU **12** Ziq always walks, goes by bus or is given a lift by his dad to school.

If he walks, the probability that he is late for school is 0.3.

If he goes by bus, the probability that he is late for school is 0.1.

Explain why it is not necessarily true that if his dad gives him a lift, the chance of his being late for school is 0.6.

15.3 Expectation

This section will show you how to:

- predict the likely number of successful events, given the number of trials and the probability of any one event

Key words

expectation

When you know the probability of an event, you can predict how many times you would expect that event to happen in a certain number of trials. This is called **expectation**.

Note that this is what you *expect*. It is not what is going to happen. If what you expected always happened, life would be very dull and boring and the National Lottery would be a waste of time.

EXAMPLE 6

A bag contains 20 balls, nine of which are black, six white and five yellow. A ball is drawn at random from the bag, its colour is noted and then it is put back in the bag. This is repeated 500 times.

a How many times would you expect a black ball to be drawn?

b How many times would you expect a yellow ball to be drawn?

c How many times would you expect a black or a yellow ball to be drawn?

a P(black ball) = $\frac{9}{20}$

Expected number of black balls = $\frac{9}{20} \times 500 = 225$

b P(yellow ball) = $\frac{5}{20} = \frac{1}{4}$

Expected number of yellow balls = $\frac{1}{4} \times 500 = 125$

c Expected number of black or yellow balls = 225 + 125 = 350

EXAMPLE 7

Four in 10 cars sold in Britain are made by Japanese companies.

a What is the probability that the next car to drive down your road will be Japanese?

b If there are 2000 cars in a multistorey car park, how many of them would you expect to be Japanese?

a P(Japanese car) = $\frac{4}{10} = \frac{2}{5} = 0.4$

b Expected number of Japanese cars in 2000 cars = $0.4 \times 2000 = 800$ cars

EXERCISE 15C

C

1 I throw an ordinary dice 150 times. How many times can I expect to get a 6?

2 I toss a coin 2000 times. How many times can I expect to get a head?

3 I draw a card from a pack of cards and replace it. I do this 520 times. How many times would I expect to get:

a a black card　　**b** a King

c a Heart　　**d** the King of Hearts?

4 The ball in a roulette wheel can land in 37 spaces which are the numbers from 0 to 36 inclusive. I always bet on the same number, 13. If I play all evening and there are altogether 185 spins of the wheel in that time, how many times could I expect to win?

5 In a bag there are 30 balls, 15 of which are red, five yellow, five green and five blue. A ball is taken out at random and then replaced. This is repeated 300 times. How many times would I expect to get:

a a red ball　　**b** a yellow or blue ball

c a ball that is not blue　　**d** a pink ball?

6 The same experiment described in question **5** is carried out 1000 times. Approximately how many times would you expect to get: **a** a green ball **b** a ball that is not blue?

C

7 A sampling bottle contains red and white balls. It is known that the probability of getting a red ball is 0.3. 1500 samples are taken. How many of them would you expect to give a white ball?

8 Josie said, "When I throw a dice, I expect to get a score of 3.5."

"Impossible," said Paul, "you can't score 3.5 with a dice."

"Do this and I'll prove it," said Josie.

a Throw an ordinary dice 60 times. Copy and fill in the table for the expected number of times each score will occur.

Score						
Expected occurrences						

b Now work out the average score that is expected over 60 throws.

c There is an easy way to get an answer of 3.5 for the expected average score. Can you see what it is?

9 The probability of some cloud types being seen on any day is given below.

Cumulus	0.3
Stratocumulus	0.25
Stratus	0.15
Altocumulus	0.11
Cirrus	0.05
Cirrcocumulus	0.02
Nimbostratus	0.005
Cumulonimbus	0.004

a What is the probability of **not** seeing one of the above clouds in the sky?

b On how many days of the year would you expect to see altocumulus clouds in the sky?

B

PS 10 Every evening Tamara and Chris cut a pack of cards to see who washes up.

If they cut a King or a Jack, Chris washes up.

If they cut a Queen, Tamara washes up.

Otherwise they wash up together.

In a year of 365 days, how many days would you expect them to wash up together?

AU 11 A market gardener is supplied with tomato plant seedlings. She knows that the probability that any plant will develop a disease is 0.003.

How will she find out the number of tomato plants that are likely to develop a disease?

12 I have 20 tickets for a raffle and I know that the probability of my winning the prize is 0.05. How many tickets were sold altogether in the raffle?

15.4 Two-way tables

This section will show you how to:

- read two-way tables and use them to work out probabilities and interpret data

Key words

two-way table

A **two-way table** is a table that links together two variables. For example, this shows how many boys and girls are in a form and whether they are left-handed or right-handed.

	Boys	Girls
Left-handed	2	4
Right-handed	10	13

This table shows the colours and makes of cars in the school car park.

	Red	Blue	White
Ford	2	4	1
Vauxhall	0	1	2
Toyota	3	3	4
Peugeot	2	0	3

One variable is written in the rows of the table and the other variable is written in the columns of the table.

EXAMPLE 8

Using the first two-way table above, answer these questions.

a If a student is selected at random from the form, what is the probability that it will be a left-handed boy?

b It is known that a student selected at random is a girl. What is the probability that she is right-handed?

a $\frac{2}{29}$ **b** $\frac{13}{17}$

EXAMPLE 9

Using the second two-way table above, answer these questions.

a What percentage of the cars in the car park are red?

b What percentage of the white cars are Vauxhalls?

a 28%. Seven out of 25 is the same as 28 out of 100.

b 20%. Two out of 10 is 20%.

EXERCISE 15D

1 The two-way table shows the age and gender of a sample of 50 students in a school.

	Age (years)					
	11	**12**	**13**	**14**	**15**	**16**
No. of boys	4	3	6	2	5	4
No. of girls	2	5	3	6	4	6

a How many students are aged 13 years or less?

b What percentage of the students in the table are 16?

c A student from the table is selected at random. What is the probability that the student will be 14 years of age? Give your answer as a fraction in its simplest form.

d There are 1000 students in the school. Use the table to estimate how many boys there are in the school altogether.

2 The two-way table shows the numbers of adults and the numbers of cars in 50 houses in one street.

		No. of adults			
		1	**2**	**3**	**4**
No. of cars	**0**	2	1	0	0
	1	3	13	3	1
	2	0	10	6	4
	3	0	1	4	2

a How many houses have exactly two adults and two cars?

b How many houses altogether have three cars?

c What percentage of the houses have three cars?

d What percentage of the houses with just one car have three adults living in the house?

3 Jane has two four-sided spinners. One has the numbers 1 to 4 and the other has the numbers 5 to 8.

Both spinners are spun together.

This two-way table shows all the ways the two spinners can land.

Some of the total scores are filled in.

1 2 4 3

Spinner A

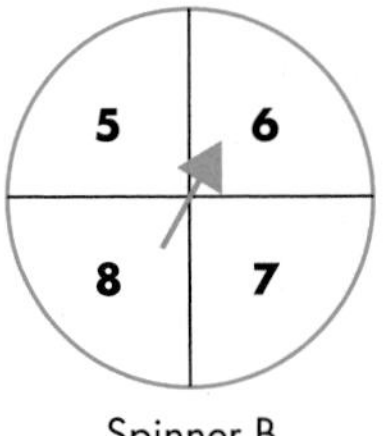

Spinner B

		Score on spinner A			
		1	**2**	**3**	**4**
Score on spinner B	**5**	6	7		
	6	7			
	7				
	8				

a Copy and complete the table to show all the possible total scores.

b How many of the total scores are 9?

c When the two spinners are spun together, what is the probability that the total score will be:

i 9 **ii** 8 **iii** a prime number?

4 The table shows information about the number of items in Flossy's music collection.

		Type of music		
		Pop	**Folk**	**Classical**
Format	**Tape**	16	5	2
	CD	51	9	13
	Mini disc	9	2	0

a How many pop tapes does Flossy have?

b How many items of folk music does Flossy have?

c How many CDs does Flossy have?

d If a CD is chosen at random from all the CDs, what is the probability that it will be a pop CD?

5 Zoe throws a fair coin and rolls a fair dice.

If the coin shows a head she records the score on the dice.
If the coin shows tails she doubles the number on the dice.

a Complete the two-way table to show Zoe's possible scores.

		No. on dice					
		1	**2**	**3**	**4**	**5**	**6**
Coin	**Head**	1	2				
	Tail	2	4				

b How many of the scores are square numbers?

c What is the probability of getting a score that is a square number?

AU **6** A gardener plants some sunflower seeds in a greenhouse and plants some in the garden. After they have fully grown, he measures the diameter of the sunflower heads. The table shows his results.

	Greenhouse	Garden
Mean diameter	16.8 cm	14.5 cm
Range of diameter	3.2 cm	1.8 cm

a The gardener, who wants to enter competitions, says, "The sunflowers from the greenhouse are better." Using the data in the table, give a reason to justify this statement.

b The gardener's wife, who does flower arranging, says, "The sunflowers from the garden are better." Using the data in the table, give a reason to justify this statement.

C

FM **7** The two-way table shows the wages for the men and women in a factory.

Wage, w (£) per week	Men	Women
£100 < w ≤ £150	3	4
£150 < w ≤ £200	7	5
£200 < w ≤ £250	23	12
£250 < w ≤ £300	48	27
£300 < w ≤ £350	32	11
More than £350	7	1

a What percentage of the men earn between £250 and £300 per week?

b What percentage of the women earn between £250 and £300 per week?

c Is it possible to work out the mean wage of the men and women? Explain your answer.

AU **8** Reyki plants some tomato plants in her greenhouse, while her husband Daniel plants some in the garden.

After the summer they compared their tomatoes.

	Garden	Greenhouse
Mean diameter (cm)	1.8	4.2
Mean number of tomatoes per plant	24.2	13.3

Use the data in the table to explain who had the better crop of tomatoes.

9 Here is a two way table for the members of a sports club.

	Men	Women	Children
Left footed			
Right footed			

Altogether there are 60 members.

4 men are left footed.

The ratio of men to women to children is 5 : 3 : 4.

There are 10 more right footed children then left footed children.

One-fifth of the members are left footed.

Copy and complete the table.

C

PS 10 Two hexagonal spinners are spun.

Spinner A is numbered 3, 5, 7, 9, 11 and 13.

Spinner B is numbered 4, 5, 6, 7, 8 and 9.

What is the probability that when the two spinners are spun, the two numbers given will multiply to a total greater than 40?

11 Here are two fair spinners.

The spinners are spun.

The two numbers obtained are added together.

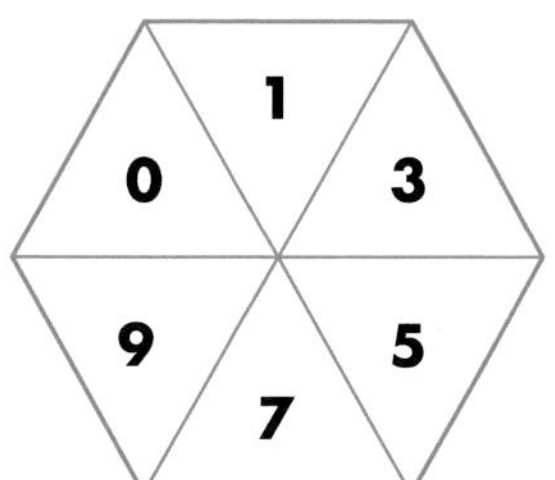

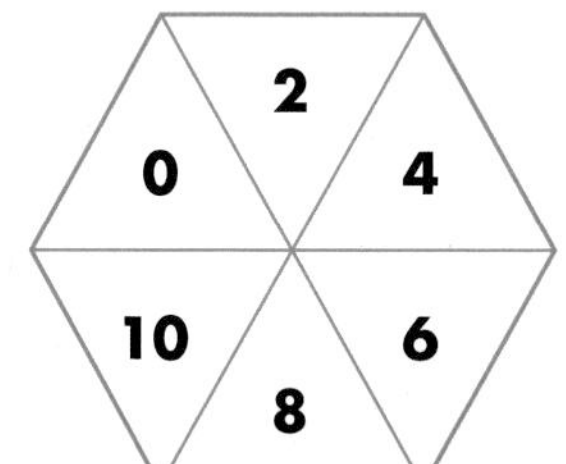

a Draw a two-way diagram showing all the possible scores.

b What is the most likely score?

c What is the probability of getting a total of 12?

d What is the probability of getting a total of 11 or more?

e What is the probability of getting a total that is an odd number?

B

PS 12 These two-way tables show information about the students in year 11.

	Vegetarian	Non-vegetarian
Girls	45	50
Boys	30	75

	Blue eyed	Brown eyed
Girls	60	35
Boys	45	60

	Vegetarian	Non-vegetarian
Blue eyed	40	65
Brown eyed	35	60

a How many students are in year 11 altogether?

b An individual student is picked at random from year 11.

What is the probability that the student is a vegetarian?

c One of the boys is picked at random.

What is the probability that he is blue eyed?

d A third of the vegetarians and two-fifths of the non-vegetarians bring a packed lunch to school.

A year 11 students leaves their packed lunch on the bus. What is the probability that it is a vegetarian lunch?

15.5 Addition rule for events

This section will show you how to:
- work out the probability of two events such as P(A) or P(B)

Key words
either

You have used the addition rule already but it has not yet been formally defined.

When two events are mutually exclusive, you can work out the probability of **either** of them occurring by adding up the separate probabilities.

EXAMPLE 10

A bag contains twelve red balls, eight green balls, five blue balls and fifteen black balls. A ball is drawn at random. What is the probability of it being:

a red **b** black **c** red or black

d not green **e** neither green nor blue?

a $\text{P(red)} = \frac{12}{40} = \frac{3}{10}$ **b** $\text{P(black)} = \frac{15}{40} = \frac{3}{8}$

c $\text{P(red or black)} = \text{P(red)} + \text{P(black)} = \frac{3}{10} + \frac{3}{8} = \frac{27}{40}$

d $\text{P(not green)} = \frac{32}{40} = \frac{4}{5}$

e $\text{P(neither green nor blue)} = \text{P(red or black)} = \frac{27}{40}$

The last part of Example 10 is another illustration of how confusing probability can be. You might think:

$$\text{P(neither green nor blue)} = \text{P(not green)} + \text{P(not blue)} = \frac{32}{40} + \frac{35}{40} = \frac{67}{40}$$

This cannot be correct, as the answer is greater than 1. In fact, the events 'not green' and 'not blue' are not mutually exclusive, as there are lots of balls that satisfy both outcomes.

EXERCISE 15E

D

1 Iqbal throws an ordinary dice. What is the probability that he throws these scores?

a 2 **b** 5 **c** 2 or 5

2 Jennifer draws a card from a pack of cards. What is the probability that she draws these?

a A Heart **b** A Club **c** A Heart or a Club

3 Jasper draws a card from a pack of cards. What is the probability that he draws one of the following numbers?

a 2 **b** 6 **c** 2 or 6

D

4 A letter is chosen at random from the letters on these cards. What is the probability of choosing each of these?

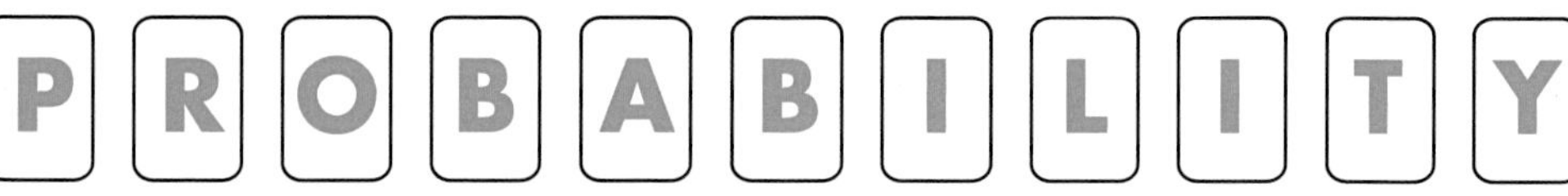

a A 'B' **b** A vowel **c** A 'B' or a vowel

5 A bag contains 10 white balls, 12 black balls and eight red balls. A ball is drawn at random from the bag. What is the probability of each of these outcomes?

a White **b** Black

c Black or white **d** Not red

e Not red or black

> **HINTS AND TIPS**
>
> You can only add fractions with the same denominator.

C

FM **6** At the School Fayre the tombola stall gives out a prize if you draw from the drum a numbered ticket that ends in 0 or 5. There are 300 tickets in the drum altogether and the probability of getting a winning ticket is 0.4.

a What is the probability of getting a losing ticket?

b How many winning tickets are there in the drum?

7 John needs his calculator for his mathematics lesson. It is always in his pocket, bag or locker.

The probability it is in his pocket is 0.35; the probability it is in his bag is 0.45. What is the probability that:

a he will have the calculator for the lesson

b it is in his locker?

8 A spinner has numbers and colours on it, as shown in the diagram. Their probabilities are given in the tables.

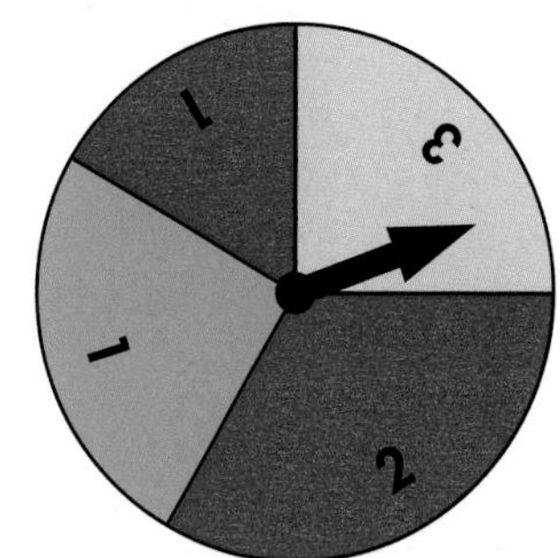

When the spinner is spun, what is the probability of each of the following?

a Red or green

b 2 or 3

c 3 or green

d 2 or green

AU

e **i** Explain why the answer to P(1 or red) is not 0.9.

ii What is the answer to P(1 or red)?

Red	0.5
Green	0.25
Blue	0.25

1	0.4
2	0.35
3	0.25

C

9 Debbie has 20 unlabelled CDs, 12 of which are rock, five are pop and three are classical. She picks a CD at random. What is the probability of these outcomes?

a Rock or pop

b Pop or classical

c Not pop

AU **10** The probability that it rains on Monday is 0.5. The probability that it rains on Tuesday is 0.5 and the probability that it rains on Wednesday is 0.5. Kelly argues that it is certain to rain on Monday, Tuesday or Wednesday because 0.5 + 0.5 + 0.5 = 1.5, which is bigger than 1 so that it is a certain event. Explain why she is wrong.

FM **11** Brian and Kathy put music onto their iPod so that at their wedding reception they had a variety of background music. They uploaded 100 different tracks onto the iPod.

40 love songs

35 musical show songs

15 classical music tracks

10 rock tracks

They set it to play the tracks continuously at random.

a What is the probability that:

i the first track played is a love song

ii the last track of the evening is either a musical show song or a classical track

iii the track when they start their meal is not a rock track?

b When they start cutting the cake they want a love song or a classical track playing. What is the probability that they will not get a track of their choice?

c The reception lasts for five and a half hours. What amount of time, in hours and minutes, would you expect the iPod to have been playing love song tracks?

PS **12** James, John and Joe play the *Count Dracula* game together every Saturday. John is always the favourite to win, with a probability of 0.75.

In the year 2009 there were 52 Saturdays and James won eight times.

What was the probability of Joe winning?

B

AU **13** A bag contains some red and some blue balls.

A ball is taken out at random and its colour noted. The ball is then replaced in the bag. Another ball is then taken out at random and its colour noted.

a Which of these **could not** be the probability of two red balls?

$\frac{9}{25}$ $\quad \frac{1}{9}$ $\quad \frac{13}{20}$

Give a reason for your choice.

b It is known that there are more blue balls than red balls in the bag.

Which of the probabilities in part **a** must be the probability of two red balls? Give a reason for your choice.

15.6 Combined events

This section will show you how to:
- work out the probability of two events occurring at the same time

Key words
probability space diagram
sample space diagram

There are many situations where two events occur together. Some examples are given below. Note that, in each case, all the possible outcomes of the events are shown in diagrams. These are called **probability space diagrams** or **sample space diagrams**.

Throwing two dice

Imagine that two fair dice, one red and one blue, are thrown. The red dice can land with any one of six scores: 1, 2, 3, 4, 5 or 6. The blue dice can also land with any one of six scores. This gives a total of 36 possible combinations. These are shown in the left-hand diagram, where each combination is given as (2, 3), etc. The first number is the score on the blue dice and the second number is the score on the red dice.

The combination (2, 3) gives a total score of 5. The total scores for all the combinations are shown in the right-hand diagram.

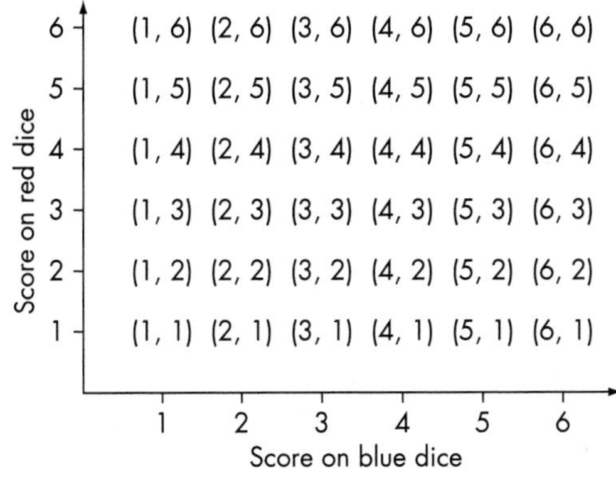

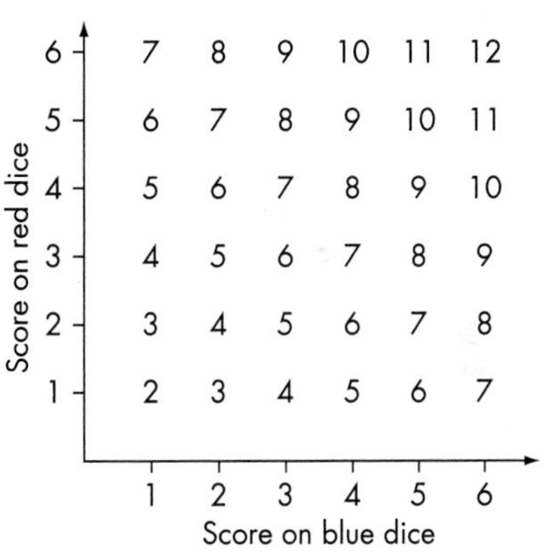

From the diagram on the right, you can see that there are two ways to get a score of 3. This gives a probability of:

$$P(3) = \frac{2}{36} = \frac{1}{18}$$

From the diagram on the left, you can see that there are six ways to get a 'double'. This gives a probability of:

$$P(\text{double}) = \frac{6}{36} = \frac{1}{36}$$

Throwing coins

Throwing one coin

There are two equally likely outcomes, head or tail. 

Throwing two coins together

There are four equally likely outcomes: 

H T

T H

T T

P(2 heads) = $\frac{1}{4}$

P(head and tail) = 2 ways out of 4 = $\frac{2}{4} = \frac{1}{2}$

Dice and coins

Throwing a dice and a coin

Outcome on coin						
H	(1, H)	(2, H)	(3, H)	(4, H)	(5, H)	(6, H)
T	(1, T)	(2, T)	(3, T)	(4, T)	(5, T)	(6, T)
	1	2	3	4	5	6

Score on dice

P (head and an even number) = 3 ways out of 12 = $\frac{3}{12} = \frac{1}{4}$

EXERCISE 15F

1 To answer these questions, use the right-hand diagram on page 389 for the total scores when two fair dice are thrown together.

- **a** What is the most likely score?
- **b** Which two scores are least likely?
- **c** Write down the probabilities of all scores from two to 12.
- **d** What is the probability of each of these scores?
  - **i** Bigger than 10
  - **ii** From 3 to 7
  - **iii** Even
  - **iv** A square number
  - **v** A prime number
  - **vi** A triangular number

2 Using the left-hand diagram on page 389 that shows, as coordinates, the outcomes when two fair dice are thrown together, what is the probability that:

- **a** the score is an even 'double'
- **b** at least one of the dice shows 2
- **c** the score on one dice is twice the score on the other dice
- **d** at least one of the dice shows a multiple of 3?

D

3 Using the left-hand diagram on page 389 that shows, as coordinates, the outcomes when two fair dice are thrown together, what is the probability that:

a both dice show a 6

b at least one of the dice shows a 6

c exactly one dice shows a 6?

4 This diagram shows the score for the event 'the difference between the scores when two fair dice are thrown'. Copy and complete the diagram.

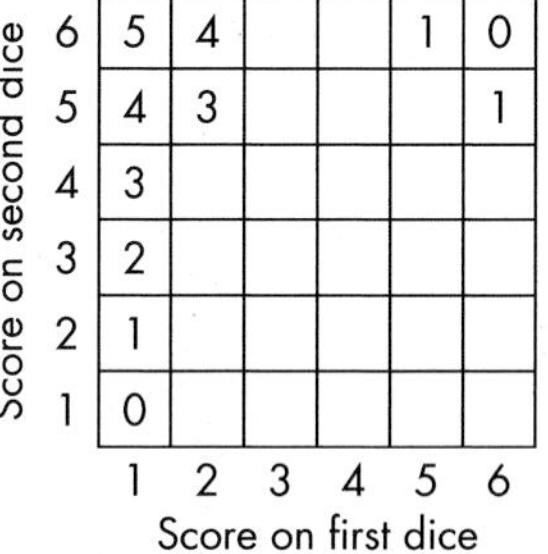

For the event described above, what is the probability of a difference of:

a 1 **b** 0 **c** 4

d 6 **e** an odd number?

5 When two fair coins are thrown together, what is the probability of each of these outcomes?

a two heads **b** A head and a tail

c At least one tail **d** No tails

Use the diagram of the outcomes when two coins are thrown together, on page 390.

C

6 Two fair five-sided spinners are spun together and the total score of the faces that they land on is worked out. Copy and complete this probability space diagram.

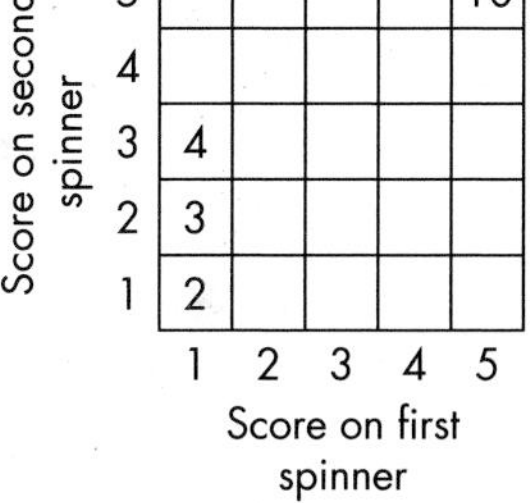

a What is the most likely score?

b When two five-sided spinners are spun together, what is the probability of:

i the total score being 5 **ii** the total score being an even number

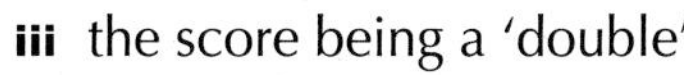

iii the score being a 'double' **iv** the score being less than 7?

7 When three fair coins are tossed together, what is the probability of:

a three heads **b** two heads and one tail

c at least one tail **d** no tails?

8 When one coin is tossed, there are two outcomes. When two coins are tossed, there are four outcomes. When three coins are tossed, there are eight outcomes.

a How many outcomes will there be when four coins are tossed?

b How many outcomes will there be when five coins are tossed?

c How many outcomes will there be when 10 coins are tossed?

d How many outcomes will there be when n coins are tossed?

9 When a dice and a coin are thrown together, what is the probability of each of the following outcomes?

a You get a head on the coin and a 6 on the dice.

b You get a tail on the coin and an even number on the dice.

c You get a head on the coin and a square number on the dice.

Use the diagram on page 171 that shows the outcomes when a dice and a coin are thrown together.

FM 10 When Mel walked into her local shopping mall, she saw a competition taking place. Mel decided to have a go.

> Roll 2 dice get a total of 11 and win a £10 note! Only £1 a go!

a Draw the sample space diagram for this event.

b What is the probability of winning a £10 note?

c How many goes should she have in order to expect to win at least once?

d If she had 40 goes, how many times could she expect to have won?

PS 11 I toss five coins. What is the probability that I will get more heads than tails?

AU 12 I roll a dice three times and add the three numbers obtained.

Explain the difficulty in drawing a sample space to show all the possible events.

PS 13 A code consists of two digits from 0 to 9.

For example 00, 01, … 12, … 46, …. 98, 99

What is the probability that the code contains the digit 9?

PS 14 Tommy has 4 black socks, 6 striped socks and 10 spotted socks

a Explain why, if Tommy takes 4 socks at random from his drawer he will be sure to get a pair.

b Tommy takes a black sock from the drawer.
Explain why, if he now takes a sock at random from the drawer the probability that it is black is $\frac{3}{19}$

c Explain why a sample space diagram cannot be used to show all the possible outcomes of taking two socks at random from the drawer.

d Use these rectangles to work out the probability that if Tommy takes two socks at random from his drawer he gets a matching pair.

	1st sock		1st black		1st stripe		1st spotted
2nd sock	20×19	2nd black	4×3	2nd stripe	6×5	2nd spotted	10×9

GRADE BOOSTER

- **D** You can calculate the probability of an outcome not happening if you know the probability it happening
- **D** You understand that the total probability of all possible outcomes in a particular situation is 1
- **D** You can predict the expected number of successes from a given number of trials if you know the probability of one success
- **D** You can extract information from two-way tables and calculate percentages and probabilities
- **D** You can draw a sample space diagram for combined events.
- **C** You can calculate relative frequency from experimental evidence and compare this with the theoretical probability
- **C** You can work out the probability of two independent events
- **C** You can work out the expected value given the probability of the event and the total population

What you should know now

- How to calculate the relative frequency of an event from experimental data
- How to work out the probability of an event using equally likely outcomes
- How to use the fact that the total probability of all events is 1 to calculate individual probabilities
- How to use a two-way table to calculate probabilities and percentages
- How to work out the probability of two independent events occurring
- How to work out an expected value
- How to use a sample space diagram to work out the probabilities of combined events

EXAMINATION QUESTIONS

1 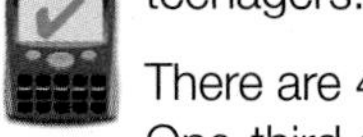 Here is some information about a group of 40 teenagers.

There are 4 more boys than girls
One-third of the girls are vegetarians.

Use this information to complete the two-way table.

	Boys	Girls	Total
Vegetarian			10
Non-vegetarian			30
Total			40

(3 marks)

AQA, November 2011, Higher 2, Question 2

2 A bag contains five discs that are numbered 1, 2, 3, 4 and 5.

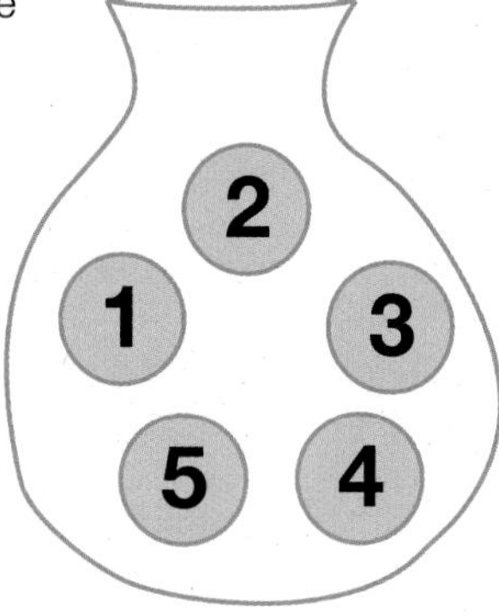

Suraj is playing a game. He takes a disc at random from the bag. He notes the number and replaces the disc. He then takes another disc at random from the bag. He notes this number. He then add the two number together to obtain a total score.

Suraj wins the game if his total score is 5, 6 or 7.

Is he more likely to win or lose this game?

You must show your working.

You may use the two-way table to help you.

+	1	2	3	4	5
1					
2					
3					
4					
5					

(4 marks)

AQA, June 2011, Foundation 1, Question 18

3  Sam and Pat are playing a game with dice.

a Sam plays the game using a **biased** dice. The probability that he throws a six is 0.1.

i Write down the probability that he does **not** throw a six. *(1 mark)*

ii Sam throws this biased dice 120 times. Work out an estimate for the number of sixes he will throw. *(2 marks)*

b Pat plays the game using a **fair** dice. She throws this fair dice 120 times. Is she likely to throw more sixes than Sam?

Give a reason for your answer. *(2 marks)*

AQA, November 2010, Foundation, Question 19

4 My wife asks me to tell her if the milk in the fridge has gone off or not. I say that it might have or it might not.

She says, "So the probability that it's off is 0.5 then."

How can I explain to her that this might not be so?

5  A home-made four-sided spinner is known to be biased.

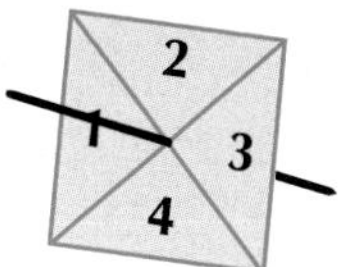

The table shows the probabilities of getting each score.

Score	1	2	3	4
Probability	$\frac{1}{10}$	$\frac{3}{10}$	$\frac{7}{20}$	$\frac{1}{4}$

a What is the probability of landing on a 1 or a 2? *(1 mark)*

b The spinner is spun 400 times. How many times would you expect it to land on a 3? *(2 marks)*

AQA, November 2011, Foundation 2, Question 27

6 A four-sided spinner is spun 1000 times.

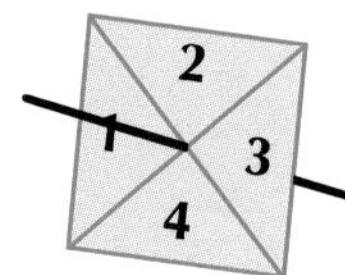

The results are shown in the table.

Number	1	2	3	4
Frequency	126	241	389	244

a **i** What is the relative frequency of 1? *(1 mark)*

ii How can you tell that the spinner is biased? *(1 mark)*

b A regular fair six-sided dice is thrown 600 times. How many time would you expect the dice to land on 6? *(1 mark)*

AQA, June 2011, Higher 2, Question 7

7 Paul has two six-sided dice. He knows that one is fair and one is biased. Describe how he can find out which is the biased dice. *(4 marks)*

AQA, June 2011, Higher 2, Question 12

8 A five-sided spinner is labelled A, B, C, D and E.

The spinner is biased.

The table shows some of the probabilities of the spinner landing on each letter.

Letter	Probability
A	0.40
B	0.25
C	
D	
E	0.05

The probability that the spinner lands on C is equal to the probability that it lands on D.

a Calculate the probability that the spinner lands on D. *(3 marks)*

b Calculate the probability that the spinner lands on either A or B. *(2 marks)*

AQA, November 2008, Module 1 Higher, Question 7

9 I have ten coins. I spin them all at the same time and count the number of heads I have. If I do this one thousand times, how many times would I expect to have ten heads?

Worked Examination Questions

FM **1** In a raffle 400 tickets have been sold. There is only one prize.

Mr Raza buys five tickets for himself and sells another 40.

Mrs Raza buys 10 tickets for herself and sells another 50.

Mrs Hewes buys eight tickets for herself and sells just 12 others.

a What is the probability of:

i Mr Raza winning the raffle

ii Mr Raza selling the winning ticket

iii Mr Raza either winning the raffle or selling the winning ticket?

b What is the probability of either Mr or Mrs Raza selling the winning ticket?

c What is the probability of Mrs Hewes not winning the lottery?

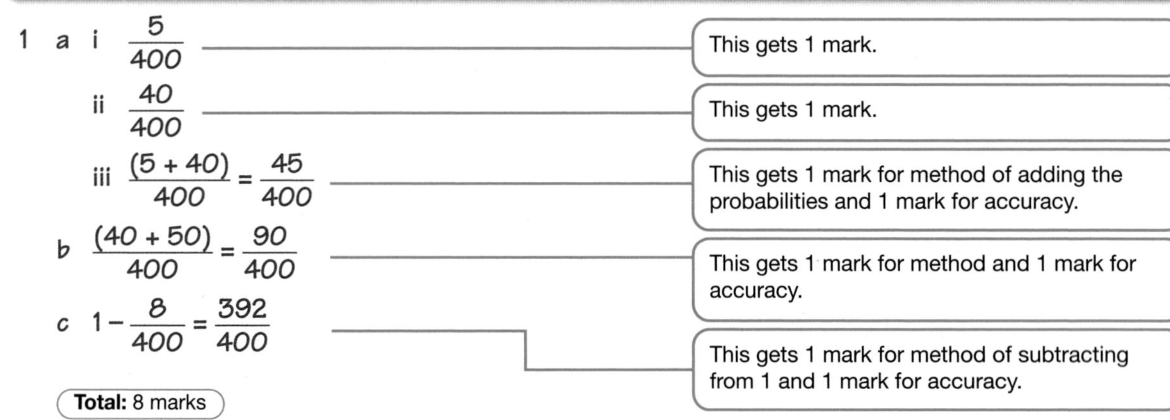

1 a i $\frac{5}{400}$ — This gets 1 mark.

ii $\frac{40}{400}$ — This gets 1 mark.

iii $\frac{(5+40)}{400} = \frac{45}{400}$ — This gets 1 mark for method of adding the probabilities and 1 mark for accuracy.

b $\frac{(40+50)}{400} = \frac{90}{400}$ — This gets 1 mark for method and 1 mark for accuracy.

c $1 - \frac{8}{400} = \frac{392}{400}$ — This gets 1 mark for method of subtracting from 1 and 1 mark for accuracy.

Total: 8 marks

Worked Examination Questions

AU **2** Harry is given a bag of coloured balls. He takes ten balls out as a sample of what might be in there and finds they are only blue and white balls.

He says, "There are no red balls in the bag."

Is he correct? Give a reason to support your answer.

2 No, he is incorrect as there might be at least one red ball in the bag but not chosen in the sample.

You need the statement 'No' and a reason similar to this one for 1 mark.

Total: 1 mark

15 Functional Maths
Fairground games

Joe had a stall at the local fair and wanted to make a reasonable profit from the game below.

In this game, the player rolls two balls down the sloping board and wins a prize if they land in slots that total more than seven.

Joe wants to know how much he should charge for each go and what the price should be. He would also like to know how much profit he is likely to make.

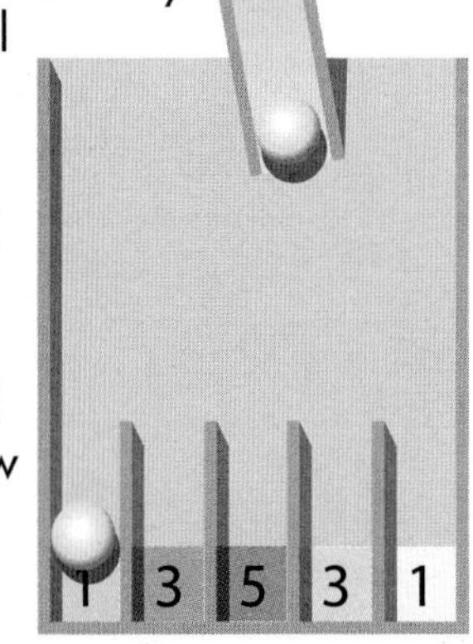

Getting started

Practise calculating probabilities using the spinner and questions below.

- What is the probability of spinning the spinner and getting:
  - a one
  - a five
  - a two
  - a number other than five?

  Give your answer as a fraction and as a decimal.
- If you spun the spinner 20 times, how many times would you expect to get:
  - a five
  - not five
  - an odd number?
- If you spun the spinner 100 times, how many times would you expect to get:
  - a two
  - not a two
  - a prime number?

Now, think about which probabilities you must calculate, in order to help Joe, and to design your own profitable game.

Your task

For this task you can work individually or in pairs.

1 Write a report for Joe which includes:

- a diagram to show the probability of each outcome
- the probability of at least three different outcomes
- the probability of winning
- at least three different ways of setting up the game, showing the cost of one go, the value of the prize and the expected profit for varying numbers of people playing the game.

Based on this information, advise Joe on how he should set up the game. Justify your decision.

2 Design your own fairground game. Describe the rules for winning, show the probability of winning using a diagram, and give costings and the projected profit.

Explain why you think your game should be used at the next local fair.

Why this chapter matters

We have already seen that patterns appear in numbers — prime numbers, square numbers and multiples — all form patterns. Number patterns are not only of mathematical value, they also make the study of nature and geometric patterns a little more interesting.

There are many mathematical patterns that appear in nature. The most famous of these is probably the Fibonacci series.

1 1 2 3 5 8 13 21 …

This is formed by adding the two previous terms to get the next term.

The sequence was discovered by the Italian, Leonardo Fibonacci, in 1202, when he was investigating the breeding patterns of rabbits!

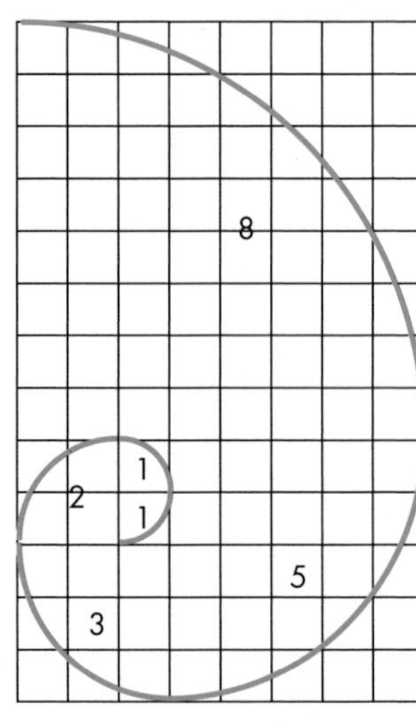

Since then, the pattern has been found in many other places in nature. The spirals found in a nautilus shell and in the seed heads of a sunflower plant also follow the Fibonacci series.

The Fibonnaci sequence has some other interesting properties.

- Divide each term by the previous term and write down the results.

  $1 \div 1 = 1,\ 2 \div 1 = 2,\ 3 \div 2 = 1.5,\ 5 \div 3 = 1.66,\ 8 \div 5 = 1.6,\ 13 \div 8 = 1.625, \ldots$
- Set up a spreadsheet to do this for 100 terms.
- Then look up the *Golden ratio* on the internet.

Fractals form another kind of pattern.

Fractals are geometric patterns that are continuously repeated on a smaller and smaller scale.

A good example of a fractal is this: start with an equilateral triangle and draw an equilateral triangle, a third the size of the original, on the middle of each side. Keep on repeating this and you will get an increasingly complex-looking shape.

The pattern shown here is called the Koch snowflake. It is named after the Swedish mathematician, Helge von Koch (1870–1924).

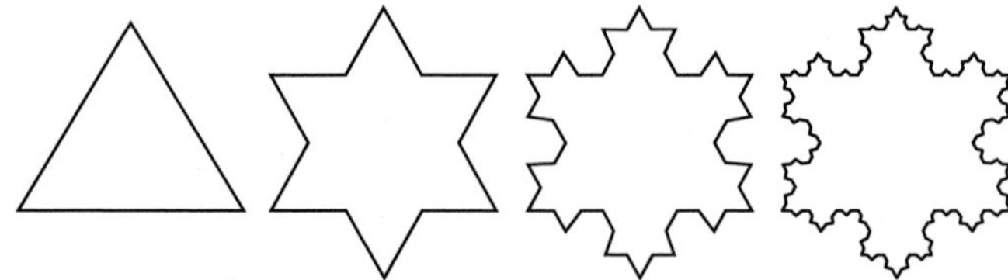

Fractals are commonly found in nature, a good example being the complex patterns found in plants, such as the leaves of a fern.

Algebra: Algebraic methods

1 Number sequences

2 Finding the nth term of a linear sequence

3 Special sequences

4 General rules from given patterns

This chapter will show you ...

- **C** How to express a rule for a sequence in words and algebraically
- **B** Some common sequences of numbers

Visual overview

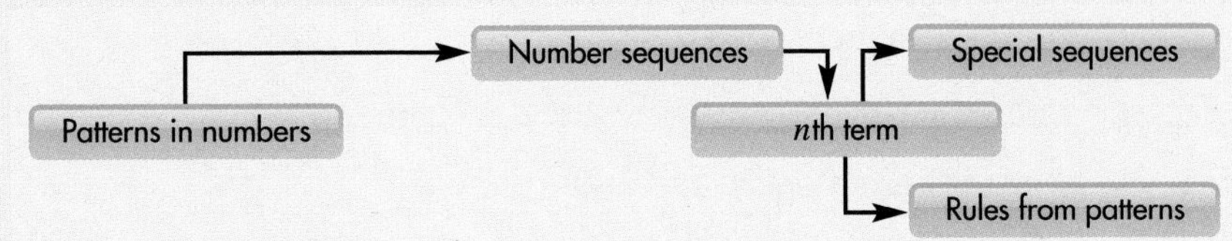

What you should already know

- How to substitute numbers into an algebraic expression **(KS3 level 5, GCSE grade E)**
- How to state a rule for a simple linear sequence in words **(KS3 level 6, GCSE grade D)**
- How to factorise simple linear expressions **(KS3 level 6, GCSE grade D)**
- How to expand a pair of linear brackets to get a quadratic equation **(KS3 level 7, GCSE grade C)**

Quick check

1 Write down the next three terms of these sequences.

a 2, 5, 8, 11, 14, ... **b** 1, 4, 9, 16, 25, 36, ...

2 Work out the value of the expression $3n - 2$ for:

a $n = 1$ **b** $n = 2$ **c** $n = 3$

16.1 Number sequences

This section will show you how to:
- recognise how number sequences are built up
- generate sequences, given the nth term

Key words
coefficient
consecutive
difference
nth term
sequence
term
term-to-term

A number **sequence** is an ordered set of numbers with a rule to find every number in the sequence. The rule that takes you from one number to the next could be a simple addition or multiplication, but often it is more tricky than that. So you need to look most carefully at the pattern of a sequence.

Each number in a sequence is called a **term** and is in a certain position in the sequence.

Look at these sequences and their rules.

3, 6, 12, 24, … doubling the last term each time … 48, 96, …

2, 5, 8, 11, … adding 3 to the last term each time … 14, 17, …

1, 10, 100, 1000, … multiplying the last term by 10 each time … 10 000, 100 000, …

1, 8, 15, 22, … adding 7 to the last term each time … 29, 36, …

These are all quite straightforward once you have looked for the link from one term to the next (**consecutive** terms). A pattern in which each term (apart from the first) is derived from the term before it is a **term-to-term** sequence.

Differences

For some sequences you need to look at the **differences** between consecutive terms to determine the pattern.

EXAMPLE 1

Find the next two terms of the sequence 1, 3, 6, 10, 15, …

Looking at the differences between each pair of consecutive terms, you notice:

```
1     3     6     10     15
   ↑     ↑     ↑      ↑
   2     3     4      5
```

So, you can continue the sequence as follows:

```
1     3     6     10     15     21     28
   ↑     ↑     ↑      ↑
   2     3     4      5    [+6]   [+7]
```

The differences usually form a number sequence of their own, so you need to find out the sequence of the differences before you can expand the original sequence.

Generalising to find the rule

When using a number sequence, you sometimes need to know, say, its 50th term, or even a later term in the sequence. To do so, you need to find the rule that produces the sequence in its general form.

Let's first look at the problem backwards. That is, take a rule and see how it produces a sequence.

EXAMPLE 2

A sequence is formed by the rule $3n + 1$, where $n = 1, 2, 3, 4, 5, 6, \ldots$. Write down the first five terms of the sequence.

Substituting $n = 1, 2, 3, 4, 5$ in turn:

$(3 \times 1 + 1), (3 \times 2 + 1), (3 \times 3 + 1), (3 \times 4 + 1), (3 \times 5 + 1), \ldots$

4 7 10 13 16

So the sequence is 4, 7, 10, 13, 16, …

Notice that the difference between each term and the next is always 3, which is the **coefficient** of n (the number attached to n). The constant term is the difference between the first term and the coefficient (in this case, $4 - 3 = 1$).

EXAMPLE 3

The ***n*th term** of a sequence is $4n - 3$. Write down the first five terms of the sequence.

Substituting $n = 1, 2, 3, 4, 5$ in turn:

$(4 \times 1 - 3), (4 \times 2 - 3), (4 \times 3 - 3), (4 \times 4 - 3), (4 \times 5 - 3)$

1 5 9 13 17

So the sequence is 1, 5, 9, 13, 17, …

Notice that the difference between each term and the next is always 4, which is the coefficient of n. The constant term is the difference between the first term and the coefficient ($1 - 4 = -3$).

EXERCISE 16A

1 Look carefully at each number sequence below. Find the next two numbers in the sequence and try to explain the pattern.

a 1, 1, 2, 3, 5, 8, 13, …

b 1, 4, 9, 16, 25, 36, …

c 3, 4, 7, 11, 18, 29, …

HINTS AND TIPS

These patterns do not go up by the same value each time so you will need to find another connection between the terms.

D

D

2 Triangular numbers are found as follows.

Find the next four triangular numbers.

1

3

6

10

3 Hexagonal numbers are found as follows.

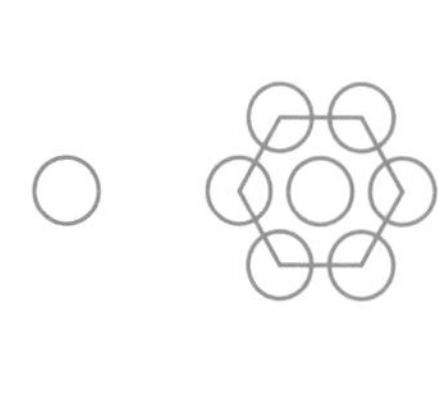
1 7

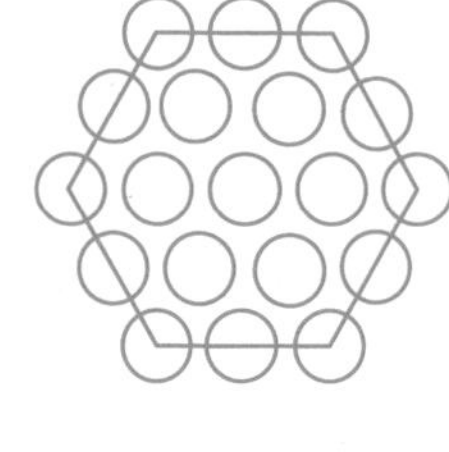
19

37

Find the next three hexagonal numbers.

C

4 The first two terms of the sequence of fractions $\frac{n-1}{n+1}$ are:

$n = 1 : \frac{1-1}{1+1} = \frac{0}{2} = 0 \qquad n = 2 : \frac{2-1}{2+1} = \frac{1}{3}$

Work out the next five terms of the sequence.

5 A sequence is formed by the rule $\frac{1}{2} \times n \times (n + 1)$ for $n = 1, 2, 3, 4, \ldots$

The first term is given by $n = 1 : \frac{1}{2} \times 1 \times (1 + 1) = 1$

The second term is given by $n = 2 : \frac{1}{2} \times 2 \times (2 + 1) = 3$

a Work out the next five terms of this sequence.

b This is a well-known sequence you have met before. What is it?

6 5! means 'factorial 5', which is $5 \times 4 \times 3 \times 2 \times 1 = 120$

In the same way 7! means $7 \times 6 \times 5 \times 4 \times 3 \times 2 \times 1 = 5040$

a Calculate 2!, 3!, 4! and 6!

b If your calculator has a factorial button, check that it gives the same answers as you get for part **a**. What is the largest factorial you can work out with your calculator before you get an error?

PS 7 The letters of the alphabet are written as the pattern:

ABBCCCDDDDEEEEEFFFFFFGGGGGGG …

so that the number of each times the letter is written matches its place in the alphabet.

So, for example, as J is the 10th letter in the alphabet, there will be 10 Js in the list.

When the pattern gets to the 26th Z, it repeats.

What letter will be the 1000th in the list?

HINTS AND TIPS

Work out how many letters there are in the sequence from ABB … to … ZZZ, then work out how many of these sequences are needed to get past 1000.

AU 8 On the first day of Christmas my true love sent to me:

a partridge in a pear tree

On the second day of Christmas my true love sent to me:

two turtle doves
and a partridge in a pear tree

and so on until…

On the twelfth day of Christmas my true love sent to me:

twelve drummers drumming
eleven pipers piping
ten lords a-leaping
nine ladies dancing
eight maids a-milking
seven swans a-swimming
six geese a-laying
five golden rings
four calling birds
three French hens
two turtle doves
and a partridge in a pear tree

How many presents were given, in total, on the 12 days of Christmas?

HINTS AND TIPS

Work out the pattern for the number of presents each day. For example, on day 1 there was 1 present, on day 2 there were 2 + 1 = 3 presents, and so on. Total the presents after each day, so after 1 day there was a total of 1 present, after 2 days a total of 4 presents, and so on. Also, try to spot any patterns.

PS 9 The first term that these two sequences have in common is 17:

8, 11, 14, 17, 20, …

1, 5, 9, 13, 17, …

What are the next two terms that the two sequences have in common?

AU 10 Two sequences are:

2, 5, 8, 11, 14, …

3, 6, 9, 12, 15, …

Will the two sequences ever have a term in common?

Show how you decided.

AU 11 The nth term of a sequence is $3n + 7$.

The nth term of another sequence is $4n - 2$.

These two sequences have several terms in common but only one term that is common *and* has the same position in the sequence.

Without writing out the sequences, show how you can tell, using the expressions for the nth term, that this is the 9th term.

C

16.2 Finding the nth term of a linear sequence

This section will show you how to:
- find the nth term of a linear sequence

Key words
nth term

A linear sequence has the same *difference* between each term and the next.

For example:

2, 5, 8, 11, 14, … difference of 3

The ***n*th term** of this sequence is given by $3n - 1$.

Here is another linear sequence:

5, 7, 9, 11, 13, … difference of 2

The nth term of this sequence is given by $2n + 3$.

So, you can see that the nth term of a linear sequence is *always* of the form $An + b$, where:

- A, the coefficient of n, is the difference between each term and the next term (consecutive terms)
- b is the difference between the first term and A.

EXAMPLE 4

Find the nth term of the sequence 5, 7, 9, 11, 13, …

The difference between consecutive terms is 2. So the first part of the nth term is $2n$.

Subtract the difference, 2, from the first term, 5, which gives $5 - 2 = 3$.

So the nth term is given by $2n + 3$.

(You can test it by substituting $n = 1, 2, 3, 4, \ldots$.)

EXAMPLE 5

Find the nth term of the sequence 3, 7, 11, 15, 19, …

The difference between consecutive terms is 4. So the first part of the nth term is $4n$.

Subtract the difference 4 from the first term 3, which gives $3 - 4 = -1$.

So the nth term is given by $4n - 1$.

EXAMPLE 6

From the sequence 5, 12, 19, 26, 33, … find the following.

a the nth term **b** the 50th term **c** the first term that is greater than 1000

a The difference between consecutive terms is 7. So the first part of the nth term is $7n$.

Subtract the difference 7 from the first term 5, which gives $5 - 7 = -2$.

So the nth term is given by $7n - 2$.

b The 50th term is found by substituting $n = 50$ into the rule, $7n - 2$.

So 50th term $= 7 \times 50 - 2 = 350 - 2$

$= 348$

c The first term that is greater than 1000 is given by:

$$7n - 2 > 1000$$

$$\Rightarrow 7n > 1000 + 2$$

$$\Rightarrow n > \frac{1002}{7}$$

$$n > 143.14$$

So the first term (which has to be a whole number) over 1000 is the 144th.

EXERCISE 16B

1 Find the next two terms and the nth term in each of these linear sequences.

a 3, 5, 7, 9, 11, … **b** 5, 9, 13, 17, 21, …

c 8, 13, 18, 23, 28, … **d** 2, 8, 14, 20, 26, …

e 5, 8, 11, 14, 17, … **f** 2, 9, 16, 23, 30, …

g 1, 5, 9, 13, 17, … **h** 3, 7, 11, 15, 19, … **i** 2, 5, 8, 11, 14, …

j 2, 12, 22, 32, … **k** 8, 12, 16, 20, … **l** 4, 9, 14, 19, 24, …

HINTS AND TIPS

Remember to look at the differences and the first term.

2 Find the nth term and the 50th term in each of these linear sequences.

a 4, 7, 10, 13, 16, … **b** 7, 9, 11, 13, 15, … **c** 3, 8, 13, 18, 23, …

d 1, 5, 9, 13, 17, … **e** 2, 10, 18, 26, … **f** 5, 6, 7, 8, 9, …

g 6, 11, 16, 21, 26, … **h** 3, 11, 19, 27, 35, … **i** 1, 4, 7, 10, 13, …

j 21, 24, 27, 30, 33, … **k** 12, 19, 26, 33, 40, … **l** 1, 9, 17, 25, 33, …

3 **a** Which term of the sequence 5, 8, 11, 14, 17, … is the first one to be greater than 100?

b Which term of the sequence 1, 8, 15, 22, 29, … is the first one to be greater than 200?

c Which term of the sequence 4, 9, 14, 19, 24, … is the closest to 500?

C

C

4 For each sequence **a** to **j**, find:

i the nth term **ii** the 100th term **iii** the term closest to 100.

a 5, 9, 13, 17, 21, …

b 3, 5, 7, 9, 11, 13, …

c 4, 7, 10, 13, 16, …

d 8, 10, 12, 14, 16, …

e 9, 13, 17, 21, …

f 6, 11, 16, 21, …

g 0, 3, 6, 9, 12, …

h 2, 8, 14, 20, 26, …

i 7, 15, 23, 31, …

j 25, 27, 29, 31, …

5 A sequence of fractions is $\frac{3}{4}, \frac{5}{7}, \frac{7}{10}, \frac{9}{13}, \frac{11}{16}, \ldots$

a Find the nth term in the sequence.

b By changing each fraction to a decimal, can you see a pattern?

c What, as a decimal, will be the value of:

i the 100th term **ii** the 1000th term?

d Use your answers to part **c** to predict what the 10 000th term and the millionth term are. (Check these on your calculator.)

6 Repeat question **5** for $\frac{3}{6}, \frac{7}{11}, \frac{11}{16}, \frac{15}{21}, \frac{19}{26}, \ldots$

B

FM 7 A haulage company uses this formula to calculate the cost of transporting n pallets.

For $n \leqslant 5$, the cost will be £$(40n + 50)$.

For $6 \leqslant n \leqslant 10$, the cost will be £$(40n + 25)$.

For $n \geqslant 11$, the cost will be £$40n$.

a How much will the company charge to transport seven pallets?

b How much will the company charge to transport 15 pallets?

c A company is charged £170 for transporting pallets. How many pallets did they transport?

d Another haulage company uses the formula £$50n$ to calculate the costs for transporting n pallets.

At what value of n do the two companies charge the same amount?

PS 8 The formula for working out a series of fractions is $\frac{2n + 1}{3n + 1}$.

a Work out the first three fractions in the series.

b **i** Work out the value of the fraction as a decimal when $n = 1\,000\,000$.

ii What fraction is equivalent to this decimal?

c How can you tell this from the original formula?

B

FM **9** This chart is used by an online CD retailer for the charges for buying n CDs including any postage and packing charges.

n	1	2	3	4	5	6	7	8	9	10	11	12	13	14	15
Charge (£)	10	18	26	34	42	49	57	65	73	81	88	96	104	112	120

a Using the charges for one to five CDs, work out an expression for the nth term.

b Using the charges for six to 10 CDs, work out an expression for the nth term.

c Using the charges for 11 to 15 CDs, work out an expression for the nth term.

d What is the basic charge for a CD?

PS **10** Look at this series of fractions.

$\frac{31}{109}, \frac{33}{110}, \frac{35}{111}, \frac{37}{112}, \frac{39}{113}, \ldots$

a Explain why the nth term of the numerators is $2n + 29$.

b Write down the nth term of the denominators.

c Explain why the terms of the series will eventually get very close to 2.

d Which term of the series has a value equal to 1?

HINTS AND TIPS

Use algebra to set up an equation.

16.3 Special sequences

This section will show you how to:

- recognise and continue some special number sequences

Key words

even, odd, powers of 2, powers of 10, prime, square, triangular

There are some number sequences that occur frequently. It is useful to know these as they are very likely to occur in examinations.

Even numbers

The even numbers are 2, 4, 6, 8, 10, 12, …

The nth term of this sequence is $2n$.

Odd numbers

The odd numbers are 1, 3, 5, 7, 9, 11, …

The nth term of this sequence is $2n - 1$.

Square numbers

The square numbers are 1, 4, 9, 16, 25, 36, …
The nth term of this sequence is n^2.

Triangular numbers

The triangular numbers are 1, 3, 6, 10, 15, 21, …
The nth term of this sequence is $\frac{1}{2}n(n + 1)$.

Powers of 2

The powers of 2 are 2, 4, 8, 16, 32, 64, …
The nth term of this sequence is 2^n.

Powers of 10

The powers of 10 are 10, 100, 1000, 10 000, 100 000, 1 000 000, …
The nth term of this sequence is 10^n.

Prime numbers

The first 20 prime numbers are 2, 3, 5, 7, 11, 13, 17, 19, 23, 29, 31, 37, 41, 43, 47, 53, 59, 61, 67, 71.

A prime number is a number that only has two factors, 1 and itself.

There is no pattern to the prime numbers so they do not have a formula for the nth term.

One important fact that you should remember is that there is only one even prime number, 2.

EXAMPLE 7

p is a prime number, q is an odd number and r is an even number.

Say if the following are always odd (O), always even (E) or could be either odd or even (X).

a pq **b** $p + q + r$ **c** pqr **d** $q^2 + r^2$

a The easiest way to do this question is to substitute numbers and see whether the outcome is odd or even.

For example, let $p = 2$ and $q = 3$, then $pq = 6$ and is even; but p could also be 3 or 5, which are odd, so $pq = 3 \times 5 = 15$ which is odd.

So pq could be either (X).

b Let $p = 2$ or 3, $q = 5$ and $r = 4$; so $p + q + r = 2 + 5 + 4 = 11$, or $3 + 5 + 4 = 12$

So $p + q + r$ could be either (X).

c Let $p = 2$ or 3, $q = 5$ and $r = 4$; so $pqr = 2 \times 5 \times 4 = 40$ or $3 \times 5 \times 4 = 60$

Both are even, so pqr is always even (E).

d Let $q = 5$ and $r = 4$; $q^2 + r^2 = 5^2 + 4^2 = 21 + 16 = 37$

This is odd, so $q^2 + r^2$ is always odd (O).

EXERCISE 16C

1 The powers of 2 are 2^1, 2^2, 2^3, 2^4, 2^5, …

This gives the sequence 2, 4, 8, 16, 32, …

The nth term is given by 2^n.

a Continue the sequence for another five terms.

b Give the nth term of these sequences.

i 1, 3, 7, 15, 31, … **ii** 3, 5, 9, 17, 33, … **iii** 6, 12, 24, 48, 96, …

2 The powers of 10 are 10^1, 10^2, 10^3, 10^4, 10^5, …

This gives the sequence 10, 100, 1000, 10 000, 100 000, …

The nth term is given by 10^n.

a Describe the connection between the number of zeros in each term and the power of the term.

b If $10^n = 1\,000\,000$, what is the value of n?

c Give the nth term of these sequences.

i 9, 99, 999, 9 999, 99 999, … **ii** 20, 200, 2000, 20 000, 200 000, …

3 a Pick any odd number. Pick any other odd number.

Add the two numbers together. Is the answer odd or even?

Complete this table.

+	**Odd**	**Even**
Odd	Even	
Even		

b Pick any odd number. Pick any other odd number.

Multiply the two numbers together. Is the answer odd or even?

Complete this table.

×	**Odd**	**Even**
Odd	Odd	
Even		

AU 4 a Write down the next two lines of this number pattern.

$$1 = 1 = 1^2$$
$$1 + 3 = 4 = 2^2$$
$$1 + 3 + 5 = 9 = 3^2$$

b Use the pattern in part **a** to write down the total of.

i $1 + 3 + 5 + 7 + 9 + 11 + 13 + 15 + 17 + 19$

ii $2 + 4 + 6 + 8 + 10 + 12 + 14$

C

C

5 The triangular numbers are 1, 3, 6, 10, 15, 21, …

a Continue the sequence for another five terms.

b The *n*th term of this sequence is given by $\frac{1}{2}n(n + 1)$.

Use the formula to find:

i the 20th triangular number

ii the 100th triangular number.

c Add consecutive terms of the triangular number sequence.

For example, $1 + 3 = 4$, $3 + 6 = 9$, …

What do you notice?

PS 6 p is an odd number, q is an even number. State if the following are odd or even.

a $p + 1$ **b** $q + 1$ **c** $p + q$

d p^2 **e** $qp + 1$ **f** $(p + q)(p - q)$

g $q^2 + 4$ **h** $p^2 + q^2$ **i** p^3

PS 7 p is a prime number, q is an even number.

State if the following are odd or even, or could be either odd or even.

a $p + 1$ **b** $p + q$ **c** p^2

d $qp + 1$ **e** $(p + q)(p - q)$ **f** $2p + 3q$

AU 8 **a** p is an odd number, q is an even number and r is an odd number.

Are the following expressions odd or even?

i $pq + r$ **ii** pqr **iii** $(p + q)^2 + r$

b x is a prime number and both y and z are odd.

Write an expression using all of x, y, and z, and no other numbers or letters, so that the answer is always even.

PS 9 A palindromic number is one that reads the same forwards as backwards, e.g. 242, 1001.

In the triangular number series 1, 3, 6, 10, 15, …, the first palindromic number is the 10th term: 55.

Find the next two palindromic triangular numbers.

AU 10 The square numbers are 1, 4, 9, 16, 25, …

The *n*th term of this sequence is n^2.

a Continue the sequence for another five terms.

b Give the *n*th term of these sequences.

i 2, 5, 10, 17, 26, … **ii** 2, 8, 18, 32, 50, … **iii** 0, 3, 8, 15, 24, …

16.4 General rules from given patterns

This section will show you how to:

- find the nth term from practical problems

Key words
difference
pattern
rule

Many problem-solving situations that you are likely to meet involve number sequences. So you need to be able to formulate general **rules** from given number **patterns**.

EXAMPLE 8

The diagram shows a pattern of squares building up.

a How many squares will be in the nth pattern?

b Which pattern has 99 squares in it?

a First, build up the following table for the patterns.

Pattern number	1	2	3	4	5
No. of squares	1	3	5	7	9

Looking at the difference between consecutive patterns, you can see it is always two squares. So, use $2n$.

Subtract the difference 2 from the first number, which gives $1 - 2 = -1$.

So the number of squares on the base of the nth pattern is $2n - 1$.

b Now find n when $2n - 1 = 99$:

$2n - 1 = 99$

$2n = 99 + 1 = 100$

$n = 100 \div 2 = 50$

The pattern with 99 squares is the 50th.

When you are trying to find a general rule from a sequence of diagrams, always set up a table to connect the pattern number with the number of the variable (squares, matches, seats, etc.) that you are trying to find the rule for. Once the table is set up, it is easy to find the nth term.

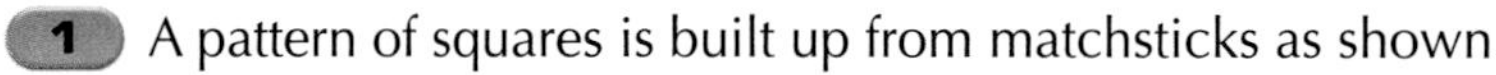

EXERCISE 16D

C

1 A pattern of squares is built up from matchsticks as shown.

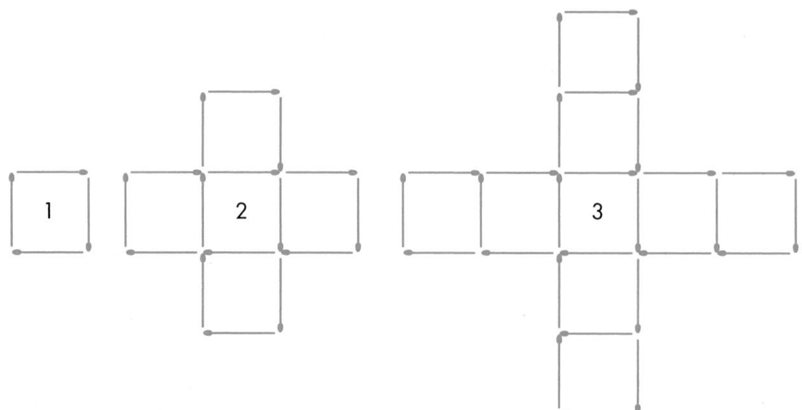

a Draw the fourth diagram.

b How many squares are in the nth diagram?

c How many squares are in the 25th diagram?

d With 200 squares, which is the biggest diagram that could be made?

> **HINTS AND TIPS**
> Write out the number sequences to help you see the patterns.

2 A pattern of triangles is built up from matchsticks.

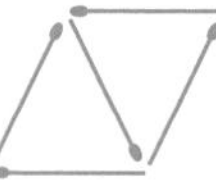
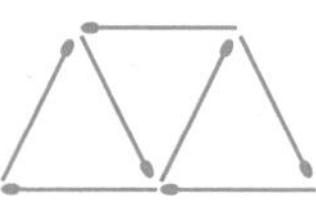
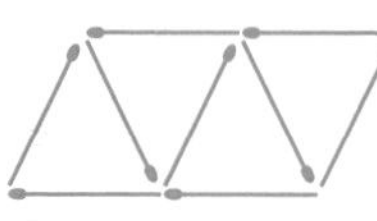

a Draw the fifth set of triangles in this pattern.

b How many matchsticks are needed for the nth set of triangles?

c How many matchsticks are needed to make the 60th set of triangles?

d If there are only 100 matchsticks, which is the largest set of triangles that could be made?

FM 3 A conference centre had tables each of which could sit six people. When put together, the tables could seat people as shown.

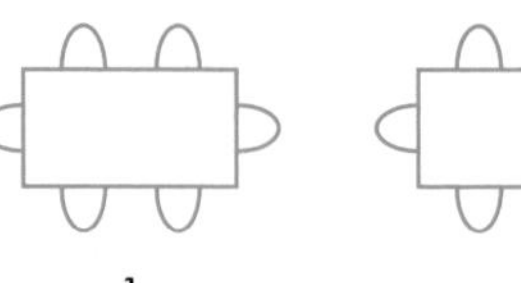
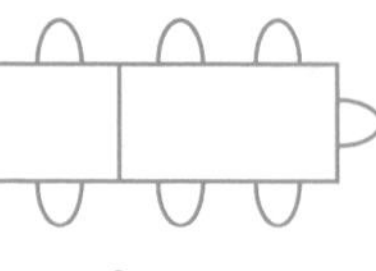
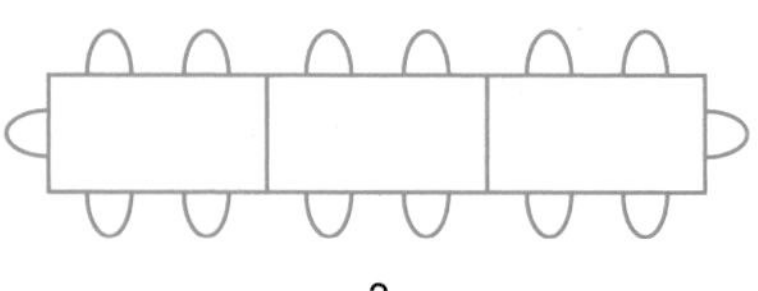

a How many people could be seated at four tables put together this way?

b How many people could be seated at n tables put together in this way?

c A conference had 50 people who wished to use the tables in this way. How many tables would they need?

4 Prepacked fencing units come in the shape shown on the right, made of four pieces of wood. When you put them together in stages to make a fence, you also need joining pieces, so the fence will start to build up as shown below.

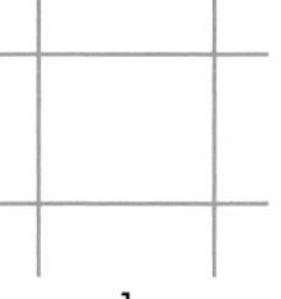

1

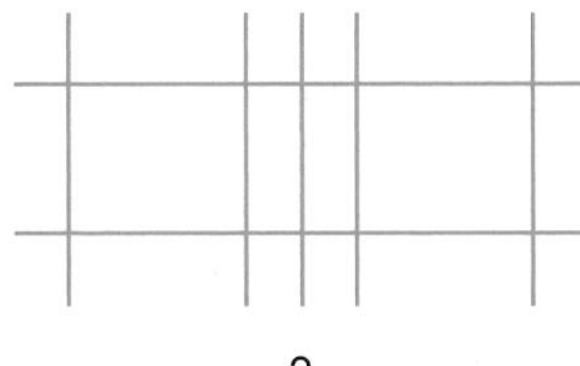

2

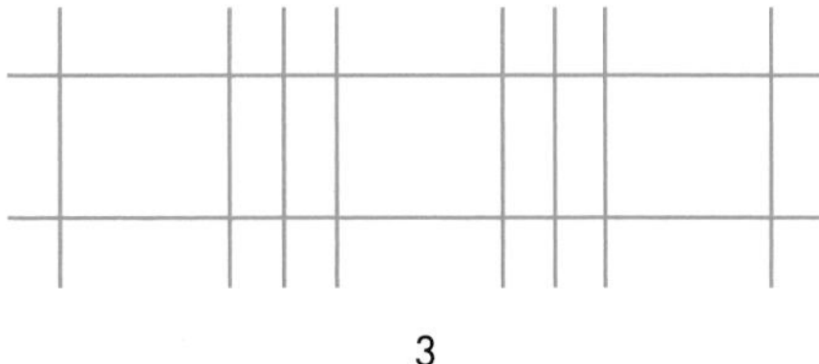

3

a How many pieces of wood would you have in a fence made up in:

i five stages

ii n stages

iii 45 stages?

b I made a fence out of 124 pieces of wood. How many stages did I use?

5 Regular pentagons of side length 1 cm are joined together to make a pattern, as shown.

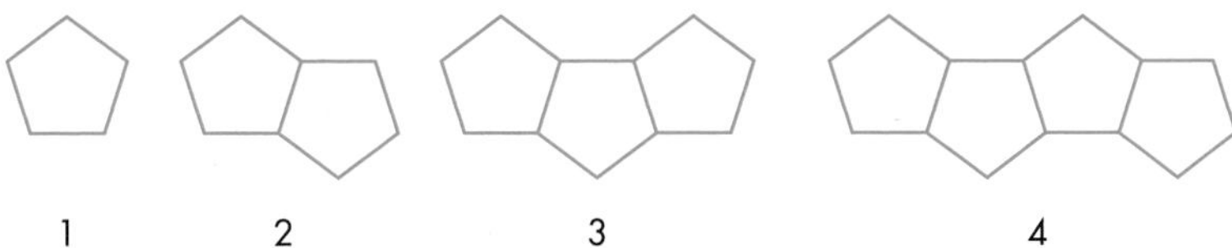

1 2 3 4

Copy this pattern and write down the perimeter of each shape.

a What is the perimeter of patterns like this made from:

i six pentagons

ii n pentagons

iii 50 pentagons?

b What is the largest number of pentagons that can be put together like this to have a perimeter less than 1000 cm?

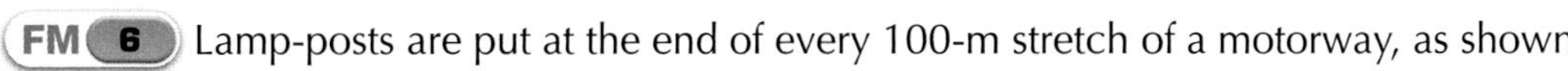

FM 6 Lamp-posts are put at the end of every 100-m stretch of a motorway, as shown.

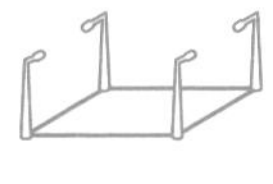

1

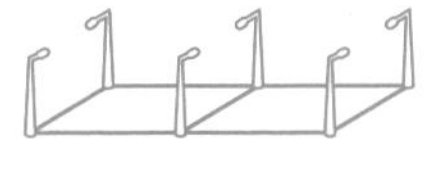

2

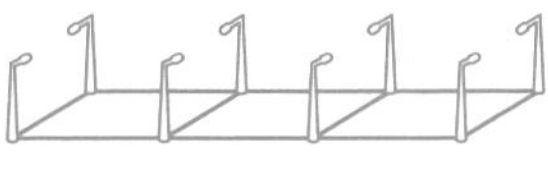

3

a How many lamp-posts are needed for:

i 900 m of this motorway

ii 8 km of this motorway?

b The M99 motorway is being built. The contractor has ordered 1598 lamp-posts. How long is this motorway?

FM 7 A school dining hall had trapezium-shaped tables. Each table could seat five people, as shown on the right. When the tables were joined together, as shown below, each table could not seat as many people.

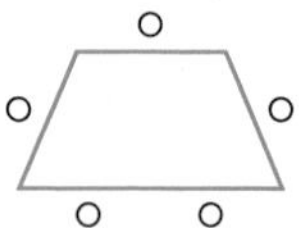

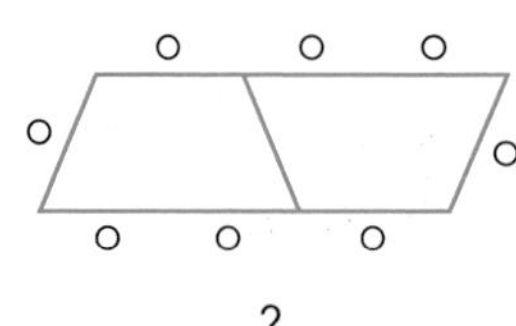

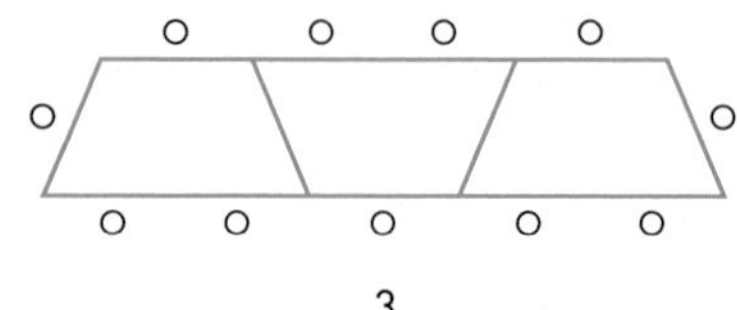

1 2 3

a In this arrangement, how many could be seated if there were:

i four tables **ii** n tables **iii** 13 tables?

b For an outside charity event, up to 200 people had to be seated. How many tables arranged like this did they need?

FM 8 When setting out tins to make a display of a certain height, you need to know how many tins to start with at the bottom.

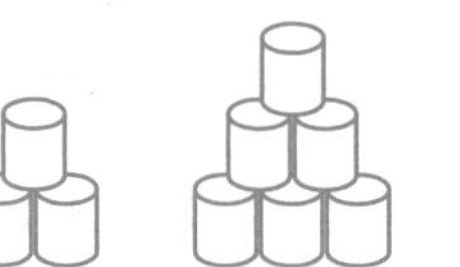

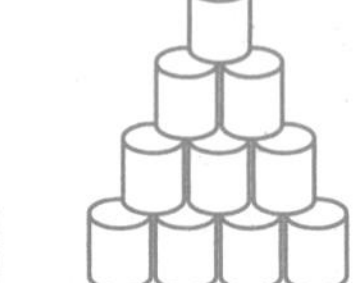

a How many tins are needed on the bottom if you wish the display to be:

i five tins high **ii** n tins high **iii** 18 tins high?

b I saw a shop assistant starting to build a display, and noticed he was starting with 20 tins on the bottom. How high was the display when it was finished?

9 **a** The values of 2 raised to a positive whole-number power are 2, 4, 8, 16, 32, …

What is the nth term of this sequence?

b A supermarket sells four different-sized bottles of water: pocket size, 100 ml; standard size, 200 ml; family size, 400 ml; giant size, 800 ml.

i Describe the number pattern that the contents follow.

ii The supermarket introduces a super giant size, which is the next-sized bottle in the pattern. How much does this bottle hold?

PS 10 Draw an equilateral triangle.

Mark the midpoints of each side and draw and shade in the equilateral triangle formed by these points.

Repeat this with the three unshaded triangles remaining.

Keep on doing this with the unshaded triangles that are left.

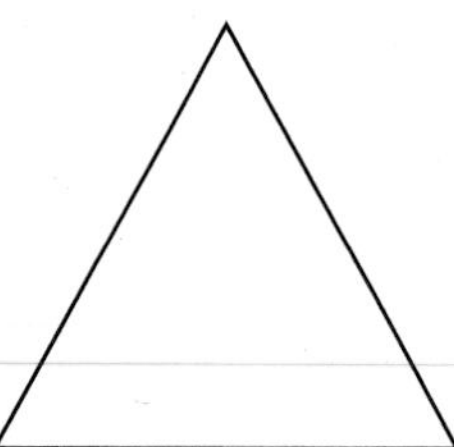

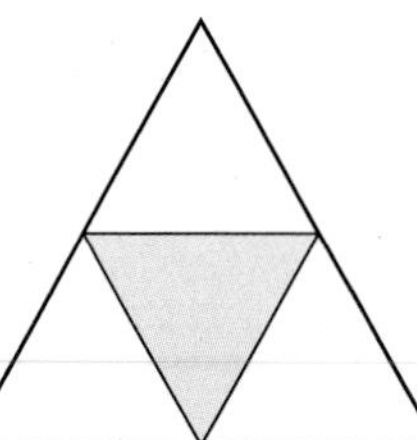

C

The pattern is called a Sierpinski triangle and is one of the earliest examples of a fractal type pattern.

The shaded areas in each triangle are $\frac{1}{4}$, $\frac{7}{16}$, $\frac{37}{64}$, $\frac{175}{256}$.

It is very difficult to work out an nth term for this series of fractions.

Use your calculator to work out the **unshaded** area, e.g. $\frac{3}{4}$, $\frac{9}{16}$,...

You should be able to write down a formula for the nth term of this pattern.

Pick a large value for n.

Will the shaded area ever cover all of the original triangle?

AU **11** Thom is building three different patterns with matches.

He builds the patterns in steps.

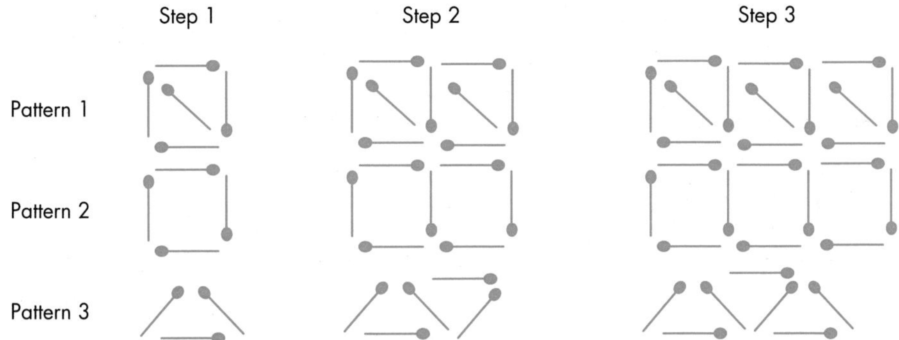

Thom has five boxes of matches that are labelled 'Average contents 42 matches'.

Will Thom have enough matches to get to step 20?

Show your working.

FM **12** For a display of grapefruit, a supermarket manager stacks them in layers, each of which is a triangle.

These are the first four layers.

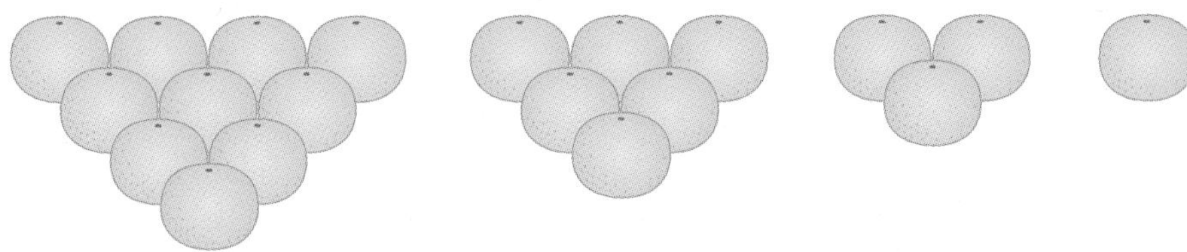

a If the display is four layers deep, how many grapefruit will be in the display?

b The manager tells her staff that there should not be any more than eight layers, as the fruit will get squashed otherwise.

What is the most grapefruit that could be stacked?

GRADE BOOSTER

- **D** You can substitute numbers into an nth-term rule
- **D** You can understand how odd and even numbers interact in addition, subtraction and multiplication problems
- **C** You can give the nth term of a linear sequence
- **C** You can give the nth term of a sequence of powers of 2 or 10

What you should know now

- How to recognise a linear sequence and find its nth term
- How to recognise a sequence of powers of 2 or 10

1 **a** Write down the nth term of the sequence.

4, 9, 14, 19, 24, … *(1 mark)*

b The nth term of a sequence is given by

$$\frac{2n-1}{n+1}$$

The first three terms are $\frac{1}{2} = 0.5$, $\frac{3}{3} = 1$ and $\frac{5}{4} = 1.25$

Show that the 6th term of the sequence is the first one that is not a terminating decimal. *(3 marks)*

AQA, June 2009, Paper 2 Higher, Question 11

2 The first ten prime numbers are 2, 3, 5, 7, 11, 13, 17, 19, 23, 29.

P is a prime number.
Q is an odd number.

State whether each of the following is always odd, always even or could be either odd or even.

a $P(Q + 1)$

b $Q - P$

3 The nth term of a sequence is given by the expression

$n^2 + 5$

Write down the first **three** terms of the sequence. *(2 marks)*

AQA, November 2008, Paper 2 Foundation, Question 19

4 **a** A sequence has nth term $4n + 1$.

i Write down the first three terms of this sequence. *(2 marks)*

ii Toms says that 2006 is a term in this sequence.

Explain why he is wrong. *(1 mark)*

b A different sequence has nth term

$(n + 3)^2 - 9$

Show that the first term of this sequence is 7. *(1 mark)*

AQA, November 2006, Paper 1 Intermediate, Question 8

5 p and q are odd numbers.

a Is $p + q$ an odd number, an even number or could it be either? *(1 mark)*

b Is pq an odd number, an even number or could it be either? *(1 mark)*

AQA, June 2005, Paper 1 Foundation, Question 24

6 Consecutive patterns are put together.

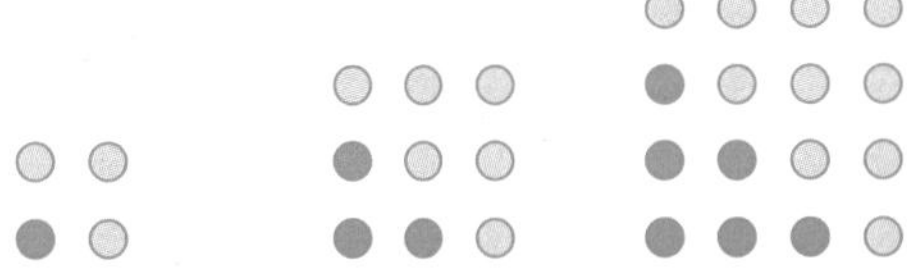

Pattern 1 and 2 Pattern 2 and 3 Pattern 3 and 4

a The numbers of counters in the combined patterns form the sequence:

4, 9, 16, …

How many counters will be in the next combined pattern in the sequence?

b What type of numbers are 4, 9, 16, … ?

c How many counters will be in the combined pattern formed by patterns 9 and 10?

7 Martin says that the square of any number is always bigger than the number. Give an example to show that Martin is wrong.

8 It is known that n is an integer.

a Explain why $2n + 1$ is always an odd number for all values of n.

b Explain why n^2 could be either odd or even.

9 Here are the nth terms of three sequences.

Sequence 1	nth term	$4n + 1$
Sequence 2	nth term	$3n + 3$
Sequence 3	nth term	$3n - 1$

For each sequence state whether the numbers in the sequence are

A Always multiples of 3

S Sometimes multiples of 3

N Never multiples of 3 *(3 marks)*

AQA, November 2006, Paper 2 Intermediate, Question 13

Worked Examination Questions

PS **1 a** Matches are used to make patterns.

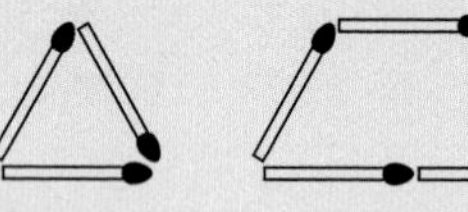

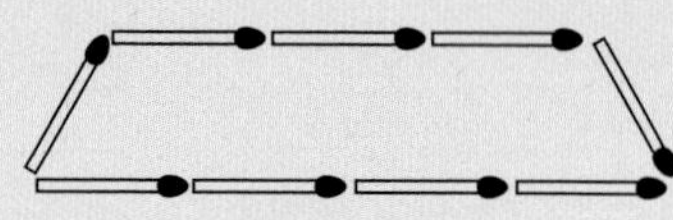

i How many matches would be needed for the 10th pattern?

ii How many matches would be needed for the nth pattern?

b The patterns are used to make a sequence of shapes.

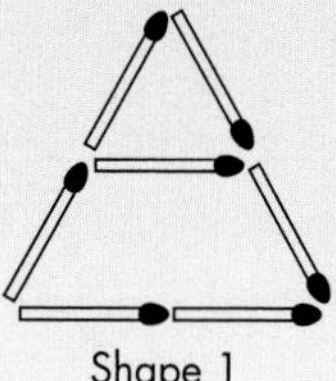

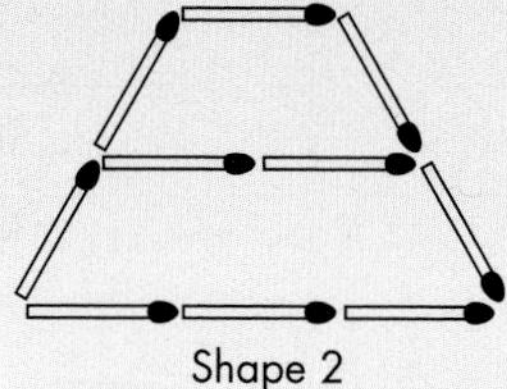

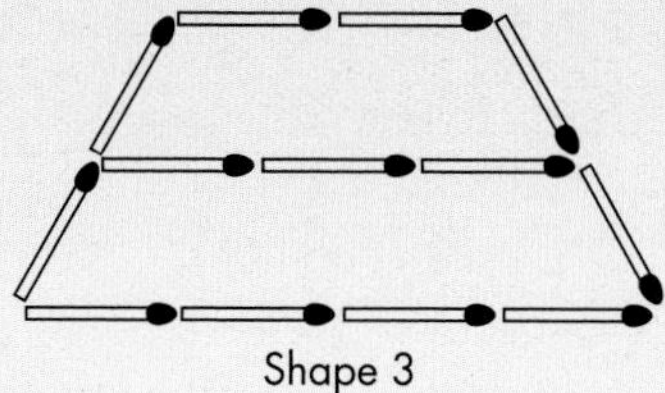

The number of matches needed to make these shapes is

7, 10, 13, 16, ….

i How many matches are needed for the nth shape?

ii What shape number could be made with 64 matches?

1 a i The number of matches needed is 21.

The number of matches needed is 3, 5, 7, 9, … which is going up by 2 each time. Continuing the sequence for 10 terms gives 3, 5, 7, 9, 11, 13, 15, 17, 19, 21. You get 1 mark for the correct answer.

ii $2n + 1$

The sequence goes up by 2 each time and the first term is 3, so the nth term is $2n + 1$. You get you 1 mark for the correct answer.

2 marks

b i $3n + 4$

The sequence goes up by 3 each time and the first term is 7 so the nth term is $3n + 4$. ($3n$ on its own would get 1 method mark. The full answer gets you 1 method mark and 1 mark for accuracy.)

ii 20th shape

$3n + 4 = 64$, then $3n = 60$, so $n = 20$. (Setting up the equation $3n + 4 = 64$ gets you 1 method mark and the answer of 20 gets 1 mark for accuracy.)

4 marks

Total: 6 marks

Worked Examination Questions

2 Tom is building fences, using posts and rails.

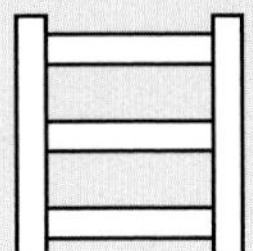

Stage 1
2 posts, 3 rows

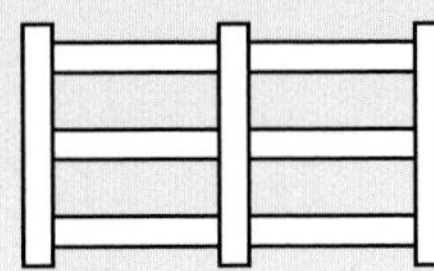
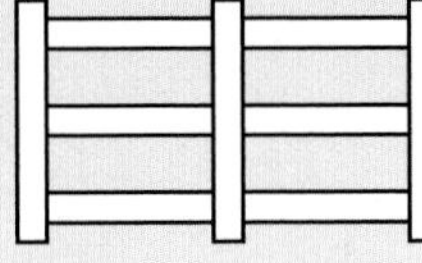

Stage 2
3 posts, 6 rows

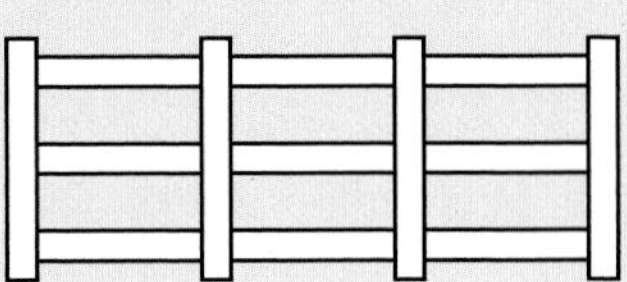

Stage 3
4 posts, 9 rows

a How many rails will there be in a fence with 6 posts?

b How many posts will be needed for a fence with 27 rails?

c Posts cost £12 and rails costs £5.
Write down a formula for the cost of a fence with n posts.

2 a 15 rails

1 mark

This answer can be found by 'counting on', i.e.
5 posts, 12 rails
6 posts, 15 rails.
This gets you 1 mark.

b 10 posts

1 mark

This could be found by 'counting on' but it is better to find a rule. The number of rails is in the 3 times table and is the multiple that is 1 less than the number of posts.
$27 = 9 \times 3$, so number of posts is $9 + 1$.
This gets you 1 mark.

c £$(27n - 15)$

2 marks

Once the formula is found in part (b) it is a matter of putting the costs and the nth terms together.
Cost $= 12n + 3 \times 5 \times (n - 1)$
You get 1 method mark for the equation.
You get 1 accuracy mark for simplifying the answer.

Total: 4 marks

It is very important that once a formula is found, it is checked. This is a fundamental part of functional maths. So take one of the given examples at the start of the question:
Stage 2 uses 3 posts and 6 rails, for which the cost is $3 \times 12 + 6 \times 5 = £66$.
The formula gives $27 \times 3 - 15 = 81 - 15 = 66$.

PS AU **3** Here are formulae for the nth terms of three sequences.

Formula 1: $4n + 1$
Formula 2: $5n - 2$
Formula 3: $5n + 10$

Say if the sequences generated by the nth terms always (A) give multiples of 5, never (N) give multiples of 5 or sometimes (S) give multiples of 5.

3 Formula 1: Sometimes (S)
Formula 2: Never (N)
Formula 3: Always (A)

Total: 3 marks

Substitute $n = 1, 2, 3$, etc. until you can be sure of the sequences.
These are:
5, 9, 13, 17, 21, 25, 29, …
3, 8, 13, 18, 23, 28, …
15, 20, 25, 30, 35, 40, …
(You get 1 mark for each part.)

16 Problem Solving
Patterns in the pyramids

Some sets of square numbers have a special connection. For example:

$$3^2 + 4^2 = 5^2$$

The set (3, 4, 5) is called a Pythagorean triple. (In Book 2, you will learn about Pythagoras' theorem, which is connected with right-angled triangles. Pythagorean triples, however, were well known before Pythagoras himself!)

Pythagorean triples were used by ancient people in the construction of great monuments such as Stonehenge and the Egyptian pyramids. They are just one example of how we can use patterns.

Getting started

Many patterns, including Pythagorean triples, can be found, using diagrams.

1 Working in pairs, start with the smallest odd number and, going up in consecutive odd numbers, draw diagrams to represent the numbers. Extend your diagrams using a rule.

Have a look at what another group has done. Have they found the same rule as yours? Is there more than one way to represent odd numbers diagrammatically?

Getting started (continued)

2 Represent a square number diagrammatically – the clue is in the name.

One representation of odd numbers can be used to show how consecutive odd numbers add together to make square numbers. See if you can find this and show it in a diagram of 3^2.

3 Use the pattern from the diagrams to complete this sequence:

$1 = 1^2 = 1$

$1 + 3 = 2^2 = 4$

$1 + 3 + 5 =$ ..... $=$ .....

$1 + 3 + 5 + 7 =$ ..... $=$ .....

$=$ ..... $=$ .....

Look for connections between the last odd number and the square number.

Use these connections to fill in the missing numbers in this sequence:

$1 + 3 + 5 +$ ..... $+ 19 =$ ..... $=$ .....

$1 + 3 + 5 +$ ..... $+$ ..... $= 12^2 = 144$

$1 + 3 + 5 +$ ..... $+ 99 =$ ..... $=$ .....

$1 + 3 + 5 +$ ..... $+$ ..... $= 200^2 = 40\,000$

Your task

It is your task to use diagrams to find patterns, and make and test generalisations.

1 Draw accurate diagrams to show that (5, 12, 13) is a Pythagorean triple. Find at least three more Pythagorean triples.

2 Here are some sequences which are seen in everyday life, for example nature and science.

1, 3, 6, 10, 28, 36, 45, ...

1, 8, 27, 64, 125, 216, 343, 512, 729, ...

2, 8, 18, 32, 50, 72, 98, 162, ...

0, 1, 1, 2, 3, 4, 8, 13, 21, 34, ...

Investigate the sequences, looking for patterns, diagrammatical representations and *n*th terms.

3 Now see if you can generate some sequences of your own. Can you think where they may apply in real-life situations?

Extension

Use the internet to find out about Pythagoras, his theorem and Euler's Formula.

Why this chapter matters

A famous saying is 'A picture is worth a thousand words' and graphs in mathematics are worth many lines of algebra as they are a visual way of showing the relationship between two variables. This chapter deals with linear graphs, which are straight line graphs.

Many years ago the city of Konigsberg (now known as Kalingrad) had seven bridges joining the four separate parts of the city.

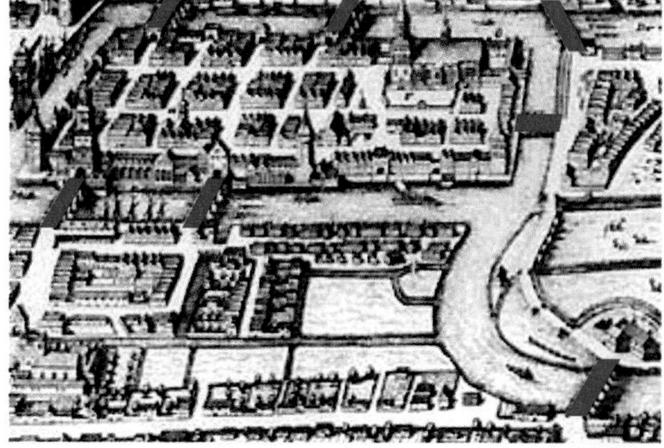

Bridges of Konigsberg

The citizens had a challenge to see if anyone could walk across all seven bridges without crossing any of them twice.

The problem is the same as trying to draw this diagram without taking your pencil off the paper.

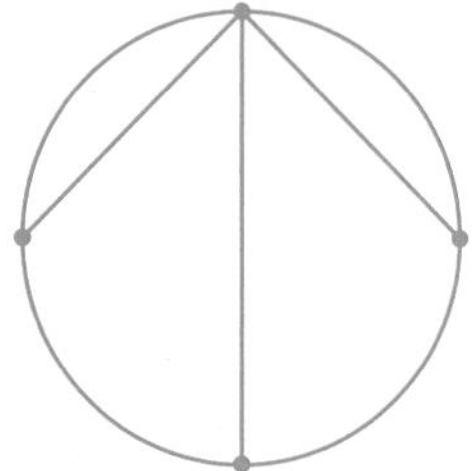

Copy the diagram and see if it can be done.

Like the citizens of Konigsberg and their problem with the bridges, you will find that it is impossible.

This problem was investigated by the Swiss mathematician Leonhard Euler (1707–1783).

Leonard Euler

He proved that it was not possible and started a new branch of mathematics called 'Graph theory'. This is not the same as the study of linear graphs, but is more concerned with where lines meet (vertices) and the lines joining them (arcs).

Copy the following diagrams and see if you can draw any of them without taking your pencil off the paper.

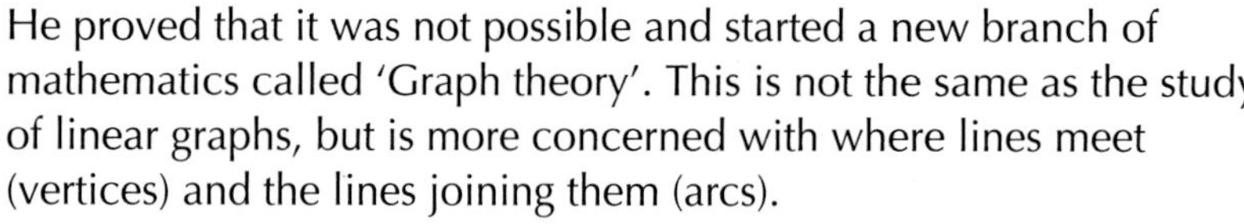

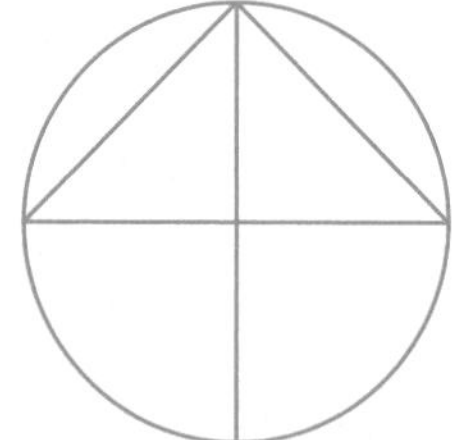

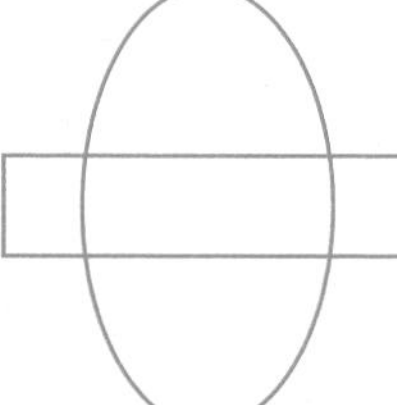

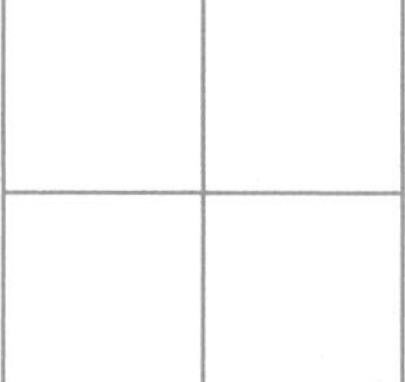

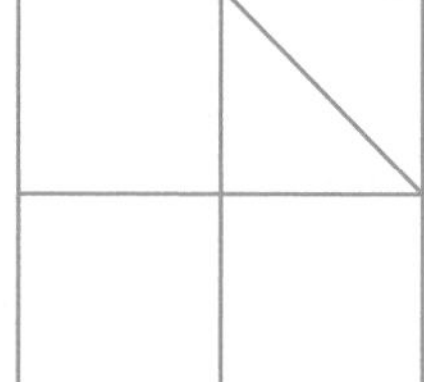

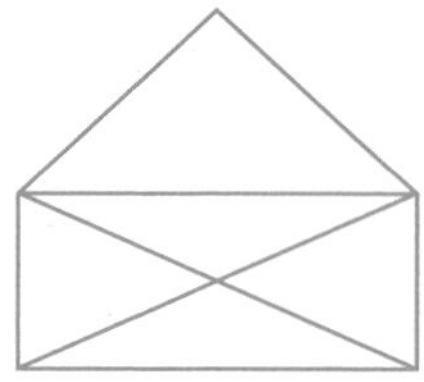

There is a way of telling if this can be done, which you can investigate at the end of this chapter.

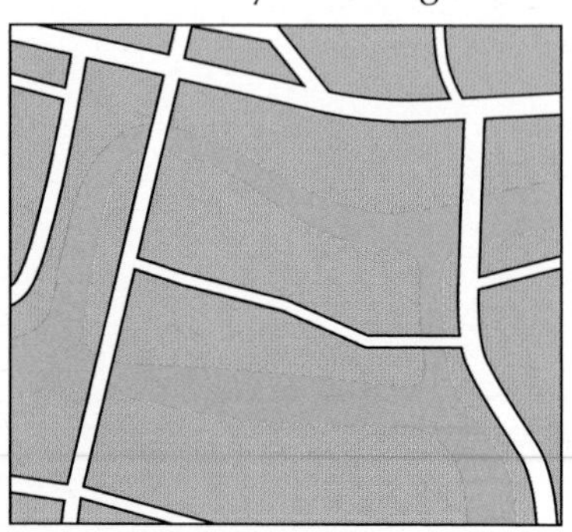

Layout of Kalingrad

In the Second World War, Konigsberg was badly damaged and many of the bridges were destroyed. This is a rough layout of Kalingrad (Konigsberg) as it is today:

Can you trace a route, starting from the island, that crosses all the remaining bridges without crossing a bridge twice?

Algebra: Linear graphs and equations

This chapter will show you ...

- **C** how to draw graphs of linear equations
- **C** how to use graphs to find the solution to linear equations
- **B** how to find the equation of a linear graph
- **B** how to read diagrams using 3-D coordinates

Visual overview

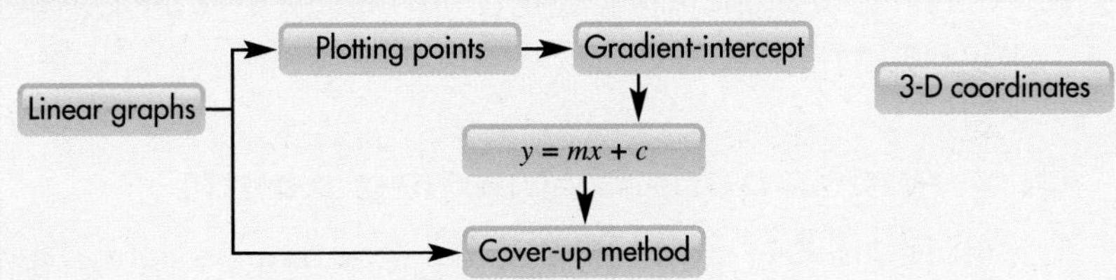

What you should already know

- How to read and plot coordinates **(KS3 level 4, GCSE grade F)**
- How to substitute into simple algebraic functions **(KS3 level 4, GCSE grade F)**
- How to plot a graph from a given table of values **(KS3 level 5, GCSE grade E)**

Quick check

1 This table shows values of $y = 2x + 3$ for $-2 \leqslant x \leqslant 5$.

x	−2	−1	0	1	2	3	4	5
y	−1	1	3	5	7	9	11	

a Complete the table for $x = 5$.

b Copy these axes and plot the points to draw the graph of $y = 2x + 3$.

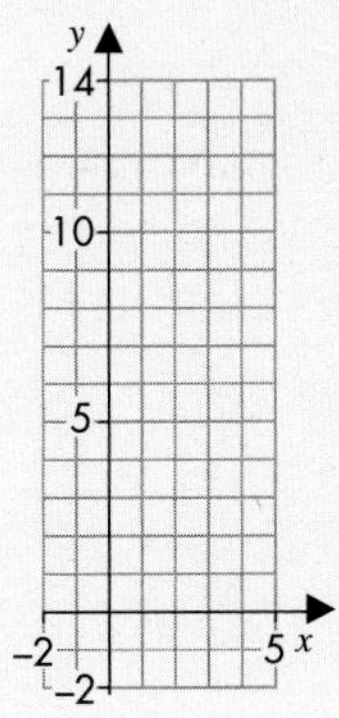

17.1 Linear graphs

This section will show you how to:
- draw linear graphs without using flow diagrams

Key words
axis (pl: axes)
linear graphs
scale

This chapter is concerned with drawing straight-line graphs. These are usually referred to as **linear graphs**.

The minimum number of points needed to draw a linear graph is two but it is better to use three or more because that gives at least one point to act as a check. There is no rule about how many points to plot but here are some tips.

- Use a sharp pencil and mark each point with an accurate cross.
- Position yourself so that your eyes are directly over the graph. If you look from the side, you will not be able to line up your ruler accurately.

Drawing graphs by finding points

This method is quite quick and does not need flow diagrams. However, if you prefer flow diagrams, use them. Work through Example 1 to see how to draw a graph by finding points.

EXAMPLE 1

Draw the graph of $y = 4x - 5$ for values of x from 0 to 5. This is usually written as $0 \leq x \leq 5$.

Choose three values for x: these should be the highest and lowest x-values and one in between.

Work out the y-values by substituting the x-values into the equation.

When $x = 0$, $y = 4(0) - 5 = -5$. This gives the point $(0, -5)$.

When $x = 3$, $y = 4(3) - 5 = 7$. This gives the point $(3, 7)$.

When $x = 5$, $y = 4(5) - 5 = 15$. This gives the point $(5, 15)$.

Keep a record of your calculations in a table.

x	0	3	5
y	–5	7	15

You now have to decide the extent (range) of the **axes**. You can find this out by looking at the coordinates that you have so far.
The smallest x-value is 0, the largest is 5.
The smallest y-value is –5, the largest is 15.

Now draw the axes, plot the points and complete the graph.
It is usually a good idea to choose 0 as one of the x-values.
In an examination, the range for the x-values will usually be given and the axes already drawn.

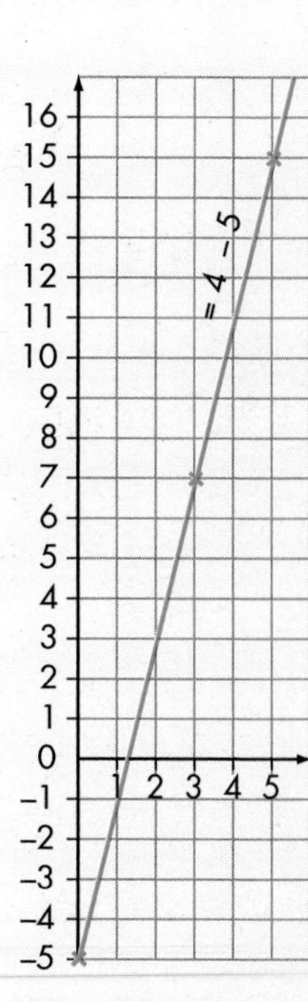

Read through these hints before drawing the linear graphs required in Exercise 17A.

- Use the highest and lowest values of x given in the range.
- Don't pick x-values that are too close together, for example, 1 and 2. Try to space them out so that you can draw a more accurate graph.
- Always label your graph with its equation. This is particularly important when you are drawing two graphs on the same set of axes.
- If you want to use a flow diagram, use one.
- Create a table of values. You will often have to complete these in your examinations.

EXERCISE 17A

D

1 Draw the graph of $y = 3x + 4$ for x-values from 0 to 5 $(0 \leqslant x \leqslant 5)$.

2 Draw the graph of $y = 2x - 5$ for $0 \leqslant x \leqslant 5$.

3 Draw the graph of $y = \frac{x}{2} - 3$ for $0 \leqslant x \leqslant 10$.

4 Draw the graph of $y = 3x + 5$ for $-3 \leqslant x \leqslant 3$.

5 Draw the graph of $y = \frac{x}{3} + 4$ for $-6 \leqslant x \leqslant 6$.

6 **a** On the same set of axes, draw the graphs of $y = 3x - 2$ and $y = 2x + 1$ for $0 \leqslant x \leqslant 5$.

b At which point do the two lines intersect?

7 **a** On the same axes, draw the graphs of $y = 4x - 5$ and $y = 2x + 3$ for $0 \leqslant x \leqslant 5$.

b At which point do the two lines intersect?

8 **a** On the same axes, draw the graphs of $y = \frac{x}{3} - 1$ and $y = \frac{x}{2} - 2$ for $0 \leqslant x \leqslant 12$.

b At which point do the two lines intersect?

9 **a** On the same axes, draw the graphs of $y = 3x + 1$ and $y = 3x - 2$ for $0 \leqslant x \leqslant 4$.

b Do the two lines intersect? If not, why not?

10 **a** Copy and complete the table to draw the graph of $x + y = 5$ for $0 \leqslant x \leqslant 5$.

x	0	1	2	3	4	5
y	5		3		1	

b Now draw the graph of $x + y = 7$ for $0 \leqslant x \leqslant 7$ on the same axes.

D

11 Ian the electrician used this formula to work out how much to charge for a job:

$C = 25 + 30H$

where C is the charge and H is how long the job takes.

John the electrician uses this formula:

$C = 35 + 27.5H$

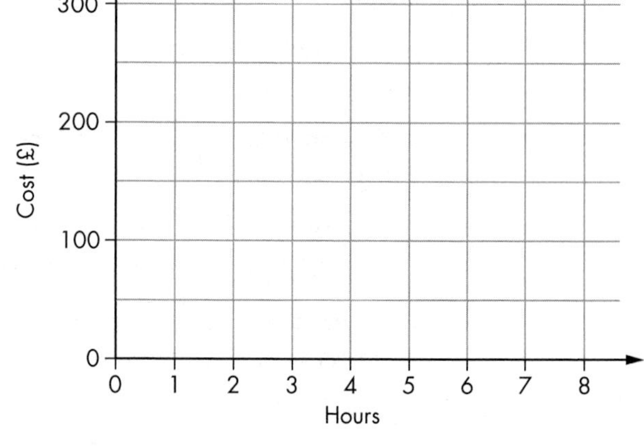

a On a copy of the grid, draw lines to represent these formulae.

FM **b** Who would you hire for a job that takes 2 hours?

C

AU **12** **a** Draw the graphs $y = 4$, $y = x$ and $x = 1$ on a copy of the grid shown on the right.

b What is the area of the triangle formed by the three lines?

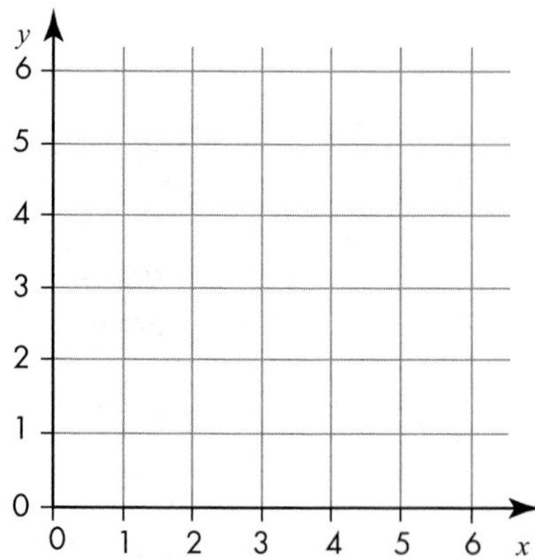

PS **13** These two graphs show y against x and y against z.

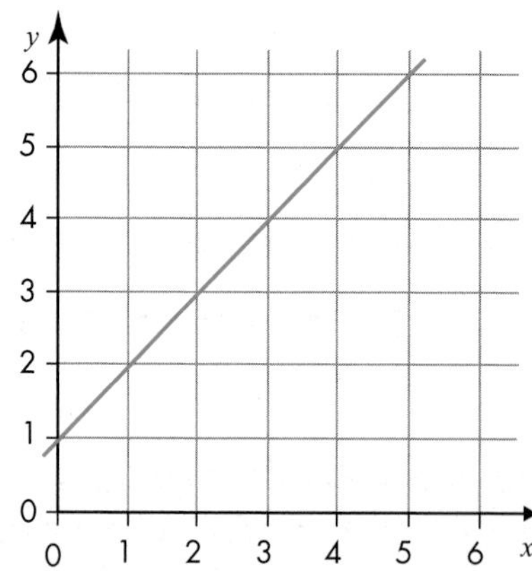

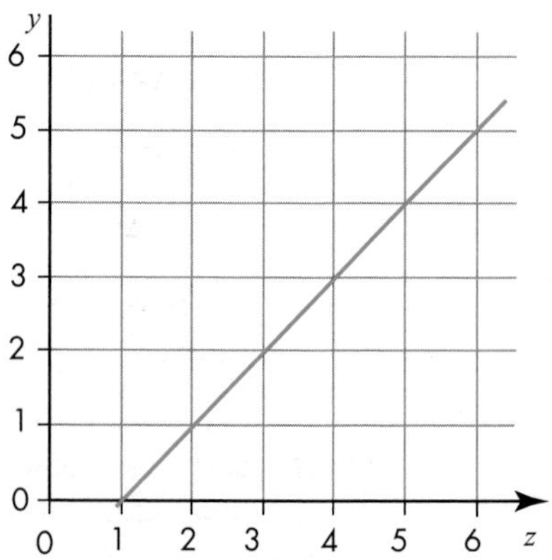

On a copy of this blank grid, show the graph of x against z.

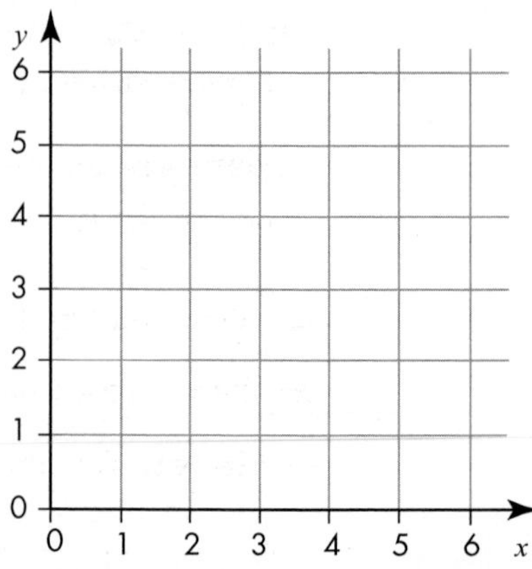

Gradient

The slope of a line is called its gradient. The steeper the slope of the line, the larger the value of the gradient.

The gradient of the line shown here can be measured by drawing, as large as possible, a right-angled triangle that has part of the line as its hypotenuse (sloping side). The gradient is then given by:

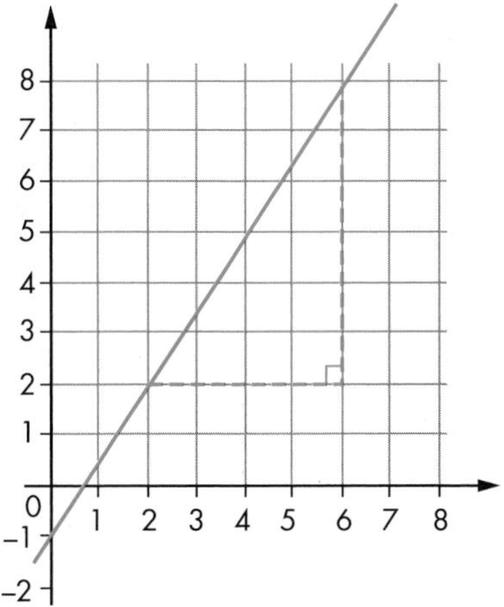

$$\text{gradient} = \frac{\text{distance measured up}}{\text{distance measured along}}$$

$$= \frac{\text{difference on } y\text{-axis}}{\text{difference on } x\text{-axis}}$$

For example, to measure the steepness of the line in the next figure, you first draw a right-angled triangle of which the hypotenuse is part of this line. It does not matter where you draw the triangle but it makes the calculations much easier if you choose a sensible place. This usually means using existing grid lines, so that you avoid fractional values.

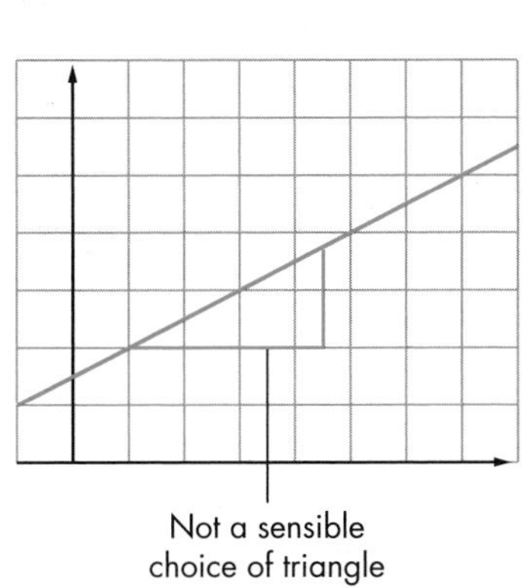

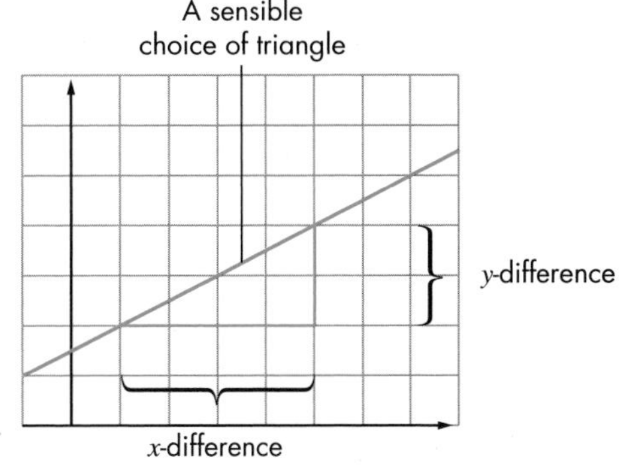

After you have drawn the triangle, you measure (or count) how many squares there are on the vertical side. This is the difference between your y-coordinates. In the case above, this is 2.

You then measure (or count) how many squares there are on the horizontal side. This is the difference between your x-coordinates. In the case above, this is 4.

To work out the gradient, you do the following calculation.

$$\text{gradient} = \frac{\text{difference of the } y\text{-coordinates}}{\text{difference of the } x\text{-coordinates}} = \frac{2}{4} = \frac{1}{2} \text{ or } 0.5$$

Note that the value of the gradient is not affected by where the triangle is drawn. As you are calculating the ratio of two sides of the triangle, the gradient will always be the same wherever you draw the triangle.

You can use the method of counting squares in cases like this, where the **scale** is one square to one unit.

Remember: When a line slopes down from left to right, the gradient is negative, so you must place a minus sign in front of the calculated fraction.

EXAMPLE 2

Find the gradient of each of these lines.

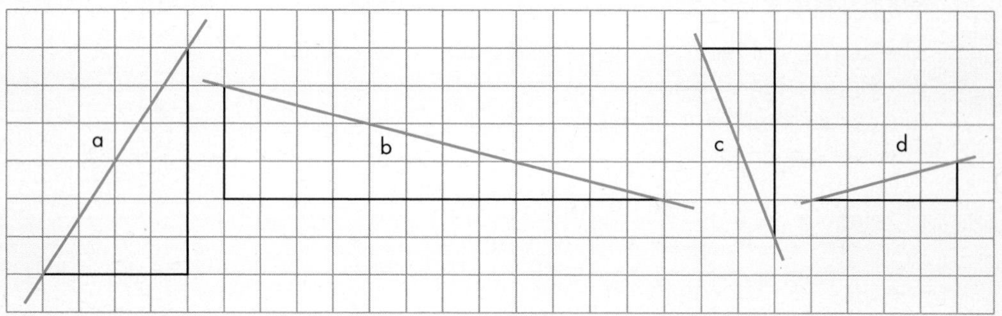

In each case, a sensible choice of triangle has already been made.

a y difference = 6, x difference = 4 Gradient $= 6 \div 4 = \frac{3}{2} = 1.5$

b y difference = 3, x difference = 12 Line slopes down from left to right, so gradient $= -(3 \div 12) = -\frac{1}{4} = -0.25$

c y difference = 5, x difference = 2 Line slopes down from left to right, so gradient $= -(5 \div 2) = -\frac{5}{2} = -2.5$

d y difference = 1, x difference = 4 Gradient $= 1 \div 4 = \frac{1}{4} = 0.25$

Drawing a line with a certain gradient

To draw a line with a certain gradient, you reverse the process described above. That is, you first draw the right-angled triangle, using the given gradient. For example, take a gradient of 2.

Start at a convenient point (A in the diagrams below). A gradient of 2 means that for an x-step of 1 the y-step must be 2 (because 2 is the fraction $\frac{2}{1}$). So, move one square across and two squares up, and mark a dot.

Repeat this as many times as you like and draw the line. You can also move one square back and two squares down, which gives the same gradient, as the third diagram shows.

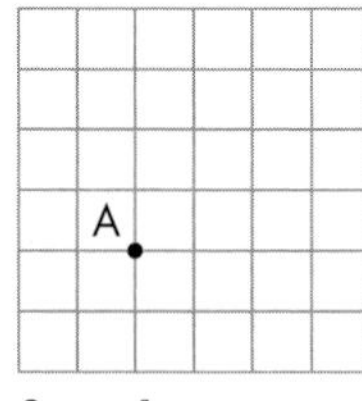

Stage 1

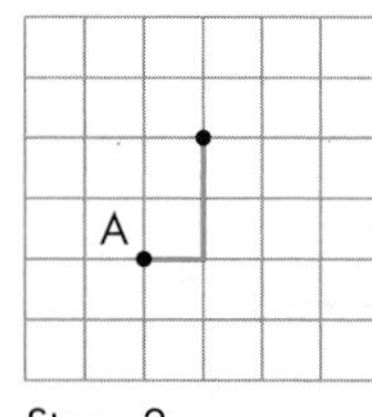

Stage 2

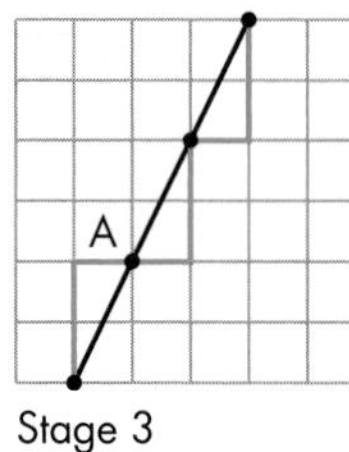

Stage 3

EXAMPLE 3

Draw lines with these gradients. **a** $\frac{1}{3}$ **b** -3 **c** $-\frac{1}{4}$

a This is a fractional gradient which has a y-step of 1 and an x-step of 3. Move three squares across and one square up every time.

b This is a negative gradient, so for every one square across, move three squares down.

c This is also a negative gradient and it is a fraction. So for every four squares across, move one square down.

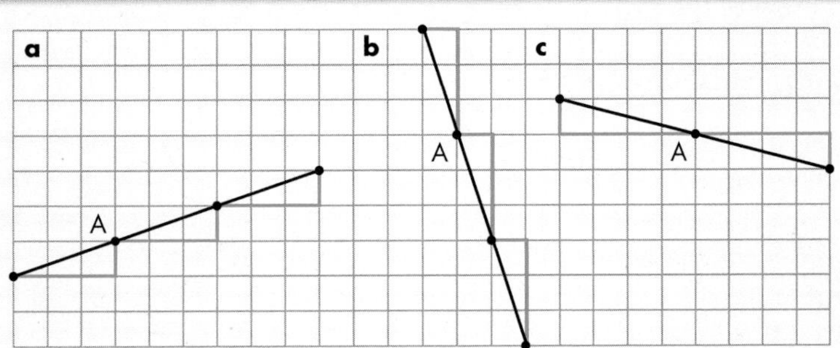

EXERCISE 17B

1 Find the gradient of each of these lines.

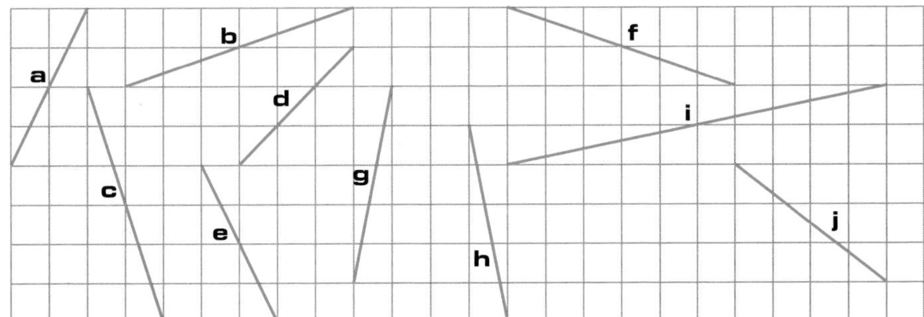

2 Draw lines with these gradients.

a 4 **b** $\frac{2}{3}$ **c** -2 **d** $-\frac{4}{5}$ **e** 6 **f** -6

3 Find the gradient of each of these lines. What is special about these lines?

a

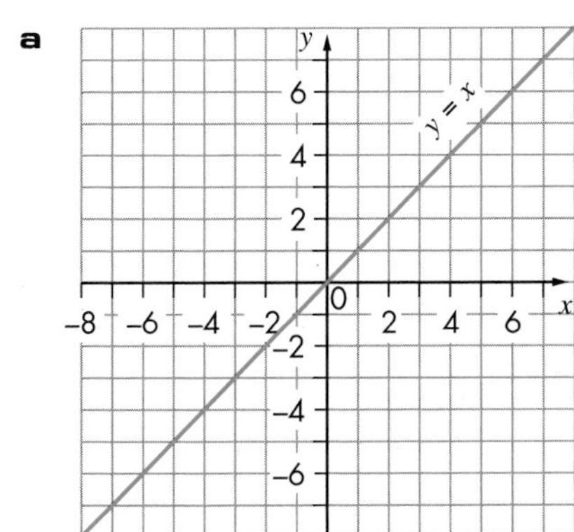

b

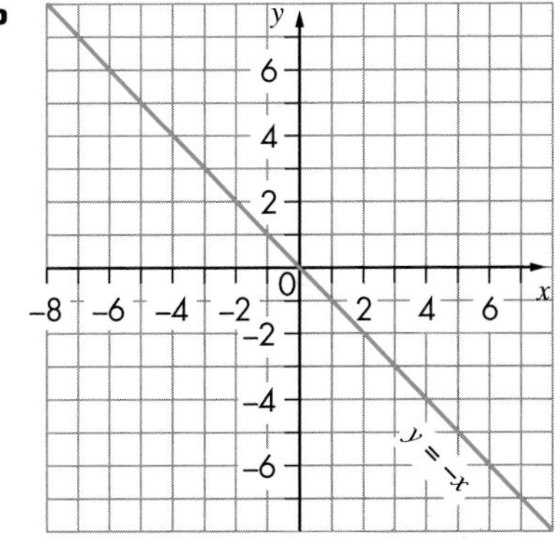

C

FM 4 This graph shows the profile of a fell race. The horizontal axis shows the distance, in miles, of the race. The vertical axis is the height above sea level throughout the race. There are 5280 feet in a mile.

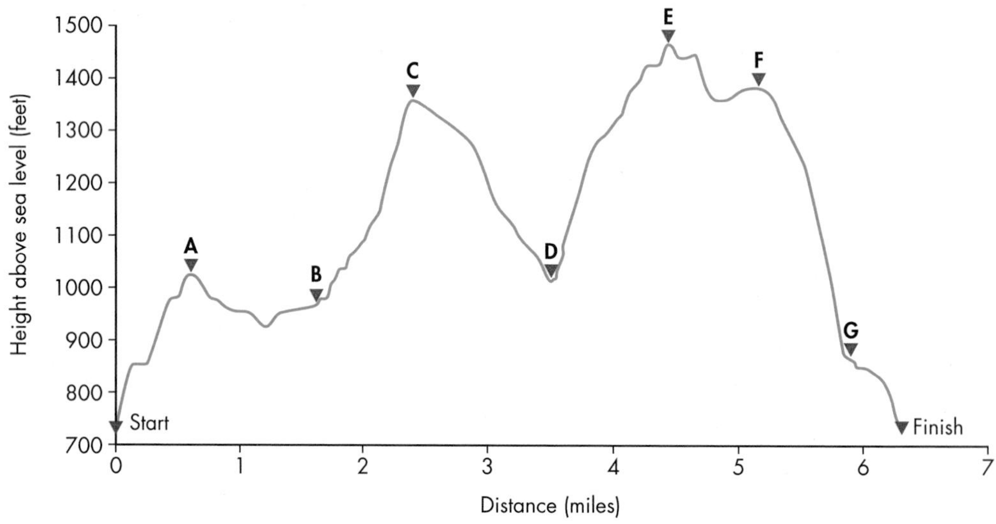

a Work out the approximate gradient of the race from the start to point A.

b The steepest part of the race is from F to G.

i How can you tell this from the graph?

ii Work out the approximate gradient from F to G.

c Fell races are classified in terms of distance and amount of ascent.

Distance:	Short (S)	Less than 6 miles
	Medium (M)	Between 6 and 12 miles
	Long (L)	Over 12 miles
Ascent	C	An average of 100 to 125 feet per mile
	B	An average of 125 to 250 feet per mile
	A	An average of 250 or more feet per mile

So, for example, an AL race would be over 12 miles and have at least 250 feet of ascent, on average, per mile.

What category is the race above?

5 The line on grid **e** is horizontal. The lines on grids **a** to **d** get nearer and nearer to the horizontal.

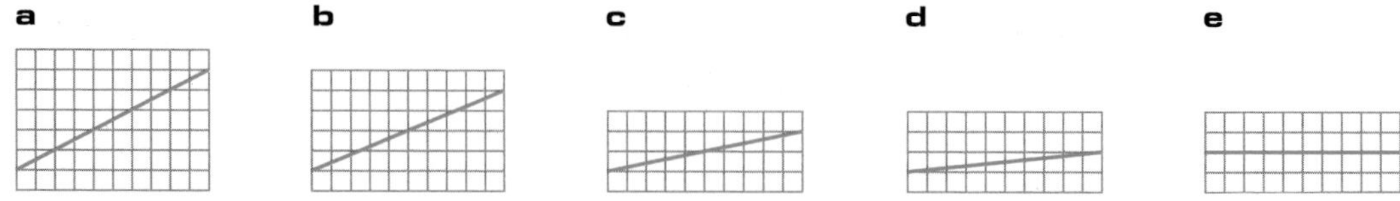

Find the gradient of each line in grids **a** to **d**. By looking at the values you obtain, what do you think the gradient of a horizontal line is?

6 The line on grid **e** is vertical. The lines on grids **a** to **d** get nearer and nearer to the vertical.

a

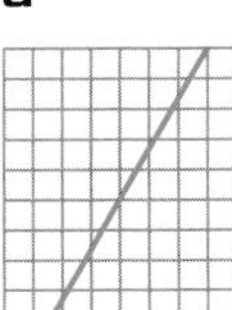

b

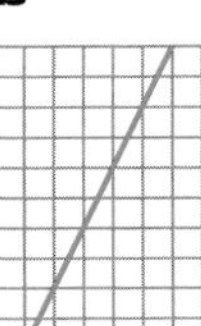

c

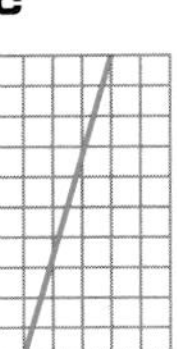

d

e

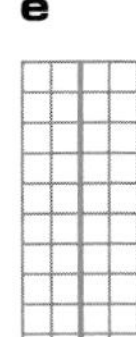

Find the gradient of each line in grids **a** to **d**. By looking at the values you obtain, what do you think the gradient of a vertical line is?

AU 7 Raisa says the gradients of these two lines are the same.

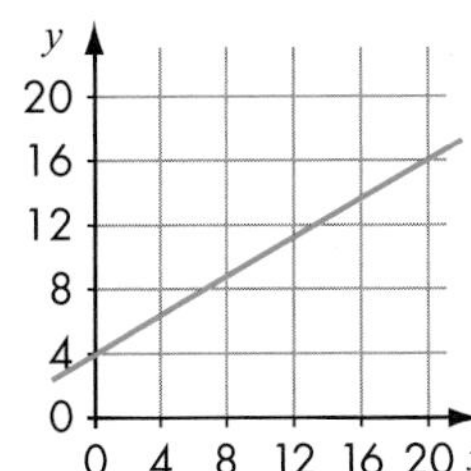

Why is Raisa wrong?

PS 8 Put the following gradients in order of steepness, starting with the shallowest.

1 horizontal, 2 vertical	2 horizontal, 5 vertical	3 horizontal, 5 vertical
4 horizontal, 6 vertical	5 horizontal, 8 vertical	6 horizontal, 11 vertical

17.2 Drawing graphs by the gradient-intercept method

This section will show you how to:

- draw graphs using the gradient-intercept method

Key words

coefficient
constant term
cover-up method
gradient-intercept
$y = mx + c$

The ideas that you have discovered in the last activity lead to another way of plotting lines, known as the **gradient-intercept** method.

EXAMPLE 4

Draw the graph of $y = 3x - 1$, using the gradient-intercept method.

- Because the **constant term** is –1, you know that the graph goes through the y-axis at –1. Mark this point with a dot or a cross (**A** on diagram **i**).
- The number in front of x (called the **coefficient** of x) gives the relationship between y and x. 3 is the coefficient of x and this tells you that the y-value is 3 times the x-value, so the gradient of the line is 3. For an x-step of one unit, there is a y-step of three. Starting at –1 on the y-axis, move one square across and three squares up and mark this point with a dot or a cross (**B** on diagram **i**).

Repeat this from every new point. You can also move one square back and three squares down. When enough points have been marked, join the dots (or crosses) to make the graph (diagram **ii**). Note that if the points are not in a straight line, you have made a mistake.

i

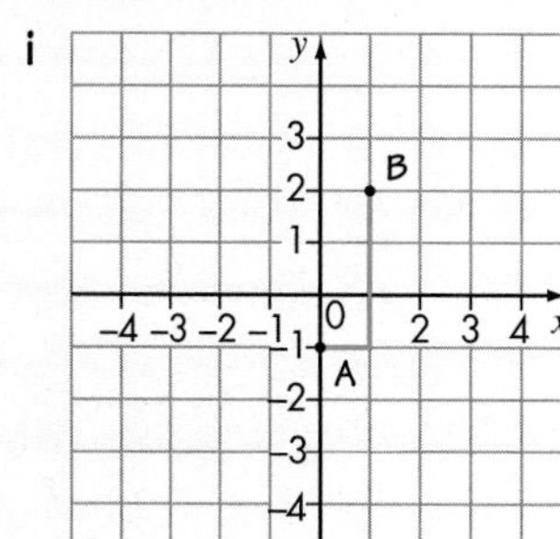

ii

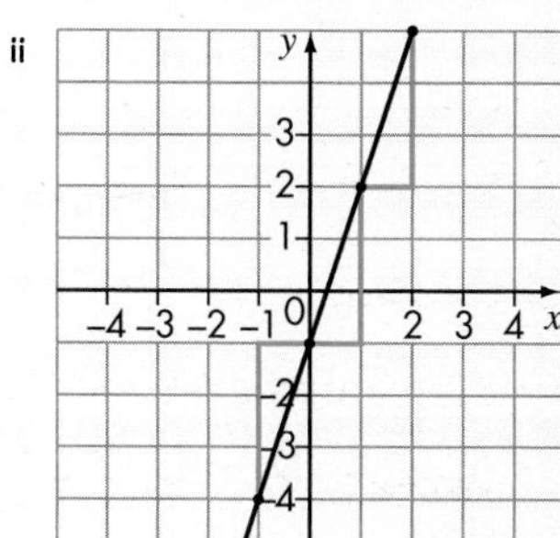

In any equation of the focus $y = mx + c$, the constant term, c, is the intercept on the y-axis and the coefficient of x, m, is the gradient of the line.

EXERCISE 17C

C

1 Draw these lines, using the gradient-intercept method. Use the same grid, taking x from –10 to 10 and y from –10 to 10. If the grid gets too 'crowded', draw another one.

a $y = 2x + 6$ **b** $y = 3x - 4$ **c** $y = \frac{1}{2}x + 5$

d $y = x + 7$ **e** $y = 4x - 3$ **f** $y = 2x - 7$

g $y = \frac{1}{4}x - 3$ **h** $y = \frac{2}{3}x + 4$ **i** $y = 6x - 5$

j $y = x + 8$ **k** $y = \frac{4}{5}x - 2$ **l** $y = 3x - 9$

2 **a** Using the gradient-intercept method, draw the following lines on the same grid. Use axes with ranges $-6 \leqslant x \leqslant 6$ and $-8 \leqslant y \leqslant 8$.

i $y = 3x + 1$ **ii** $y = 2x + 3$

b Where do the lines cross?

3 **a** Using the gradient-intercept method, draw the following lines on the same grid. Use axes with ranges $-14 \leqslant x \leqslant 4$ and $-2 \leqslant y \leqslant 6$.

i $y = \dfrac{x}{3} + 3$

ii $y = \dfrac{x}{4} + 2$

b Where do the lines cross?

4 **a** Using the gradient-intercept method, draw the following lines on the same grid. Use axes with ranges $-4 \leqslant x \leqslant 6$ and $-6 \leqslant y \leqslant 8$.

i $y = x + 3$

ii $y = 2x$

b Where do the lines cross?

AU **5** Here are the equations of three lines.

A: $y = 3x - 1$ B: $2y = 6x - 4$ C: $y = 2x - 2$

a State a mathematical property that lines A and B have in common.

b State a mathematical property that lines B and C have in common.

c Which of the following points is the intersection of lines A and C?

$(1, -4)$ $(-1, -4)$ $(1, 4)$

PS **6** **a** What is the gradient of line A?

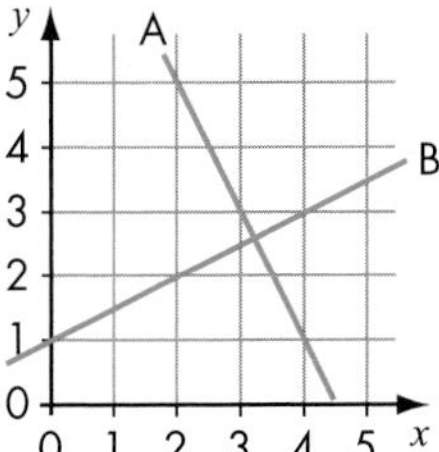

b What is the gradient of line B?

c What angle is there between lines A and B?

d What relationship do the gradients of A and B have with each other?

e Another line C has a gradient of 3.

What is the gradient of a line perpendicular to C?

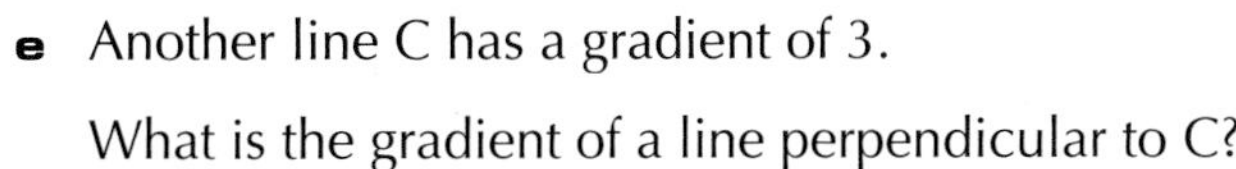

Cover-up method for drawing graphs

The x-axis has the equation $y = 0$. This means that all points on the x-axis have a y-value of 0.

The y-axis has the equation $x = 0$. This means that all points on the y-axis have an x-value of 0.

You can use these facts to draw any line that has an equation of the form:

$ax + by = c$

EXAMPLE 5

Draw the graph of $4x + 5y = 20$.

Because the value of x is 0 on the y-axis, you can solve the equation for y:

$$4(0) + 5y = 20$$
$$5y = 20$$
$$\Rightarrow y = 4$$

Hence, the line passes through the point (0, 4) on the y-axis (diagram **A**).

Because the value of y is 0 on the x-axis, you can also solve the equation for x:

$$4x + 5(0) = 20$$
$$4x = 20$$
$$\Rightarrow x = 5$$

Hence, the line passes through the point (5, 0) on the x-axis (diagram **B**). You need only two points to draw a line. (Normally, you would like a third point but in this case you can accept two.) Draw the graph by joining the points (0, 4) and (5, 0) (diagram **C**).

A
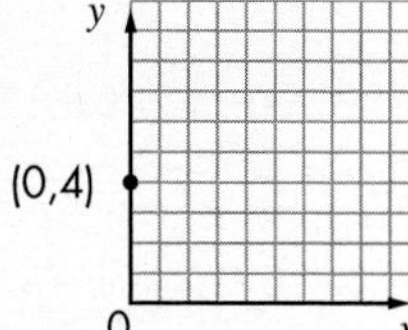

B
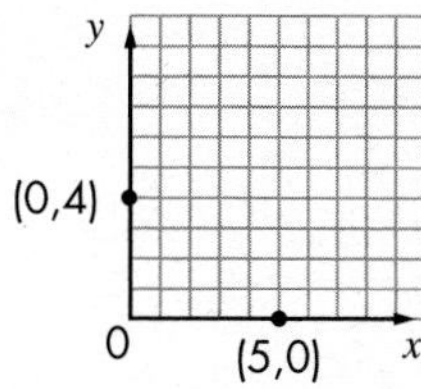

C
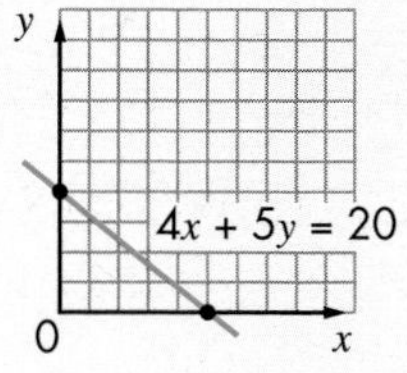

This type of equation can be drawn very easily, without much working at all, using the **cover-up method**.

Start with the equation: $4x + 5y = 20$

Cover up the x-term: [thumb] $+ 5y = 20$

Solve the equation (when $x = 0$): $y = 4$

Now cover up the y-term: $4x +$ [thumb] $= 20$

Solve the equation (when $y = 0$): $x = 5$

This gives the points (0, 4) on the y-axis and (5, 0) on the x-axis.

EXAMPLE 6

Draw the graph of $2x - 3y = 12$.

Start with the equation: $2x - 3y = 12$

Cover up the x-term: [thumb] $- 3y = 12$

Solve the equation (when $x = 0$): $y = -4$

Now cover up the y-term: $2x +$ 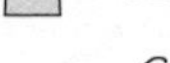 $=12$

Solve the equation (when $y = 0$): $x = 6$

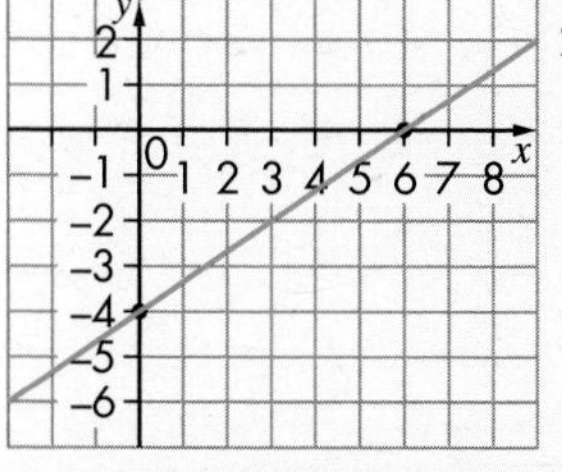

This gives the points (0, –4) on the y-axis and (6, 0) on the x-axis.

EXERCISE 17D

B

1 Draw these lines using the cover-up method. Use the same grid, taking x from –10 to 10 and y from –10 to 10. If the grid gets too 'crowded', draw another.

a $3x + 2y = 6$ **b** $4x + 3y = 12$ **c** $4x - 5y = 20$

d $x + y = 10$ **e** $3x - 2y = 18$ **f** $x - y = 4$

g $5x - 2y = 15$ **h** $2x - 3y = 15$ **i** $6x + 5y = 30$

j $x + y = -5$ **k** $x + y = 3$ **l** $x - y = -4$

2 **a** Using the cover-up method, draw the following lines on the same grid. Use axes with ranges $-2 \leqslant x \leqslant 6$ and $-2 \leqslant y \leqslant 6$.

i $2x + y = 4$

ii $x - 2y = 2$

b Where do the lines cross?

3 **a** Using the cover-up method, draw the following lines on the same grid.

Use axes with ranges $-2 < x < 6$ and $-3 < y < 6$.

i $x + 2y = 6$

ii $2x - y = 2$

b Where do the lines cross?

4 **a** Using the cover-up method, draw the following lines on the same grid.

Use axes with ranges $-6 \leqslant x \leqslant 8$ and $-2 \leqslant y \leqslant 8$.

i $x + y = 6$

ii $x - y = 2$

b Where do the lines cross?

AU **5** Here are the equations of three lines.

A: $2x + 6y = 12$ B: $x - 2y = 6$ C: $x + 3y = -9$

a State a mathematical property that lines A and B have in common.

b State a mathematical property that lines B and C have in common.

c State a mathematical property that lines A and C have in common.

d The line A crosses the y-axis at (0, 2).

The line C crosses the x-axis at (–9, 0).

Find values of a and b so that this line passes through these two points.

$ax + by = 18$

B

PS 6 The diagram shows an octagon ABCDEFGH.

The equation of the line through A and B is $y = 3$.

The equation of the line through B and C is $x + y = 4$.

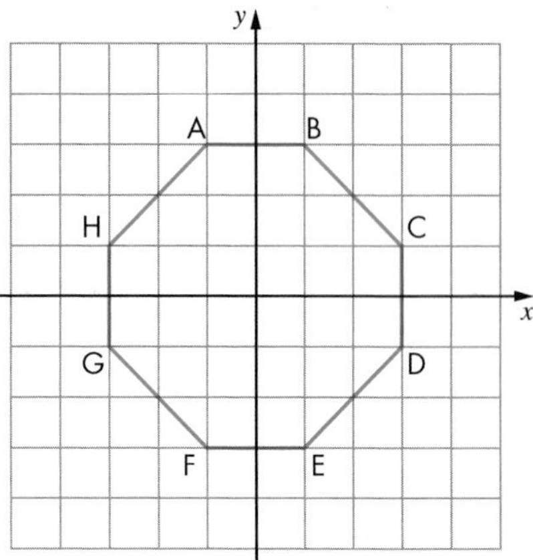

a Write down the equation of the lines through:

i C and D **ii** D and E **iii** E and F

iv F and G **v** G and H **vi** H and A

b The gradient of the line through F and B is 3.

Write down the gradient of the lines through:

i A and E **ii** G and C **iii** H and D

17.3 Finding the equation of a line from its graph

This section will show you how to:
- find the equation of a line, using its gradient and intercept

Key words
coefficient
gradient
intercept

The equation $y = mx + c$

When a graph can be expressed in the form $y = mx + c$, the **coefficient** of x, m, is the **gradient**, and the constant term, c, is the **intercept** on the y-axis.

This means that if you know the gradient, m, of a line and its intercept, c, on the y-axis, you can write down the equation of the line immediately.

For example, if $m = 3$ and $c = –5$, the equation of the line is $y = 3x – 5$.

All linear graphs can be expressed in the form $y = mx + c$.

This gives a method of finding the equation of any line drawn on a pair of coordinate axes.

EXAMPLE 7

Find the equation of the line shown in diagram A.

A

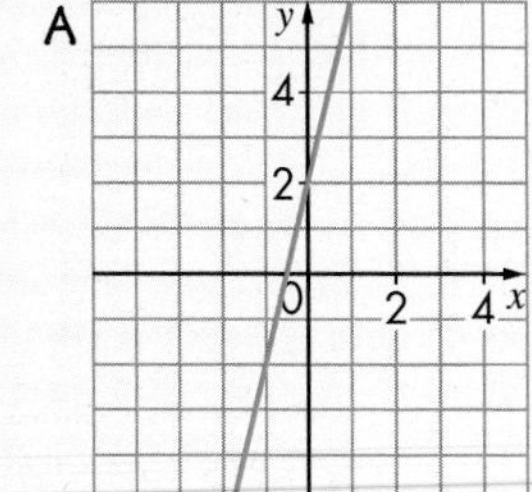

B

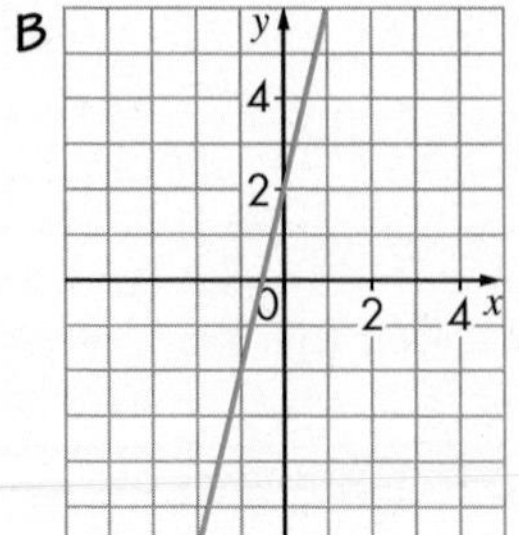

C

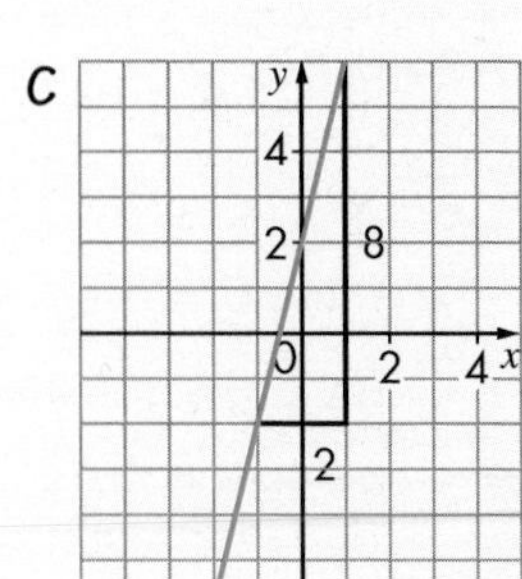

First, find where the graph crosses the y-axis (diagram **B**).

So $c = 2$

Next, measure the gradient of the line (diagram **C**).

y-step = 8

x-step = 2

gradient = $8 \div 2 = 4$

So $m = 4$

Finally, write down the equation of the line: $y = 4x + 2$

EXERCISE 17E

1 Give the equation of each of these lines, all of which have positive gradients. (Each square represents one unit.)

a

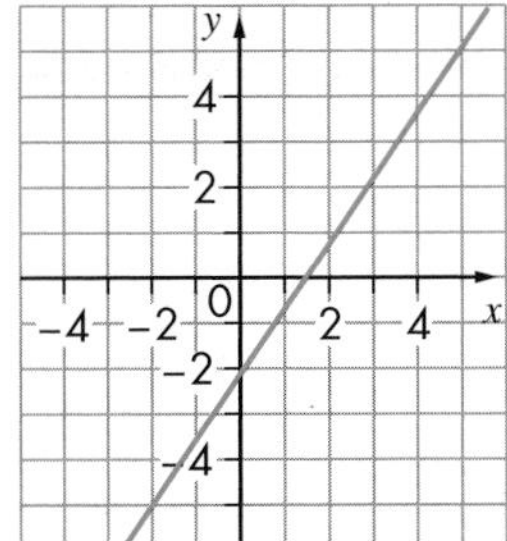

b

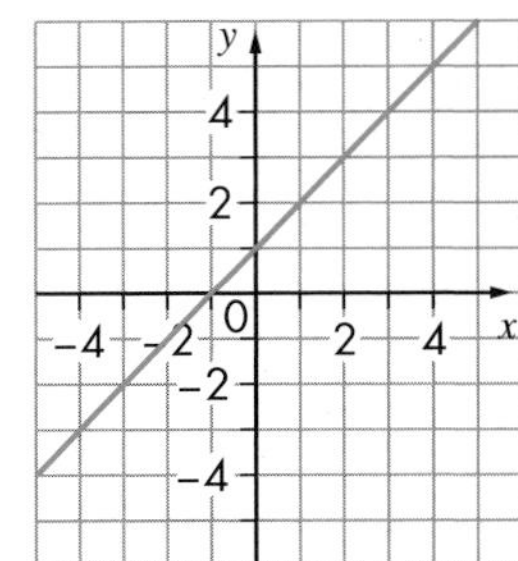

c

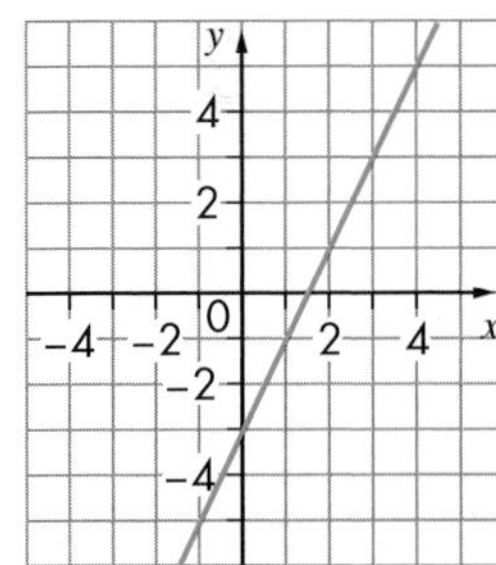

d

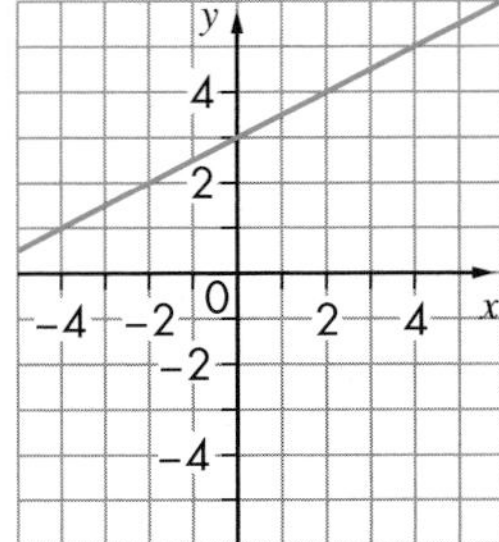

e

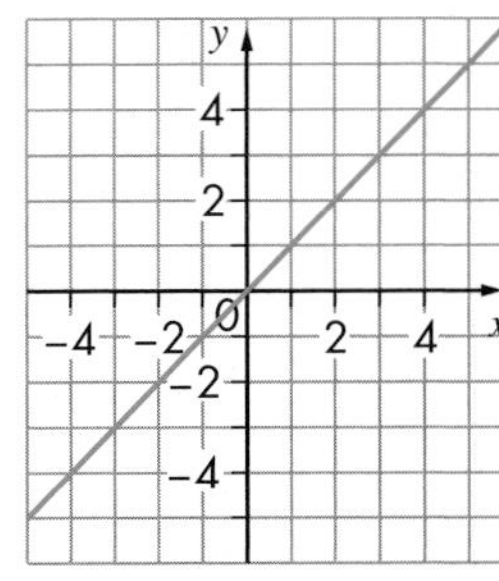

f

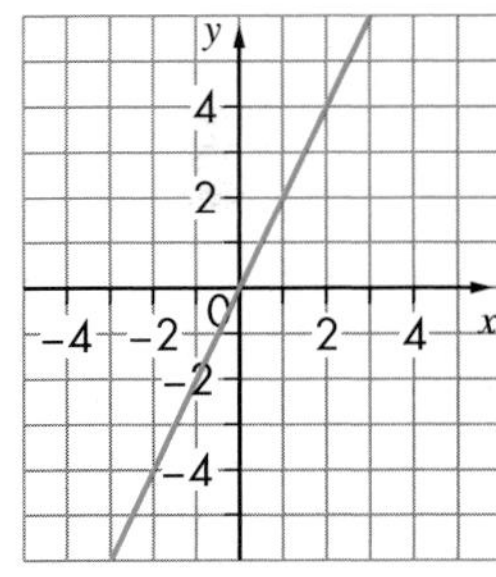

PS 2 In each of these grids, there are two lines. (Each square represents one unit.)

a

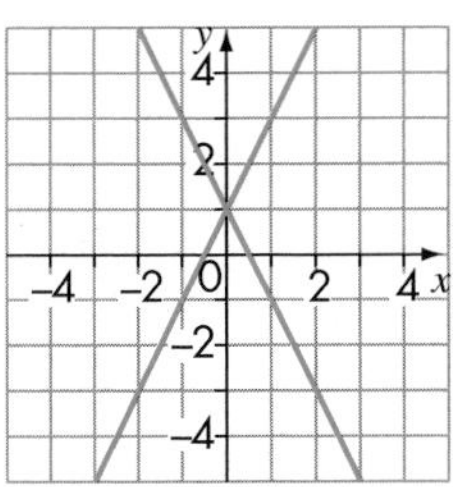

b

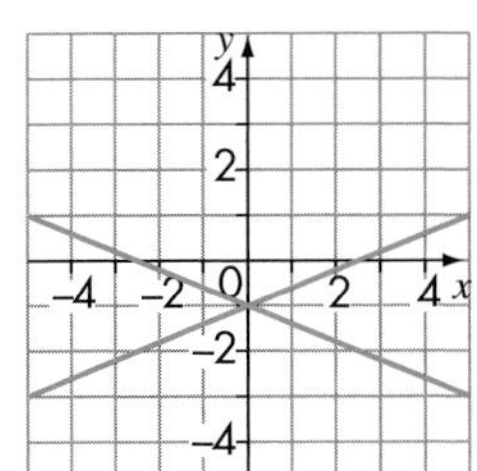

c

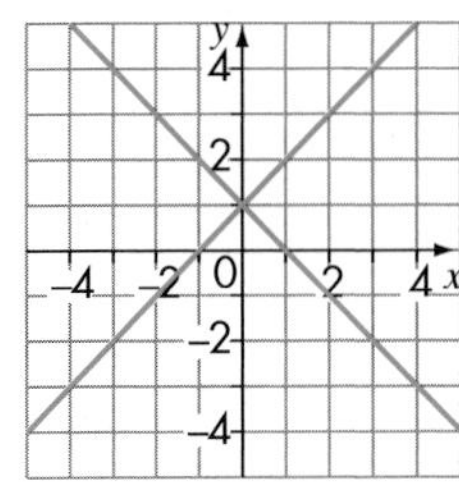

For each grid:

i find the equation of each of the lines

ii describe any symmetries that you can see

iii describe any connection between the gradients of each pair of lines.

B

AU 3 A straight line passes through the points (1, 3) and (2, 5).

a Explain how you can tell that the line also passes through (0, 1).

b Explain how you can tell that the line has a gradient of 2.

c Work out the equation of the line that passes through (1, 5) and (2, 8).

4 Give the equation of each of these lines, all of which have negative gradients. (Each square represents one unit.)

a

b

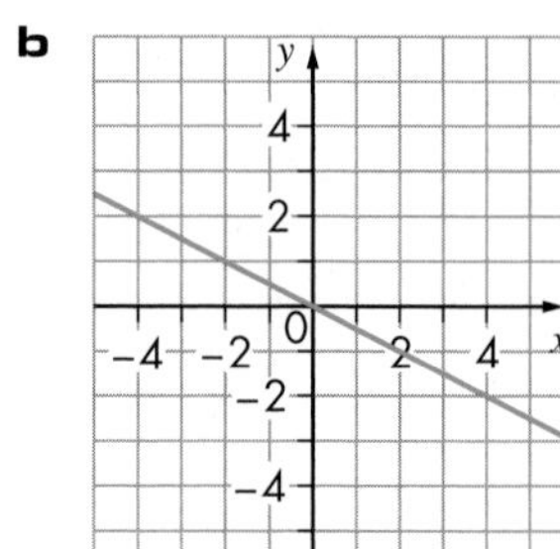

c

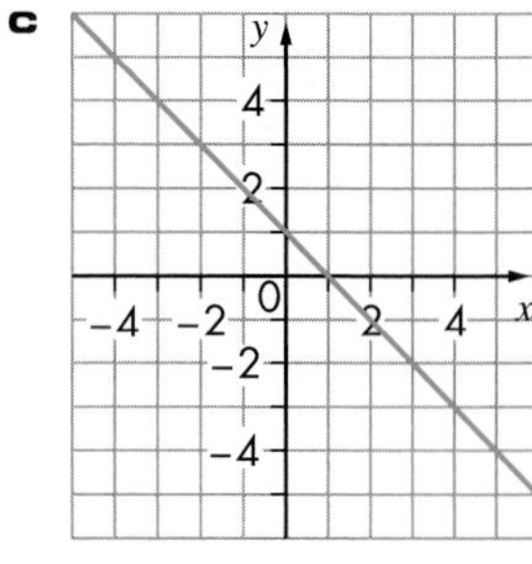

d

e

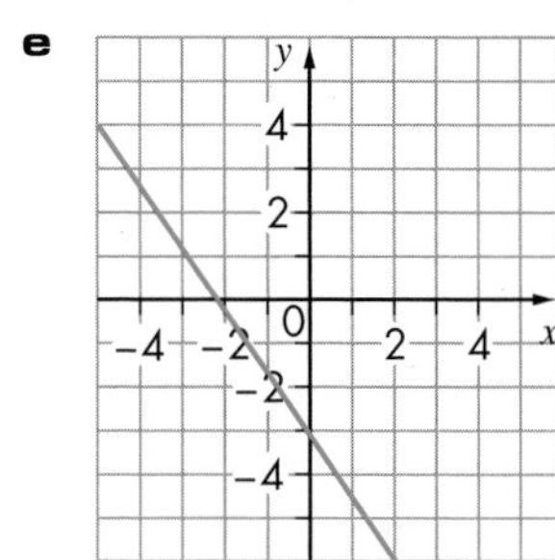

PS 5 In each of these grids, there are three lines. One of them is $y = x$. (Each square represents one unit.)

a

b

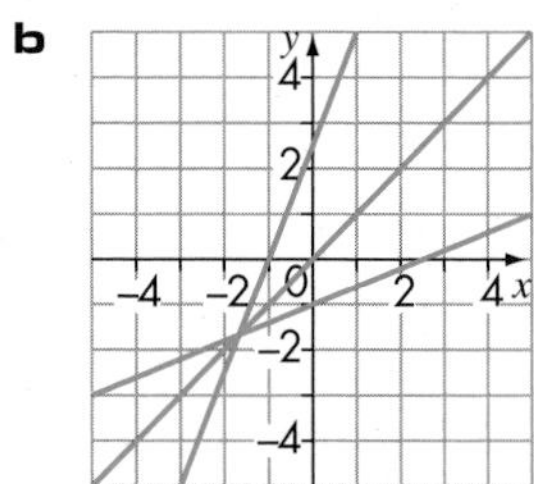

c

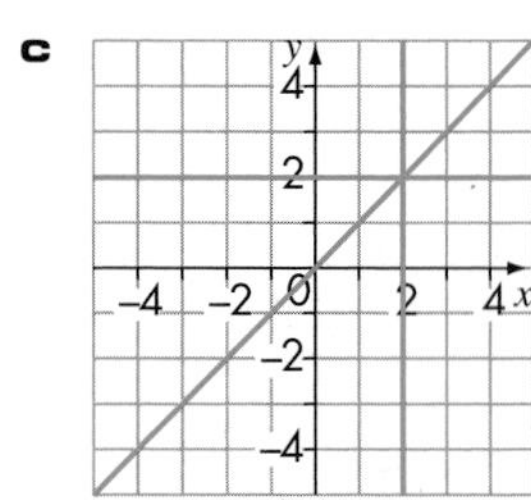

For each grid:

i find the equation of each of the other two lines

ii describe any symmetries that you can see

iii describe any connection between the gradients of each group of lines.

17.4 3-D coordinates

You have already worked with coordinates in two dimensions, so you may remember that these are always expressed as a pair of values, for example, (4, –3) where the first value is the x-coordinate and the second value is the y-coordinate. This is the ordered pair (x, y).

In three dimensions, which is the actual world we live in, coordinates are expressed as a triple of values, for example, (2, 3, –1). The first and second values are still x and y but the third represents the z-coordinate. This is the ordered triple (x, y, z).

It is not possible to show three dimensions exactly on the flat page of a book. Therefore, to draw three dimensions, certain conventions are used. The axes obey what is known as the right-hand rule. This means that if you hold out your right hand with the thumb, index finger and middle finger at right angles to one another, the thumb represents the positive x-axis, the index finger is the positive y-axis and the middle finger is the positive z-axis.

The normal convention is to draw a set of axes, like this.

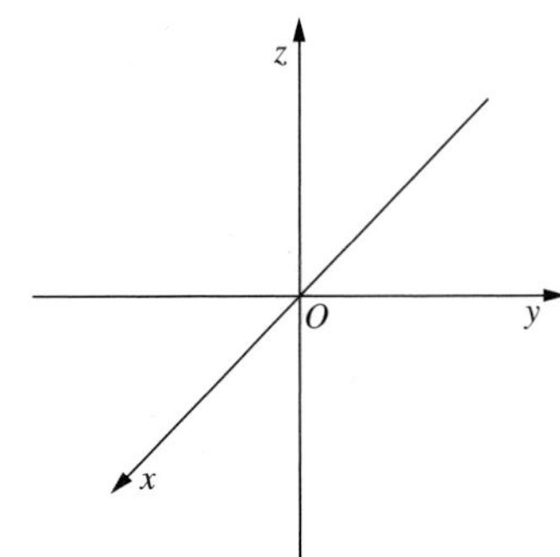

The positive y-axis is to the right, the positive z-axis is vertical and the positive x-axis is out of the page. The point where all three axes meet is called the origin and is marked with an O, as in 2D coordinates.

To mark a point in 3D, you move in the x-direction first, then in the y-direction, then in the z-direction.

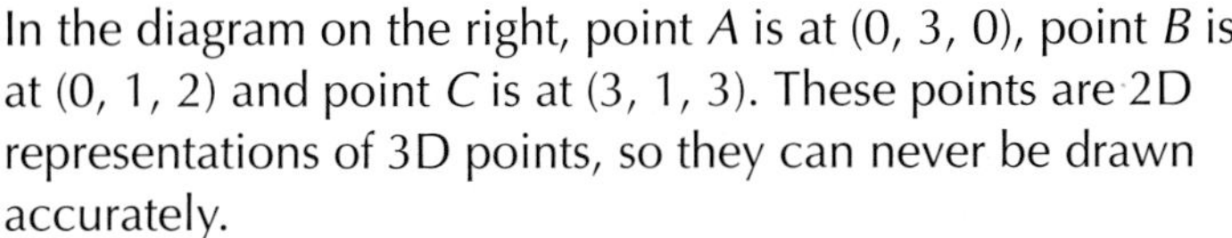

In the diagram on the right, point A is at (0, 3, 0), point B is at (0, 1, 2) and point C is at (3, 1, 3). These points are 2D representations of 3D points, so they can never be drawn accurately.

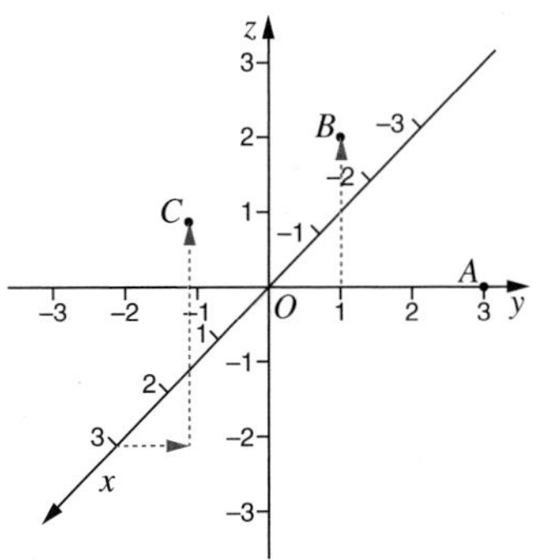

To make things easier, most questions use positive values and usually use the axes so that at least one coordinate is zero.

EXAMPLE 8

In the diagram below, $ABCDEFGH$ is a cuboid with a square end, with sides of two units and a length of four units.

The vertex A is placed at the origin and the sides of the cuboid are aligned along the axes.

1 Write down the coordinates of the following points.

a C

b H

2 Write down the coordinates of the midpoint of CH.

EXAMPLE 8 (continued)

3 Work out the length of the diagonal AG.

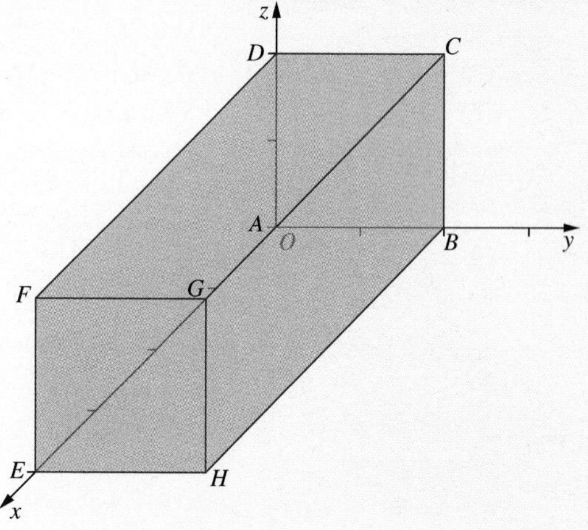

a **i** C is at the point (0, 2, 2). **ii** H is at the point (4, 2, 0).

b The midpoint of CH is at (2, 2, 1).

c The length of AG is the same problem as finding the length of the diagonal of a cuboid with sides of 2 cm, 2 cm and 4 cm. G is the point (4, 2, 2).

Using Pythagoras in 3D gives:

$AG = \sqrt{(4^2 + 2^2 + 2^2)} = \sqrt{24} = 4.9$ units

EXAMPLE 9

In a game of 3-D Noughts and Crosses, two white, two grey and two black balls have been placed as shown in the diagram.

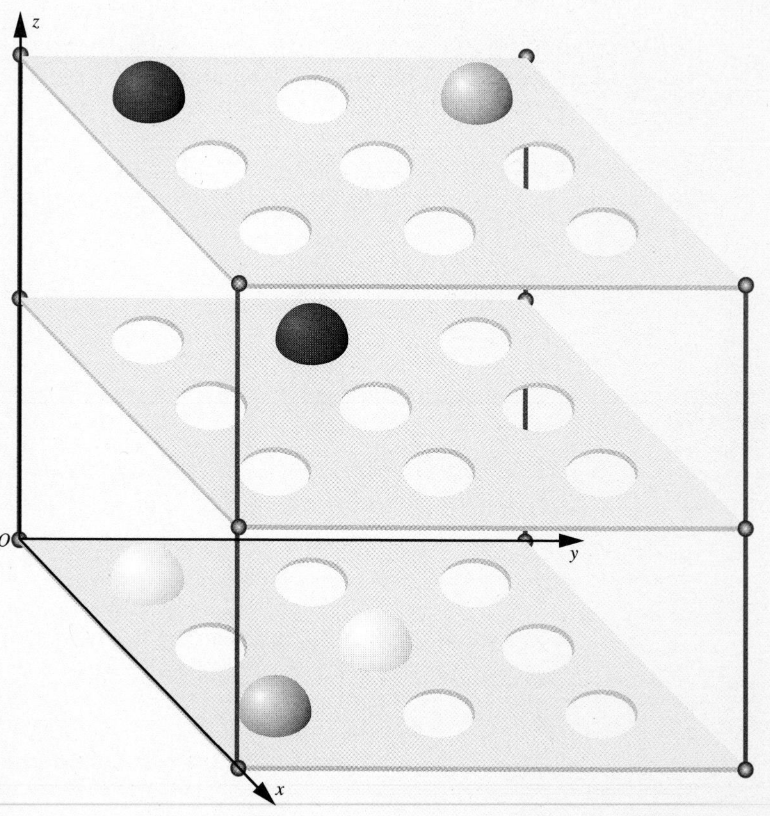

EXAMPLE 9 (continued)

For example, the white balls are at (1, 1, 1) and (2, 2, 1).

Write down where the next white, grey and black balls should be placed to give a winning line of three balls.

Answers

The white ball should be placed at (3, 3, 1).

The grey balls are at (3, 1, 1) and (1, 3, 3). The next grey ball should be placed at (2, 2, 2).

The black balls are at (1, 1, 3) and (1, 2, 2). The next black ball should be placed at (1, 3, 1).

EXERCISE 17F

1 The diagram shows a cube *ABCDEFGH* of side three units with *A* at the origin and the sides of the cube along the axes.

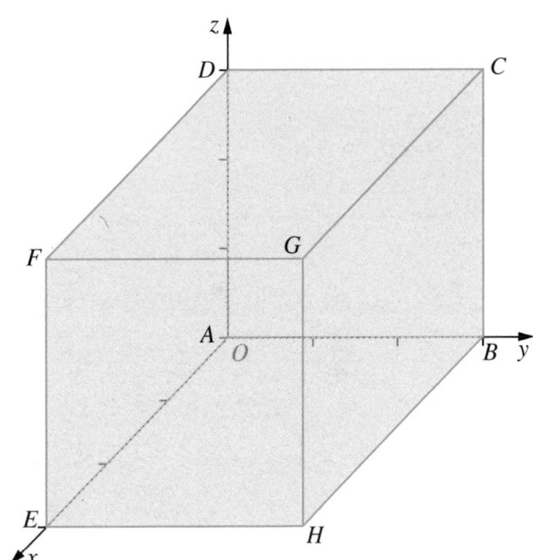

a Write down the coordinates of all the vertices: *A, B, C, D, E, F, G, H*.

b Write down the coordinates of the midpoints of the following faces.

i *ABCD* **ii** *GCBH*

c Write down the coordinates of the centre of the cube.

d Work out the length of each of the following.

i *ED*

ii *AG*

2 The diagram shows a triangular prism *ABCDEF*.

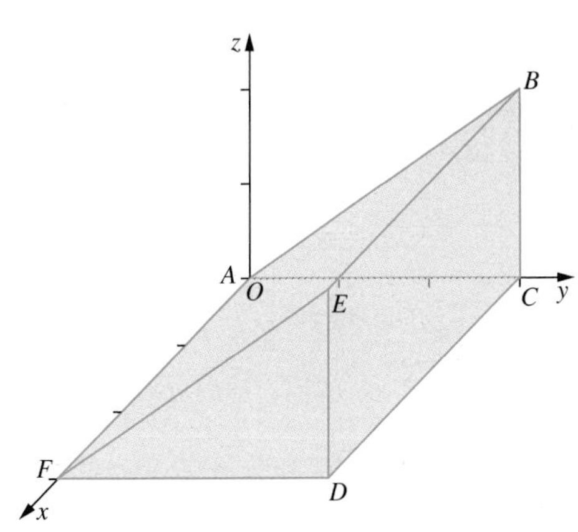

a Write down the coordinates of the points

i *C* **ii** *D* **iii** *F*

b The prism is reflected in the plane *ABC*. What are the coordinates of the reflections of the following points?

i *B* **ii** *E* **iii** *D*

c Work out the angle *BFC*.

A

A*

B

3 Write down the midpoint of each of the following pairs of points.

a (1, 1, 1) and (3, 3, 3)

b (1, 2, 1) and (3, 2, 5)

c (0, 1, –2) and (0, 3, 0)

d (–6, 2, 0) and (–2, 4, 6)

e (8, –4, 2) and (6, 4, 4)

f (4, –6, 0) and (4, 0, 4)

C

4 The diagram shows a cuboid ABCDEFGH.

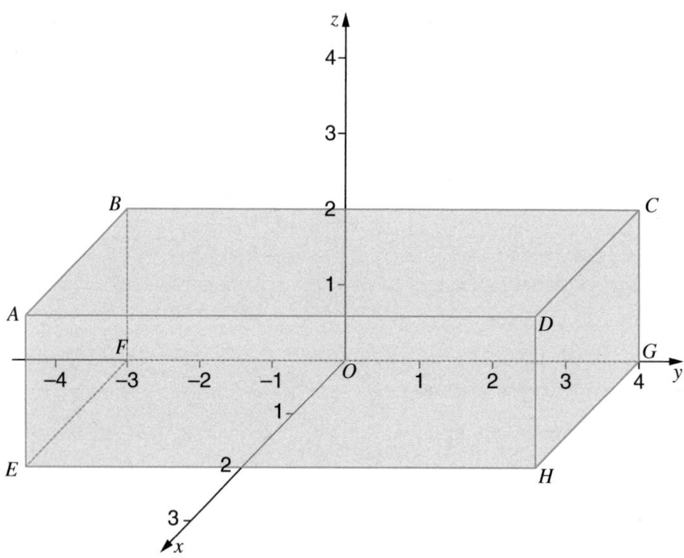

a Write down the coordinates of C.

b Write down the coordinates of E.

c Work out the volume of the cuboid.

GRADE BOOSTER

- **D** You can draw straight lines by plotting points
- **C** You can draw straight lines using the gradient-intercept method
- **C** You can read a 3-D coordinate from a diagram.
- **C** You can work out the midpoint of two points on a 3-D diagram.
- **B** You can find the equation of a line form its graph.
- **B** You can find points on a 3-D coordinate diagram from, for example, the reflection of another point.
- **B** You can use Pythagoras theorem to work out a length on a 3-D coordinate diagram where the points have a common coordinate.
- **A** You can use Pythagoras theorem to work out a length on a 3-D coordinate diagram where the points do not have a common coordinate.

What you should know now

- How to draw linear graphs
- How to work out the equation of a line from its graph.
- How to read 3-D coordinates.
- How to interpret a 3-D coordinate diagram to solve problems using, for example, Pythagoras theorem.

EXAMINATION QUESTIONS

1 **a** Draw the graph of $y = 2x + 3$ for values of x from 0 to 5. Use a grid with axes covering $0 \leqslant x \leqslant 6$ and $0 \leqslant y \leqslant 14$.

b Use your graph to solve $6.5 = 2x + 3$.

2 The diagram shows a sketch of the graph of $y = 3x + 1$.

Copy the diagram, and draw and label sketch graphs of these.

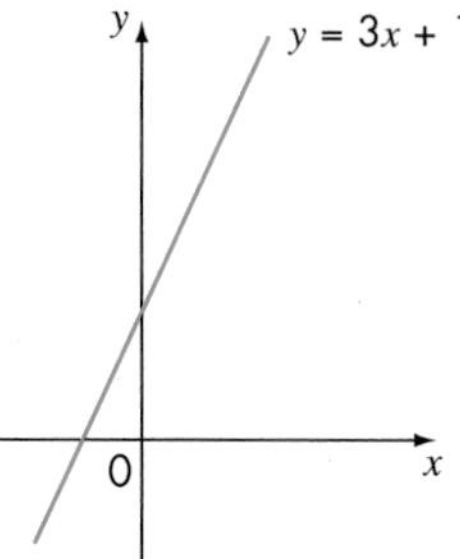

a $y = 1$

b $y = x + 1$

3 **a** Find the equation of the line AB.

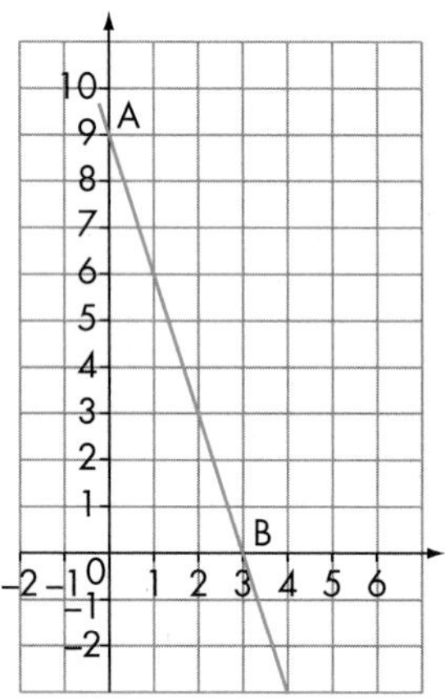

b Give the y-coordinate of the point on the line with an x-coordinate of 6. *(2 marks)*

c Write down the gradient of a line perpendicular to AB. *(1 mark)*

AQA, November 2004, Paper 2 Higher, Question 7

4 The line $y = x - 1$ crosses the line $y = 5$ at the point P.

Work out the coordinates of the point P. You may use a grid to help you.

(3 marks)

AQA, November 2011, Higher 1, Question 6

5 Draw the graph of $y = 3x - 1$ for $-4 \leqslant x \leqslant 4$. Use an x axis from –4 to + 4 and a y axis from –14 to + 14

AQA, June 2011, Higher 2, Question 6

6 Work out the area enclosed by the lines

$y = 3 \qquad x = -2 \qquad y = x$

AQA, November 2010, Higher 2, Question 10

7 **a** Draw the line $y = 3x + 2$ for $-4 \leqslant x \leqslant 4$. Use a grid with an x-axis from –4 to 4 and a y-axis from –16 to +16

b What is the equation of the line L?

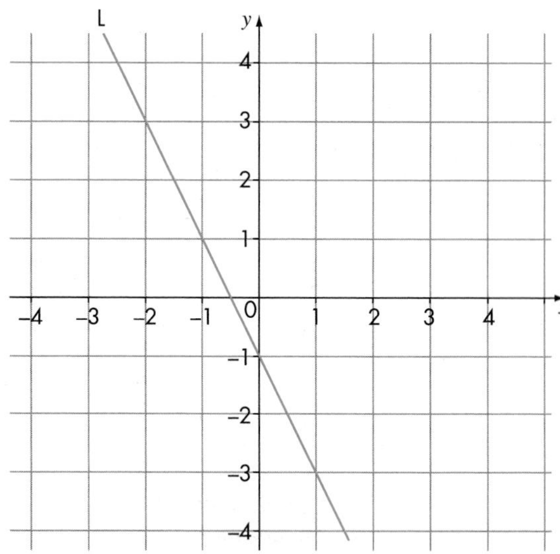

(3 marks)

AQA, June 2010, Higher 2, Question 13

8

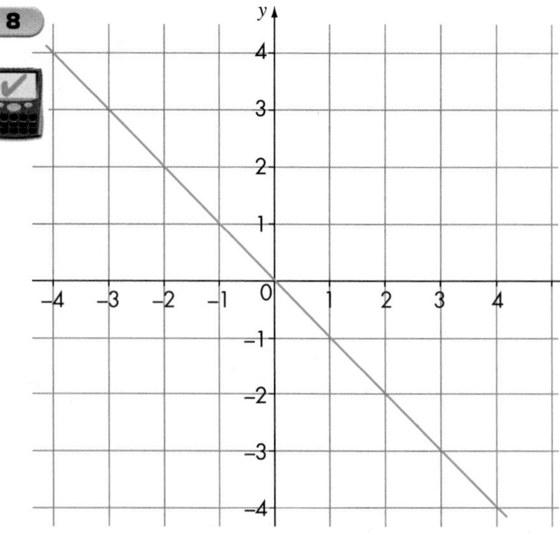

a Pat says the graph shows the line $y = x$.

Explain why Pat is wrong. *(1 mark)*

b Find the equation of the line that passes through the points (–4, –3), (0, 1) and (3, 4). *(2 marks)*

Worked Examination Questions

1 The line $y = 2x - 1$ crosses the line $y = 3$ at point P.

Work out the coordinates of P.

You may use graph paper to help you.

1.

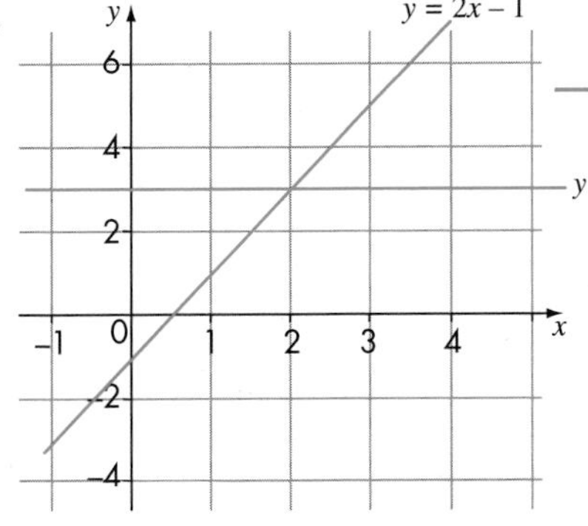

Drawing axes and marking them with a y-axis that goes form at least –3 to 4 and drawing one of the lines will get 1 method mark. Drawing both lines will get an extra accuracy mark.

P = (2, 3)

Writing down the intersection point gets 1 accuracy mark. This question could be done without drawing a graph. The y coordinate is obviously 3 so substitute this into $y = 2x - 1$ and solve for x.

Total: 3 marks

2 Work out the equation if the line L.

2. Gradient = $\frac{5}{4}$

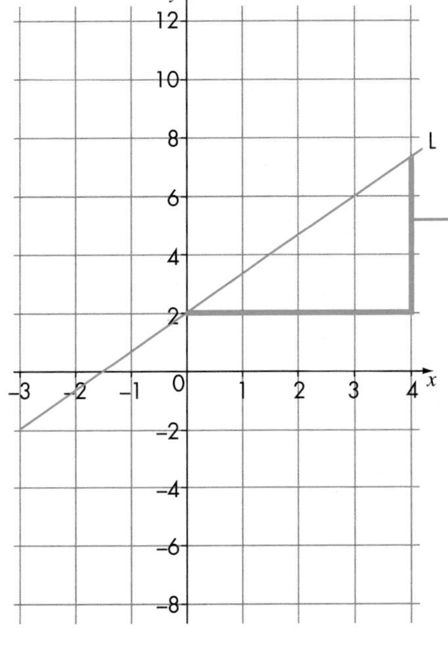

Drawing an appropriate right angled triangle on the grid gets 1 method mark. Writing down the y distance divided by the x distance gets I accuracy mark. This is best done as a fraction. Be careful with scales!

Intercept = (0, 2)

$y = \frac{5}{4}x + 2$

The intercept can be written down or it can be part of the final answer. The final equation gets 1 accuracy mark. Sometimes the answer is asked for in the form $ax + by = c$. In this case that would be $5x - 4y = -2$

Total: 3 marks

17 Problem Solving
Traverse the network

Graph theory is the study of mathematical structures that model the relationship between objects from one collection. Just think about the 3D molecular structures that you see in chemistry – these are an example of how graph theory can be used.

Graph theory is also used to create networks. You can often 'traverse' a network. Effectively, this means that you can draw them without taking your pen off the page. In this task you will look for a rule for working out when it is possible to traverse a network.

Getting started

Draw these networks, without taking your pen off the page if possible.

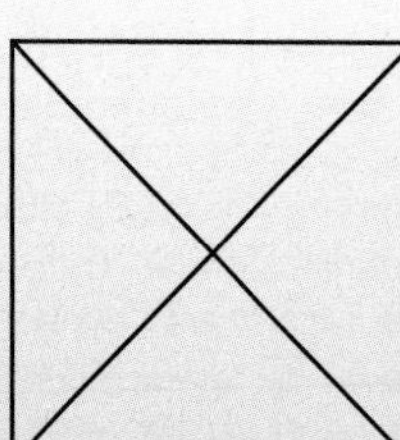

Facts

Arcs and nodes

A **node** is a point where **arcs** meet. This shape has 5 nodes and 10 arcs.

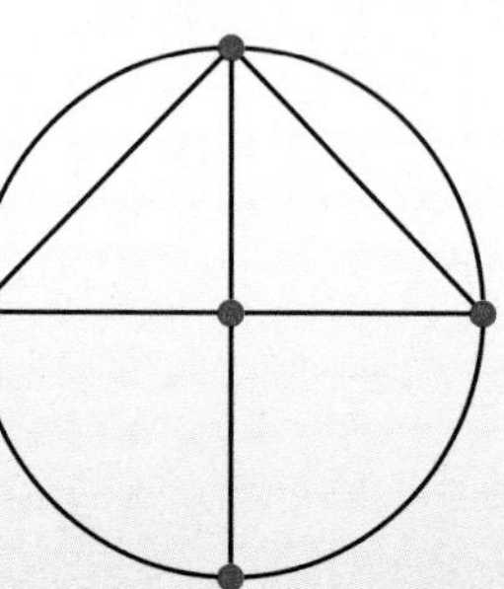

The degree of a node

The degree of a node is how many arcs intersect at that node.

This shape has one node of degree 5, one node of degree 3 and three nodes of degree 4

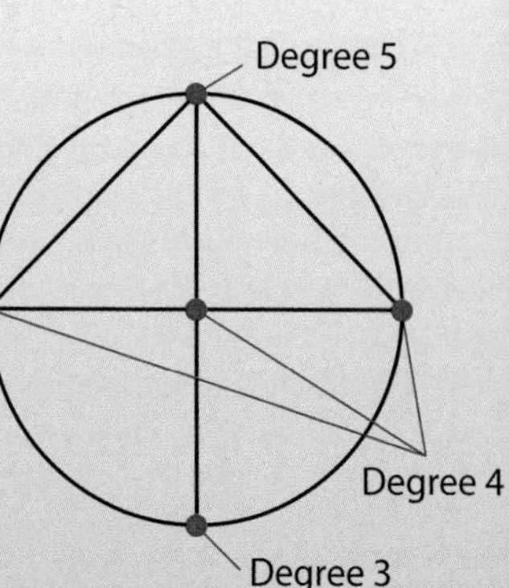

Your task

1 Investigate the number of odd nodes and even nodes that each of these shapes has, and whether they are traversable networks. Find the rule that makes a network traversable.

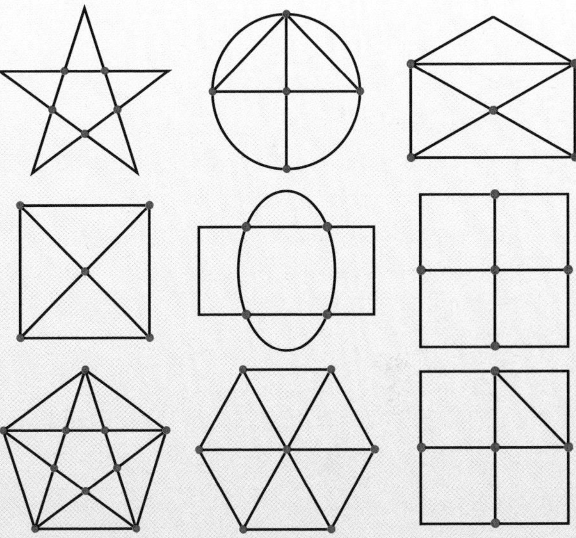

Prove your rule by creating networks of your own.

2 This is a plan of a building

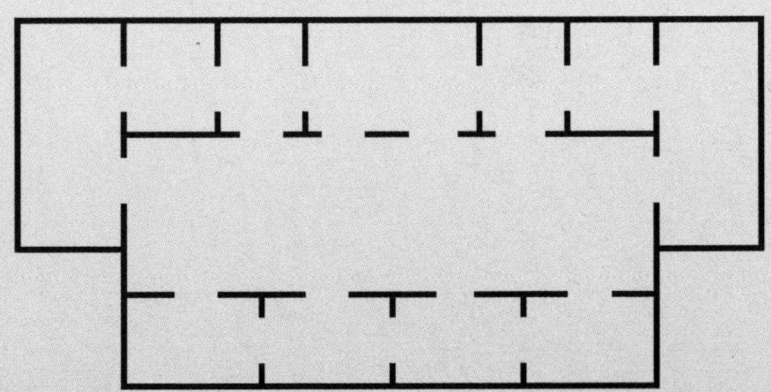

Using your rule, trace a route through the house so that you pass through each door once, and only once.

Extension

Research the Königsberg Bridge problem (which you first learnt about on page 416), and Leonhard Euler. Use your research to check the rule you found in your task and to find out more about networks.

Why this chapter matters

Have you ever wondered why the lights on a car shine forward in a fairly tight beam?
Why isn't the side of the road also lit up?
It is the same for a torch or a searchlight
The answer is of course mathematical.

Before you read the reason below try this exercise

Trace this curve, the tangents drawn at 3 places and the point called F.

From F draw a line to where the tangent touches the curve. At that point draw the perpendicular to the tangent (this is called the normal) and then draw a line on the other side of the normal at the same angle as the line from F. This has been done for point T_1.

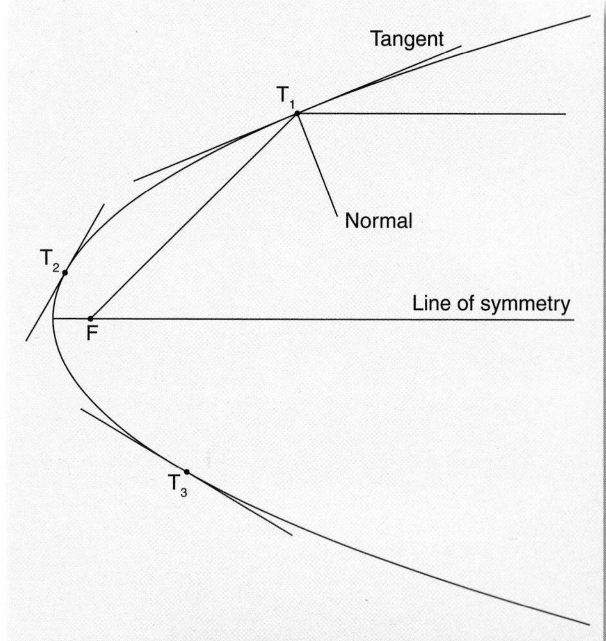

What you should find.
The three lines from the tangent should all be parallel to the line of symmetry.

Why does it work?
The curve is a parabola. This is a quadratic curve which you will be learning about in this chapter.

This is the shape of all car headlights. The point F is called the focus. This is where the bulb is placed in a car headlight. Wherever a beam of light from the focus meets the parabola it is always reflected parallel to the line of symmetry.

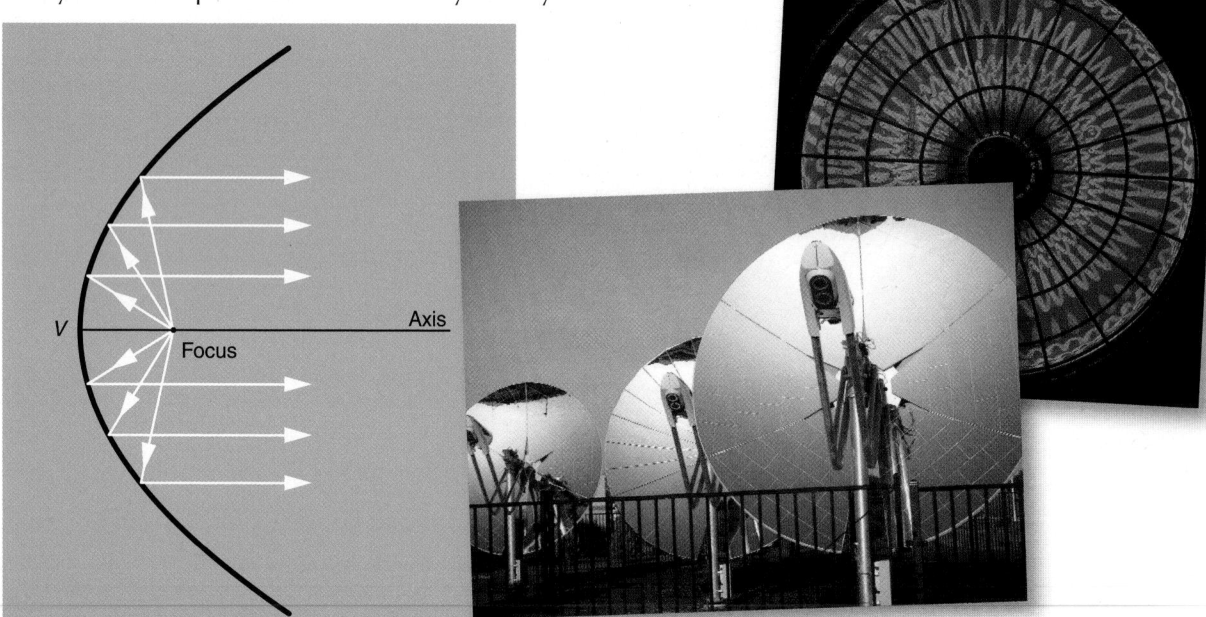

The same principal works in heaters and in reverse in satellite dishes.

Algebra: More graphs and equations

1 Quadratic graphs

2 The significant points of a quadratic graph

This chapter will show you ...

C how to draw quadratic graphs

A how to recognise and find the significant points of a quadratic graph

Visual overview

What you should already know

- How to draw linear graphs **(KS3 level 6, GCSE grade D)**
- How to find the equation of a graph, using the gradient-intercept method **(KS3 level 8, GCSE grade B)**

Quick check

1 Draw the graph of $y = 3x - 1$ for values of x from -2 to $+3$.

2 Give the equation of the graph shown.

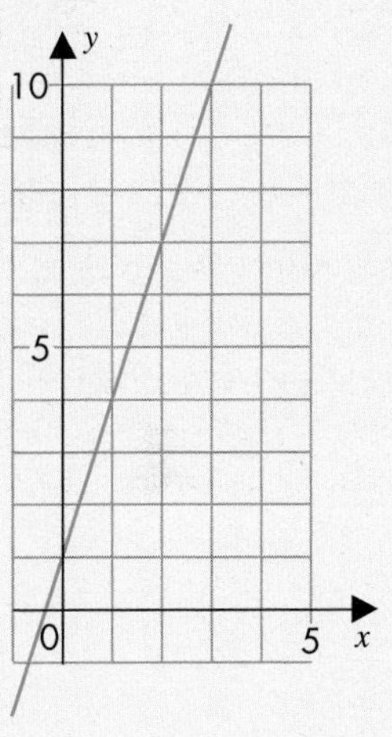

18.1 Quadratic graphs

This section will show you how to:

- draw and read values from quadratic graphs

Key words
parabola
quadratic

A **quadratic** graph has a term in x^2 in its equation. All of the following are quadratic equations and each would produce a quadratic graph.

$y = x^2$ $\qquad$ $y = x^2 + 5$ $\qquad$ $y = x^2 - 3x$

$y = x^2 + 5x + 6$ $\qquad$ $y = 3x^2 - 5x + 4$

EXAMPLE 1

Draw the graph of $y = x^2 + 5x + 6$ for $-5 \leq x \leq 3$.

Make a table, as shown below. Work out the values in each row (x^2, $5x$, 6) separately, adding them together to obtain the values of y. Then plot the points from the table.

x	−5	−4	−3	−2	−1	0	1	2	3
y^2	25	16	9	4	1	0	1	4	9
$+5x$	−25	−20	−15	−10	−5	0	5	10	15
+6	6	6	6	6	6	6	6	6	6
y	6	2	0	0	2	6	12	20	30

Note that in an examination paper you may be given only the first and last rows, with some values filled in. For example,

x	−5	−4	−3	−2	−1	0	1	2	3
y	6		0		2				30

In this case, you would either construct your own table, or work out the remaining y-values with a calculator.

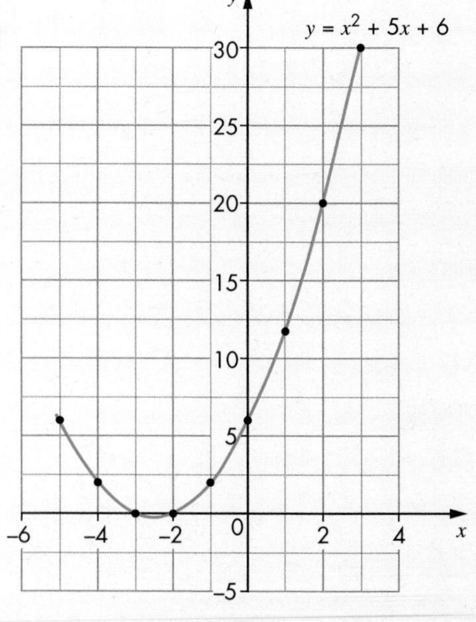

EXAMPLE 2

a Complete the table for $y = 3x^2 - 5x + 4$ for $-1 \leqslant x \leqslant 3$, then draw the graph.

x	–5	–0.5	0	0.5	1	1.5	2	2.5	3
y	12		0	2.25	2			10.25	16

b Use your graph to find the value of y when $x = 2.2$.

c Use your graph to find the values of x that give a y-value of 9.

a The table gives only some values. So you either set up your own table with $3x^2$, $-5x$ and $+4$, or calculate each y-value. For example, on the majority of scientific calculators, the value for –0.5 will be worked out as:

Check that you get an answer of 7.25

If you want to make sure that you are doing the correct arithmetic with your calculator, try some values for x for which you know the answer. For example, try $x = 0.5$, and see whether your answer is 2.25

The complete table should be:

x	–1	–0.5	0	0.5	1	1.5	2	2.5	3
y	12	7.25	4	2.25	2	3.25	6	10.25	16

The graph is shown on the right.

b To find the corresponding y-value for any value of x, you start on the x-axis at that x-value, go up to the curve, across to the y-axis and read off the y-value. This procedure is marked on the graph with arrows.

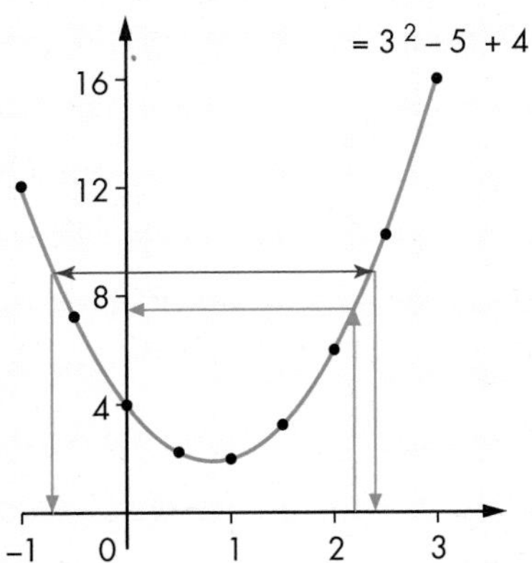

Always show these arrows because even if you make a mistake and misread the scales, you may still get a mark.

When $x = 2.2$, $y = 7.5$.

c This time start at 9 on the y-axis and read off the two x-values that correspond to a y-value of 9. Again, this procedure is marked on the graph with arrows.

When $y = 9$, $x = -0.7$ or $x = 2.4$.

A quadratic curve drawn correctly will always give a smooth curve, called a **parabola**.

Drawing accurate graphs

Although it is difficult to draw accurate curves, examiners work to a tolerance of only 1 mm. Here are some of the more common ways in which marks are lost in an examination.

- When the points are too far apart, a curve tends to 'wobble'.
- Drawing curves in small sections leads to 'feathering'.
- The place where a curve should turn smoothly is drawn 'flat'.
- A line is drawn through a point that, clearly, has been incorrectly plotted.

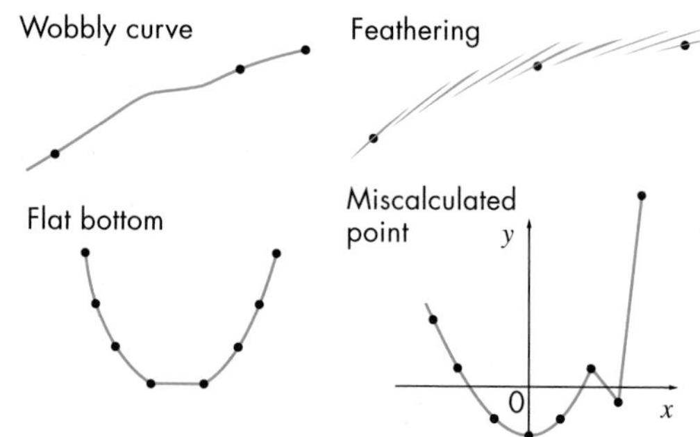

Here are some tips which will make it easier for you to draw smooth, curved lines.

- If you are *right-handed*, turn your paper or exercise book round so that you draw from left to right. Your hand is steadier this way than when you are trying to draw from right to left or away from your body. If you are *left-handed*, you should find drawing from right to left the more accurate way.
- Move your pencil over the points as a practice run without drawing the curve.
- Do one continuous curve and only stop at a plotted point.
- Use a *sharp* pencil and do not press too heavily, so that you may easily rub out mistakes.

Normally, in an examination, grids are provided with the axes clearly marked, so the examiner can place a transparent master over a graph and see immediately whether any lines are badly drawn or points are misplotted. Remember: a tolerance of 1 mm is all that you are allowed.

You do not need to work out all values in a table. You need only to work out the y-value. The other rows in the table are just working lines to break down the calculation. Learn how to calculate y-values with a calculator as there is no credit given for setting up tables in examinations.

EXERCISE 18A

In this exercise, suitable ranges are suggested for the axes. You can use any type of graph paper.

C

1 **a** Copy and complete the table or use a calculator to work out values for the graph of $y = 3x^2$ for values of x from –3 to 3.

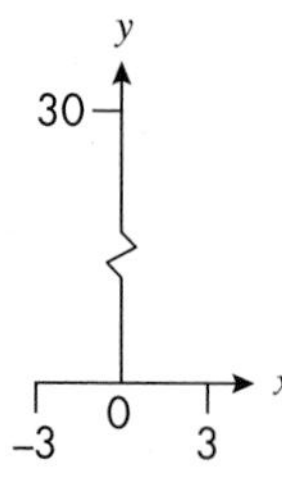

x	–3	–2	–1	0	1	2	3
y	27		3			12	

b Use your graph to find the value of y when $x = -1.5$.

c Use your graph to find the values of x that give a y-value of 10.

2 **a** Copy and complete the table or use a calculator to work out values for the graph of $y = x^2 + 2$ for values of x from –5 to 5.

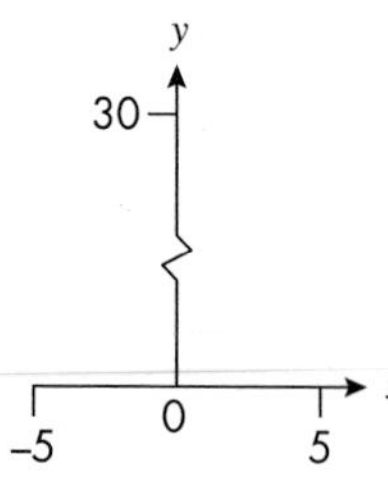

x	–5	–4	–3	–2	–1	0	1	2	3	4	5
$y = x^2 + 2$	27		11					6			

b Use your graph to find the value of y when $x = -2.5$.

c Use your graph to find the values of x that give a y-value of 14.

3 **a** Copy and complete the table or use a calculator to work out values for the graph of $y = x^2 - 2x - 8$ for values of x from –5 to 5.

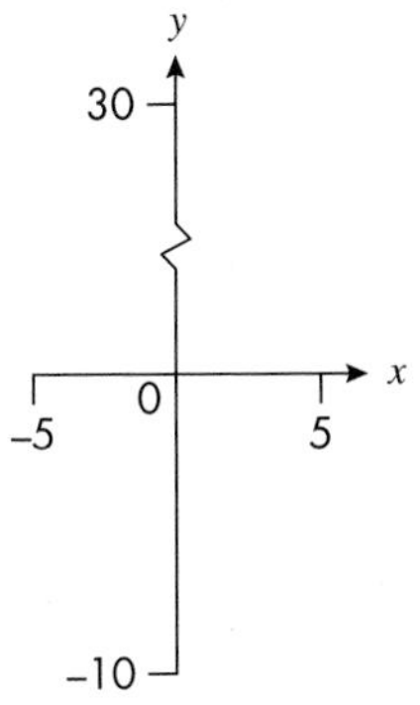

x	–5	–4	–3	–2	–1	0	1	2	3	4	5
x^2	25		9					4			
$-2x$	10							–4			
-8	–8							–8			
y	27							–8			

b Use your graph to find the value of y when $x = 0.5$.

c Use your graph to find the values of x that give a y-value of –3.

4 **a** Copy and complete the table or use a calculator to work out the values for the graph of $y = x^2 + 2x - 1$ for values of x from –3 to 3.

x	–3	–2	–1	0	1	2	3
x^2	9				1	4	
$+2x$	–6		–2			4	
-1	–1	–1				–1	
y	2					7	

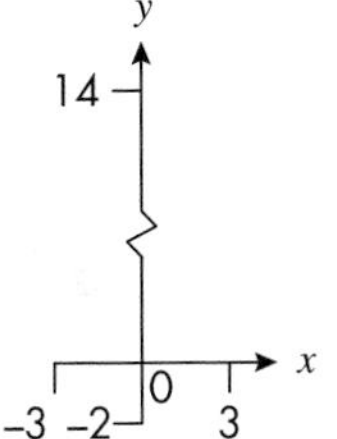

b Use your graph to find the y-value when $x = -2.5$.

c Use your graph to find the values of x that give a y-value of 1.

d On the same axes, draw the graph of $y = \frac{x}{2} + 2$.

e Where do the graphs $y = x^2 + 2x - 1$ and $y = \frac{x}{2} + 2$ cross?

5 **a** Copy and complete the table or use a calculator to work out values for the graph of $y = x^2 - x + 6$ for values of x from –3 to 3.

x	–3	–2	–1	0	1	2	3
x^2	9				1	4	
$-x$	3					–2	
$+6$	6					6	
y	18					8	

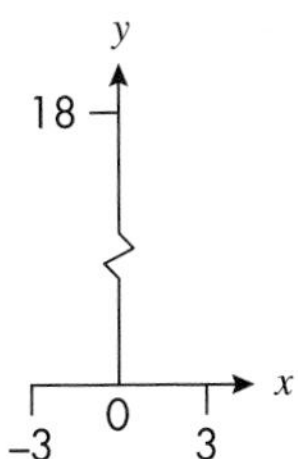

b Use your graph to find the y-value when $x = 2.5$.

c Use your graph to find the values of x that give a y-value of 8.

d Copy and complete the table or use a calculator to draw the graph of $y = x^2 + 5$ on the same axes.

x	–3	–2	–1	0	1	2	3
y	14		6				14

e Where do the graphs $y = x^2 - x + 6$ and $y = x^2 + 5$ cross?

C

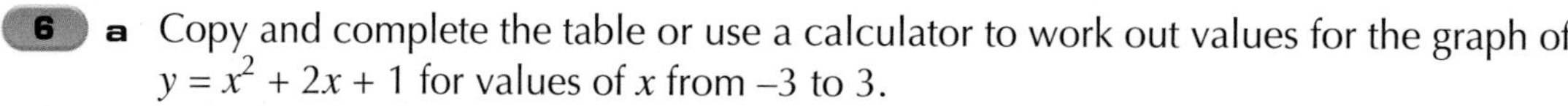

6 **a** Copy and complete the table or use a calculator to work out values for the graph of $y = x^2 + 2x + 1$ for values of x from –3 to 3.

x	–3	–2	–1	0	1	2	3
x^2	9				1	4	
$+2x$	–6					4	
$+1$	1					1	
y	4						

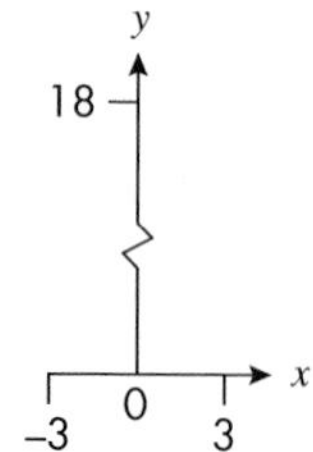

b Use your graph to find the y-value when $x = 1.7$.

c Use your graph to find the values of x that give a y-value of 2.

d On the same axes, draw the graph of $y = 2x + 2$.

e Where do the graphs $y = x^2 + 2x + 1$ and $y = 2x + 2$ cross?

7 **a** Copy and complete the table or use a calculator to work out values for the graph of $y = 2x^2 – 5x – 3$ for values of x from –2 to 4.

x	–2	–1.5	–1	–0.5	0	0.5	1	1.5	2	2.5	3	3.5	4
y	15	9			–3	–5				–3			9

b Where does the graph cross the x-axis?

PS **8** The diagram shows a side elevation of a cone with a cut parallel to one side.

The cone is divided into horizontal sections.

A plan view of the cone is shown.

Construction lines have been drawn to link the elevation and the plan.

Two of the intersecting points have been drawn on the plan.

Two points have also been drawn where the construction lines from the side elevation intersect with the construction lines from the plan.

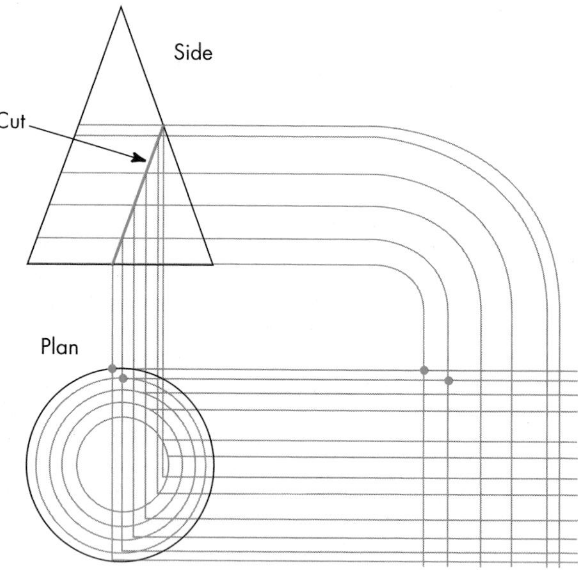

a Plot the rest of the points on the plan and join them with a smooth curve to see the plan view of the parabola.

b Plot the rest of the points on the intersecting lines and join them with a smooth curve to see the parabola.

PS 9 Copy the grid onto centimetre-squared paper.

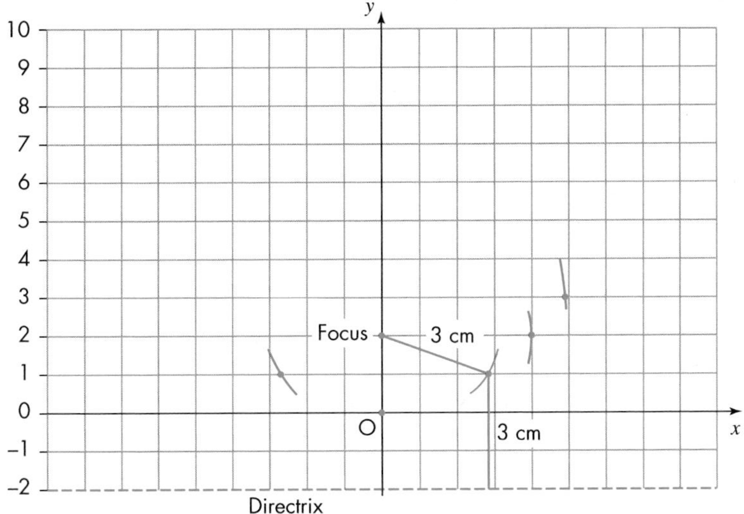

Mark a point at (0, 2). This is the focus.

Draw the line $y = -2$. This is the directrix.

A property of all parabolas is that all the points on a parabola are the same distance from the focus and the directrix.

The origin is 2 units away from both and this will be the lowest point of the parabola.

Set a pair of compasses to a radius of 3 cm. Using the focus as the centre, draw arcs on both sides to intersect with the line $y = 1$, which is 3 cm from the directrix.

Now set the compasses at 4 cm and draw arcs from the focus to intersect with $y = 2$.

Repeat with the compasses set to 5 cm, 6 cm, etc.

Once you have drawn all the points, join them with a smooth curve to show a parabola.

The parabola drawn has the equation $y = \frac{1}{8}x^2$.

AU 10 Here are the equations of three quadratic equations.

Parabola A: $y = 2x^2$

Parabola B: $y = -x^2$

Parabola C: $y = x^2 + 2$

Give a reason why each line may be the odd one out.

18.2 The significant points of a quadratic graph

This section will show you how to:

- recognise and calculate the significant points of a quadratic graph

Key words
intercept
maximum
minimum
roots
vertex

A quadratic graph has four points that are of interest to a mathematician. These are the points A, B, C and D on the diagram. The x-values at A and B are called the **roots**, and are where the graph crosses the x-axis. C is the point where the graph crosses the y-axis (the **intercept**) and D is the **vertex**, which is the lowest or highest point of the graph.

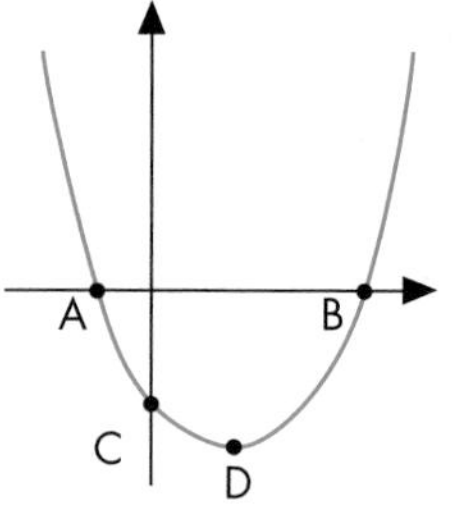

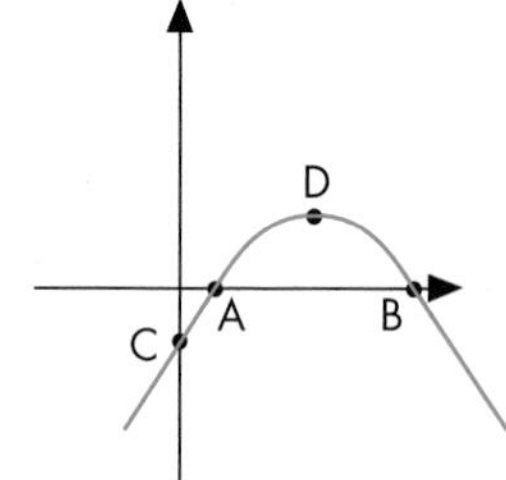

The roots

If you look at your answer to question **7** in Exercise 18A, you will see that the graph crosses the x-axis at $x = -0.5$ and $x = 3$. Since the x-axis is the line $y = 0$, the y-value at any point on the x-axis is zero. So, you have found the solution to the equation:

$0 = 2x^2 - 5x - 3$ that is $2x^2 - 5x - 3 = 0$

Equations of this type are known as *quadratic equations*.

You can solve quadratic equations by finding the values of x that make them true. Such values are called the roots of the equation. On the graph, these occur where the curve cuts the x-axis. So the roots of the quadratic equation $2x^2 - 5x - 3 = 0$ are -0.5 and 3.

Let's check these values.

For $x = 3.0$ $\quad 2(3)^2 - 5(3) - 3 = 18 - 15 - 3 = 0$

For $x = 0.5$ $\quad 2(-0.5)^2 - 5(-0.5) - 3 = 0.5 + 2.5 - 3 = 0$

You can find the roots of a quadratic equation by drawing its graph and finding where the graph crosses the x-axis.

EXAMPLE 3

a Draw the graph of $y = x^2 - 3x - 4$ for $-2 \leqslant x \leqslant 5$.

b Use your graph to find the roots of the equation $x^2 - 3x - 4 = 0$.

a Set up a table.

x	−2	−1	0	1	2	3	4	5
y^2	4	1	0	1	4	9	16	25
$-3x$	6	3	0	−3	−6	−9	−12	−15
-4	−4	−4	−4	−4	−4	−4	−4	−4
y	6	0	−4	−6	−6	−4	0	6

Draw the graph.

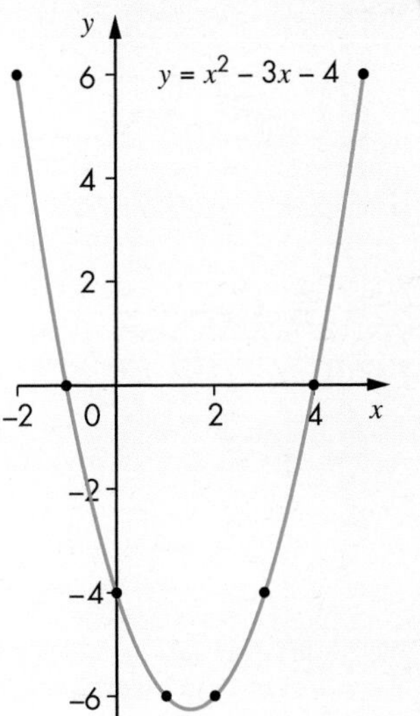

b The points where the graph crosses the x-axis are −1 and 4.

So, the roots of $x^2 - 3x - 4 = 0$ are $x = -1$ and $x = 4$.

Note that sometimes the quadratic graph may not cross the x-axis. In this case there are no roots. This was dealt with in Chapter 10 of Book 1.

The y-intercept

If you look at all the quadratic graphs you have drawn so far you will see a connection between the equation and the point where the graph crosses the y-axis. Very simply, the constant term of the equation $y = ax^2 + bx + c$ (that is, the value c) is where the graph crosses the y-axis. The intercept is at $(0, c)$.

The vertex

The lowest (or highest) point of a quadratic graph is called the *vertex*.

If it is the highest point, it is called the **maximum**.

If it is the lowest point, it is called the **minimum**.

It is difficult to find a general rule for this point, but the x-coordinate is always half-way between the roots. The easiest way to find the y-value is to substitute the x-value into the original equation.

Another way to find the vertex is to complete the square.

EXAMPLE 4

a Write the equation $x^2 - 3x - 4 = 0$ in the form $(x - p)^2 - q = 0$.

b What is the least value of the graph $y = x^2 - 3x - 4$?

a $(x - p)^2 - q = x^2 - 2px + p^2 - q = x^2 - 3x - 4$

So $p = 1\frac{1}{2}$ and $p^2 - q = -4$, $q = p^2 + 4 = (1\frac{1}{2})^2 + 4 = 6\frac{1}{4}$

b Looking at the graph drawn in Example 3 you can see that the minimum point is at $(1\frac{1}{2}, -6\frac{1}{4})$, so the least value is $-6\frac{1}{4}$.

You should be able to see the connection between the vertex point and the equation written in the 'completing the square' form.

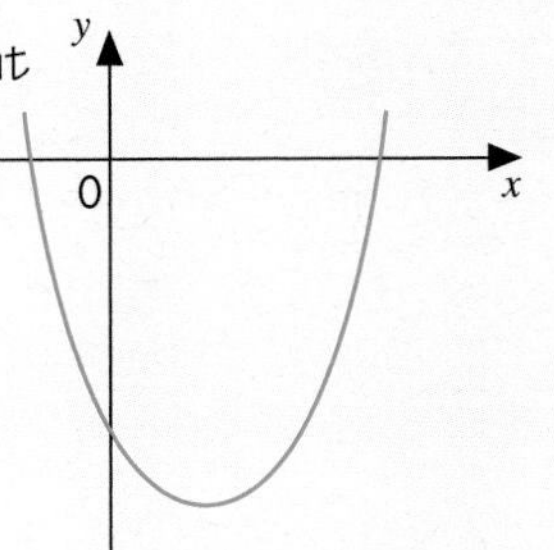

As a general rule when a quadratic is written in the form $(x - p)^2 + q$ then the minimum point is (p, q). Note the sign change of p.

Note: If the x^2 term is negative then the graph will be inverted and the vertex will be a maximum.

EXERCISE 18B

B

1 **a** Copy and complete the table to draw the graph of $y = x^2 - 4$ for $-4 \leqslant x \leqslant 4$.

x	–4	–3	–2	–1	0	1	2	3	4
y	12			–3				5	

b Use your graph to find the roots of $x^2 - 4 = 0$.

2 **a** Copy and complete the table and draw the graph of $y = x^2 - 9$ for $-4 \leqslant x \leqslant 4$.

x	–4	–3	–2	–1	0	1	2	3	4
y	7				–9			0	

b Use your graph to find the roots of $x^2 - 9 = 0$.

PS **3** **a** Look at the equations of the graphs you drew in questions **1** and **2**. Is there a connection between the numbers in each equation and its roots?

b Before you draw the graphs in parts **c** and **d**, try to predict what their roots will be.

c Copy and complete the table and draw the graph of $y = x^2 - 1$ for $-4 \leqslant x \leqslant 4$.

x	–4	–3	–2	–1	0	1	2	3	4
y	15				–1			8	

d Copy and complete the table and draw the graph of $y = x^2 - 5$ for $-4 \leqslant x \leqslant 4$.

x	–4	–3	–2	–1	0	1	2	3	4
y	11		–1					4	

e Were your predictions correct?

4 **a** Copy and complete the table and draw the graph of $y = x^2 + 4x$ for $-5 \leqslant x \leqslant 2$.

x	–5	–4	–3	–2	–1	0	1	2
x^2	25			4			1	
$+4x$	–20			–8			4	
y	5			–4			5	

b Use your graph to find the roots of the equation $x^2 + 4x = 0$.

5 **a** Copy and complete the table and draw the graph of $y = x^2 - 6x$ for $-2 \leqslant x \leqslant 8$.

x	–2	–1	0	1	2	3	4	5	6	7	8
x^2	4			1			16				
$-6x$	12			–6			–24				
y	16			–5			–8				

b Use your graph to find the roots of the equation $x^2 - 6x = 0$.

B

6 **a** Copy and complete the table and draw the graph of $y = x^2 + 3x$ for $-5 \leqslant x \leqslant 3$.

x	–5	–4	–3	–2	–1	0	1	2	3
y	10			–2				10	

b Use your graph to find the roots of the equation $x^2 + 3x = 0$.

PS **7** **a** Look at the equations of the graphs you drew in questions **4**, **5** and **6**. Is there a connection between the numbers in each equation and the roots?

b Before you draw the graphs in parts **c** and **d**, try to predict what their roots will be.

c Copy and complete the table and draw the graph of $y = x^2 - 3x$ for $-2 \leqslant x \leqslant 5$.

x	–2	–1	0	1	2	3	4	5
y	10			–2				10

d Copy and complete the table and draw the graph of $y = x^2 + 5x$ for $-6 \leqslant x \leqslant 2$.

x	–6	–5	–4	–3	–2	–1	0	1	2
y	6			–6				6	

e Were your predictions correct?

8 **a** Copy and complete the table and draw the graph of $y = x^2 - 4x + 4$ for $-1 \leqslant x \leqslant 5$.

x	–1	0	1	2	3	4	5
y	9				1		

b Use your graph to find the roots of the equation $x^2 - 4x + 4 = 0$.

c What happens with the roots?

9 **a** Copy and complete the table and draw the graph of $y = x^2 - 6x + 3$ for $-1 \leqslant x \leqslant 7$.

x	–1	0	1	2	3	4	5	6	7
y	10			–5			–2		

b Use your graph to find the roots of the equation $x^2 - 6x + 3 = 0$.

A

10 **a** Copy and complete the table and draw the graph of $y = 2x^2 + 5x - 6$ for $-5 \leqslant x \leqslant 2$.

x	–5	–4	–3	–2	–1	0	1	2
y								

b Use your graph to find the roots of the equation $2x^2 + 5x - 6 = 0$.

PS 11 Look back at questions **1** to **7**.

a Write down the point of intersection of the graph with the y-axis for each one.

b Write down the coordinates of the minimum point (vertex) of each graph for each one.

c Explain the connection between these points and the original equation.

12 **a** Write the equation $y = x^2 - 4x + 4$ in the form $y = (x - p)^2 + q$.

b Write down the minimum value of the equation $y = x^2 - 4x + 4$.

13 **a** Write the equation $y = x^2 - 6x + 3$ in the form $y = (x - p)^2 + q$.

b Write down the minimum value of the equation $y = x^2 - 6x + 3$.

14 **a** Write the equation $y = x^2 - 8x + 2$ in the form $y = (x - p)^2 + q$.

b Write down the minimum value of the equation $y = x^2 - 8x + 2$.

15 **a** Write the equation $y = -x^2 + 2x - 6$ in the form $y = -(x - p)^2 + q$.

b Write down the maximum value of the equation $y = -x^2 + 2x - 6$.

PS 16 Look at your answers to questions **12** to **13**.

a What is the connection between the maximum or minimum point and the values in the equation when written as $(x - a)^2 + b$?

b Without drawing the curve, predict the minimum point of the graph:

$y = x^2 + 10x - 3$.

AU 17 Masood draws a quadratic graph which has a minimum point at (3, –7).

He forgets to label it and later cannot remember what the quadratic function was.

He knows it is of the form $y = x^2 + px + q$.

Can you help him?

AU 18 **a** The graph $y = x^2 + 4x + 2$ has a minimum point at (–2, 2).

Write down the minimum point of the graph $y = x^2 + 4x - 3$.

b The graph $y = x^2 - 2ax + b$ has a minimum point at $(a, b - a^2)$.

Write down the minimum points of:

i $y = x^2 - 2ax + 2b$

ii $y = x^2 - 4ax + b$

GRADE BOOSTER

C You can draw quadratic graphs, using a table of values

B You can solve quadratic equations from their graphs

A You can find the maximum and minimum values of a quadratic graph.

What you should know now

- How to draw non-linear graphs
- How to find the roots of a quadratic equation from its graph
- How to find the significant points of a quadratic equation from its graph.

1 **a** Copy and complete the table of values for $y = x^2 - x - 5$. *(2 marks)*

x	−2	−1	0	1	2	3	4
y	1		−5	−5	−3	1	

b Draw, using x-axis range −2 to 4 and y-axis −7 to 8, the graph of $y = x^2 - x - 5$ for values of x from −2 to 4. *(2 marks)*

c An approximate solution of the equation $x^2 - x - 5 = 0$ is $x = 2.8$.

Explain how you can find this from the graph. *(1 mark)*

AQA, May 2008, Higher 1, Question 10

2 The graph of $y = x^2 - x - 3$ is shown on the grid.

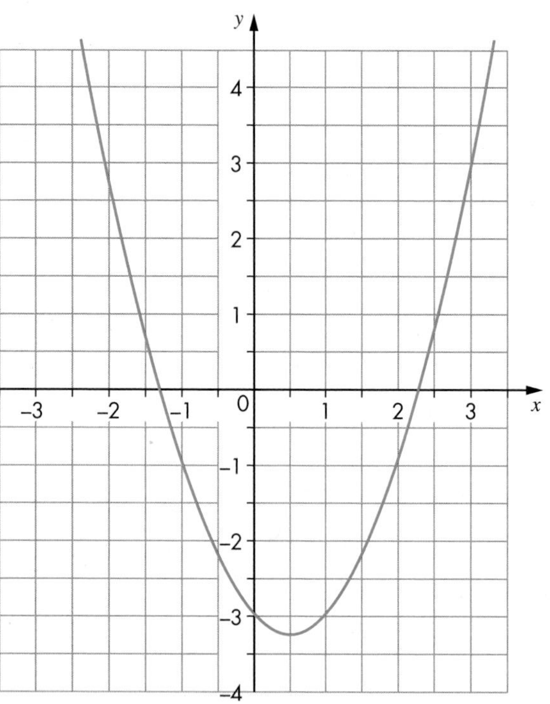

a Write down the value of y when $x = 1.5$ *(1 mark)*

b Write down the value of x when $y = -1$ *(2 marks)*

AQA, November 2011, Foundation 2, Question 28

3 **a** Complete the table of values for $y = x^2 - 4x + 2$. *(2 marks)*

x	−1	0	1	2	3	4	5
y		2	−1		−1	2	7

b On a grid with an x-axis from −1 to 5 and a y-axis from −3 to 8. Draw the graph of $y = x^2 - 4x + 2$ for values of x from −1 to 5. *(2 marks)*

c Use the graph to solve the equation $x^2 - 4x + 2 = 0$. *(1 mark)*

AQA, June 2010, Higher 1, Question 12

4 The graph shows the equation $y = x^2 + 3x - 2$

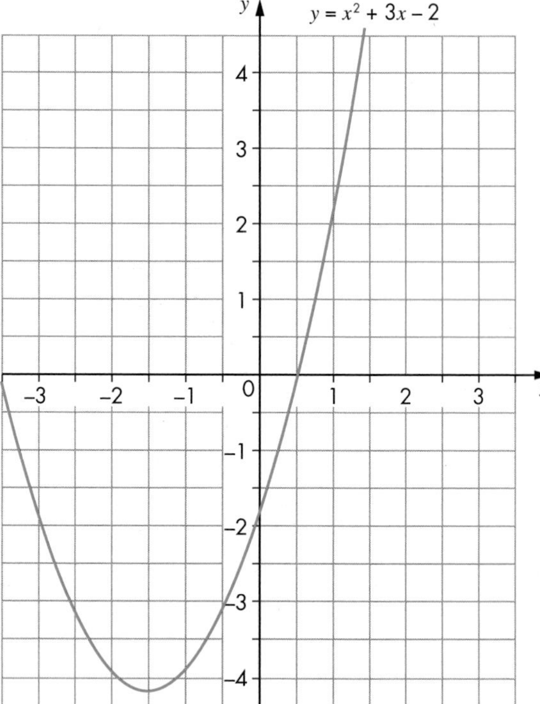

a Use the graph to find the positive solution of the equation $x^2 + 3x - 2 = 0$

b What is the minimum value of y?

c Deduce the minimum value of $y = x^2 + 3x - 4$

Worked Examination Questions

1 The grid below shows the graph of $y = x^2 - x - 6$.

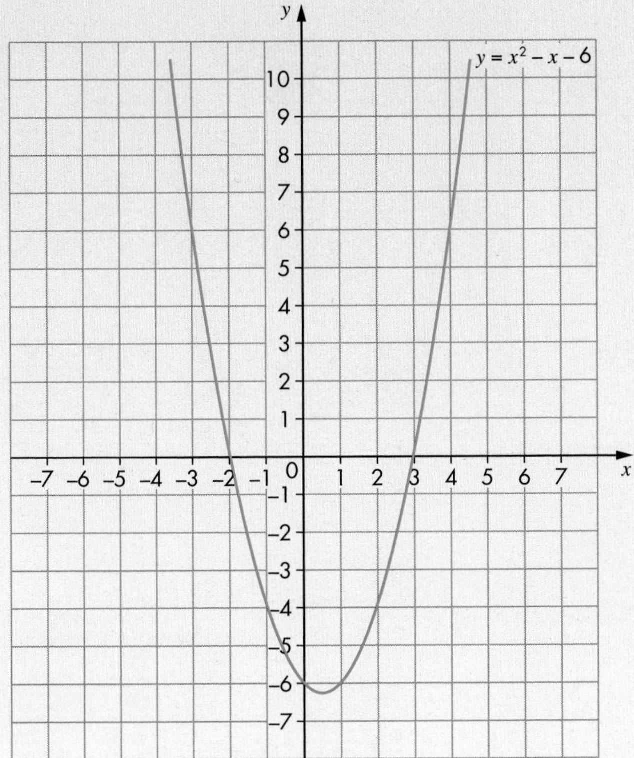

a Deduce the coordinates of the minimum point of the graph $y = x^2 - x - 12$.

b Use the graph to find the approximate solutions to the simultaneous equations $y = x^2 - x - 6$ and $y = x + 3$.

1 a (0.5, –12.25)

The minimum point of the given graph is (0.5, –6.25) and the required graph is 6 lower than the given graph, as the constant terms have a difference of 6.

b (3.2, 6.2) and (–1.2, 2.2)

Draw the line $y = x + 3$ and read off the points of intersection. This scores 1 mark each.

Total: 3 marks

18 Problem Solving
Quadratics in bridges

The forms of many suspension bridges are based on quadratic functions. Their shape is a classic quadratic curve.

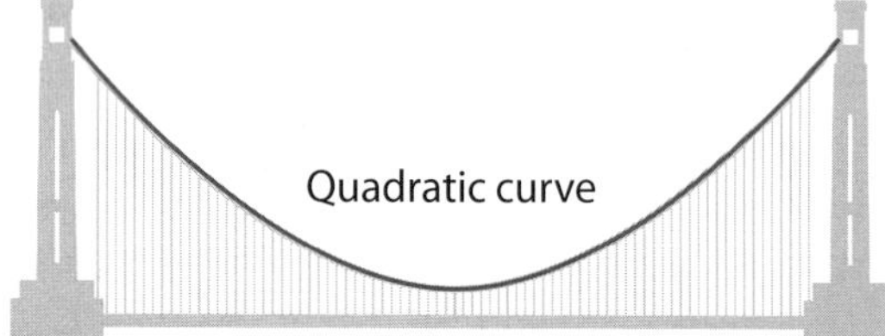

In this task you will investigate the quadratic functions that can be used to describe stable suspension bridges.

Getting started

Use these questions to familiarise yourself with how quadratic expressions can be used to represent bridges.

- What would you need to know, to be able to describe the shape of a bridge, in terms of a quadratic formula?
- Think about bridges and other landmarks in your local area. How many of these are based on quadratic equations and form parabolas. Use images to illustrate your findings.

Your task

Below you can see the dimensions of the Clifton Suspension Bridge. Use these dimensions to construct a diagram of the bridge.

Then, using your diagram, estimate the quadratic equation for the curve of the bridge. Represent this equation appropriately.

Write a report explaining the mathematical process that you used to solve this problem. State any assumptions that you made and explain whether you could have found different answers if you had changed your assumptions.

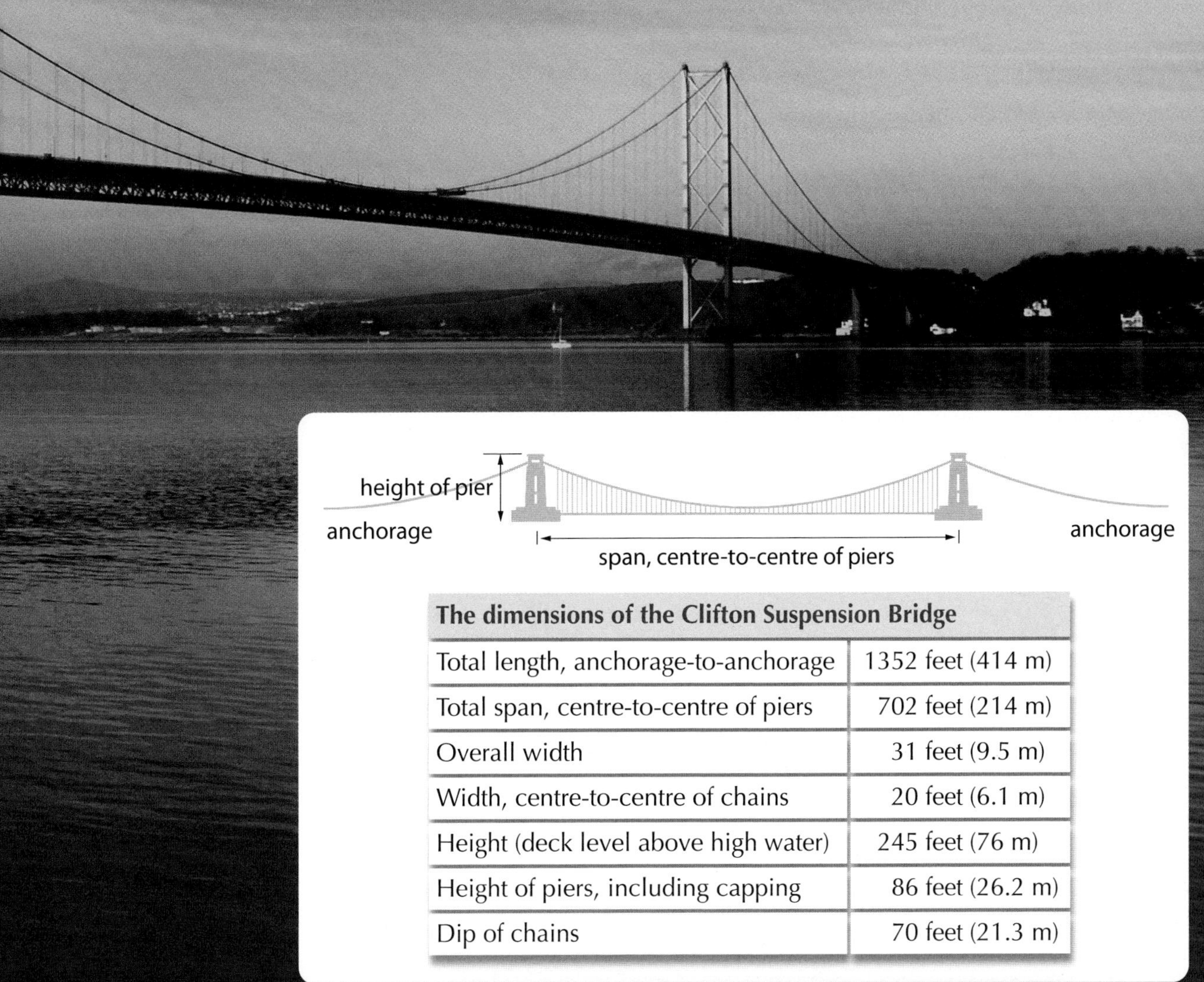

The dimensions of the Clifton Suspension Bridge	
Total length, anchorage-to-anchorage	1352 feet (414 m)
Total span, centre-to-centre of piers	702 feet (214 m)
Overall width	31 feet (9.5 m)
Width, centre-to-centre of chains	20 feet (6.1 m)
Height (deck level above high water)	245 feet (76 m)
Height of piers, including capping	86 feet (26.2 m)
Dip of chains	70 feet (21.3 m)

Why this chapter matters

The theory of linear programming has been used by many companies to reduce their costs and increase productivity.

The theory of linear programming, which uses inequalities in two dimensions, was developed at the start of the Second World War in 1939.

It was used to work out ways to get armaments as efficiently as possible and to increase the effectiveness of resources. It was so powerful an analytical tool that the Allies did not want the Germans to know about it, so it was not made public until 1947.

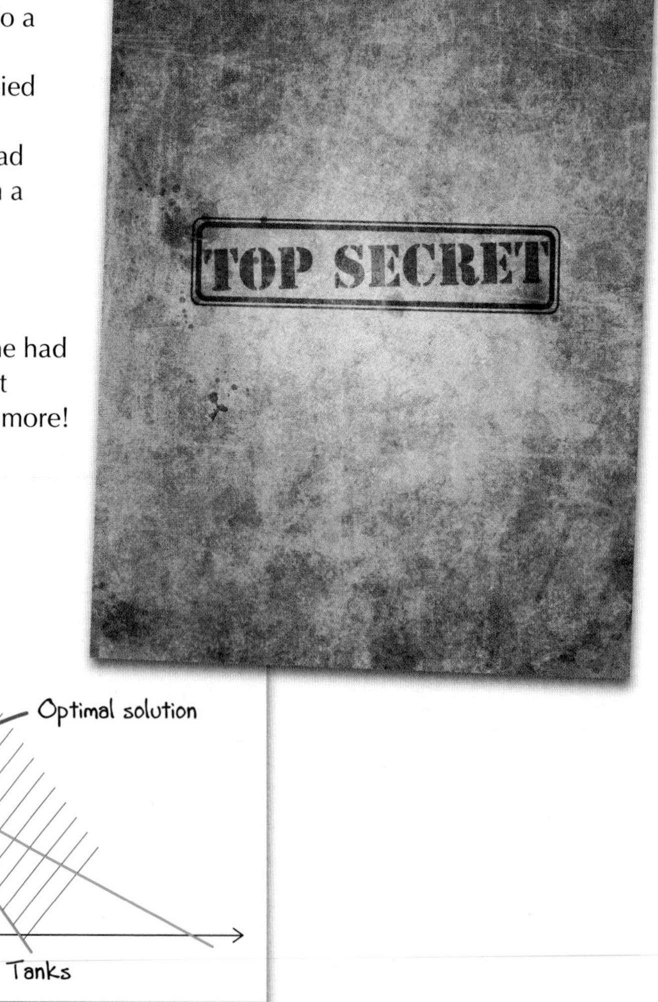

George Dantzig, who was one of the inventors of linear programming, came late to a lecture at University one day and saw two problems written on the blackboard. He copied them, thinking they were the homework assignment. He solved both problems, but had to apologise to the lecturer as he found them a little harder than the usual homework, so he took a few days to solve them and was late handing them in.

The lecturer was astonished. The problems he had written on the board were not homework but examples of 'impossible problems'. Not any more!

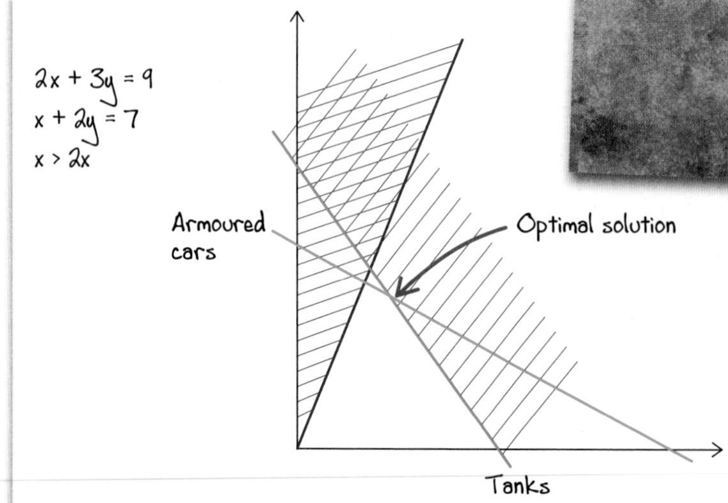

Chapter 19

Algebra: Inequalities and regions

This chapter will show you ...

C how to solve a linear inequality

B how to find a region on a graph that obeys a linear inequality in two variables

Visual overview

Linear inequalities → Inequalities in two variables

What you should already know

- How to solve linear equations **(KS3 level 6, GCSE grade D)**
- How to draw linear graphs **(KS3 level 6, GCSE grade D)**

Quick check

1 Solve these equations.

a $\frac{2x + 5}{3} = 7$ **b** $2x - 7 = 13$

2 On a grid with x- and y-axes from 0 to 10, draw the graphs of these equations.

a $y = 3x + 1$ **b** $2x + 3y = 12$

19.1 Solving inequalities

This section will show you how to:

- solve a simple linear inequality

Key words
inclusive inequality
inequality
number line
strict inequality

Inequalities behave similarly to equations, which you have already met. In the case of linear inequalities, you use the same rules to solve them as you use for linear equations. There are four inequality signs, $<$ which means 'less than', $>$ which means 'greater than', $\leqslant$ which means 'less than or equal to' and $\geqslant$ which means 'greater than or equal to'.

Be careful. Never replace the inequality sign with an equals sign or you could end up getting no marks in an examination.

EXAMPLE 1

Solve $2x + 3 < 14$

Rewrite this as: $2x < 14 - 3$

$2x < 11$

Divide both sides by 2: $\frac{2x}{2} < \frac{11}{2}$

$\Rightarrow x < 5.5$

This means that x can take any value below 5.5 but *not* the value 5.5.

$<$ and $>$ are called **strict inequalities.**

Note: The inequality sign given in the problem is the sign to use in the answer.

EXAMPLE 2

Solve $\frac{x}{2} + 4 \geqslant 13$

Solve just like an equation but leave the inequality sign in place of the equals sign.

Subtract 4 from both sides: $\frac{x}{2} \geqslant 9$

Multiply both sides by 2: $x \geqslant 18$

This means that x can take any value above and including 18.

$\leqslant$ and $\geqslant$ are called **inclusive inequalities.**

EXAMPLE 3

Solve $\frac{3x + 7}{2} < 14$

Rewrite this as: $3x + 7 < 14 \times 2$

That is: $3x + 7 < 28$

$\Rightarrow \quad 3x < 28 - 7$

$\Rightarrow \quad 3x < 21$

$\Rightarrow \quad x < 21 \div 3$

$\Rightarrow \quad x < 7$

EXAMPLE 4

Solve $1 < 3x + 4 \leqslant 13$

Divide the inequality into two parts, and treat each part separately.

$1 < 3x + 4$	$3x + 4 \leqslant 13$
$\Rightarrow \quad 1 - 4 < 3x$	$\Rightarrow \quad 3x \leqslant 13 - 4$
$\Rightarrow \quad -3 < 3x$	$\Rightarrow \quad 3x \leqslant 9$
$\Rightarrow \quad -\frac{3}{3} < x$	$\Rightarrow \quad x \leqslant \frac{9}{3}$
$\Rightarrow \quad -1 < x$	$\Rightarrow \quad x \leqslant 3$

Hence, $-1 < x \leqslant 3$

EXERCISE 19A

1 Solve the following linear inequalities.

a $x + 4 < 7$ **b** $t - 3 > 5$ **c** $p + 2 \geqslant 12$

d $2x - 3 < 7$ **e** $4y + 5 \leqslant 17$ **f** $3t - 4 > 11$

g $\frac{x}{2} + 4 < 7$ **h** $\frac{y}{5} + 3 \leqslant 6$ **i** $\frac{t}{3} - 2 \geqslant 4$

j $3(x - 2) < 15$ **k** $5(2x + 1) \leqslant 35$ **l** $2(4t - 3) \geqslant 34$

2 Write down the largest integer value of x that satisfies each of the following.

a $x - 3 \leqslant 5$, where x is positive

b $x + 2 < 9$, where x is positive and even

c $3x - 11 < 40$, where x is a square number

d $5x - 8 \leqslant 15$, where x is positive and odd

e $2x + 1 < 19$, where x is positive and prime

C

C

3 Write down the smallest integer value of x that satisfies each of the following.

a $x - 2 \geqslant 9$, where x is positive
b $x - 2 > 13$, where x is positive and even
c $2x - 11 \geqslant 19$, where x is a square number

FM 4 Ahmed went to town with £20 to buy two CDs. His bus fare was £3. The CDs were both the same price. When he reached home he still had some money in his pocket. What was the most each CD could cost?

HINTS AND TIPS

Set up an inequality and solve it.

AU 5 **a** Explain why you cannot make a triangle with three sticks of length 3 cm, 4 cm and 8 cm.

b Three sides of a triangle are x, $x + 2$ and 10 cm.

x is a whole number.

What is the smallest value x can take?

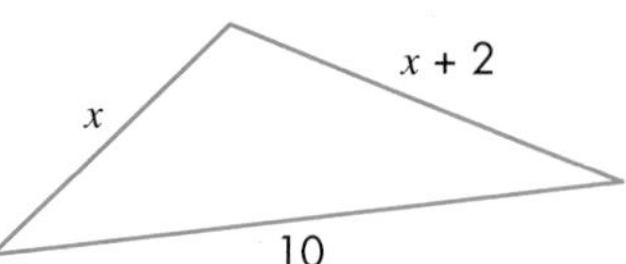

B

PS 6 Five cards have inequalities and equations marked on them.

$x > 0$	$x < 3$	$x \geqslant 4$	$x = 2$	$x = 6$

The cards are shuffled and then turned over, one at a time.
If two consecutive cards have any numbers in common, then a point is scored.
If they do not have any numbers in common, then a point is deducted.

a The first two cards below score –1 because $x = 6$ and $x < 3$ have no numbers in common. Explain why the total for this combination scores 0.

$x = 6$	$x < 3$	$x > 0$	$x = 2$	$x \geqslant 4$

b What does this combination score?

$x > 0$	$x = 6$	$x \geqslant 4$	$x = 2$	$x < 3$

c Arrange the cards to give a maximum score of 4.

7 Solve the following linear inequalities.

a $4x + 1 \geqslant 3x - 5$
b $5t - 3 \leqslant 2t + 5$
c $3y - 12 \leqslant y - 4$
d $2x + 3 \geqslant x + 1$
e $5w - 7 \leqslant 3w + 4$
f $2(4x - 1) \leqslant 3(x + 4)$

8 Solve the following linear inequalities.

a $\dfrac{x + 4}{2} \leqslant 3$
b $\dfrac{x - 3}{5} > 7$
c $\dfrac{2x + 5}{3} < 6$
d $\dfrac{4x - 3}{5} \geqslant 5$
e $\dfrac{2t - 2}{7} > 4$
f $\dfrac{5y + 3}{5} \leqslant 2$

9 Solve the following linear inequalities.

a $7 < 2x + 1 < 13$
b $5 < 3x - 1 < 14$
c $-1 < 5x + 4 \leqslant 19$
d $1 \leqslant 4x - 3 < 13$
e $11 \leqslant 3x + 5 < 17$
f $-3 \leqslant 2x - 3 \leqslant 7$

The number line

The solution to a linear inequality can be shown on the **number line** by using the following conventions.

A strict inequality does not include the boundary point but an inclusive inequality does include the boundary point.

Below are five examples.

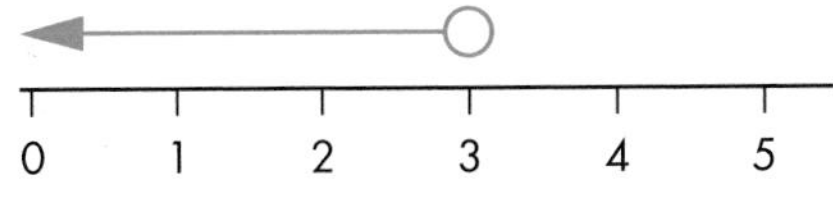

represents $x < 3$

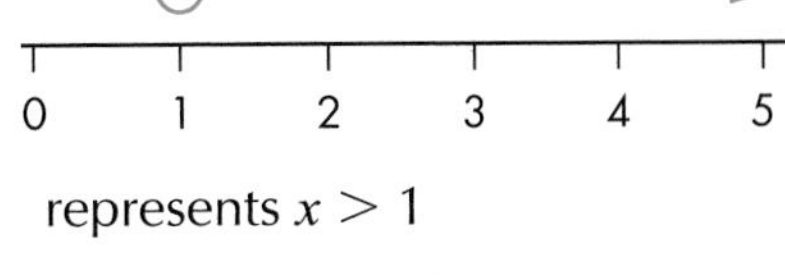

represents $x > 1$

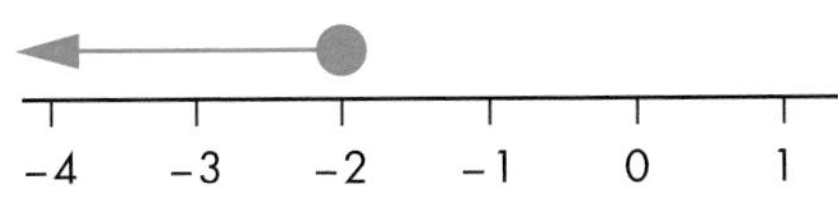

represents $x \leqslant -2$

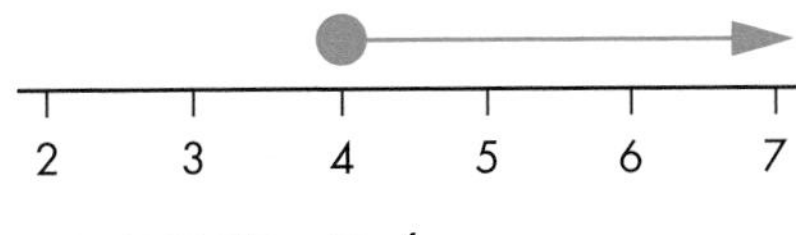

represents $x \geqslant 4$

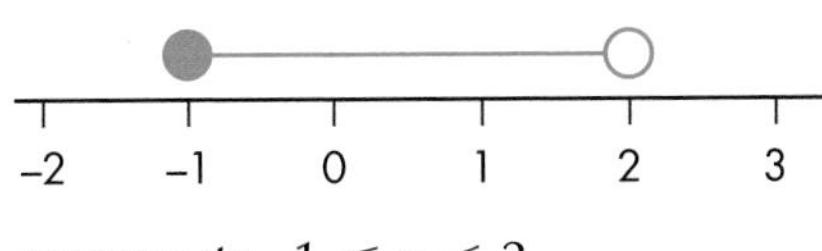

represents $-1 \leqslant x < 2$

This is a 'between' inequality. It can be written as $x \geqslant -1$ and $x < 2$, but the notation $-1 \leqslant x < 2$ is much neater.

EXAMPLE 5

a Write down the inequality shown by this diagram.

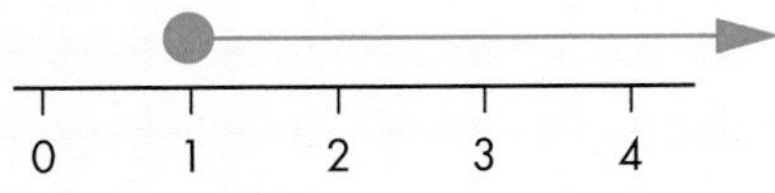

b **i** Solve the inequality $2x + 3 < 11$.

ii Mark the solution on a number line.

c Write down the integers that satisfy both the inequalities in **a** and **b**.

a The inequality shown is $x \geqslant 1$.

b **i** $2x + 3 < 11$

$\Rightarrow 2x < 8$

$\Rightarrow x < 4$

ii

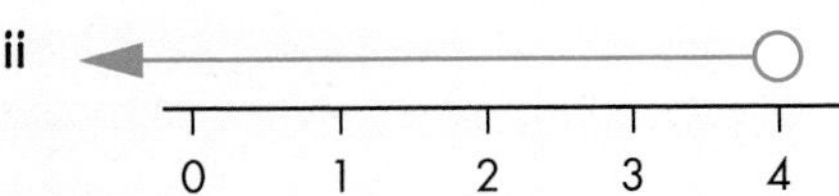

c The integers that satisfy both inequalities are 1, 2 and 3.

EXERCISE 19B

C

1 Write down the inequality that is represented by each diagram below.

a

0 1 2 3 4

b

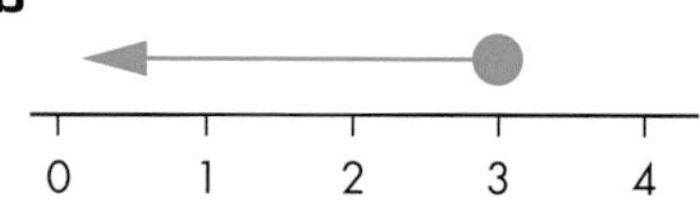

c

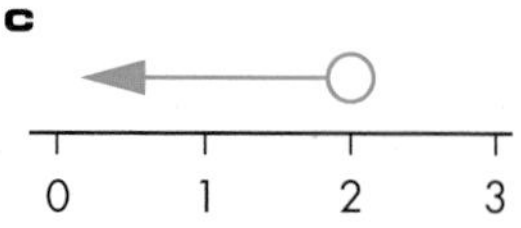

d

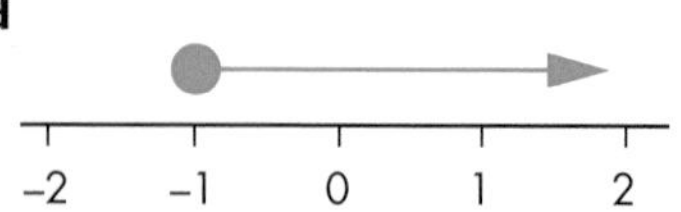

e

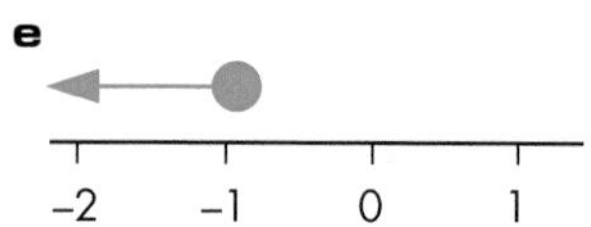

f

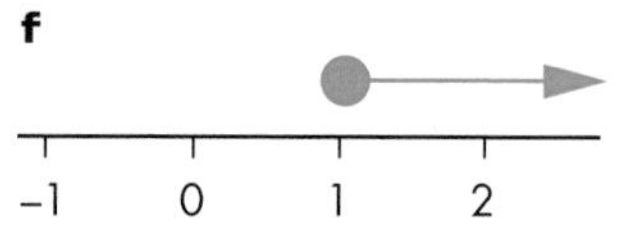

2 Draw diagrams to illustrate these inequalities.

a $x \leqslant 3$ **b** $x > -2$ **c** $x \geqslant 0$ **d** $x < 5$

e $x \geqslant -1$ **f** $2 < x \leqslant 5$ **g** $-1 \leqslant x \leqslant 3$ **h** $-3 < x < 4$

3 Solve the following inequalities and illustrate their solutions on number lines.

a $x + 4 \geqslant 8$ **b** $x + 5 < 3$ **c** $4x - 2 \geqslant 12$ **d** $2x + 5 < 3$

e $2(4x + 3) < 18$ **f** $\dfrac{x}{2} + 3 \leqslant 2$ **g** $\dfrac{x}{5} - 2 > 8$ **h** $\dfrac{x}{3} + 5 \geqslant 3$

FM 4 Max went to the supermarket with £1.20. He bought three apples costing x pence each and a chocolate bar costing 54p. When he got to the till, he found he didn't have enough money.

Max took one of the apples back and paid for two apples and the chocolate bar. He counted his change and found he had enough money to buy a 16p chew.

a Explain why $3x + 54 > 120$ and solve the inequality.

b Explain why $2x + 54 \leqslant 104$ and solve the inequality.

c Show the solution to both of these inequalities on a number line.

d What is the possible price of an apple?

AU 5 On copies of the number lines below, draw two inequalities so that only the integers {–1, 0, 1, 2} are common to both inequalities.

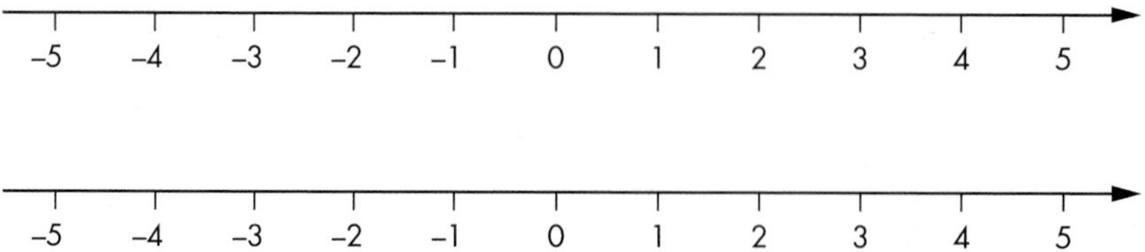

C

PS **6** What numbers are being described?

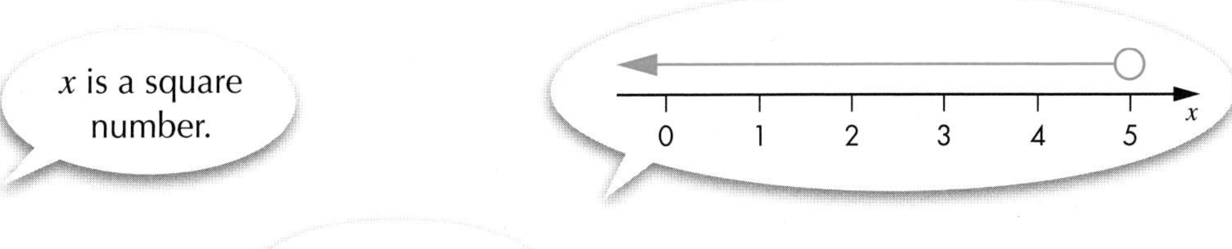

$2x + 3 > 5$

B

7 Solve the following inequalities and illustrate their solutions on number lines.

a $\frac{2x+5}{3} > 3$ **b** $\frac{3x+4}{2} \geqslant 11$ **c** $\frac{2x+8}{3} \leqslant 2$ **d** $\frac{2x-1}{3} \geqslant -3$

19.2 Graphical inequalities

This section will show you how to:
- show a graphical inequality
- how to find regions that satisfy more than one graphical inequality

Key words
boundary
included
origin
region

A linear inequality can be plotted on a graph. The result is a **region** that lies on one side or the other of a straight line. You will recognise an inequality by the fact that it looks like an equation but instead of the equals sign it has an inequality sign: $<$, $>$, $\leqslant$, or $\geqslant$.

The following are examples of linear inequalities that can be represented on a graph.

$y < 3$ $x > 7$ $-3 \leqslant y < 5$ $y \geqslant 2x + 3$ $2x + 3y < 6$ $y \leqslant x$

The method for graphing an inequality is to draw the **boundary** line that defines the inequality. This is found by replacing the inequality sign with an equals sign. When a strict inequality is stated ($<$ or $>$), the boundary line should be drawn as a *dashed* line to show that it is not included in the range of values. When $\leqslant$ or $\geqslant$ is used to state the inequality, the boundary line should be drawn as a *solid* line to show that the boundary is **included**.

After the boundary line has been drawn, shade the *required region*.

To confirm on which side of the line the region lies, choose any point that is not on the boundary line and test it in the inequality. If it satisfies the inequality, that is the side required. If it doesn't, the other side is required.

Work through the six inequalities in the following example to see how the procedure is applied.

EXAMPLE 6

Show each of the following inequalities on a graph.

a $y \leqslant 3$ **b** $x > 7$ **c** $-3 \leqslant y < 5$

d $y \leqslant 2x + 3$ **e** $2x + 3y < 6$ **f** $y \leqslant x$

a Draw the line $y = 3$. Since the inequality is stated as $\leqslant$, the line is *solid*. Test a point that is not on the line. The **origin** is always a good choice if possible, as 0 is easy to test.

Putting 0 into the inequality gives $0 \leqslant 3$. The inequality is satisfied and so the region containing the origin is the side we want.

Shade it in.

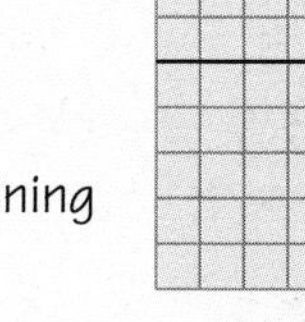

≤ 3

b Since the inequality is stated as >, the line is *dashed*. Draw the line $x = 7$.

Test the origin (0, 0), which gives $0 > 7$. This is not true, so you want the other side of the line from the origin.

Shade it in.

> 7

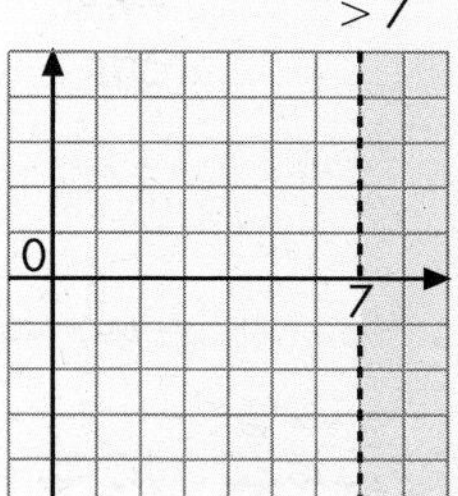

c Draw the lines $y = -3$ (solid for $\leqslant$) and $y = 5$ (dashed for <).

Test a point that is not on either line, say (0, 0). Zero is between −3 and 5, so the required region lies between the lines.

Shade it in.

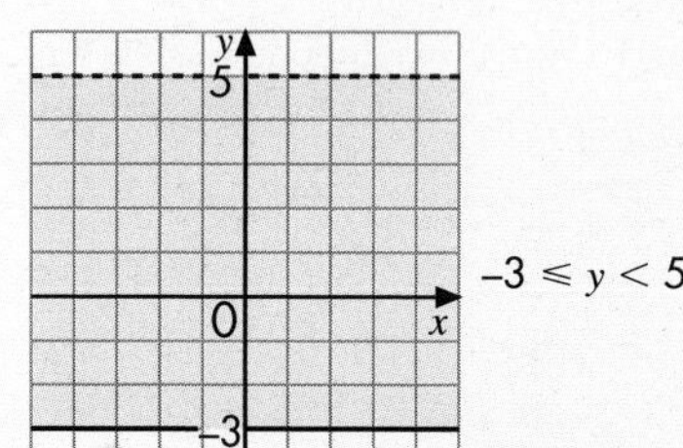

d Draw the line $y = 2x + 3$. Since the inequality is stated as $\leqslant$, the line is solid.

Test a point that is not on the line, (0, 0). Putting these x- and y-values in the inequality gives $0 \leqslant 2(0) + 3$, which is true. So the region that includes the origin is what you want.

Shade it in.

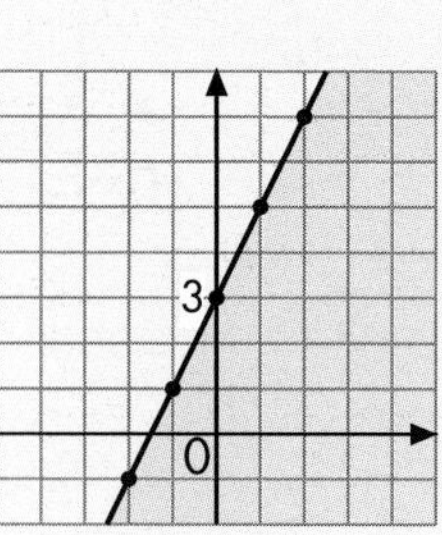

≤ 2 + 3

e Draw the line $2x + 3y = 6$. Since the inequality is stated as <, the line is dashed.

Test a point that is not on the line, say (0, 0). Is it true that $2(0) + 3(0) < 6$? The answer is yes, so the origin is in the region that you want.

Shade it in.

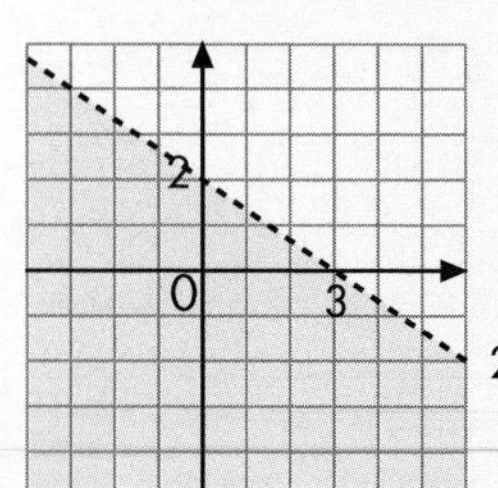

2 + 3 < 6

f Draw the line $y = x$. Since the inequality is stated as $\leqslant$, the line is solid.

This time the origin is on the line, so pick any other point, say (1, 3). Putting $x = 1$ and $y = 3$ in the inequality gives $3 \leqslant 1$. This is not true, so the point (1, 3) is not in the region you want.

Shade in the other side to (1, 3)

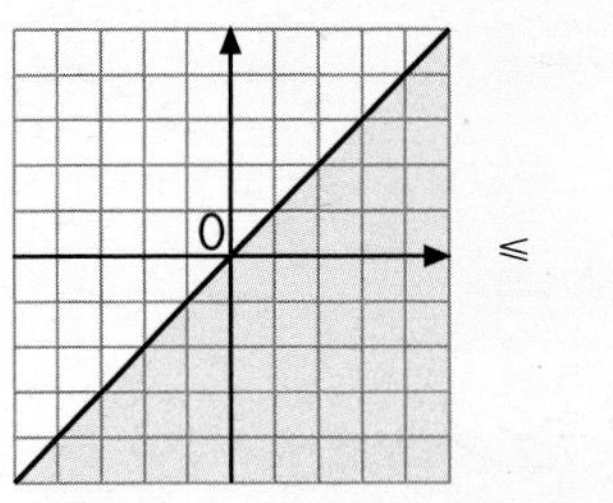

More than one inequality

When you have to show a region that satisfies more than one inequality, it is clearer to *shade* the regions *not required,* so that the *required region* is left *blank.*

EXAMPLE 7

a On the same grid, show the regions that represent the following inequalities by shading the unwanted regions.

i $x > 2$ **ii** $y \geqslant x$ **iii** $x + y < 8$

b Are these points

i (3, 4) **ii** (2, 6) **iii** (3, 3)

in the region that satisfies all three inequalities?

a **i**

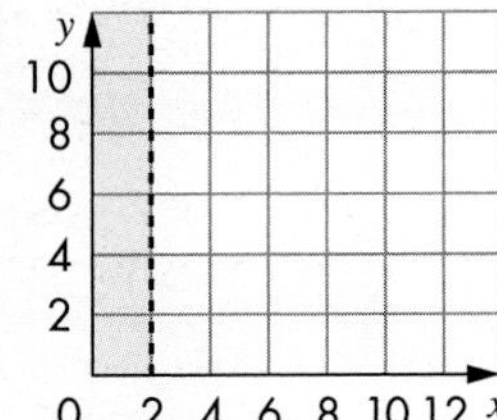

ii

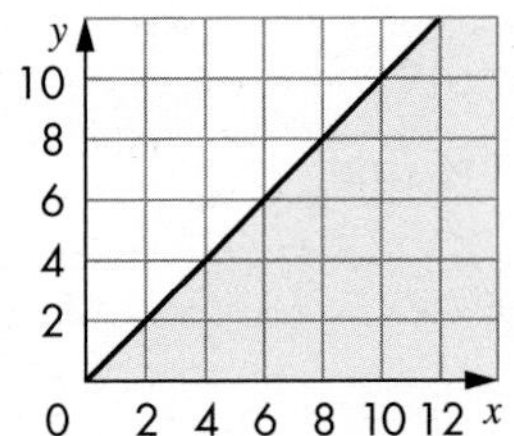

iii

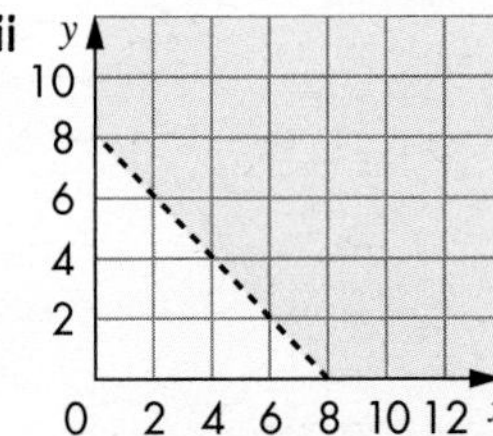

i The region $x > 2$ is shown unshaded in diagram **i**. The boundary line is $x = 2$ (dashed).

ii The region $y \geqslant x$ is shown unshaded in diagram **ii**. The boundary line is $y = x$ (solid).

iii The region $x + y < 8$ is shown unshaded in diagram **iii**.
The boundary line is $x + y = 8$ (dashed). The regions have first been drawn separately so that each may be clearly seen. The diagram on the right shows all three regions on the same grid. The white triangular area defines the region that satisfies all three inequalities.

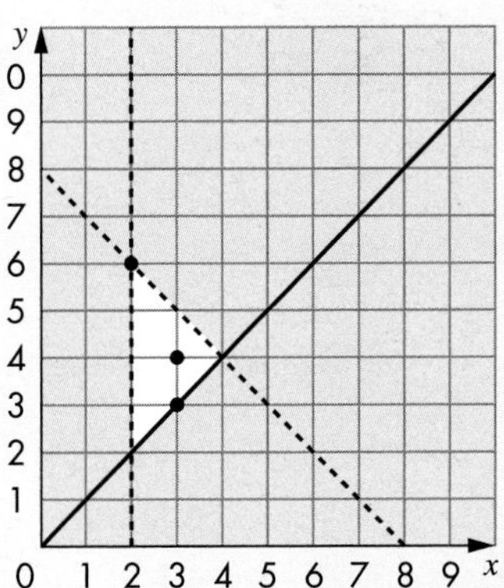

b **i** The point (3, 4) is clearly within the region that satisfies all three inequalities.

ii The point (2, 6) is on the boundary lines $x = 2$ and $x + y = 8$. As these are dashed lines, they are not included in the region defined by all three inequalities. So, the point (2, 6) is not in this region.

iii The point (3, 3) is on the boundary line $y = x$. As this is a solid line, it is included in the region defined by all three inequalities. So, the point (3, 3) is included in this region.

EXERCISE 19C

C

1 **a** Draw the line $x = 2$ (as a solid line). **b** Shade the region defined by $x \leqslant 2$.

2 **a** Draw the line $y = -3$ (as a dashed line). **b** Shade the region defined by $y > -3$.

3 **a** Draw the line $x = -2$ (as a solid line).

b Draw the line $x = 1$ (as a solid line) on the same grid.

c Shade the region defined by $-2 \leqslant x \leqslant 1$.

4 **a** Draw the line $y = -1$ (as a dashed line).

b Draw the line $y = 4$ (as a solid line) on the same grid.

c Shade the region defined by $-1 < y \leqslant 4$.

B

5 **a** On the same grid, draw the regions defined by these inequalities.

i $-3 \leqslant x \leqslant 6$ **ii** $-4 < y \leqslant 5$

b Are the following points in the region defined by both inequalities?

i (2, 2) **ii** (1, 5) **iii** (–2, –4)

6 **a** Draw the line $y = 2x - 1$ (as a dashed line).

b Shade the region defined by $y < 2x - 1$.

7 **a** Draw the line $3x - 4y = 12$ (as a solid line).

b Shade the region defined by $3x - 4y \leqslant 12$.

8 **a** Draw the line $y = \frac{1}{2}x + 3$ (as a solid line).

b Shade the region defined by $y \geqslant \frac{1}{2}x + 3$.

> **HINTS AND TIPS**
>
> In exams it is always made clear which region is to be labelled or shaded. Make sure you do as the question asks, and label or shade as required, otherwise you could lose a mark.

9 Shade the region defined by $y < -3$.

10 **a** Draw the line $y = 3x - 4$ (as a solid line).

b Draw the line $x + y = 10$ (as a solid line) on the same diagram.

c Shade the diagram so that the region defined by $y \geqslant 3x - 4$ is left *unshaded.*

d Shade the diagram so that the region defined by $x + y \leqslant 10$ is left *unshaded.*

e Are the following points in the region defined by both inequalities?

i (2, 1) **ii** (2, 2) **iii** (2, 3)

11 **a** Draw the line $y = x$ (as a solid line).

b Draw the line $2x + 5y = 10$ (as a solid line) on the same diagram.

c Draw the line $2x + y = 6$ (as a dashed line) on the same diagram.

d Shade the diagram so that the region defined by $y \geqslant x$ is left *unshaded.*

e Shade the diagram so that the region defined by $2x + 5y \geqslant 10$ is left *unshaded.*

f Shade the diagram so that the region defined by $2x + y < 6$ is left *unshaded.*

g Are the following points in the region defined by these inequalities?

i (1, 1) **ii** (2, 2) **iii** (1, 3)

12 **a** On the same grid, draw the regions defined by the following inequalities. (Shade the diagram so that the overlapping region is left blank.)

i $y > x - 3$ **ii** $3y + 4x \leqslant 24$ **iii** $x \geqslant 2$

b Are the following points in the region defined by all three inequalities?

i (1, 1) **ii** (2, 2) **iii** (3, 3) **iv** (4, 4)

AU 13 The graph shows three points (1, 2), (1, 3) and (2, 3).

Write down three inequalities that between them surround these three grid intersection points and *no others.*

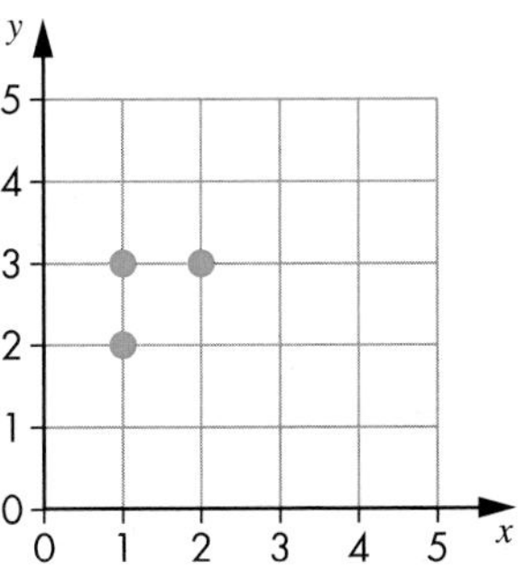

PS 14 If $x + y > 40$, which of the following may be true (M), must be false (F) or must be true (T)?

a $x > 40$ **b** $x + y \leqslant 20$ **c** $x - y = 10$

d $x \leqslant 5$ **e** $x + y = 40$ **f** $y > 40 - x$

g $y = 2x$ **h** $x + y \geqslant 39$

AU 15 Explain how you would find which side of the line represents the inequality $y < x + 2$.

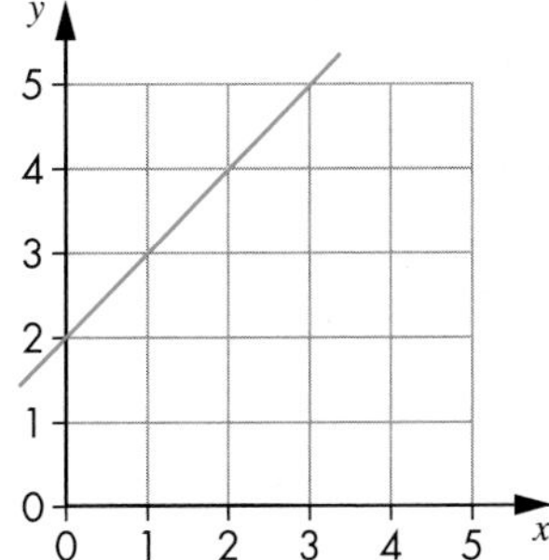

AU 16 The region marked R is the overlap of the inequalities:
the inequality $y < x + 2$.

$x + y \geqslant 3$ $y \leqslant \frac{1}{2}x + 3$ $y \geqslant 5x - 15$

a For which point in the region R is the value of the function $2x - y$ the greatest ?

b For which point in the region R is the value of the function $x - 3y$ the least?

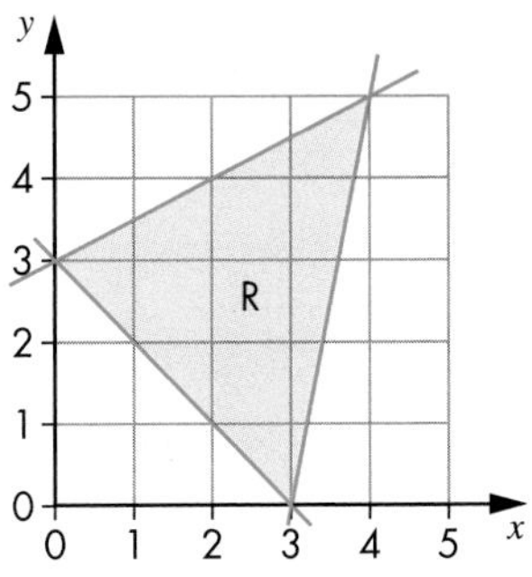

GRADE BOOSTER

C You can solve inequalities such as $3x + 2 < 5$ and represent the solution on a number line

B You can represent a region that satisfies a linear inequality graphically, and solve more complex linear inequalities

B You can represent a region that simultaneously satisfies more than one linear inequality graphically

What you should know now

- How to solve simple inequalities
- How to create algebraic inequalities from verbal statements
- How to represent linear inequalities on a graph
- How to depict a region satisfying more than one linear inequality

1 Solve the inequality:

$5x + 3 \leqslant 10$ *(2 marks)*

AQA, June 2006, Paper 2 Higher, Question 1

2 **a** Solve the inequality:

$3(x - 2) \leqslant 9$ *(3 marks)*

b The inequality $x \leqslant 3$ is shown on the number line below.

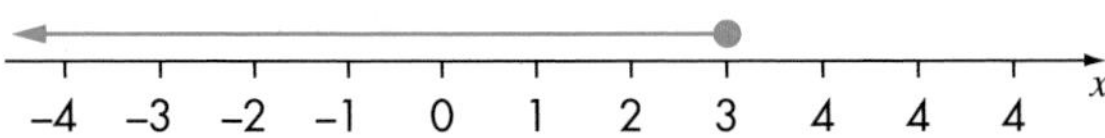

Draw another inequality on the number line so that only the following integers satisfy

$\{-2, -1, 0, 1, 2, 3\}$ *(1 mark)*

AQA, November 2006, Paper 2 Higher, Question 4

3 **a** Solve the inequality:

$2x + 3 \geqslant 4x + 5$ *(2 marks)*

b Which inequality is represented by the shaded region? *(1 mark)*

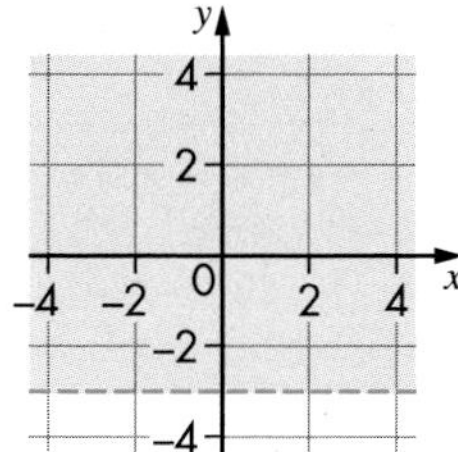

c The line $2x - 3y = 12$ is drawn on the grid:

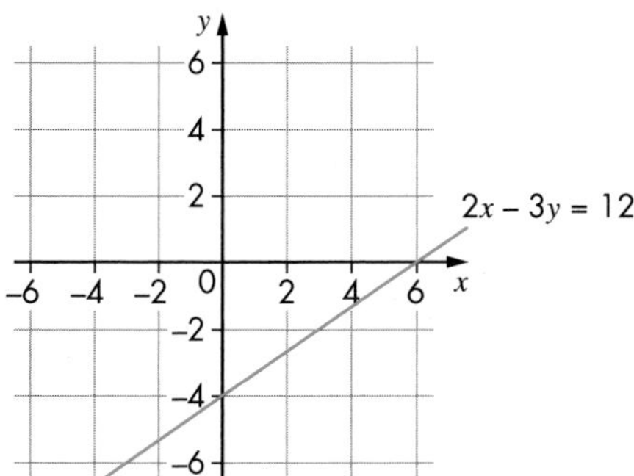

Shade the side of the line that represents

$2x - 3y \leqslant 12.$

Explain how you know which side to shade.

(1 mark)

AQA, June 2008, Paper 2 Higher, Question 15

4 **a** Which of these inequalities is shown shaded on the grid below?

$y > 2 \quad y \geqslant 2 \quad x > 2 \quad x \geqslant 2$

(1 mark)

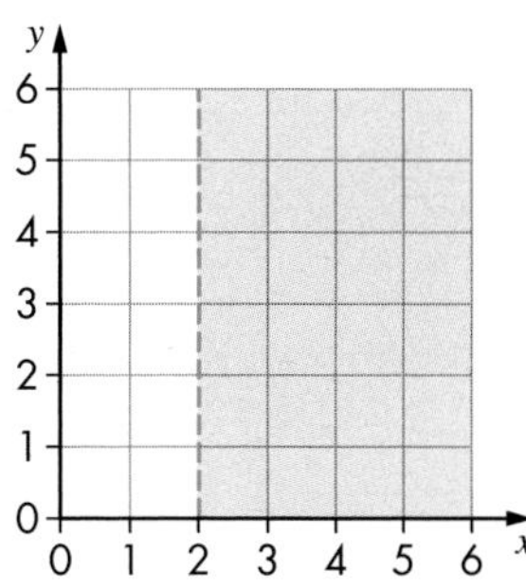

b On a grid like the one above, draw lines to find the region satisfied by the three inequalities:

$y > 2$

$y < x + 1$

$x + y < 5$

Label the region with the letter R. *(3 marks)*

AQA, November 2007, Paper 2 Higher, Question 12

5 Match each of the shaded regions to one of these inequalities.

A $y \leqslant -x + 2$ **D** $y \geqslant 2x - 4$

B $y \leqslant x + 2$ **E** $y \leqslant 2x - 4$

C $y \geqslant -2x + 4$ *(4 marks)*

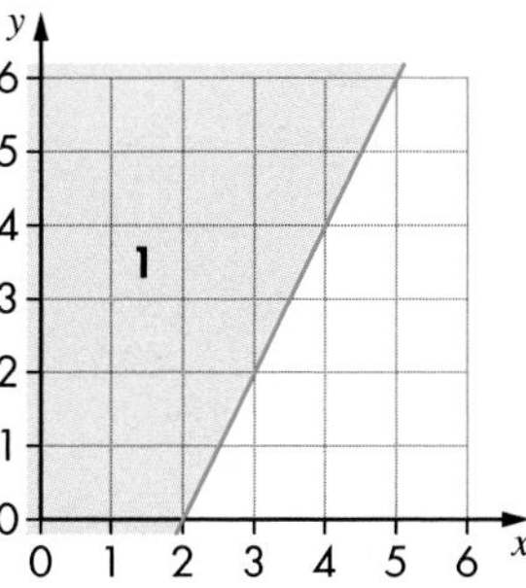

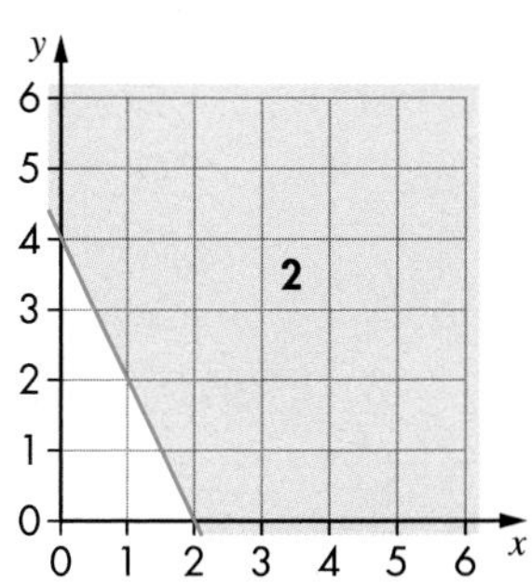

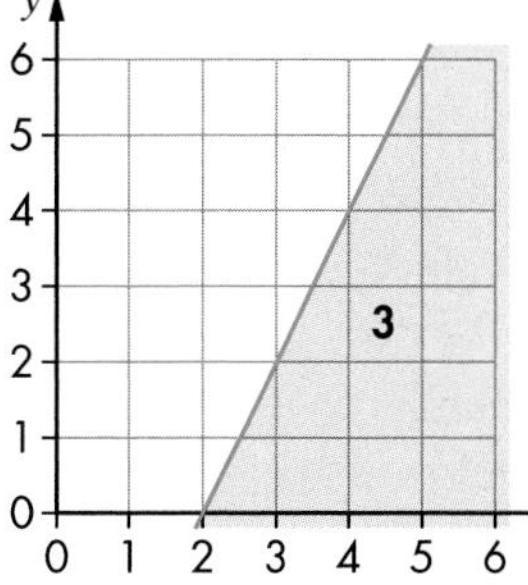

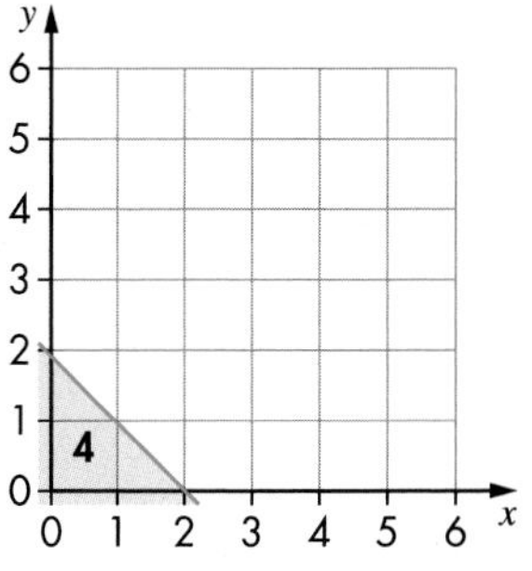

AQA, June 2006, Paper 1 Higher, Question 8

Worked Examination Questions

1 a On the number lines show these inequalities.

i $-2 \leqslant n < 4$

ii $n < 2$

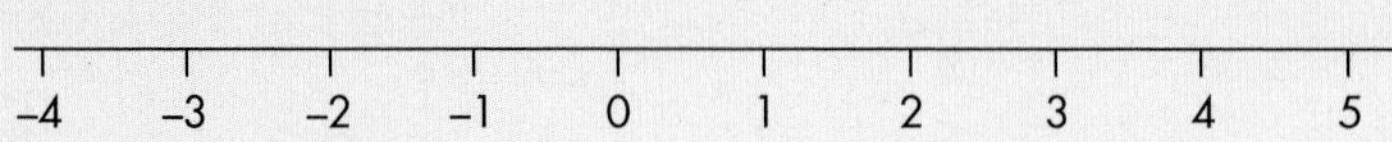

b n is an integer. Find the values of n that satisfy both inequalities in part **a**.

c Solve these inequalities.

i $3x + 8 > 2$

ii $3(x - 4) \leqslant \frac{1}{2}(x + 1)$

1 a i

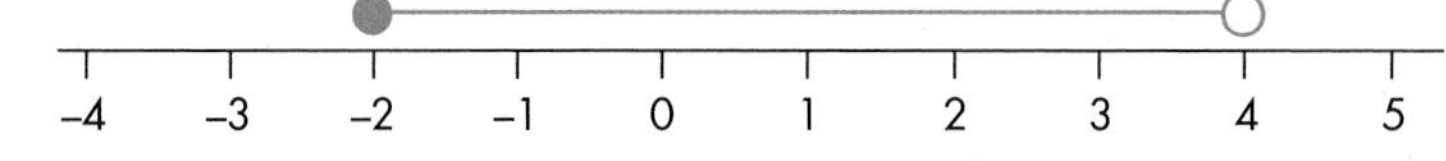

ii

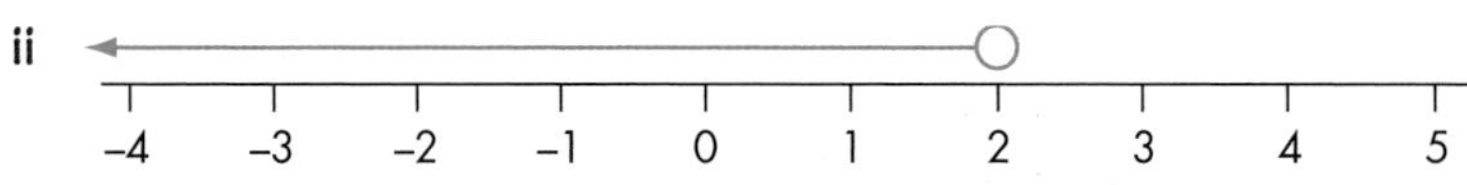

Remember that a strict inequality has an open circle to show the boundary and an inclusive inequality has a solid circle to show the boundary. These get 1 mark each.

b $\{-2, -1, 0, 1\}$

The integers that satisfy both inequalities are in the overlap of both lines. This gets 1 mark.

c i $3x + 8 > 2$

$3x > -6$

$x > -2$

As when solving an equation do the same thing to both sides. First subtract 8, then divide by 3. This gets 1 mark for method and 1 mark for accuracy.

ii $3(x - 4) \leqslant \frac{1}{2}(x + 1)$

$6(x - 4) \leqslant x + 1$

$6x - 24 < x + 1$

$5x < 25$

$x < 5$

First multiply by 2 to get rid of the fraction, then expand the brackets. This gets 1 mark for method. Then collect all the x terms on the left-hand side and the number terms on the right-hand side. This gets 1 mark for method.
Then simplify and divide by 5.
This gets 1 mark for accuracy.

Total: 7 marks

FM 2 A school uses two coach firms, Excel and Storm, to take students home from school. An Excel coach holds 40 students and a Storm coach holds 50 students. 1500 students need to be taken home by coach. If E Excel coaches and S Storm coaches are used, explain why:

$4E + 5S \geqslant 150$

2 E Excel coaches take $40E$ and S Storm coaches take $50S$ students. Together they take $40E + 50S$ students.

Write down the number of students and the total carried by each company's coaches.
This gets 1 mark.

There are 1500 students to use the coaches, so $40E + 50S$ must be at least 1500.
$40E + 50S \geqslant 1500$, then cancel by 10.

Explain that this must be at least the total number of students to be carried, and that the equation will cancel by 10.
This gets 1 mark for method and 1 mark for accuracy.

Total: 3 marks

Worked Examination Questions

PS **3** A bookshelf holds P paperback and H hardback books. The bookshelf can hold a total of 400 books. Which of the following may be true?

a $P + H < 300$ **b** $P \geqslant H$ **c** $P + H > 500$

3 $P + H < 300$ and $P \geqslant H$.

Both of these inequalities could be true. The bookshelf doesn't have to be full and there could be more paperbacks than hardbacks. The final inequality cannot be true as there can only be a maximum of 400 books.
These get 1 mark each.

Total: 2 marks

EQ **4** The region R is shown shaded below.

Write down three inequalities which together describe the shaded region.

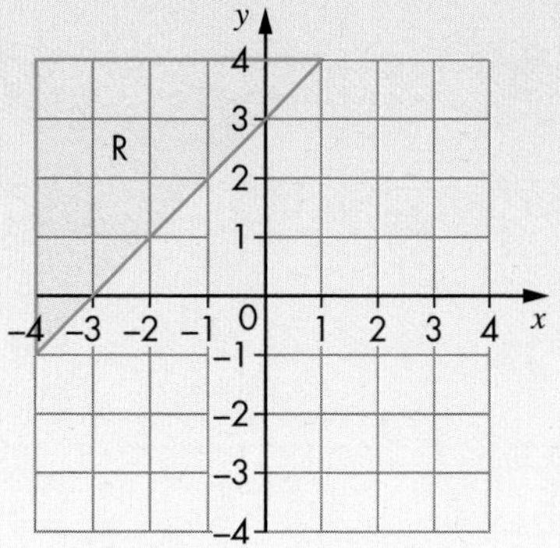

4 $y \leqslant 4, x \geqslant -4$ and $y \geqslant x + 3$

Work out the equation of each boundary line and then decide if points off the line are greater or less than the boundary.
These get 1 mark each.

Total: 3 marks

19

Functional Maths
Linear programming

Linear programming is a mathematical method that uses two-dimensional inequalities. It takes into account maximum and minimum values, and constraints to find the optimum solution to a problem. It is often used by shops to work out stock levels, and reduce cost to increase profit.

Getting started

A boy goes to the fair with £6.00 in his pocket. He only likes rides on the big wheel and eating hot-dogs. A big wheel ride costs £1.50 and a hot-dog costs £2.00. He has W big wheel rides and D hot-dogs.

a Explain why:

i $W \leq 4$

ii $D \leq 3$

iii $3W + 4D \leq 12$

b If he cannot eat more than two hot-dogs before feeling full, write down an inequality that must be true.

c Which of these combinations of big wheel rides and hot-dogs are possible if they obey all of the above conditions?

i two big wheel rides and one hot-dog

ii three big wheel rides and two hot-dogs

iii two big wheel rides and two hot-dogs

iv one big wheel ride and one hot-dog

Your task

A shop stocks only sofas and beds.

A sofa takes up 3 m^2 of floor area and is worth £500. A bed takes up 4 m^2 of floor area and is worth £300.

The shop has 48 m^2 of floor space for stock.

The shop stocks at least 3 sofas and 2 beds at any one time. The insurance policy will allow a total of only £6000 of stock to be in the shop at any one time.

The shop stocks x beds and y sofas.

Use this information to investigate the number of sofas and beds the shop should stock.

Extension

Give some limiting factors of your own and display them diagrammatically.

ANSWERS

Answers to Chapter 1

Quick check

1 **a** 468 **b** 366 **c** 54 **d** 300 **e** 102 **f** 95

2 **a** 3841 **b** 41 **c** 625

3 **a** 17 **b** 25 **c** 5

1.1 Solving real-life problems

Exercise 1A

1 **a** 6000
b 5 cans cost £1.95, so 6 cans cost £1.95.
$32 = 5 \times 6 + 2$. Cost is £10.53.

2 **a** 288
b 16

3 **a** 38
b Coach price for adults = £8, coach price for juniors = £4, money for coaches raised by tickets = £12 400, cost of coaches = £12 160, profit = £240

4 £34.80

5 (18.81...) Kirsty can buy 18 models.

6 (7.58...) Eunice must work for 8 weeks.

7 £8.40 per year, 70p per copy

8 £450

9 15

10 3 weeks

11 £248.75

12 Gavin pays $2296.25 - 1840 =$ £456.25

1.2 Multiplication and division with decimals

Exercise 1B

1 **a** 0.028 **b** 0.09 **c** 0.192 **d** 3.0264 **e** 7.134 **f** 50.96 **g** 3.0625 **h** 46.512

2 **a** 35, 35.04, 0.04 **b** 16, 18.24, 2.24 **c** 60, 59.67, 0.33 **d** 180, 172.86, 7.14 **e** 12, 12.18, 0.18 **f** 24, 26.016, 2.016 **g** 40, 40.664, 0.664 **h** 140, 140.58, 0.58

3 **a** 572 **b** **i** 5.72 **ii** 1.43 **iii** 22.88

4 **a** Incorrect as should end in the digit 2
b Incorrect since $9 \times 5 = 45$, so answer must be less than 45

5 $26.66 \div 3.1$ (answer 8.6) since approximately $27 \div 3 = 9$

6 **a** 18 **b** 140 **c** 1.4 **d** 12 **e** 21.3 **f** 6.9 **g** 2790 **h** 12.1 **i** 18.9

7 **a** 280 **b** 12 **c** 0.18 **d** 450 **e** 0.62 **f** 380 **g** 0.26 **h** 240 **i** 12

8 750

9 300

10 **a** 27 **b** **i** 27 **ii** 0.027 **iii** 0.27

11 £54.20

12 Mark bought a DVD, some jeans and a pen.

1.3 Approximation of calculations

Exercise 1C

1 **a** 50 000 **b** 60 000 **c** 30 000 **d** 90 000 **e** 90 000 **f** 0.5 **g** 0.3 **h** 0.006 **i** 0.05 **j** 0.0009 **k** 10 **l** 90 **m** 90 **n** 200 **o** 1000

2 **a** 56 000 **b** 27 000 **c** 80 000 **d** 31 000 **e** 14 000 **f** 1.7 **g** 4.1 **h** 2.7 **i** 8.0 **j** 42 **k** 0.80 **l** 0.46 **m** 0.066 **n** 1.0 **o** 0.0098

3 **a** 60 000 **b** 5300 **c** 89.7 **d** 110 **e** 9 **f** 1.1 **g** 0.3 **h** 0.7 **i** 0.4 **j** 0.8 **k** 0.2 **l** 0.7

4 **a** 65, 74 **b** 95, 149 **c** 950, 1499

5 Any correct multiplication such as $200 \times 6\,000$, 1000×1200 etc.

6 Elsecar 750, 849, Hoyland 1150, 1249, Barnsley 164 500, 165 499

7 15, 16 or 17

8 1, because there could be 450 then 449

9 Donte has rounded to 2 significant figures or nearest 10 000

Exercise 1D

1 **a** 60 000 **b** 120 000 **c** 10 000 **d** 15 **e** 140 **f** 100 **g** 200 **h** 0.028 **i** 0.09 **j** 400 **k** 8000 **l** 0.16 **m** 45 **n** 0.08 **o** 0.25 **p** 4 000 000 **q** 360 000

2 **a** 5 **b** 50 **c** 25 **d** 600 **e** 3000 **f** 5000 **g** 2000 **h** 2000 **i** 400 **j** 8000 **k** 4 000 000 **l** 3 200 000

3 **a** 54 400 **b** 16 000

4 $30 \times 90\,000 = 2\,700\,000$
$600 \times 8000 = 4\,800\,000$
$5000 \times 4000 = 20\,000\,000$
$200\,000 \times 700 = 140\,000\,000$

5 1400 million

Exercise 1E

1 **a** 35000 **b** 15000 **c** 960 **d** 5 **e** 1200 **f** 500
2 **a** 39700 **b** 17000 **c** 933 **d** 4.44 **e** 1130 **f** 550
3 **a** 4000 **b** 10 **c** 1 **d** 20 **e** 3 **f** 18
4 **a** 4190 **b** 8.79 **c** 1.01 **d** 20.7 **e** 3.07 **f** 18.5
5 **a** £3000 **b** £2000 **c** £1500 **d** £700
6 **a** £15000 **b** £18000 **c** £17500
7 £20000
8 8p
9 \$1000
10 **a** 40 miles per hour **b** 10 gallons **c** £70
11 **a** 80000 **b** 2000 **c** 1000 **d** 30000 **e** 5000 **f** 200000 **g** 75 **h** 140 **i** 100 **j** 3000
12 **a** 86900 **b** 1760 **c** 1030 **d** 29100 **e** 3930 **f** 237000 **g** 84.8 **h** 163 **i** 96.9 **j** 2440
13 Approximately 500
14 1000 or 1200
15 400 or 500
16 **a** **i** 27.571 428 57 **ii** 27.6
b **i** 16.896 516 39 **ii** 16.9
c **i** 704.419 889 5 **ii** 704

Exercise 1F

1 **a** 1.74 m **b** 6 minutes **c** 240 g **d** 83°C **e** 35000 people **f** 15 miles **g** 14 m^2
2 82°F, 5km, 110 min, 43000 people, 6.2 seconds, 67th, 1788, 15 practice walks, 5 seconds
The answers will depend on the approximations made. Your answers should be to the same order as these.
3 40
4 300 miles
5 40 × £20 = £800
6 40 minutes
7 60 stamps
8 270 fans
9 80000 kg (80 tonnes)
10 22.5° C − 18.2° C = 4.3 Celsius degrees
11 149000000 ÷ 300000 = 496.66 ≈ 500 seconds
12 Macau's population density is approximately 710000 times the population density of Greenland.

Examination questions

1 6 weeks
2 17 boxes
3 13
4 **a** 30.946 944 26
b 30.95
5 **a** 3.586 440 678
b 3.59
6 4200
7 Briony
8 20 cartridges
9 £0.79 x 500 = £395.
£14.95 x 24 = £358.80
so two year contract is cheaper.
10 $20^2 \div (5 \times 10) = 400 \div 50 = 8$
11 **a** 0.8735818
b 0.874
12 Sal £5 Bill £7
13 18km ≈ 3 hours 36 mins.
1400m climbed ≈ 2 hour 20 mins.
3h 36m + 2h 20m = 5h 56m which is about 6 hours.

Answers to Chapter 2

Quick check

1 **a** 4427 **b** 36 **c** 36
2 Answers will vary
3 **a** 64 **b** 144 **c** 13

2.1 Multiples, factors, prime numbers, powers and roots

Exercise 2A

1 **a** 12 **b** 9 **c** 6 **d** 13 **e** 15 **f** 14 **g** 16 **h** 10 **i** 18 **j** 17 **k** 8 (or 16) **l** 21
2 4 packs of sausages and 5 packs of buns (or multiples of these)
3 24 seconds
4 30 seconds
5 12 minutes; Debbie will have run 4 laps; Fred will have run 3 laps.
6 1 + 3 + 5 + 7 + 9 = 25, 1 + 3 + 5 + 7 + 9 + 11 = 36, 1 + 3 + 5 + 7 + 9 + 11 + 13 = 49, 1 + 3 + 5 + 7 + 9 + 11 + 13 + 15 = 64
7 **a** −2 **b** −5 **c** −7 **d** −1 **e** −9 **f** −11 **g** −12 **h** −20 **i** −30 **j** −13
8 **a** 1 **b** 3 **c** 4 **d** 2 **e** 10 **f** −2 **g** −1 **h** 20 **i** 40 **j** −4
9 **a** 1, 3, 6, 10, 15, 21, 28, 36, 45, 55, 66, 78, 91, 105
b Adding consecutive pairs gives you square numbers.
10

	Square number	Factor of 56
Cube number	64	8
Multiple of 7	49	28

11 **a** These numbers of dots can be arranged in a triangle pattern.
b 21, 28, 36, 45, 55
12 2, 3 and 12
13 **a** 1, 64, 729, 4096, 15 625
b 1, 8, 27, 64, 125
c $\sqrt{a^3} = a \times \sqrt{a}$
d Square numbers
14 **a** 0.2 **b** 0.5 **c** 0.6 **d** 0.9 **e** 1.2 **f** 0.8 **g** 1.1 **h** 1.5
15 The answers will depend on the approximations made. Your answers should be to the same order as these.
a 109 **b** 1807 **c** 197

2.2 Prime factors, LCM and HCF

Exercise 2B

1 **a** $84 = 2 \times 2 \times 3 \times 7$ **b** $100 = 2 \times 2 \times 5 \times 5$
c $180 = 2 \times 2 \times 3 \times 3 \times 5$ **d** $220 = 2 \times 2 \times 5 \times 11$
e $280 = 2 \times 2 \times 2 \times 5 \times 7$
f $128 = 2 \times 2 \times 2 \times 2 \times 2 \times 2 \times 2$
g $50 = 2 \times 5 \times 5$

2 **a** $84 = 2^2 \times 3 \times 7$ **b** $100 = 2^2 \times 5^2$
c $180 = 2^2 \times 3^2 \times 5$ **d** $220 = 2^2 \times 5 \times 11$
e $280 = 2^3 \times 5 \times 7$ **f** $128 = 2^7$
g $50 = 2 \times 5^2$

3 $1, 2, 3, 2^2, 5, 2 \times 3, 7, 2^3, 3^2, 2 \times 5, 11, 2^2 \times 3, 13, 2 \times 7, 3 \times 5, 2^4, 17, 2 \times 3^2, 19, 2^2 \times 5, 3 \times 7, 2 \times 11, 23, 2^3 \times 3, 5^2, 2 \times 13, 3^3, 2^2 \times 7, 29, 2 \times 3 \times 5, 31, 2^5, 3 \times 11, 2 \times 17, 5 \times 7, 2^2 \times 3^2, 37, 2 \times 19, 3 \times 13, 2^3 \times 5, 41, 2 \times 3 \times 7, 43, 2^2 \times 11, 3^3 \times 5, 2 \times 23, 47, 2^4 \times 3, 7^2, 2 \times 5^2$

4 **a** 2 is always the only prime factor
b 64, 128 **c** 81, 243, 729
d 256, 1024, 4096
e $3, 3^2, 3^3, 3^4, 3^5, 3^6, 4, 4^2, 4^3, 4^4, 4^5, 4^6$

5 **a** $2 \times 2 \times 3 \times 5$ **b** $2^2 \times 3 \times 5$
c $120 = 2^3 \times 3 \times 5$, $240 = 2^4 \times 3 \times 5$, $480 = 2^5 \times 3 \times 5$

6 **a** $7^2 \times 11^2 \times 13^2$ **b** $7^3 \times 11^3 \times 13^3$
c $7^{10} \times 11^{10} \times 13^{10}$

7 Because 3 is not a factor of 40 so it does not divide exactly

Exercise 2C

1 **a** 20 **b** 56 **c** 6 **d** 28 **e** 10 **f** 15
g 24 **h** 30

2 They are the two numbers multiplied together.

3 **a** 8 **b** 18 **c** 12 **d** 30

4 No. The numbers have a common factor. Multiplying them together would mean using this factor twice, thus increasing the size of the common multiple. It would not be the least common multiple.

5 **a** 168 **b** 105 **c** 84 **d** 84 **e** 96 **f** 54
g 75 **h** 144

6 3 packs of cheese slices and 4 packs of bread rolls

7 **a** 8 **b** 7 **c** 4 **d** 14 **e** 4 **f** 9
g 5 **h** 4 **i** 3 **j** 16 **k** 5 **l** 18

8 **a** ii and iii **b** iii

9 18 and 24

2.3 Negative numbers

Exercise 2D

1 **a** −15 **b** −14 **c** −24 **d** 6 **e** 14
f 2 **g** −2 **h** −8 **i** −4 **j** 3
k −24 **l** −10 **m** −18 **n** 16 **o** 36

2 **a** −9 **b** 16 **c** −3 **d** −32 **e** 18
f 18 **g** 6 **h** −4 **i** 20 **j** 16
k 8 **l** −48 **m** 13 **n** −13 **o** −8

3 **a** −2 **b** 30 **c** 15 **d** −27 **e** −7

4 **a** −9 **b** 3 **c** 1

5 **a** 16 **b** −2 **c** −12

6 $-1 \times 12, 1 \times -12, -2 \times 6, 2 \times -6, -3 \times 4, 3 \times -4,$

7 Any appropriate divisions

8 **a** −24 **b** 24 degrees **c** 3×-6

9 $13 \times -6, -15 \times 4, -72 \div 4, -56 \div -8$

Exercise 2E

1 **a** −4 **b** −6 **c** 4 **d** 45 **e** 6 **f** 6

2 **a** 38 **b** 24 **c** −3 **d** −6 **e** −1 **f** 2
g −25 **h** 25 **i** 0 **j** −20 **k** 4 **l** 0

3 **a** $(3 \times -4) + 1 = -11$ **b** $-6 \div (-2 + 1) = 6$
c $(-6 \div -2) + 1 = 4$ **d** $4 + (-4 \div 4) = 3$
e $(4 + -4) \div 4 = 0$ **f** $(16 - -4) \div 2 = 10$

4 **a** 49 **b** −1 **c** −5 **d** −12

5 **a** 40 **b** 1 **c** 78 **d** 4

6 Possible answer: $3 \times -4 \div 2$

7 Possible answer: $(2 - 4) \times (7 - 3)$

8 $(5 + 6) - (7 \div 8) \times 9$

9 **a** −15

10 **a** −16 **b** −36 **c** ±10i **d** ±12i **e** 1
f −1 **g** −i **h** −1 **i** −2.5 **j** 16

Examination questions

1 **a** $2 \times 3 \times 7$
b 84

2 **a** 90 **b** 240 **c** 6

3 $2^3 \times 5$

4 Possible answer: 18 and 36 or 4 and 9

5 5 times

6 120

7 **a** $x = 5$
b $2 \times 3 \times 5 \times 5$

8 **a** $p = 2, q = 5$
b 10

9 **a** $a = 2, b = 5$ (or vice versa)
b There are two possible solutions, 1 and 6 or 2 and 3.

10 **a** (3, 8) **b** Any pair from (12, 15) or (15, 18)

11 **a** $7^2 - 5^2 = 24$. Many other answers.
b $7^2 - 6^2 = 13$. Many other answers.

12 −2

13 **a** Could be either.
b n = 31, p = 5. Many other answers.

14 −4

15 $8^2 = 64$, $9^2 = 81$ so $\sqrt{72}$ must be between 8 and 9.

16 −10

17 **a** 4 **b** 24

Answers to Chapter 3

Quick check

1 **a** $\frac{2}{5}$ **b** $\frac{3}{8}$ **c** $\frac{3}{7}$

2

Fraction	Percentage	Decimal
$\frac{3}{4}$	75%	0.75
$\frac{2}{5}$	40%	0.4
$\frac{11}{20}$	55%	0.55

3 **a** £23 **b** £4.60 **c** 23p

3.1 One quantity as a fraction of another

Exercise 3A

1 **a** $\frac{1}{3}$ **b** $\frac{1}{5}$ **c** $\frac{2}{5}$ **d** $\frac{5}{24}$ **e** $\frac{2}{5}$
f $\frac{1}{6}$ **g** $\frac{2}{7}$ **h** $\frac{1}{3}$

2 $\frac{3}{5}$

3 $\frac{12}{31}$

4 20 weeks.

5 Jon saves $\frac{30}{90} = \frac{1}{3}$
Matt saves $\frac{35}{100}$, which is greater than $\frac{1}{3}$, so Matt saves the greater proportion of his earnings.

6 $\frac{13}{20} = \frac{65}{100}$, $\frac{16}{25} = \frac{64}{100}$, so first mark is better.

7 $\frac{1}{8}$

8 $\frac{5}{12}$

9 $\frac{1}{5}$

10 $\frac{3}{20}$

11 $\frac{3}{10}$

12 32 or 36

3.2 Adding and subtracting fractions

Exercise 3B

1 **a** $\frac{8}{15}$ **b** $\frac{7}{12}$ **c** $\frac{3}{10}$ **d** $\frac{11}{12}$ **e** $\frac{1}{10}$ **f** $\frac{1}{8}$
g $\frac{1}{12}$ **h** $\frac{1}{3}$ **i** $\frac{7}{9}$ **j** $\frac{5}{8}$ **k** $\frac{3}{8}$ **l** $\frac{1}{15}$

2 **a** $3\frac{31}{45}$ **b** $4\frac{47}{60}$ **c** $\frac{41}{72}$ **d** $\frac{29}{48}$ **e** $1\frac{43}{48}$ **f** $1\frac{109}{120}$
g $1\frac{23}{30}$ **h** $1\frac{31}{84}$

3 $\frac{1}{20}$

4 **a** $\frac{1}{6}$ **b** 30

5 No, one eighth is left, which is 12.5 cl, so enough for one cup but not two cups.

6 260

7 Three-quarters of 68

8 He has added the numerators and added the denominators instead of using a common denominator. Correct answer is $3\frac{7}{12}$.

9 Possible answer: The denominators are 4 and 5. I first find a common denominator. The lowest common denominator is 20 because 4 and 5 are both factors of 20. So I am changing the fractions to twentieths. One-quarter is the same as five-twentieths (multiplying numerator and denominator by 5). Two-fifths is the same as eight-twentieths (multiplying numerator and denominator by 4). Five-twentieths plus eight-twentieths = thirteen-twentieths.

10 £51

11 10 minutes

3.3 Increasing and decreasing quantities by a percentage

Exercise 3C

1 **a** 1.1 **b** 1.03 **c** 1.2 **d** 1.07 **e** 1.12
2 **a** £62.40 **b** 12.96 kg **c** 472.5 g **d** 599.5 m
e £38.08 **f** £90 **g** 391 kg **h** 824.1 cm
i 253.5 g **j** £143.50 **k** 736 m **l** £30.24
3 £29 425 – 7% pay rise
4 1 690 200
5 **a** Bob: £17 325, Anne: £18 165, Jean: £20 475, Brian: £26 565
b 5% of different amounts is not a fixed amount. The more pay to start with, the more the increase (5%) will be.
6 £411.95
7 193 800
8 575 g
9 918
10 60
11 TV: £294, microwave: £86.40, CD: £138, stereo: £35.40
12 £10
13 **c** Both same as $1.05 \times 1.03 = 1.03 \times 1.05$
14 **d** Shop A as $1.04 \times 1.04 = 1.0816$, so an 8.16% increase.
15 £540.96
16 Calculate the VAT on certain amounts, and $\frac{1}{6}$ of that amount. Show the error grows as the amount increases. After €600 the error is greater than €10, so the method works to within €10 with prices up to €600.

Exercise 3D

1 **a** 0.92 **b** 0.85 **c** 0.75 **d** 0.91 **e** 0.88
2 **a** £9.40 **b** 23 kg **c** 212.4 g **d** 339.5 m
e £4.90 **f** 39.6 m **g** 731 m **h** 83.52 g
i 360 cm **j** 117 min **k** 81.7 kg **l** £37.70
3 £5525
4 **a** 52.8 kg **b** 66 kg **c** 45.76 kg
5 Mr Speed: £176, Mrs Speed: £297.50, James: £341, John: £562.50
6 448
7 705
8 £18 975
9 **a** 66.5 mph
10 No, as the total is £101. She will save £20.20, which is less than the £25 it would cost to join the club.
11 **a** 524.8 units
b Less gas since 18% of the smaller amount of 524.8 units (94.464 units) is less than 18% of 640 units (115.2 units). I used 619.264 units.
12 TV £227.04, DVD player £172.80
13 10% off £50 is £45; 10% off £45 is £40.50; 20% off £50 is £40
14 £765
15 $1.10 \times 0.9 = 0.99$ (99%)
16 Offer A gives 360 grams for £1.40, i.e. 0.388 pence per gram. Offer B gives 300 grams for £1.12, i.e 0.373 pence per gram, so Offer B is the better offer.
Or Offer A is 360 for 1.40 = 2.6 g/p, offer B is 300 for 1.12 = 2.7 g/p, so offer B is better.

3.4 Expressing one quantity as a percentage of another

Exercise 3E

1 **a** 25% **b** 60.6% **c** 46.3% **d** 12.5%
e 41.7% **f** 60% **g** 20.8% **h** 10%
i 1.9% **j** 8.3% **k** 45.5% **l** 10.5%
2 32%
3 6.5%
4 33.7%
5 **a** 49.2% **b** 64.5% **c** 10.6%
6 17.9%
7 4.9%
8 90.5%
9 **a** Brit Com: 20.9%, USA: 26.5%, France: 10.3%, Other 42.3%
b Total 100%, all imports
10 Stacey had the greater percentage increase.
Stacey: $(20 - 14) \times 100 \div 14 = 42.9\%$
Calum: $(17 - 12) \times 100 \div 12 = 41.7\%$
11 Yes, as 38 out of 46 is over 80% (82.6%)

3.5 Compound interest and repeated percentage change

Exercise 3F

1 **a** **i** 10.5 g
ii 11.03 g
iii 12.16 g
iv 14.07 g
b 9 days
2 12 years
3 **a** £14 272.27 **b** 20 years
4 **a** **i** 2550
ii 2168
iii 1331
b 7 years
5 **a** £6800 **b** £5440 **c** £3481.60
6 **a** **i** 1.9 million litres
ii 1.6 million litres
iii 1.2 million litres
b 10th August
7 **a** **i** 51 980
ii 84 752
iii 138 186
b 2021
8 **a** 21 years **b** 21 years
9 3 years
10 30 years
11 $1.1 \times 1.1 = 1.21$ (21% increase)
12 Bradley Bank account is worth £1032, Monastery Building Society account is worth £1031.30, so Bradley Bank by 70p
13 4 months: fish weighs $3 \times 1.1^4 = 4.3923$ kg; crab weighs $6 \times 0.9^4 = 3.9366$ kg
14 4 weeks
15 20

3.6 Reverse percentage (working out the original quantity)

Exercise 3G

1 **a** 800 g **b** 250 m **c** 60 cm **d** £3075 **e** £200 **f** £400
2 80
3 T shirt £8.40, Tights £1.20, Shorts £5.20, Sweater £10.75, Trainers £24.80, Boots £32.40
4 £833.33
5 £300
6 240
7 £350
8 4750 blue bottles
9 £23.10
10 300 cm^3
11 8 cm
12 5 cm
13 **a** £1600
b With 10% cut each year he earns £1440 × 12 + £1296 × 12 = £17 280 + £15 552 = £32 832 With immediate 14% cut he earns £1376 × 24 = £33 024, so correct decision
14 **a** 30% **b** 15%
15 Less by $\frac{1}{4}$%
16 £900
17 Calculate the pre-VAT price for certain amounts, and $\frac{5}{6}$ of that amount. Show the error grows as the amount increases. Up to €280 the error is less than €5.
18 £1250
19 £1250
20 Baz has assumed that 291.2 is 100% instead of 112%. He rounded his wrong answer to the correct answer of £260.

3.7 Ratio

Exercise 3H

1 $\frac{7}{10}$
2 $\frac{2}{5}$
3 **a** $\frac{2}{5}$ **b** $\frac{3}{5}$
4 **a** $\frac{7}{10}$ **b** $\frac{3}{10}$
5 Amy $\frac{3}{5}$, Katie $\frac{2}{5}$
6 **a** Fruit crush $\frac{5}{32}$, lemonade $\frac{27}{32}$
b The second recipe.
7 $13\frac{1}{2}$ litres
8 **a** $\frac{1}{2}$ **b** $\frac{7}{20}$ **c** $\frac{3}{20}$
9 James $\frac{1}{2}$, John $\frac{3}{10}$, Joseph $\frac{1}{5}$
10 Sugar $\frac{5}{22}$, flour $\frac{3}{11}$, margarine $\frac{2}{11}$, fruit $\frac{7}{22}$
11 3 : 1
12 $\frac{1}{7}$
13 1 : 1 : 1

Exercise 3I

1 **a** 160 g : 240 g
b 80 kg : 200 kg
c 150 : 350
d 950 m : 50 m
e 175 min : 125 min
f £20 : £30 : £50
g £36 : £60 : £144
h 50 g : 250 g : 300 g
2 **a** 175 **b** 30%
3 **a** 40% **b** 300 kg
4 21 horses
5 **a** No, Yes, No, No, Yes
b Possible answers: W26, H30; W31, H38; W33, H37
6 **a** 1 : 400 000 **b** 1 : 125 000
c 1 : 250 000 **d** 1 : 25 000
e 1 : 20 000 **f** 1 : 40 000
7 **a** 1 : 1 000 000 **b** 47 km
c 0.8 cm
8 **a** 1 : 250 000 **b** 2 km
c 4.8 cm
9 **a** 1 : 20 000 **b** 0.54 km
c 40 cm
10 **a** 4 : 3 **b** 90 miles
c Both arrive at the same time.
11 0.4 metres
12 **a** 1 : 1.6 **b** 1 : 3.25
c 1 : 1.125 **d** 1 : 1.44
e 1 : 5.4 **f** 1 : 1.5
g 1 : 4.8 **h** 1 : 42
i 1 : 1.25

Exercise 3J

1 **a** 3 : 2
b 32
c 80
2 **a** 100
b 160
3 0.4 litres
4 102
5 1000 g
6 Jamie has 1.75 pints, so he has enough.
7 8100
8 5.5 litres
9 **a** 14 min
b 75 min ($1\frac{1}{4}$ h)
10 **a** 11 pages
b 32%
11 Kevin £2040, John £2720
12 C, F, T, T
13 51
14 100
15 40cc
16 **a** 160 cans
b 48 cans
17 **a** Lemonade 20 litres, ginger 0.5 litres
b This one, in part **a** there are 50 parts in the ratio 40 : 9 : 1, so ginger is $\frac{1}{50}$ of total amount; in part **b** there are 13 parts in the ratio 10 : 2 : 1, so ginger is $\frac{1}{13}$ of total amount. $\frac{1}{13} > \frac{1}{50}$
18 225 kg
19 54

Examination questions

1 £332.80
2 **a** 48.1 seconds
b **i** 44.1 seconds
ii Di (40.23 seconds)
iii Di
3 £141
4 £2200 per month
5 £220
6 $\frac{9}{40}$
7 $4\frac{1}{12}$ pints
8 2 tins
9 **a** $\frac{312}{77}, \frac{54}{17}, \frac{22}{7}, \frac{221}{71}$
b $\frac{22}{7}$
10 Yes, investment will be worth £4008.46
11 8% decrease
12 Estimate: 80% (78.4%)
13 **a** No, only enough for 6 days:
$5 \div \frac{4}{5} = 6\frac{1}{4}$ or $5 \div \frac{2}{5} = 12.5$,
so $12\frac{1}{2}$ meals
b $2\frac{2}{3}$
14 £7375.53
15 194.6%
16 **a** 90%
b £152000
17 **a** 18 adults, 108 children
b 1 : 4
18 4 more red balls
19 Yes, $100 \times 0.96^9 = 69.3$ kg
20 392 500 square kilometres
21 Jill is correct $0.4 \times 0.75 = 0.3$, so 30% of the original price is equal to 70% off.
22 Not correct, since $0.64^5 = 0.107$, lost 100% − 10.7% = 89.3% of its original contents
23 20%
24 60 men (and 50 women)

Answers to Chapter 4

Quick check

1 **a** $\frac{3}{5}$ **b** $\frac{1}{5}$ **c** $\frac{1}{3}$ **d** $\frac{16}{25}$ **e** $\frac{2}{5}$ **f** $\frac{3}{4}$ **g** $\frac{1}{3}$
2 **a** £12 **b** £33 **c** 175 litres **d** 15 kg **e** 40 m **f** £35 **g** 135 g **h** 1.05 litres

4.1 Speed, time and distance

Exercise 4A

1 18 mph
2 280 miles
3 52.5 mph
4 11.50 am
5 500 seconds
6 **a** 75 mph **b** 6.5 hours
c 175 miles **d** 240 km
e 64 km/h **f** 325 km
g 4.3 h (4 h 18 min)
7 **a** 7.75 h **b** 52.9 mph
8 **a** 2.25 h **b** 99 miles
9 **a** 1.25 h **b** 1 h 15 min
10 **a** 48 mph **b** 6 h 40 min
11 **a** 120 km **b** 48 km/h
12 **a** 30 min **b** 6 mph
13 **a** 10 m/s **b** 3.3 m/s
c 16.7 m/s **d** 41.7 m/s
e 20.8 m/s
14 **a** 90 km/h **b** 43.2 km/h
c 14.4 km/h **d** 108 km/h
e 1.8 km/h
15 **a** 64.8 km/h **b** 28 s
c 8.07
16 **a** 6.7 m/s **b** 66 km
c 5 minutes **d** 133.3
17 7 minutes
18 **a** 20 mph **b** 07.30

4.2 Direct proportion problems

Exercise 4B

1 60 g
2 £5.22
3 45
4 £6.72
5 **a** £312.50 **b** 8
6 **a** 56 litres **b** 350 miles
7 **a** 300 kg **b** 9 weeks
8 40 seconds
9 **a** **i** 100 g, 200 g, 250 g, 150 g
ii 150 g, 300 g, 375 g, 225 g
iii 250 g, 500 g, 625 g, 375 g
b 24
10 Peter: £2.30 ÷ 6 = 38.33p each; I can buy four packs (24 sausages) from him (£9.20)
Paul: £3.50 ÷ 10 = 35p each; I can only buy two packs (20 sausages) from him (£7)
I should use Peter's shop to get the most sausages for £10
11 11 minutes 40 seconds + 12 minutes = 23 minutes 40 seconds
12 Possible answer:
30 g plain flour (rounding to nearest 10 g)
60 ml whole milk (rounding to nearest 10 ml)
1 egg (need an egg)
1 g salt (nearest whole number)
10 ml beef dripping or lard (rounding to nearest 10 ml)
13 30 litres

4.3 Best buys

Exercise 4C

1 **a** £4.50 for a 10-pack
b £1.08 for 6
c £2.45 for 1 litre
d Same value
e 29p for 250 g
f £1.39 for a pack of 6
g £4 for 3

2 **a** Large jar as more g per £
b 600 g tin as more g per p
c 5 kg bag as more kg per £
d 75 ml tube as more ml per £
e Large box as more g per £
f Large box as more g per £
g 400 ml bottle as more ml per £

3 **a** £5.11
b Large tin (small £5.11/l, medium £4.80/l, large £4.47/l)

4 **a** 95p **b** Family size

5 Bashir's

6 Mary

7 Kelly

4.4 Density

Exercise 4D

1 0.75 g/cm^3

2 $8\frac{1}{3}$ g/cm^3

3 32 g

4 120 cm^3

5 156.8 g

6 3200 cm^3

7 2.72 g/cm^3

8 36 800 kg

9 1.79 g/cm^3 (3 sf)

10 1.6 g/cm^3

11 First statue is the fake as density is approximately 26 g/cm^3

12 Second piece by 1 cm^3

13 0.339 m^3

Examination questions

1 4 minutes

2 **a** 75%
b 36 000 litres

3 8 mph

4 **a** £105
b 70%

5 4.17 kg

6 **a** 140 km
b 100 km/h

7 11 : 8

8 0.16 km^2

9 Small 3.33p per ml, Large 3.125p per ml, so large best value.

Answers to Chapter 5

Quick check

1 **a** 90 mm^2 **b** 40 cm^2 **c** 21 m^2

2 120 cm^3

5.1 Circumference and area of a circle

Exercise 5A

1 **a** 8 cm, 25.1 cm, 50.3 cm^2
b 5.2 m, 16.3 m, 21.2 m^2
c 6 cm, 37.7 cm, 113 cm^2
d 1.6 m, 10.1 m, 8.04 m^2

2 **a** 5π cm **b** 8π cm
c 18π m **d** 12π cm

3 **a** 25π cm^2 **b** 36π cm^2
c 100π cm^2 **d** 0.25π m^2

4 8.80 m

5 4 complete revolutions

6 1p : 3.1 cm^2, 2p : 5.3 cm^2, 5p : 2.3 cm^2, 10p : 4.5 cm^2

7 0.83 m

8 38.6 cm

9 Claim is correct (ratio of the areas is just over 1.5 : 1)

10 **a** 18π cm^2 **b** 4π cm^2

11 9π cm^2

12 28.3 m^2

13 Diameter of tree is 9.96 m

14 45 complete revolutions

5.2 Area of a trapezium

Exercise 5B

1 **a** 30 cm^2 **b** 77 cm^2
c 24 cm^2 **d** 42 cm^2
e 40 m^2 **f** 6 cm
g 3 cm **h** 10 cm

2 **a** 27.5 cm, 36.25 cm^2
b 33.4 cm, 61.2 cm^2
c 38.5 m, 90 m^2

3 The area of the parallelogram is $(a + b)h$.
This is the same as two trapezia

4 Two of 20 cm^2 and two of 16 cm^2

5 **a** 57 m^2
b 702.5 cm^2
c 84 m^2

6 47 m^2

7 4, because the total area doubled is about 32 m^2

8 80.2%

9 1 100 000 km^2

10 160 cm^2

5.3 Sectors

Exercise 5C

1 **a** **i** 5.59 cm **ii** 22.3 cm^2
b **i** 8.29 cm **ii** 20.7 cm^2
c **i** 16.3 cm **ii** 98.0 cm^2
d **i** 15.9 cm **ii** 55.6 cm^2
2 2π cm, 6π cm^2
3 **a** 73.8 cm **b** 20.3 cm
4 **a** 107 cm^2
b 173 cm^2
5 43.6 cm
6 **a** $\frac{180}{\pi}$
b If arc length is 10 cm, distance along chord joining the two points of the sector on the circumference will be less than 10 cm, so angle at centre will be less than 60°
7 $(36\pi - 72)$ cm^2
8 36.5 cm^2
9 16 cm (15.7)
10 **a** 13.9 cm
b 7.07 cm^2

5.4 Volume of a prism

Exercise 5D

1 **a** **i** 21 cm^2 **ii** 63 cm^3
b **i** 48 cm^2 **ii** 432 cm^3
c **i** 36 m^2 **ii** 324 m^3
2 **a** 432 m^3 **b** 225 m^3 **c** 1332 m^3
3 **a** A cross-section parallel to the side of the pool always has the same shape
b About $3\frac{1}{2}$ hours
4 7.65 m^3
5 **a** 21 cm^3, 210 cm^3
b 54 cm^2, 270 cm^2
6 146 cm^3
7 78 m^3 (78.3 m^3)
8 327 litres
9 1.02 tonnes
10 672 cm^2

5.5 Cylinders

Exercise 5E

1 **a** **i** 226 cm^3
ii 207 cm^2
b **i** 14.9 cm^3
ii 61.3 cm^2
c **i** 346 cm^3
ii 275 cm^2
d **i** 1060 cm^3
ii 636 cm^2
2 **a** **i** 72π cm^3
ii 48π cm^2
b **i** 112π cm^3
ii 56π cm^2
c **i** 180π cm^3
ii 60π cm^2
d **i** 600π m^3
ii 120π m^2
3 £80
4 1.23 tonnes
5 665 cm^3
6 Label should be less than 10.5 cm wide so that it fits the can and does not overlap the rim and more than 23.3 cm long to allow an overlap
7 332 litres
8 There is no right answer. Students could start with the dimensions of a real can. Often drinks cans are not exactly cylindrical. One possible answer is height of 6.6 cm and diameter of 8 cm
9 1.71 g/cm^3
10 7.78 g/cm^3
11 About 127 cm
12 A diameter of 10 cm and a length of 5 cm give a volume close to 400 cm^3 (0.4 litres)
13 B has volume 200π, C has volume 400π, D has volume 337.5π, so order is B, D, C.
14 $\pi r^2 h = r^3$ so $\pi h = r$, $h = r \div \pi \approx 0.32r$.

Examination questions

1 **a** 66.5 cm^2 **b** 855.5 cm^2
2 **a** 320π cm^3 **b** 4
3 11 777 cm^3
4 9.08 cm
5 480π cm^3
6 $\frac{3}{8}$

Answers to Chapter 6

Quick check

1 **a** $2x + 12$ **b** $4x - 12$ **c** $12x - 6$
2 **a** $5y$ **b** $4x - 3$ **c** $-x - 4$
3 **a** $6x$ **b** $8y^2$ **c** $2c^3$
4 **a** $x = 1$ **b** $x = 3$ **c** $x = 9$ **d** $x = 8$ **e** $x = 24$ **f** $x = 15$

6.1a Basic algebra: Substitution

Exercise 6A

1 **a** 13 **b** −3 **c** 5
2 **a** 2 **b** 8 **c** −10
3 **a** 6 **b** 3 **c** −2
4 **a** −7 **b** −10 **c** 6.5
5 **a** −4.8 **b** 48 **c** 32
6 **a** 1.4 **b** 1.4 **c** −0.4
7 **a** 13 **b** 74 **c** 17
8 **a** 75 **b** 22.5 **c** −135
9 **a** 2.5 **b** −20 **c** 2.5
10 **a** £4 **b** 13 km
c No, 5 miles is 8 km so fare would be £6.50
11 **a** $\frac{150}{n}$ **b** £925
12 **a** $2 \times 8 + 6 \times 11 - 3 \times 2 = 76$
b $5 \times 2 - 2 \times 11 + 3 \times 8 = 12$
13 **a** £477.90
b £117.90 still owed (debit)
14 **a** One odd one even value, different from each other.
b Any valid combination, e.g. $x = 1$, $y = 2$
15 Any values such that $2lw = bh$
16 **a** **i** Odd **ii** Odd **iii** Even **iv** Odd
b Any valid expression such as $xy + z$
17 **a** £20
b **i** −£40
ii Delivery cost will be zero.
c 40 miles
18 A expression, B formula, C identity, D equation
19 **a** First term is cost of petrol, each mile is a tenth of £0.98. Second term is the hire cost divided by the miles.
b 29.8p per mile

6.1b Basic algebra: Expansion

Exercise 6B

1 **a** $6 + 2m$ **b** $10 + 5l$ **c** $12 - 3y$
d $20 + 8k$ **e** $6 - 12f$ **f** $10 - 6w$
g $10k + 15m$ **h** $12d - 8n$ **i** $t^2 + 3t$
j $k^2 - 3k$ **k** $4t^2 - 4t$ **l** $8k - 2k^2$
m $8g^2 + 20g$ **n** $15h^2 - 10h$ **o** $y^3 + 5y$
p $h^4 + 7h$ **q** $k^3 - 5k$ **r** $3t^3 + 12t$
s $15d^3 - 3d^4$ **t** $6w^3 + 3tw$ **u** $15a^3 - 10ab$
v $12p^4 - 15mp$ **w** $12h^3 + 8h^2g$ **x** $8m^3 + 2m^4$
2 **a** $5(t - 1)$ and $5t - 5$
b Yes, as $5(t - 1)$ when $t = 4.50$ is $5 \times 3.50 = £17.50$.
3 He has worked out 3×5 as 8 instead of 15 and he has not multiplied the second term by 3. Answer should be $15x - 12$.
4 **a** $3(2y + 3)$
b $2(6z + 4)$ or $4(3z + 2)$

6.1c Basic algebra: Simplification

Exercise 6C

1 **a** $7t$ **b** $9d$ **c** $3e$
d $2t$ **e** $5t^2$ **f** $4y^2$
g $5ab$ **h** $3a^2d$
2 **a** $2x$ and $2y$ **b** a and $7b$
3 **a** $3x - 1 - x$ **b** $10x$
c 25 cm
4 **a** $22 + 5t$ **b** $21 + 19k$
c $22 + 2f$ **d** $14 + 3g$
5 **a** $2 + 2h$ **b** $9g + 5$
c $17k + 16$ **d** $6e + 20$
6 **a** $4m + 3p + 2mp$
b $3k + 4h + 5hk$
c $12r + 24p + 13pr$
d $19km + 20k - 6m$
7 **a** $9t^2 + 13t$ **b** $13y^2 + 5y$
c $10e^2 - 6e$ **d** $14k^2 - 3kp$
8 **a** $17ab + 12ac + 6bc$
b $18wy + 6ty - 8tw$
c $14mn - 15mp - 6np$
d $8r^3 - 6r^2$
9 **a** $5(f + 2s) + 2(2f + 3s) = 9f + 16s$
b £$(270f + 480s)$
c £42 450 − £30 000 = £12 450
10 For x-coefficients, 3 and 1 or 1 and 4; for y-coefficients, 5 and 1 or 3 and 4 or 1 and 7
11 $5(3x + 2) - 3(2x - 1) = 9x + 13$

6.2 Factorisation

Exercise 6D

1 **a** $6(m + 2t)$ **b** $3(3t + p)$ **c** $4(2m + 3k)$
d $4(r + 2t)$ **e** $m(n + 3)$ **f** $g(5g + 3)$
g $2(2w - 3t)$ **h** $y(3y + 2)$ **i** $t(4t - 3)$
j $3m(m - p)$ **k** $3p(2p + 3t)$ **l** $2p(4t + 3m)$
m $4b(2a - c)$ **n** $5bc(b - 2)$ **o** $2b(4ac + 3de)$
p $2(2a^2 + 3a + 4)$ **q** $3b(2a + 3c + d)$ **r** $t(5t + 4 + a)$
s $3mt(2t - 1 + 3m)$ **t** $2ab(4b + 1 - 2a)$ **u** $5pt(2t + 3 + p)$

2 **a** Mary has taken out a common factor.
b Because the bracket adds up to £10.
c £30

3 **a**, **d**, **f** and **h** do not factorise.
b $m(5 + 2p)$
c $t(t - 7)$
e $2m(2m - 3p)$
g $a(4a - 5b)$
i $b(5a - 3bc)$

4 **a** Bernice
b Aidan has not taken out the largest possible common factor. Craig has taken m out of both terms but there isn't an m in the second term.

5 There are no common factors.

6 $4x^3 - 12x$, $2x - 6$

6.3a Solving linear equations: Fractional equations

Exercise 6E

1 **a** 30 **b** 21 **c** 72 **d** 12
e 6 **f** $10\frac{1}{2}$ **g** −10 **h** 7
i 11 **j** 2 **k** 7 **l** $2\frac{4}{5}$
m 1 **n** $11\frac{1}{2}$ **o** $\frac{1}{5}$

2 Any valid equations

3 **a** Amanda
b First line: Betsy adds 4 instead of multiplying by 5.
Second line: Betsy adds 5 instead of multiplying by 5.
Fourth line: Betsy subtracts 2 instead of dividing by 2.

6.3b Solving linear equations: Brackets

Exercise 6F

1 **a** 3 **b** 7 **c** 5 **d** 3
e 4 **f** 6 **g** 8 **h** 1
i $1\frac{1}{2}$ **j** $2\frac{1}{2}$ **k** $\frac{1}{2}$ **l** $1\frac{1}{5}$
m 2 **n** −2 **o** −1 **p** −2
q −2 **r** −1

2 Any values that work, e.g. $a = 2$, $b = 3$ and $c = 30$.

3 55

6.3c Solving linear equations: Variables on both sides

Exercise 6G

1 **a** $x = 2$ **b** $y = 1$ **c** $a = 7$
d $t = 4$ **e** $p = 2$ **f** $k = -1$
g $m = 3$ **h** $s = -2$

2 $3x - 2 = 2x + 5$, $x = 7$

3 **a** $d = 6$ **b** $x = 11$ **c** $y = 1$
d $h = 4$ **e** $b = 9$ **f** $c = 6$

4 $6x + 3 = 6x + 10$; $6x - 6x = 10 - 3$; $0 = 7$, which is obviously false. Both sides have $6x$, which cancels out.

5 $8x + 7 + x + 4 = 11x + 5 - x - 4$, $x = 10$

6 Check students' explanations.

6.4 Setting up equations

Exercise 6H

1 90p

2 **a** 1.5 **b** 2

3 **a** 1.5 cm **b** 6.75 cm^2

4 17

5 8

6 **a** $8c - 10 = 56$
b £8.25

7 **a** B: 450 cars, C: 450 cars, D: 300 cars
b 800
c 750

8 Length is 5.5 m, width is 2.5 m and area is 13.75 m^2. Carpet costs £123.75

9 3 years

10 9 years

11 3 cm

12 5

13 **a** $4x + 40 = 180$
b $x = 35°$

14 **a** $\frac{x + 10}{5} = 9.50$
b £37.50

15 **a** 15
b −1
c $2(n + 3)$, $2(n + 3) - 5$
d $2(n + 3) - 5 = n$, $2n + 6 - 5 = n$, $2n + 1 = n$, $n = -1$

16 No, as $x + x + 2 + x + 4 + x + 6 = 360$ gives $x = 87°$ so the consecutive numbers (87, 89, 91, 93) are not even but odd

17 $4x + 18 = 3x + 1 + 50$, $x = 33$
Large bottle 1.5 litres, small bottle 1 litre

6.5 Trial and improvement

Exercise 6I

1 a 4 and 5 b 4 and 5 c 2 and 3
2 $x = 3.5$
3 $x = 3.7$
4 $x = 2.5$
5 $x = 1.5$
6 a $x = 2.4$ b $x = 2.8$ c $x = 3.2$
7 a Area $= x(x + 5) = 100$
b width = 7.8 cm, length = 12.8 cm
8 $x = 5.8$
9 Volume $= x \times 2x(x + 8) = 500$, $x^3 + 8x^2 = 250$, $4 \Rightarrow 192$, $5 \Rightarrow 325$, $4.4 \Rightarrow 240.064$, $4.5 \Rightarrow 253.125$, $4.45 \Rightarrow 246.541125$, so dimensions are 4.5 cm, 9 cm and 12.5 cm

Examination questions

1 32
2 a i $y^2 + 4y - 5$
ii Odd × odd + 4 × odd − odd = odd + even − odd = even
b $2y(x - 3y)$
3 $x^2 + x - 12$
4 a Identity, (Formula), Equation, Expression
b Add any even number, multiply by any odd number
5 a 3.5 b 2.2 c 1.5
6 a i y^9
ii y^5
iii y^{14}
b i y^{14}
ii y^5
7 6
8 a $5x(x + 4)$
9 a 20
b $\frac{1}{3}$
c $2ab(3b - 1)$
10 $5x + 16$
11 1.25
12 a 5 b 6
13 2.4
14 a 5 b 27
15 a $x(x + 7)$ b $15x + 40$ c $4x + 9$

Answers to Chapter 7

Quick check

1 a $2x + 12$ b $4x - 12$ c $12x - 6$
2 a $5y$ b $4x - 3$ c $-x - 4$
3 a $6x$ b $8y^2$ c $2c^3$
4 a $x = 1$ b $x = 3$ c $x = 9$ d $x = 8$ e $x = 24$ f $x = 15$

7.1a Simultaneous equations: Elimination and substitution methods

Exercise 7A

1 a $x = 4, y = 1$ b $x = 1, y = 4$ c $x = 3, y = 1$
d $x = 5, y = -2$ e $x = 7, y = 1$ f $x = 5, y = \frac{1}{2}$
g $x = 4\frac{1}{2}, y = 1\frac{1}{2}$ h $x = -2, y = 4$ i $x = 2\frac{1}{4}, y = -1\frac{1}{2}$
j $x = 2\frac{1}{4}, y = 6\frac{1}{2}$ k $x = 4, y = 3$ l $x = 5, y = 3$
2 a 3 is the first term. The next term is $3 \times a + b$, which equals 14.
b $14a + b = 47$ c $a = 3, b = 5$ d 146, 443

7.1b Simultaneous equations: Balancing coefficients in one equation only

Exercise 7B

1 a $x = 2, y = -3$ b $x = 7, y = 3$ c $x = 4, y = 1$
d $x = 2, y = 5$ e $x = 4, y = -3$ f $x = 1, y = 7$
g $x = 2\frac{1}{2}, y = 1\frac{1}{2}$ h $x = -1, y = 2\frac{1}{2}$ i $x = 6, y = 3$
j $x = \frac{1}{2}, y = -\frac{3}{4}$ k $x = -1, y = 5$
l $x = 1\frac{1}{2}, y = \frac{3}{4}$ m $x = 1\frac{1}{2}, y = \frac{3}{4}$
2 a They are the same equation. Divide the first by 2 and it is the second, so they have an infinite number of solutions.
b Double the second equation to get $6x + 2y = 14$ and subtract to get $9 = 14$. The left-hand sides are the same if the second is doubled so they cannot have different values.

7.1c Simultaneous equations: Balancing coefficients in both equations

Exercise 7C

1 **a** $x = 5, y = 1$ **b** $x = 3, y = 8$ **c** $x = 9, y = 1$
d $x = 7, y = 3$ **e** $x = 4, y = 2$ **f** $x = 6, y = 5$
g $x = 3, y = -2$ **h** $x = 2, y = \frac{1}{2}$ **i** $x = -2, y = -3$
j $x = -1, y = 2\frac{1}{2}$ **k** $x = 2\frac{1}{2}, y = -\frac{1}{2}$ **l** $x = -1\frac{1}{2}, y = 4\frac{1}{2}$
m $x = -\frac{1}{2}, y = -6\frac{1}{2}$ **n** $x = 3\frac{1}{2}, y = 1\frac{1}{2}$ **o** $x = -2\frac{1}{2}, y = -3\frac{1}{2}$

2 (1, −2) is the solution to equations A and C; (−1, 3) is the solution to equations A and D; (2, 1) is the solution to B and C; (3, −3) is the solution to B and D.
3 Intersection points are (0, 6), (1, 3) and (2, 4). Area is 2 cm^2.
4 Intersection points are (0, 3), (6, 0) and (4, −1). Area is 6 cm^2.

7.2 Rearranging formulae

Exercise 7D

1 $k = \frac{T}{3}$
2 $y = X + 1$
3 $p = 3Q$
4 $r = \frac{A - 9}{4}$
5 $n = \frac{W + 1}{3}$
6 **a** $m = p - t$
b $t = p - m$
7 $m = gv$
8 $m = \sqrt{t}$
9 $r = \frac{C}{2}$
10 $b = \frac{A}{h}$
11 $l = \frac{P - 2w}{2}$
12 $p = \sqrt{m - 2}$
13 **a** $-40 - 32 = -72$, $-72 \div 9 = -8$, $5 \times -8 = -40$
b $68 - 32 = 36$, $36 \div 9 = 4$, $4 \times 5 = 20$
c $F = \frac{9}{5}C + 32$
14 **a** $5x = 9y + 75$, $y = \frac{5x - 75}{9}$
b 25p
15 Average speeds: outward journey = 72 kph, return journey = 63 kph, taking 2 hours. He was held up for 15 minutes.
16 **a** $a = \frac{v - u}{t}$
b $t = \frac{v - u}{a}$
17 $d = \sqrt{\frac{4A}{\pi}}$
18 **a** $n = \frac{W - t}{3}$
b $t = W - 3n$
19 **a** $y = \frac{x + w}{5}$
b $w = 5y - x$
20 $p = \sqrt{\frac{k}{2}}$
21 **a** $t = u^2 - v$
b $u = \sqrt{v + t}$
22 **a** $m = k - n^2$
b $n = \sqrt{k - m}$
23 $r = \sqrt{\frac{T}{5}}$
24 **a** $w = K - 5n^2$
b $n = \sqrt{\frac{K - w}{5}}$

Examination questions

1 $x = 8, y = -2$
2 **a** $4a + 3 = 2b + 5$, $4a - 2b = 2$, $2a - b = 1$
b $a = 2.25, b = 3.5$
3 **a** $2y + x + 1 + 11 = 11 + 2x + 2y + y$, $x + 1 = 2x + y$, $1 = x + y$
b $2y + x + 1 + 11 = x + 1 + 2x + y$, $2y + 11 = 2x + y$, $2x - y = 11$
c $x = 4, y = -3$
d Clockwise from 11: (11), 2, −3, 8, 5, −6
4 pen = 1p, ruler = 21p
5 $b = \frac{2A - ah}{h}$
6 $x = \frac{y - 7}{3}$
7 $x = -1, y = -4$
8 **a** $x = wy + t$ **b** $x = 4, y = 5$
9 $p = \sqrt{n - 6}$
10 $x = 4, y = 1.5$

Answers to Chapter 8

Quick check

1 5.3
2 246.5
3 0.6
4 2.8
5 16.1
6 0.7

8.1 Pythagoras' theorem

Exercise 8A

1 10.3 cm
2 5.9 cm
3 8.5 cm
4 20.6 cm
5 18.6 cm
6 17.5 cm
7 5 cm
8 13 cm
9 10 cm
10 The square in the first diagram and the two squares in the second have the same area

8.2 Finding a shorter side

Exercise 8B

1 **a** 15 cm **b** 14.7 cm **c** 6.3 cm **d** 18.3 cm
2 **a** 20.8 m **b** 15.5 cm **c** 15.5 m **d** 12.4 cm
3 **a** 5 m **b** 6 m **c** 3 m **d** 50 cm
4 There are infinite possibilities, e.g. any multiple of 3, 4, 5 such as 6, 8, 10; 9, 12, 15; 12, 16, 20; multiples of 5, 12, 13 and of 8, 15, 17
5 42.6 cm

8.3 Applying Pythagoras' theorem in real situations

Exercise 8C

1 No. The foot of the ladder is about 6.6 m from the wall
2 2.06 m
3 11.3 m
4 About 17 minutes, assuming it travels at the same speed.
5 127 m − 99.6 m = 27.4 m
6 4.58 m
7 **a** 3.87 m
b 1.74 m
8 3.16 m
9 13 units
10 **a** 4.85 m
b 4.83 m (There is only a small difference.)
11 Yes, because $24^2 + 7^2 = 25^2$
12 6 cm
13 Greater than 20 cm (no width) and less than 28.3 cm (a square)

Exercise 8D

1 **a** 32.2 cm^2
b 2.83 cm^2
c 50.0 cm^2
2 22.2 cm^2
3 15.6 cm^2
4 **a**

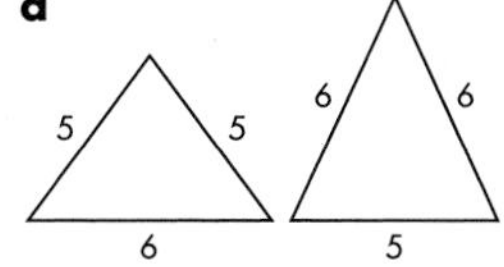

b The areas are 12 cm^2 and 13.6 cm^2 respectively, so triangle with 6 cm, 6 cm, 5 cm sides has the greater area
5 **a**

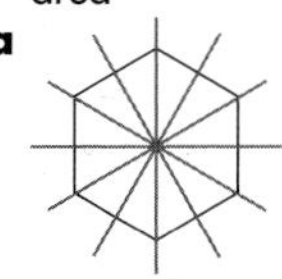

b 166.3 cm^2
6 259.8 cm^2
7 **a** No, areas vary from 24.5 cm^2 to 27.7 cm^2
b No, equilateral triangle gives the largest area
c The closer the isosceles triangle gets to an equilateral triangle the larger its area becomes
8 19.8 or 20 m^2
9 48 cm^2
10 **a** 10 cm **b** 26 cm **c** 9.6 cm

8.4 Pythagoras' theorem in three dimensions

Exercise 8E

1 **a** **i** 14.4 cm **ii** 13 cm **iii** 9.4 cm
b 15.2 cm
2 No, 6.6 m is longest length
3 **a** 20.6 cm **b** 15.0 cm
4 21.3 cm
5 **a** 8.49 m **b** 9 m
6 17.3 cm
7 20.6 cm
8 **a** 11.3 cm **b** 7 cm **c** 8.06 cm
9 **a** 50.0 cm **b** 54.8 cm **c** 48.3 cm **d** 27.0 cm

Examination questions

1 110 cm^2
2 10.8 cm
3 12 cm^2
4 24.1 cm
5 13.6 cm (3 sf)
6 Height^2 = $6^2 - 2^2 = 32$.
Height = $\sqrt{32}$:
Area = $0.5 \times 4 \times \sqrt{32} = 2\sqrt{32} = 2\sqrt{16 \times 2}$
$= 2 \times 4 \times \sqrt{2} = 8\sqrt{2}$

7 **a** AX and BY are both at right angles to AB so they are parallel. Hence ABYX has one pair of parallel sides.
b Form a right angled triangle with XY as the hypotenuse. This has sides XY = 8cm and a short side of 2 cm.
$8^2 - 2^2 = 60$. $\sqrt{60} = 7.7459 = 7.75$ to 3 sf.

Answers to Chapter 9

Quick check

1 $a = 50°$
2 $b = 140°$
3 $c = d = 65°$

9.1 Special triangles and quadrilaterals

Exercise 9A

1 $a = b = 70°, c = 50°, d = 80°,$
$e = 55°, f = 70°, g = h = 57.5°$
2
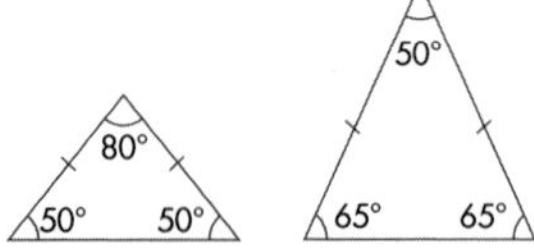

3 **a** $a = 110°, b = 55°$
b $c = e = 105°, d = 75°$
c $f = 135°, g = 25°$
d $e = f = 94°$
e $j = l = 105°, k = 75°$
f $m = o = 49°, n = 131°$
4 40°, 40°, 100°
5 $a = b = 65°, c = d = 115°,$
$e = f = 65°, g = 80°, h = 60°,$
$i = 60°, j = 60°, k = 20°$
6 **a** $x = 25°, y = 15°$
b $x = 7°, y = 31°$
c $x = 60°, y = 30°$
7 **a** $x = 50°$: 60°, 70°, 120°, 110° – possibly trapezium
b $x = 60°$: 50°, 130°, 50°, 130° – parallelogram or isosceles trapezium
c $x = 30°$: 20°, 60°, 140°, 140° – possibly kite
d $x = 20°$: 90°, 90°, 90°, 90° – square or rectangle
8 52°
9 Both 129°
10 $y = 360° - 4x$
11 **a** 65°
b Trapezium, angle A + angle D = 180° and angle B + angle C = 180°

9.2 Angles in polygons

Exercise 9B

1 **a** 1440° **b** 2340° **c** 17 640° **d** 7740°
2 **a** 150° **b** 162° **c** 140° **d** 174°
3 **a** 9 **b** 15 **c** 102 **d** 50
4 **a** 15 **b** 36 **c** 24 **d** 72
5 **a** 12 **b** 9 **c** 20 **d** 40
6 **a** 130° **b** 95° **c** 130°
7 **a** 50° **b** 40° **c** 59°
8 Hexagon
9 100°
10 141°
11 **a** Octagon **b** 89°
12 **a** **i** 71° **ii** 109° **iii** Equal
b If S = sum of the two opposite interior angles, then $S + I = 180°$ (angles in a triangle), and we know $E + I = 180°$ (angles on a straight line), so $S + I = E + I$, therefore $S = E$
13 $a = 144°$
14 Three angles are 135° and two angles are 67.5°.
15 88°; $\frac{1440° - 5 \times 200°}{5}$
16 **a** 36° **b** 10
17 8
18 45°

Examination questions

1 **a**
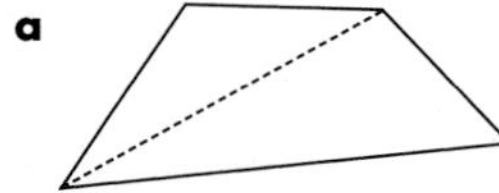
$2 \times 180° = 360°$
b **i** $6x - 18 = 270$ **ii** $x = 48°$, largest angle is 132°
2 **a** 360 ÷ 6 or angle in an equilateral triangle
b 720 **c** 20
3 72°
4 28°
5 **a** c **b** d **c** g
6 **a** A **b** 36°
7 **a** 25° **b** 50° alternate to $2x$

Answers to Chapter 10

Quick check

1 **a** 6 cm **b** 7.5 cm **c** 11 cm

2 **a** 30° **b** 135°

10.1 Constructing triangles

Exercise 10A

1 **a** BC = 2.9 cm, ∠B = 53°, ∠C = 92°
b EF = 7.4 cm, ED = 6.8 cm, ∠E = 50°
c ∠G = 105°, ∠H = 29°, ∠I = 46°
d ∠J = 48°, ∠L = 32°, JK = 4.3 cm
e ∠N = 55°, ON = OM = 7 cm
f ∠P = 51°, ∠R = 39°, QP = 5.7 cm

2 **a** Students can check one another's triangles
b ∠ABC = 44°, ∠BCA = 79°, ∠CAB = 57°

3 **a** 5.9 cm
b 18.8 cm^2

4 BC = 2.6 cm, 7.8 cm

5 **a** 4.5
b 11.25 cm^2

6 **a** 4.3 cm
b 34.5 cm^2

7 **a** Right-angled triangle constructed with sides 3, 4, 5 and 4.5, 6, 7.5, and scale marked 1 cm : 1 m
b Right-angled triangle constructed with 12 equally-spaced dots

8 An equilateral triangle of side 4 cm

9 Even with all three angles, you need to know at least one length

10.2 Bisectors

Exercise 10B

1–9 Practical work; check students' constructions

10 Leicester

11 The centre of the circle

12 Start with a base line AB; then construct a perpendicular to the line from point A. At point B, construct an angle of 60°. Ensure that the line for this 60° angle crosses the perpendicular line; where they meet will be the final point C

13–15 Practical work; check students' constructions

10.3 Defining a locus

Exercise 10C

1 Circle with radius:
a 2 cm **b** 4 cm **c** 5 cm

2 **a** **b** **c**

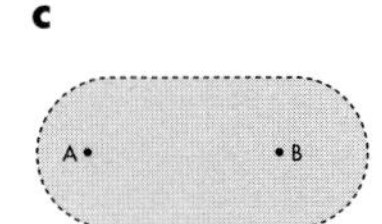

3 **a** Circle with radius 4 m
b

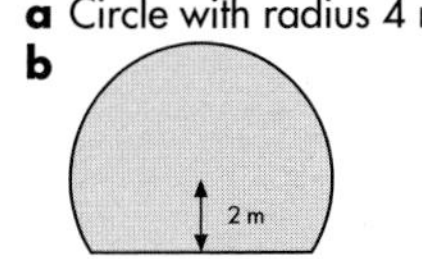
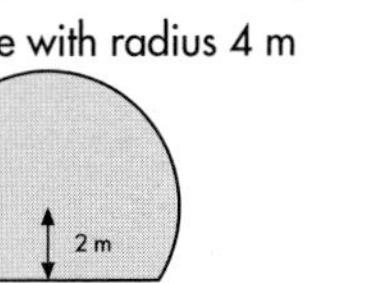

4 **a** **b** **c** **d** **e** **f**

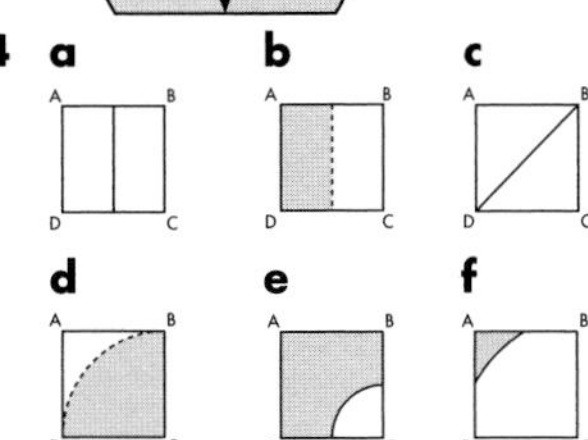

5

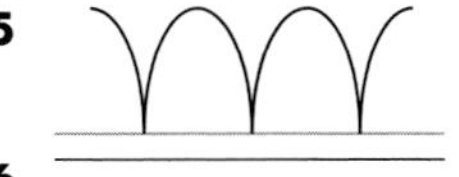

6

7 Construct the bisector of angle BAC and the perpendicular bisector of the line AC

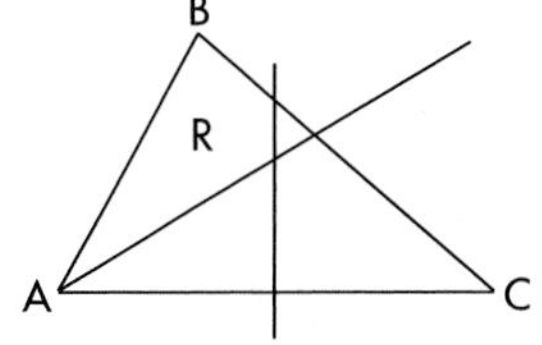

8

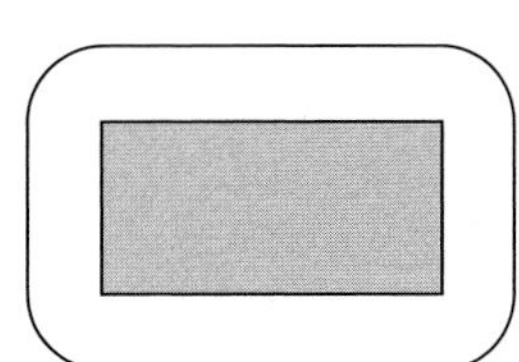

10.4 Loci problems

Exercise 10D

1 Fence Stake Fence

2

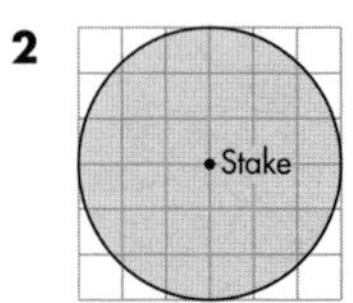

3

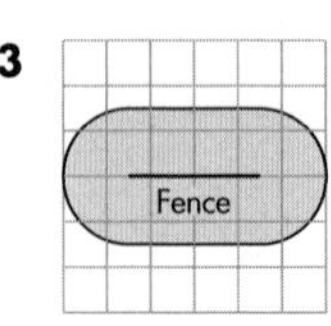

4

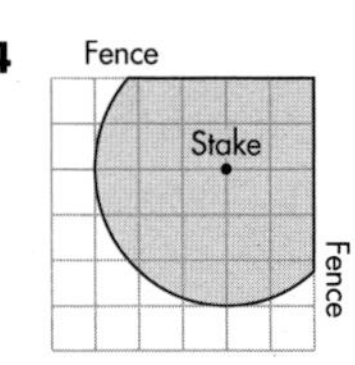

5

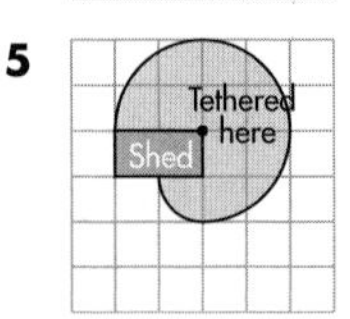

6

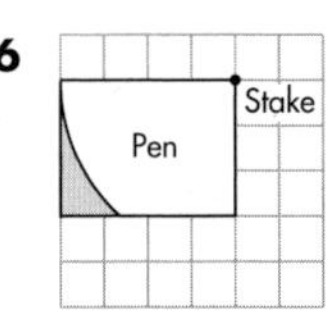

7

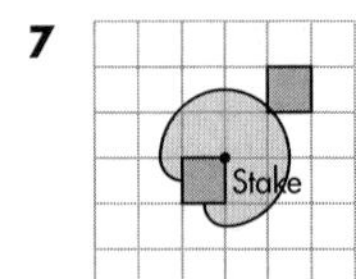

8 **a** Sketch should show a circle of radius 6 cm around London and one of radius 4 cm around Glasgow
b No
c Yes

9 **a** Yes
b Sketch should show a circle of radius 4 cm around Leeds and one of radius 4 cm around Exeter. The area where they overlap should be shaded

10 **a** This is the perpendicular bisector of the line from York to Birmingham. It should pass just below Manchester and just through the top of Norwich
b Sketch should show a circle of radius 7 cm around Glasgow and one of radius 5 cm around London
c The transmitter can be built anywhere on line constructed in part **a** that is within the area shown in part **b**

11 Sketch should show two circles around Birmingham, one of radius 3 cm and one of radius 5 cm. The area of good reception is the area between the two circles

12 Sketch should show a circle of radius 6 cm around Glasgow, two circles around York, one of radius 4 cm and one of radius 6 cm and a circle around London of radius 8 cm. The small area in the Irish Sea that is between the two circles around York and inside both the circle around Glasgow and the circle around London is where the boat can be

13 Sketch should show two circles around Newcastle upon Tyne, one of radius 4 cm and one of radius 6 cm, and two circles around Bristol, one of radius 3 cm and one of radius 5 cm. The area that is between both pairs of circles is the area that should be shaded

14 Sketch should show the perpendicular bisector of the line running from Newcastle upon Tyne to Manchester and that of the line running from Sheffield to Norwich. Where the lines cross is where the oil rig is located

15 Sketch should show the perpendicular bisector of the line running from Glasgow to Norwich and that of the line running from Norwich to Exeter. Where the lines cross is where Fred's house is

16 Sketch should show the bisectors of the angles made by the piers and the sea wall at points A and B. These are the paths of each boat

17 Leeds

18 On a map, draw a straight line from Newcastle to Bristol, construct the line bisector, then the search will be anywhere on the sea along that line

Examination questions

1 Check students' drawings, top angle should be 110°

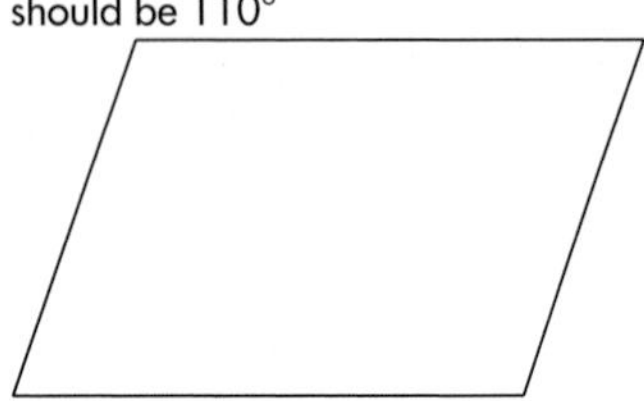

2 Check students' drawings, top angles should be 90° and 130°

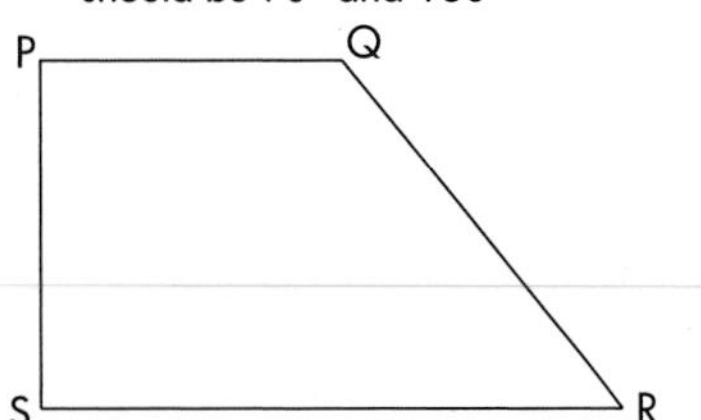

3 **a**

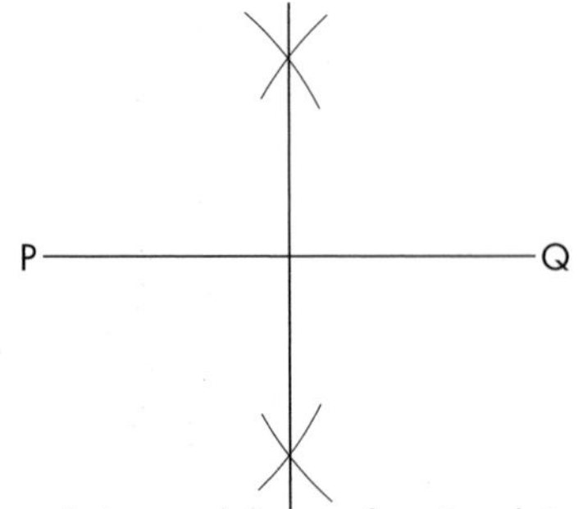

b An equal distance from P and Q

4

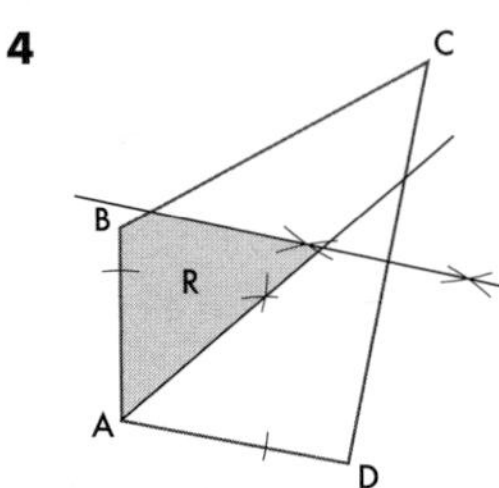

5

6 a

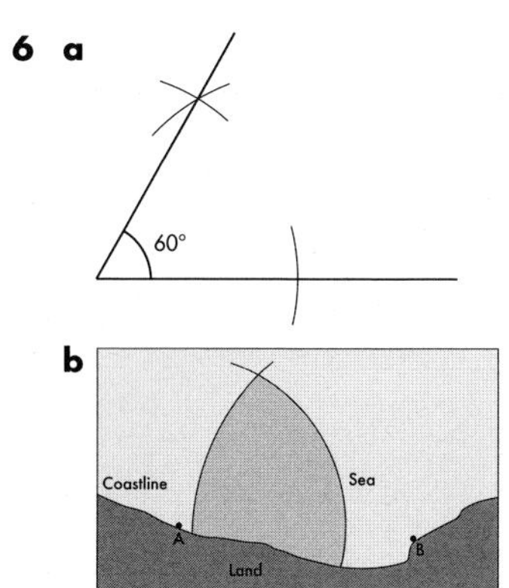

b

7 a

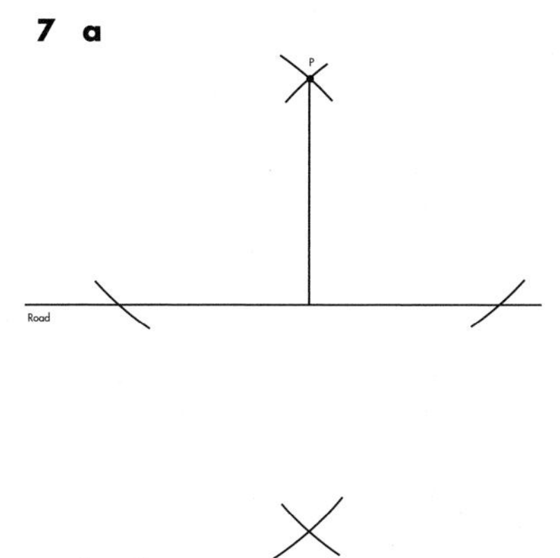

b 3 km

Answers to Chapter 11

Quick check

Trace shape **a** and check whether it fits exactly on top of the others.
You should find that shape **b** is not congruent to the others.

11.1 Congruent triangles

Exercise 11A

1 **a** SAS **b** SSS **c** ASA **d** RHS **e** SSS **f** ASA

2 **a** SSS. A to R, B to P, C to Q
b SAS. A to R, B to Q, C to P

3 **a** 60° **b** 80° **c** 40° **d** 5 cm

4 **a** 110° **b** 55° **c** 85° **d** 110° **e** 4 cm

5 SSS or RHS

6 SSS or SAS or RHS

7 For example, use $\triangle ADE$ and $\triangle CDG$. AD = CD (sides of large square), DE = DG (sides of small square), $\angle ADE = \angle CDG$ (angles sum to 90° with $\angle ADG$), so $\triangle ADE \equiv \triangle CDG$ (SAS), so AE = CG

8 AB and PQ are the corresponding sides to the 42° angle, but they are not equal in length

11.2 Translations

Exercise 11B

1 **a** **i** $\begin{pmatrix}1\\3\end{pmatrix}$ **ii** $\begin{pmatrix}4\\2\end{pmatrix}$ **iii** $\begin{pmatrix}2\\-1\end{pmatrix}$
iv $\begin{pmatrix}5\\1\end{pmatrix}$ **v** $\begin{pmatrix}-1\\6\end{pmatrix}$ **vi** $\begin{pmatrix}4\\6\end{pmatrix}$

b **i** $\begin{pmatrix}-1\\-3\end{pmatrix}$ **ii** $\begin{pmatrix}3\\-1\end{pmatrix}$ **iii** $\begin{pmatrix}1\\-4\end{pmatrix}$
iv $\begin{pmatrix}4\\-2\end{pmatrix}$ **v** $\begin{pmatrix}-2\\3\end{pmatrix}$ **vi** $\begin{pmatrix}3\\3\end{pmatrix}$

c **i** $\begin{pmatrix}-4\\-2\end{pmatrix}$ **ii** $\begin{pmatrix}-3\\1\end{pmatrix}$ **iii** $\begin{pmatrix}-2\\-3\end{pmatrix}$
iv $\begin{pmatrix}1\\-1\end{pmatrix}$ **v** $\begin{pmatrix}-5\\4\end{pmatrix}$ **vi** $\begin{pmatrix}0\\4\end{pmatrix}$

d **i** $\begin{pmatrix}3\\2\end{pmatrix}$ **ii** $\begin{pmatrix}-4\\2\end{pmatrix}$ **iii** $\begin{pmatrix}5\\-4\end{pmatrix}$
iv $\begin{pmatrix}-2\\-7\end{pmatrix}$ **v** $\begin{pmatrix}5\\0\end{pmatrix}$ **vi** $\begin{pmatrix}1\\-5\end{pmatrix}$

2

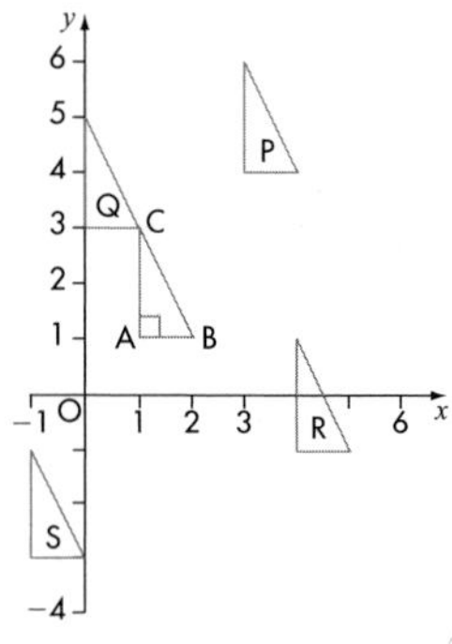

3 **a** $\begin{pmatrix}-3\\-1\end{pmatrix}$ **b** $\begin{pmatrix}4\\-4\end{pmatrix}$ **c** $\begin{pmatrix}-5\\-2\end{pmatrix}$

d $\begin{pmatrix}4\\7\end{pmatrix}$ **e** $\begin{pmatrix}-1\\5\end{pmatrix}$ **f** $\begin{pmatrix}1\\6\end{pmatrix}$

g $\begin{pmatrix}-4\\4\end{pmatrix}$ **h** $\begin{pmatrix}-4\\-7\end{pmatrix}$

4 $10 \times 10 = 100$ (including $\begin{pmatrix}0\\0\end{pmatrix}$)

5 Check students' designs for a *Snakes and ladders* board.

6 $\begin{pmatrix}-x\\-y\end{pmatrix}$

7 $\begin{pmatrix}-300\\-500\end{pmatrix}$

8 $\begin{pmatrix}-1\\4\end{pmatrix}$

11.3 Reflections

Exercise 11C

1

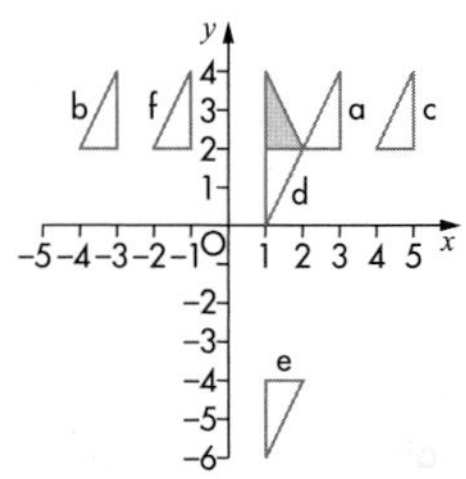

2 a–e

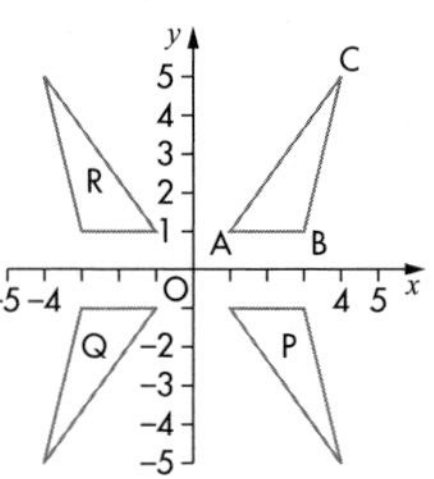

f Reflection in the y-axis

3 a–b

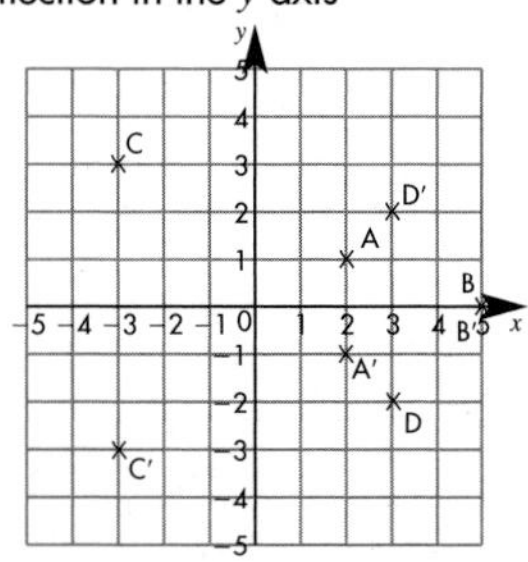

c y-value changes sign

d $(a, -b)$

4 a–b

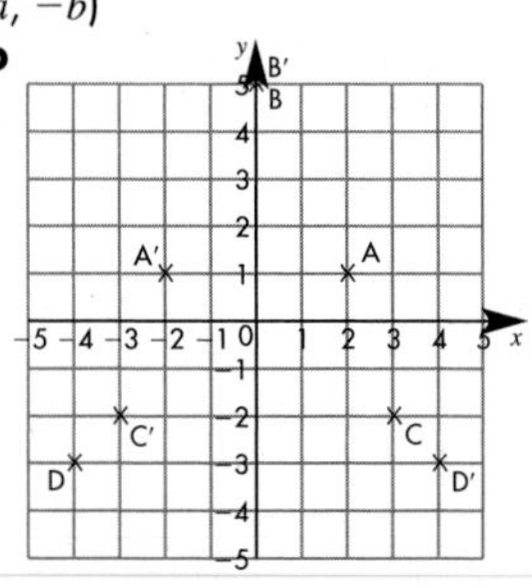

c x-value changes sign

d $(-a, b)$

5 Possible answer: Take the centre square as ABCD then reflect this square each time in the line, AB, then BC, then CD and finally AD

6 $x = -1$

7 Possible answer:

8

9 a–i

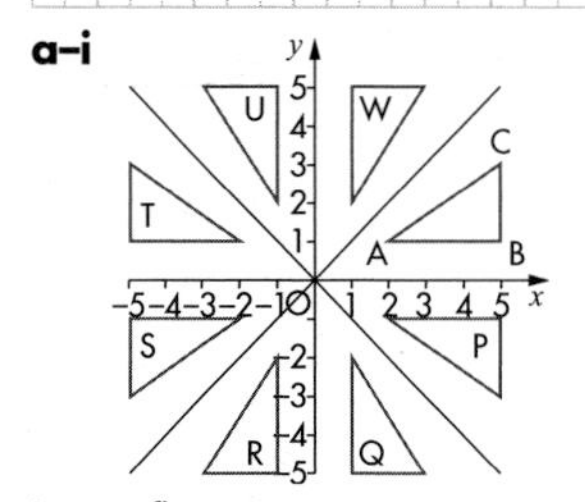

i A reflection in $y = x$

10

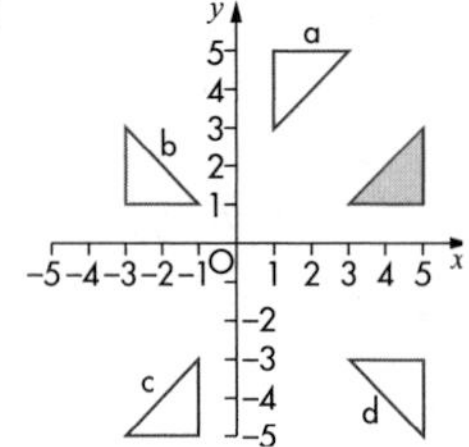

11 a–c

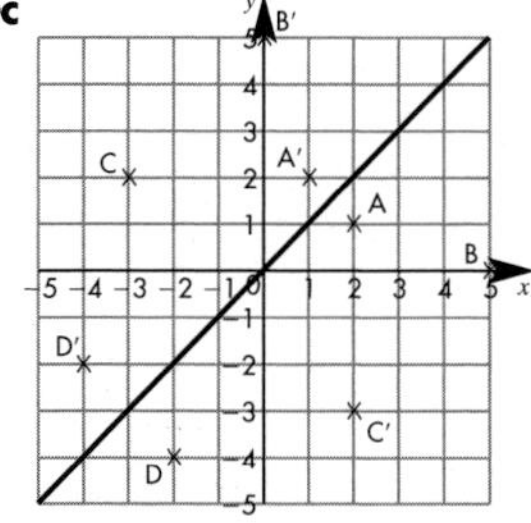

d Coordinates are reversed: x becomes y and y becomes x

e (b, a)

12 a–c

d

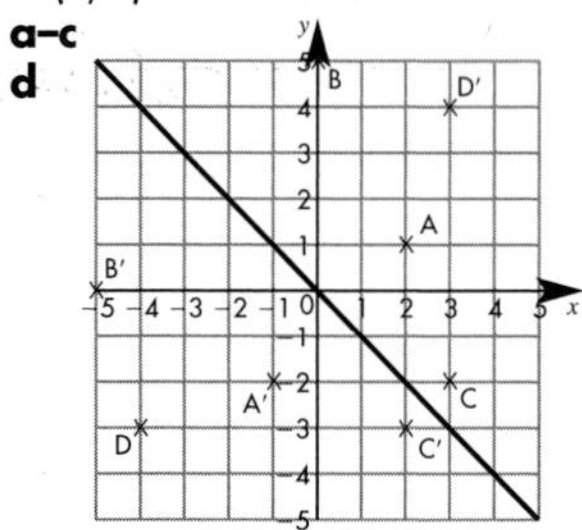

Coordinates are reversed and change sign, x becomes $-y$ and y becomes $-x$

e $(-b, -a)$

11.4 Rotations

Exercise 11D

1 **a**

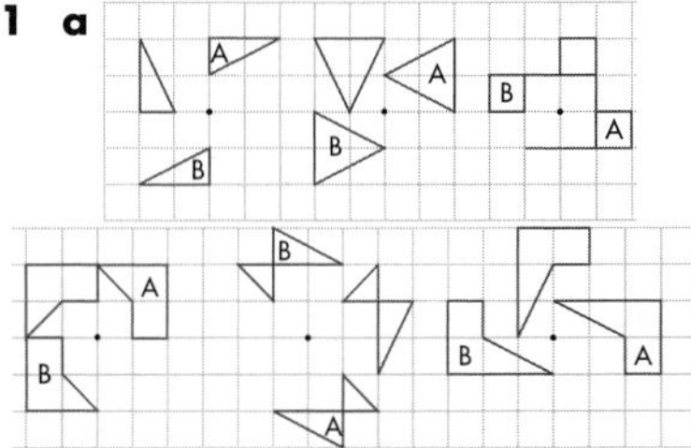

b **i** Rotation 90° anticlockwise
ii Rotation 180°

2

3 Possible answer: If ABCD is the centre square, rotate about A 90° anticlockwise, rotate about new B 180°, now rotate about new C 180°, and finally rotate about new D 180°

4

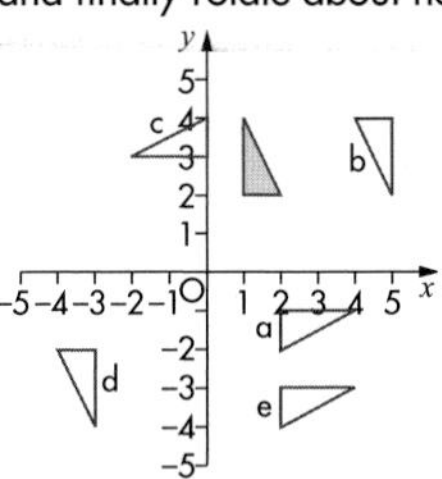

5 **a** 90° anticlockwise
b 270° anticlockwise
c 300° clockwise
d 260° clockwise

6 **a b c** **i**

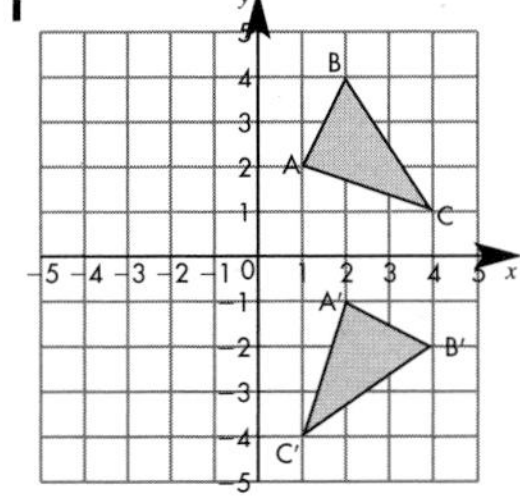

ii A′(2, −1), B′(4, −2), C′(1, −4)
iii Original coordinates (x, y) become $(y, -x)$
iv Yes

7 **i**

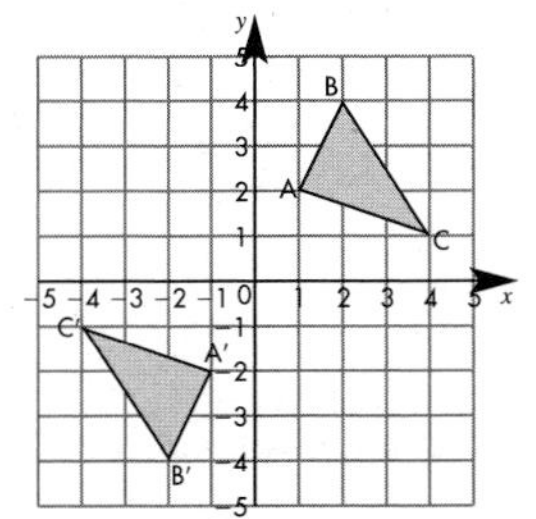

ii A′(−1, −2), B′(−2, −4), C′(−4, −1)
iii Original coordinates (x, y) become $(-x, -y)$
iv Yes

8 **i**

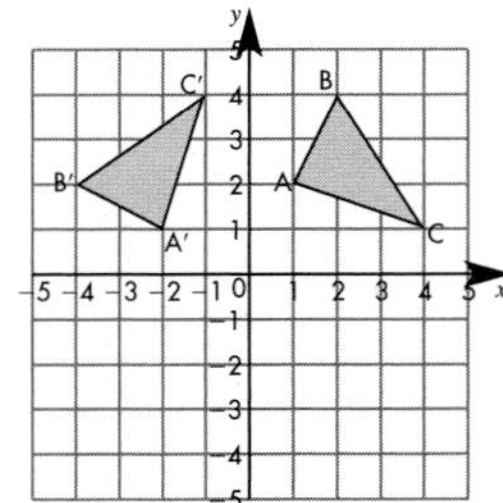

ii A′(−2, 1), B′(−4, 2), C′(−1, 4)
iii Original coordinates (x, y) become $(-y, x)$
iv Yes

9 Show by drawing a shape or use the fact that (a, b) becomes $(a, -b)$ after reflection in the x-axis, and $(a, -b)$ becomes $(-a, -b)$ after reflection in the y-axis, which is equivalent to a single rotation of 180°.

10 Show by drawing a shape or use the fact that (a, b) becomes (b, a) after reflection in the line $y = x$, and (b, a) becomes $(-a, -b)$ after reflection in the line $y = -x$, which is equivalent to a single rotation of 180°.

11 **a**

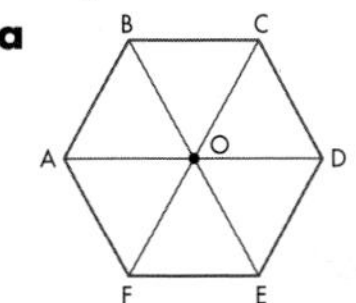

b **i** Rotation 60° clockwise about O
ii Rotation 120° clockwise about O
iii Rotation 180° about O
iv Rotation 240° clockwise about O

c **i** Rotation 60° clockwise about O
ii Rotation 180° about O

12 Rotation 90° anticlockwise about (3, −2)

11.5 Enlargements

Exercise 11E

1

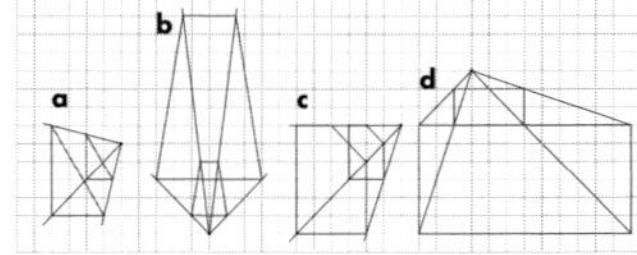

2 **a**

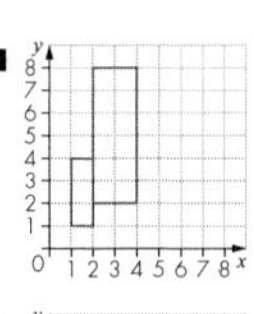

b

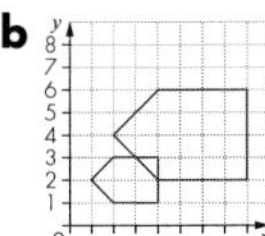

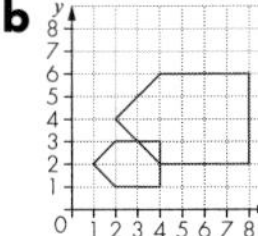

c

3 **a**

b

4

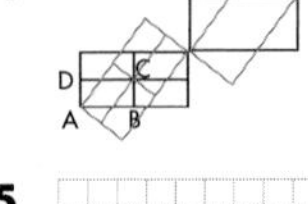

5

6 **a**

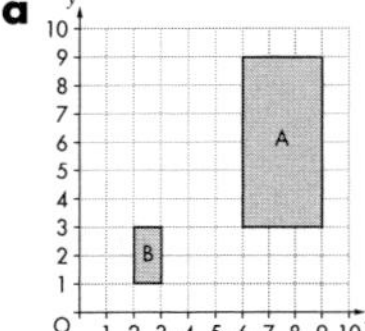

b 3 : 1
c 3 : 1
d 9 : 1

7

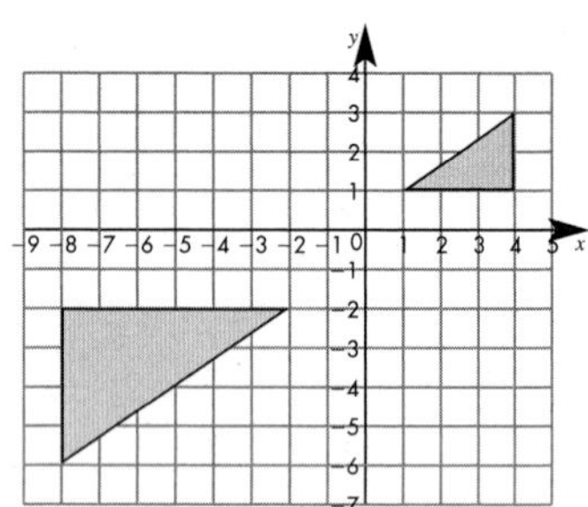

8 a–c

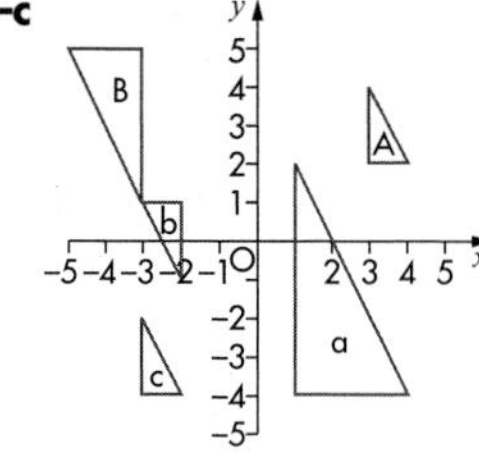

d Scale factor $-\frac{1}{2}$, centre (1, 3)

e Scale factor -2, centre (1, 3)

f Scale factor -1, centre $(-2.5, -1.5)$

g Scale factor -1, centre $(-2.5, -1.5)$

h Same centres, and the scale factors are reciprocals of each other

9 Enlargement, scale factor -2, about (1, 3)

11.6 Combined transformations

Exercise 11F

1 $(-4, -3)$

2 **a** $(-5, 2)$

b Reflection in y-axis

3 A: translation $\begin{pmatrix} 1 \\ -2 \end{pmatrix}$, B: reflection in y-axis, C: rotation 90°clockwise about (0, 0), D: reflection in $x = 3$, E: reflection in $y = 4$, F: enlargement by scale factor 2, centre (0, 1)

4 **a** T_1 to T_2: rotation 90°clockwise about (0, 0)

b T_1 to T_6: rotation 90°anticlockwise about (0, 0)

c T_2 to T_3: translation $\begin{pmatrix} 2 \\ 2 \end{pmatrix}$

d T_6 to T_2: rotation 180°about (0, 0)

e T_6 to T_5: reflection in y-axis

f T_5 to T_4: translation $\begin{pmatrix} 4 \\ 0 \end{pmatrix}$

5 **a–d**

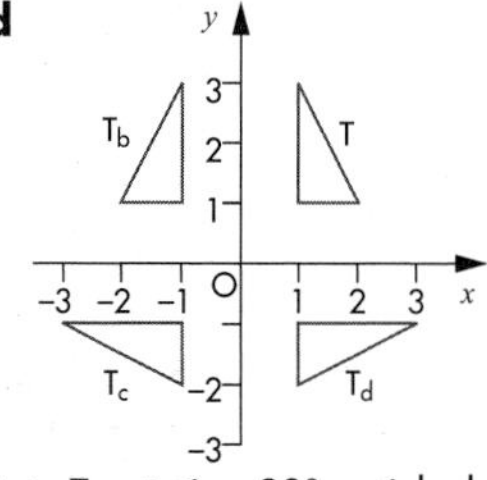

e T_d to T: rotation 90° anticlockwise about (0, 0)

6 (3, 1)

7 Reflection in x-axis, translation $\begin{pmatrix} 0 \\ -5 \end{pmatrix}$, rotation 90°clockwise about (0, 0)

8 Translation $\begin{pmatrix} 0 \\ -8 \end{pmatrix}$, reflection in x-axis, rotation 90°clockwise about (0, 0)

9 **a**

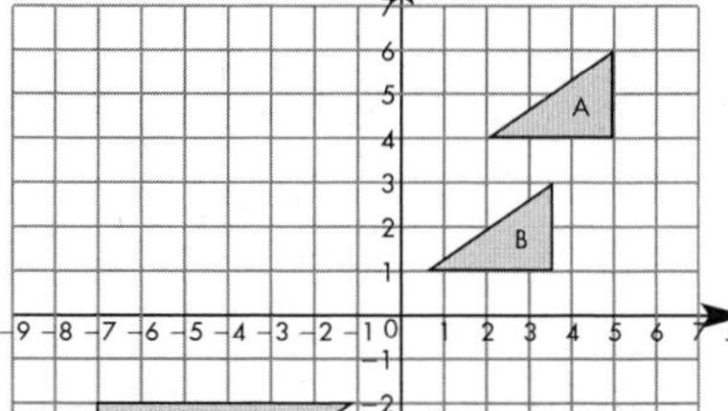

b Enlargement of scale factor $-\frac{1}{2}$ about (1, 2)

Examination questions

1

2 **a** Rotation 90° clockwise about (0, 0)

b $\begin{pmatrix} -5 \\ -4 \end{pmatrix}$

3 **a** $x = 5$

b $\begin{pmatrix} 5 \\ 0 \end{pmatrix}$

c 180° about (5, 2.5)

4 a, b

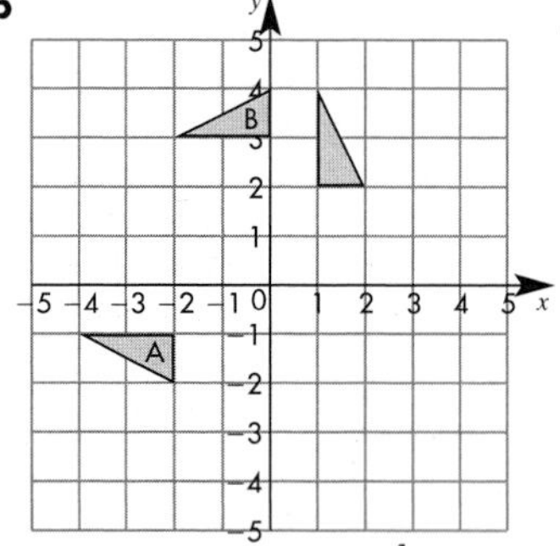

5 **a** Enlargement scale factor of $\frac{1}{3}$ about $(-4, 5)$

b

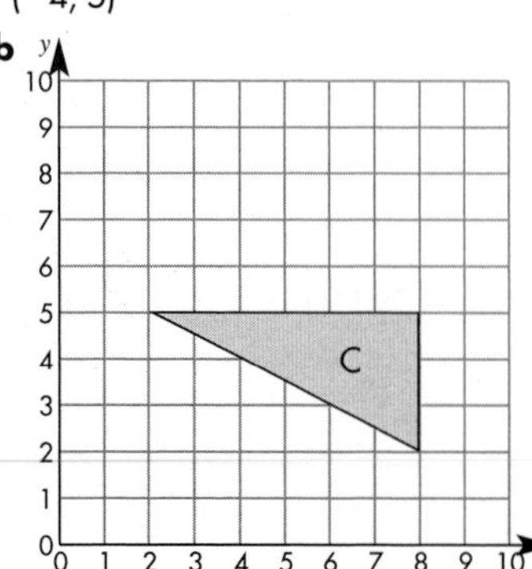

6 **a** Reflection in the line $y = x$

b Rotation of 90° anticlockwise about (1, 1)

c Enlargement scale factor of -2 about (0, 0)

7 For example, AB = BC (isosceles triangle), AM = MC (M is midpoint of AC), BM is common, so $\triangle$ABM $\equiv$ $\triangle$CBM (SSS)

8 For example, $\angle$MYZ = $\angle$NZY (given), YZ is common, MY = NZ (symmetry of isosceles triangle), so $\triangle$YMZ $\equiv$ $\triangle$ZNY (SAS)

9 **a** BE = DF (given), AD = BC (opposite sides of a parallelogram), angle ADC = angle EBC (opposite angles in a parallelogram), Hence congruent due to SAS. There are other ways to prove this.

b Angle BEF = angle AFD so EC and AF parallel as EB and DF are parallel.

Answers to Chapter 12

Quick check

1 **a** 7 **b** 6 **c** 8 **d** 6

12.1 Averages

Exercise 12A

1 Mode
2 Three possible answers: 12, 14, 14, 16, 18, 20, 24; or 12, 14, 14, 16, 18, 22, 24; or 12, 14, 14, 16, 20, 22, 24
3 53
4 **a** Median (mean could be unduly influenced by results of very able and/or very poor candidates)
b Median (mean could be unduly influenced by pocket money of students with very rich or generous parents)
c Mode (numerical value of shoe sizes irrelevant, just want most common size)
d Median (mean could be distorted by one or two extremely short or tall performers)
e Mode (the only way to get an 'average' of non-numerical values)
f Median (mean could be unduly influenced by very low weights of premature babies)
5 The mean is 31.5 which rounds up to 32, so the statement is correct (though the mode and median are 31).
6 **a** **i** £20000 **ii** £28000 **iii** £34000
b A 6% rise would increase the mean salary to £36040, a £1500 pay increase would produce a mean of £35500.
7 **a** Median **b** Mode **c** Mean
8 Tom – mean, David – median, Mohamed – mode
9 **a** 9 **b** 7
10 **a** $6x$ **b** 8
11 $2x - 4$
12 $2x + 3$
13 11.6
14 42.7 kg
15 24
16 **a** Possible answer: 1, 6, 6, 6, 6
b Possible answer: 2, 5, 5, 6, 7
17 Boss chose the mean while worker chose the mode.
18 $5x = 17 + x + y$, $4x = 17 + y$, $y = 4x - 17$
19 **a** $a = 2y - x$ **b** $b = \frac{x}{4} + \frac{3y}{4}$
20 B = 66, C = 69, D = 93, E = 69

12.2 Frequency tables

Exercise 12B

1 **a** **i** 7 **ii** 6 **iii** 6.4
b **i** 8 **ii** 8.5 **iii** 8.2
2 **a** 668 **b** 1.9 **c** 0 **d** 328
3 **a** 2.2, 1.7, 1.3 **b** Better dental care
4 **a** 50 **b** 2 **c** 2.8
5 **a** Roger 5, Brian 4 **b** Roger 3, Brian 8
c Roger 5.4, Brian 4.5 **d** Roger 5, Brian 4
e Roger, smaller range **f** Brian, better mean
6 **a** 40 **b** 7 **c** 3 **d** 2
e 2.5 **f** 2.5 **g** 2.4
7 **a** 2 **b** 1.9 **c** 49%
8 5
9 The total frequency could be an even number where the two middle numbers have an odd difference.
10 **a** 34
b $x + 80 + 3y + 104 = 266$, so $x + 3y = 266 - 184 = 82$
c $x = 10$, $y = 24$
d 2.5
11 The mean for 2009 = 396 ÷ 12 = 33, Range = 38
12 **a** 7 **b** 4 **c** 1.75
13 Any of (a, b), (1, 8), (3, 7), (5, 6), (7, 5), (9, 4), (11, 3), (13, 2), (15, 1)

12.3 Grouped data

Exercise 12C

1 **a** **i** $30 < x \leqslant 40$ **ii** 29.5
b **i** $0 < y \leqslant 100$ **ii** 158.3
c **i** $5 < z \leqslant 10$ **ii** 9.43
d **i** 7–9 **ii** 8.41
2 **a** $100\text{ g} < w \leqslant 120\text{ g}$ **b** 10.86 kg **c** 108.6 g
3 **a** $175 < h \leqslant 200$ **b** 31%
c 193.3 hours **d** No the mean was under 200.
4 **a** Yes, average distance is 11.7 miles per day.
b Because shorter runs will be done faster which will affect the average
c Yes, because the shortest could be 1 mile, the longest 25 miles
5 24
6 Soundbuy; average increases are Soundbuy 17.7p, Springfields 18.7p, Setco 18.2p
7 **a** 160 **b** 52.6 minutes
c Modal group **d** 65%
8 The first 5 and the 10 are the wrong way round.
9 Find the midpoint of each group, multiply that by the frequency and add those products. Divide that total by the total frequency.
10 **a** Yes, as total in first two columns is 50, so median is between 39 and 40
b He could be correct, as the biggest possible range is 69 − 20 = 49, and the lowest is 60 − 29 = 31.

12.4 Frequency diagrams

Exercise 12D

1 **a** 36
b Pie charts with these angles: 50°, 50°, 80°, 60°, 60°, 40°, 20°
c Check students' bar charts
d Bar chart, because easier to make comparisons

2 **a** Pie charts with these angles: 124°, 132°, 76°, 28°
b Split of total data seen at a glance

3 **a** 55° **b** 22 **c** 33%

4 **a** Pie charts with these angles:
Strings: 36°, 118°, 126°, 72°, 8°
b Brass: 82°, 118°, 98°, 39°, 23°
Overall, the brass candidates did better, as a smaller proportion got the lowest grade. A higher proportion of strings candidates scored the top two grades.

5 The sector for 'Don't know' has an angle of 360° − (80° + 90° + 150°) = 40°, as a fraction of the whole circle, the 'Don't know' sector is $\frac{40}{360} = \frac{1}{9}$, or as a percentage, 11%

6 Identify the possible ways in which students might come to school and, on the morning in question, use a tally chart to record how each arrives.

Exercise 12E

1 **a**

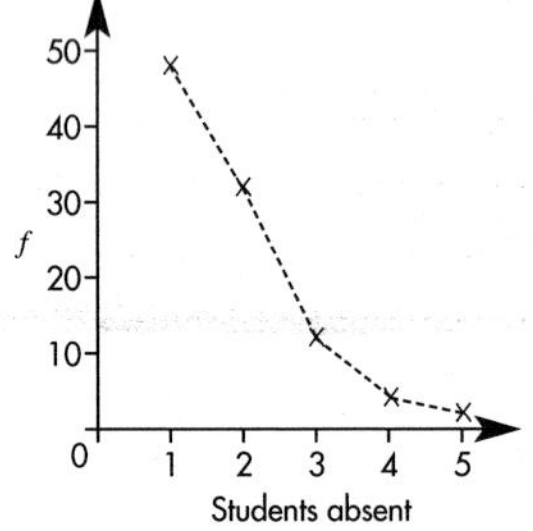

b 1.7

2 **a**

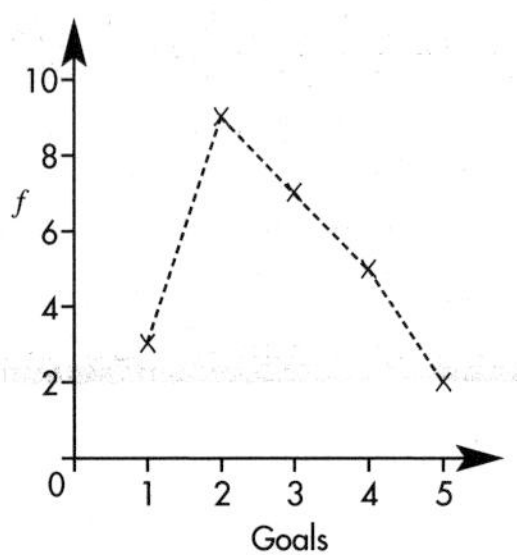

b 2.8

3 **a** **i** 17, 13, 6, 3, 1 **ii** £1.45
b **i**

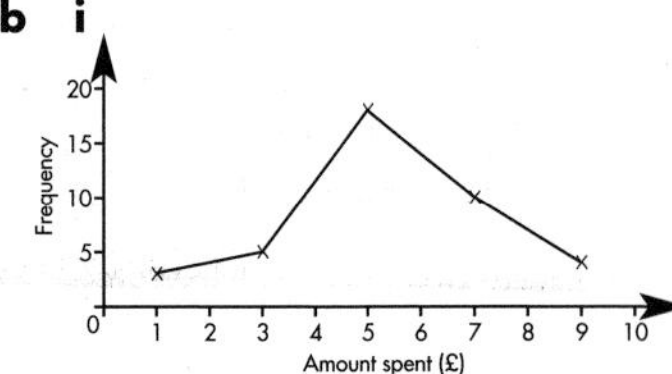

ii £5.35
c Much higher mean. Early morning, people just want a paper or a few sweets. Later people are buying food for the day.

4 **a**

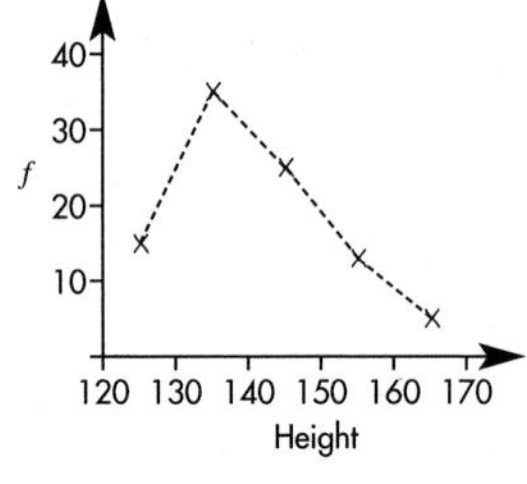

b

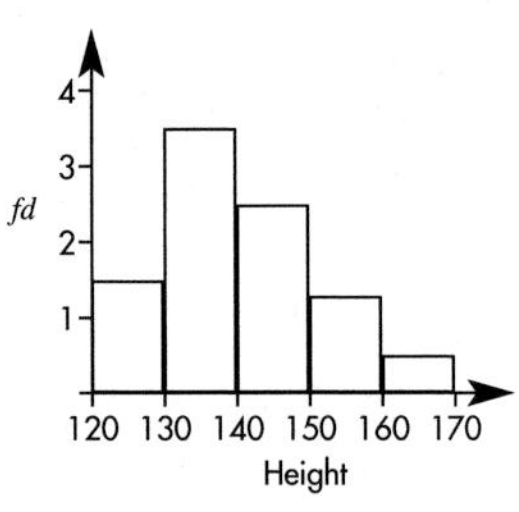

c 140.4 cm

5 **a**

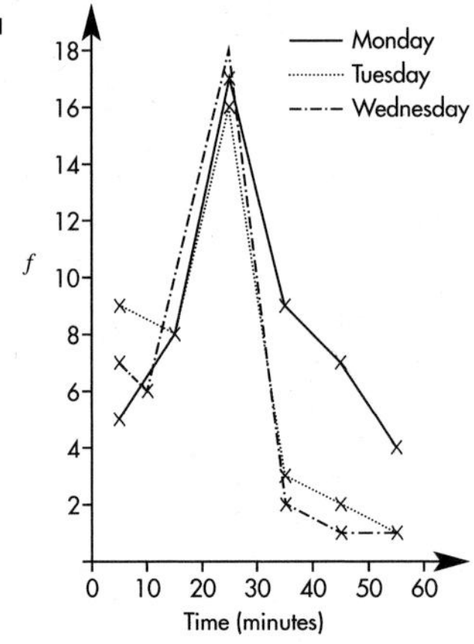

b Monday 28.4 min, Tuesday 20.9 min, Wednesday 21.3 min
c There are more patients on a Monday, and so longer waiting times, as the surgery is closed during the weekend.

6 **a**

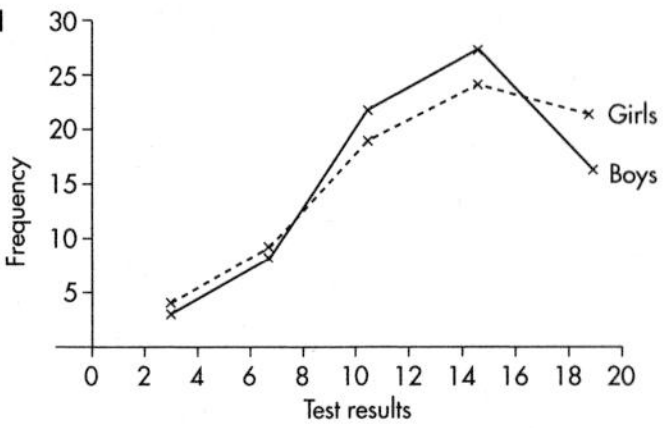

Girls
Boys
Frequency
Test results

b Boys 12.9, girls 13.1, and so the girls did slightly better than the boys.

7 2.17 hours

8 That is the middle value of the time group 0 to 1 minute. It would be very unusual for most of them to be exactly in the middle at 30 seconds.

12.5 Histograms with bars of unequal width

Exercise 12F

1 The respective frequency densities on which each histogram should be based are:
a 2.5, 6.5, 6, 2, 1, 1.5 **b** 4, 27, 15, 3
c 17, 18, 12, 6.67 **d** 0.4, 1.2, 2.8, 1 **e** 9, 21, 13.5, 9

2 a

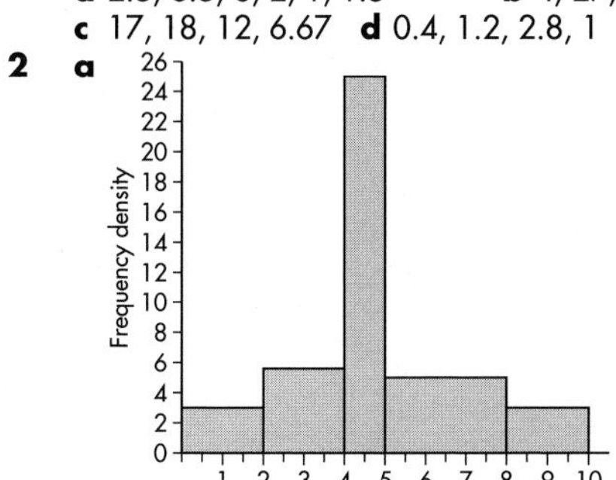

b

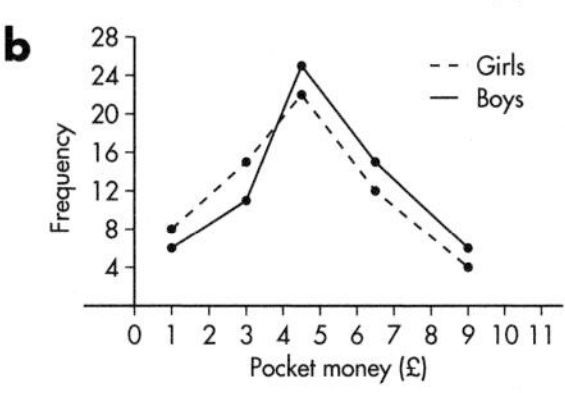

c Girls £4.36, boys £4.81. Boys get more pocket money than girls do.

3

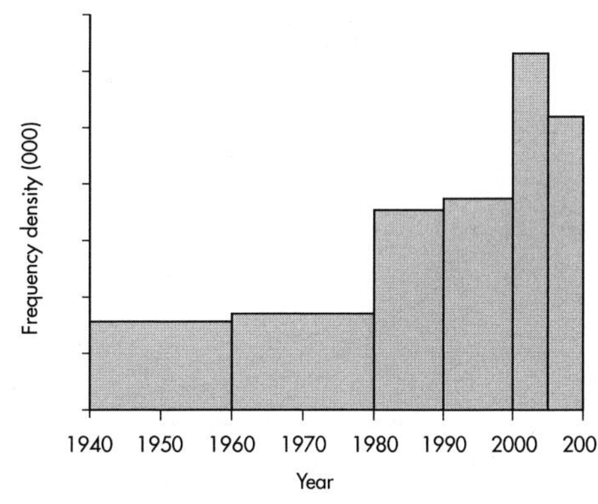

4 a 775 **b** 400

5 Divide the frequency of the class interval by the width of the class interval.

6 a i

Age, y (years)	$9 < y \leqslant 10$	$10 < y \leqslant 12$	$12 < y \leqslant 14$	$14 < y \leqslant 17$	$17 < y \leqslant 19$	$19 < y \leqslant 20$
Frequency	4	12	8	9	5	1

ii 10–12 **iii** 13 **iv** 11, 16, 5 **v** 13.4

b i

Temperature, t (°C)	$10 < t \leqslant 11$	$11 < t \leqslant 12$	$12 < t \leqslant 14$	$14 < t \leqslant 16$	$16 < t \leqslant 19$	$19 < t \leqslant 21$
Frequency	15	15	50	40	45	15

ii 12–14°C **iii** 14.5°C **iv** 12°C, 17°C, 5°C **v** 14.8°C

c i

Weight, w (kg)	$50 < w \leqslant 70$	$70 < w \leqslant 90$	$90 < w \leqslant 100$	$100 < w \leqslant 120$	$120 < w \leqslant 170$
Frequency	160	200	120	120	200

ii 70–90 kg and 120–170 kg **iii** 93.33 kg **iv** 74 kg, 120 kg, 46 kg **v** 99.0 kg

7 a 7.33 hours **b** 8.44 hours **c** 7 hours

8 a

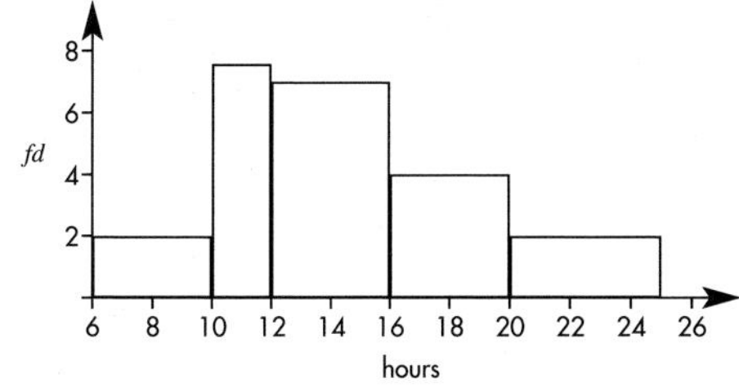

b 14.2 kg **c** 14.7 kg **d** 33 plants

9 a

Speed, v (mph)	$0 < v \leqslant 40$	$40 < v \leqslant 50$	$50 < v \leqslant 60$	$60 < v \leqslant 70$	$70 < v \leqslant 80$	$80 < v \leqslant 100$
Frequency	80	10	40	110	60	60

b 360 **c** 64.5 mph **d** 59.2 mph

10 a 102 **b** 35 **c** 104 **d** 75

11 0.45

12.6 Surveys

Exercise 12G

1–5 Check students' answers and designs, which will vary.

6 **a** Possible answer: Question – Which of the following foods would you normally eat for your main meal of the day?

Name	Sex	Chips	Beef burgers	Vegetables	Pizza	Fish

b Yes, as a greater proportion of girls ate healthy food

7 Possible answer: Question – What kind of tariff do you use on your mobile phone?

Name	Pay as you go		Contract	
	200 or over free texts	Under 200 free texts	200 or over free texts	Under 200 free texts

(Any sheet in which choices that can distinguish one from the other have to be made will be accepted.)

8 Possible answers: shop names, year of student, tally space, frequency

12.7 Questionnaires

Exercise 12H

1 **a** It is a leading question, and no option to disagree with the statement.
b Unbiased, and the responses do not overlap.

2 **a** Responses overlap.
b Provide options: up to £2, more than £2 and up to £5, more than £5 and up to £10, more than £10.

3–6 Check students' questionnaires.

7 **a** This is a leading question with no possibility of showing disagreement.
b This is a clear direct question that has an answer, and good responses as only one selection can be made.
c Check students' questions.

8 Possible questionnaire: Do you have a back problem?
☐ Yes ☐ No
Tick the diagram/text that best illustrates/describes how you sit.
☐ shoulders back awkwardly, curved spine
☐ slumped, straining lower back
☐ caved chest, pressure on spine
☐ balanced, head and spine aligned

9 The groups overlap, and the 'less than £15' is also in the 'less than £25'.

12.8 The data-handling cycle

Exercise 12I

1 **a** secondary data **b** primary data
c primary or secondary **d** primary or secondary
e primary data **f** primary data

2 Students' answers will vary

3 For example, Kath may carry out a survey among her friends or class-mates.

12.9 Other uses of statistics

Exercise 12J

1 Price 78p, 80.3p, 84.2p, 85p, 87.4p, 93.6p

2 **a** £1 = $1.88
b Greatest drop was from June to July.
c There is no trend in the data.

3 **a** 9.7 million **b** 4.5 years
c 12 million **d** 10 million

4 £74.73

5 General cost of living in 2009 dropped to 98% of the costs in 2008.

6 **a** Holiday month
b **i** 140 thousand
ii 207 thousand (an answer of 200–210 thousand over the 3 months is acceptable)

12.10 Sampling

Exercise 12K

1–4 Check students' answers as they will vary.

5 **a** How many times, on average, do you visit a fast-food outlet in a week?

☐ Never ☐ 1 or 2 times

☐ 3 or 4 times ☐ More than 4 times

b

	Boys	Girls
Y9	11	8
Y10	10	11
Y11	10	10

6 555

7 Find the approximate proportion of men and women, girls and boys, then decide on a sample size. Work out the proportion of men in the whole group and find that proportion of the sample size to give the number of men in the sample. Similarly work out the number of women, boys and girls.

8 **a** There are many possible correct answers. Below are two examples.

How far from Meadowhall do you live?

☐ Less than 5 miles

☐ Between 5 and 10 miles inclusive

☐ More than 10 miles

When you visit Meadowhall, approximately how much do you usually spend?

☐ £50 or less

☐ Between £50 and £100

☐ £100 or more

b $Y7 = 143 \times \frac{100}{670} = 21$ $Y10 = 131 \times \frac{100}{670} = 20$

$Y8 = 132 \times \frac{100}{670} = 20$ $Y11 = 108 \times \frac{100}{670} = 16$

$Y9 = 156 \times \frac{100}{670} = 23$

9 **a** There are many more girls than boys.

b 20

Examination questions

1 **a** **i** 8 **ii** 23 **iii** 19

b There is no space for 0 hours and 6 hours appears in two groups.

2 **a** 1.7

b Which days of the week are you prepared to car share?

☐ None ☐ Mon ☐ Tues ☐ Wed ☐ Thu ☐ Fri

3 28.2 min

4 **a** 4060 ÷ 100

b **i** 125 **ii** 140.6 cm

c **i**

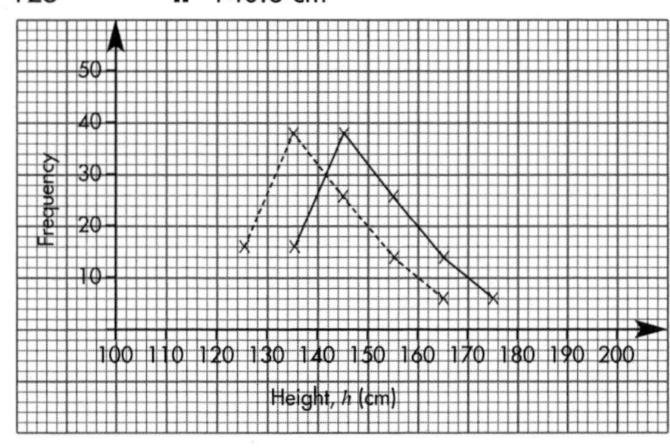

ii On average, the boys are about 10 cm taller than the girls. The range of the heights in both groups is the same, 40 cm.

5 **a**

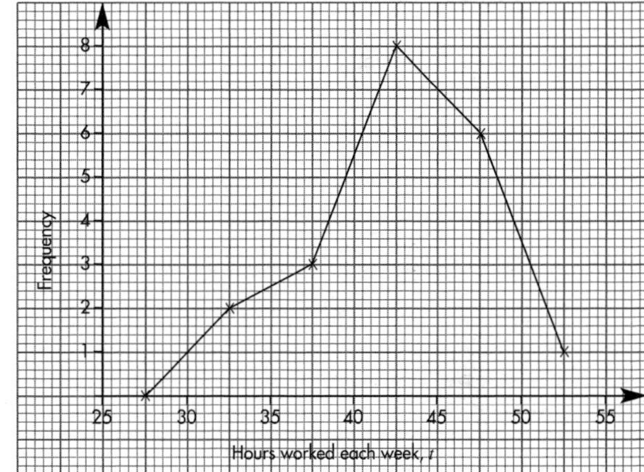

b 60

c It is what they thought and not what they measured.

6 **a** Level 5 **b** 3.2

c Because of all the zeros at the lower grades in German

7 Because over half the students have more than £10 pocket money, so the mean must be more than £10

8 24.6

9 **a** Polygon from (5, 6), (15, 10), (25, 20), (35, 8), (45, 6)

b Mean for girls is lower. Girls times are more consistent.

10 **a** 25 **b** 50 **c** 20 years

11 70

12 a

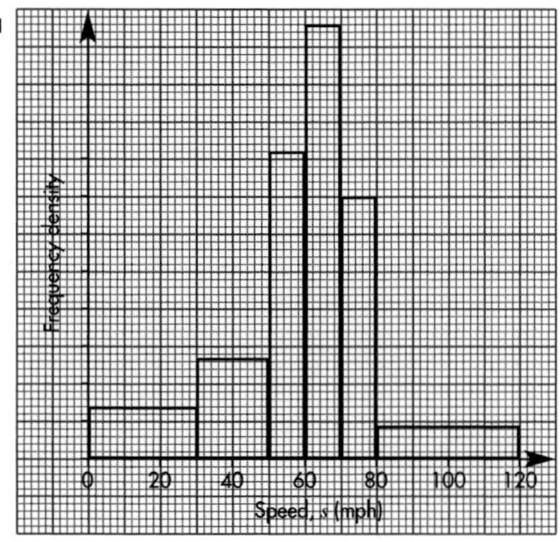

b 57

13 a

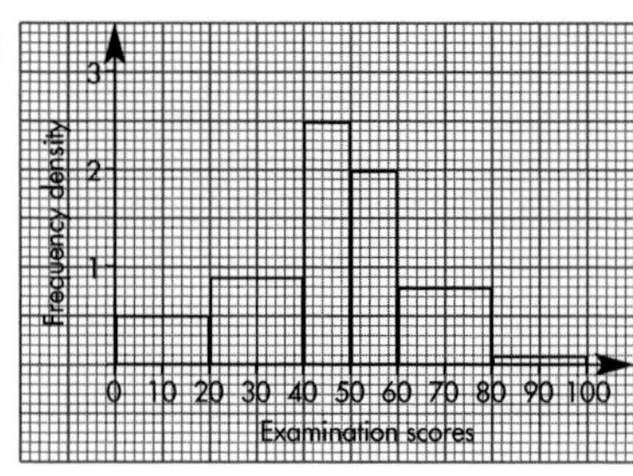

b 38

14 NUT 1040, ATL 680, NATFHE 280

15 12 − 9 = 3 more

16 Single 4, Couple 11, Family 15

Answers to Chapter 13

Quick check

1 A (3, 0), B (1, 4), C (4, 5)

2 **a** 18 **b** 265

13.1 Straight-line distance–time graphs

Exercise 13A

1 **a** **i** 2 h **ii** 3 h **iii** 5 h
b **i** 40 km/h **ii** 120 km/h **iii** 40 km/h
c 6.30 am

2 **a** **i** 125 km **ii** 125 km/h
b **i** Between 2 pm and 3 pm **ii** About $12\frac{1}{2}$ km/h

3 **a** 30 km **b** 40 km **c** 100 km/h

4 **a** **i** 263 m/min (3 sf) **ii** 15.8 km/h (3 sf)
b **i** 500 m/min
c Paul by 1 minute

5 **a** Patrick ran quickly at first, then had a slow middle section but he won the race with a final sprint. Araf ran steadily all the way and came second. Sean set off the slowest, speeded up towards the end but still came third.
b **i** 1.67 m/s **ii** 6 km/h

6 There are three methods for doing this question. This table shows the first, which is writing down the distances covered each hour.

Time	9am	9:30	10:00	10:30	11:00	11:30	12:00	12:30
Walker	0	3	6	9	12	15	18	21
Cyclist	0	0	0	0	7.5	15	22.5	30

The second method is algebra:

Walker takes T hours until overtaken, so $T = \frac{D}{6}$; Cyclist takes $T - 1.5$ to overtake, so $T - 1.5 = \frac{D}{15}$.

Rearranging gives $15T - 22.5 = 6T$, $9T = 22.5$, $T = 2.5$.

The third method is a graph:

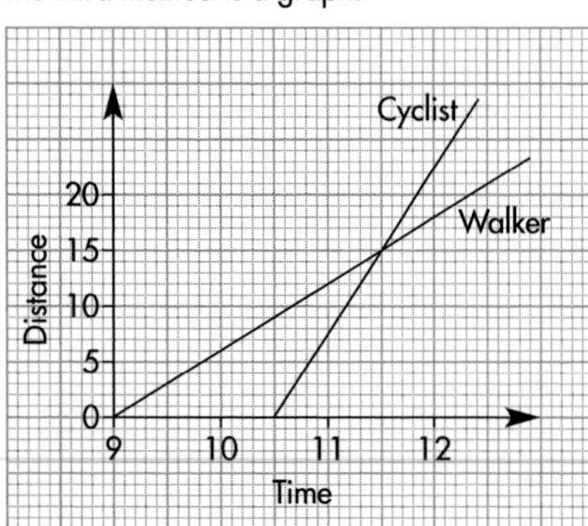

All methods give the same answer of 11:30 when the cyclist overtakes the walker.

Exercise 13A continued

7 **a** **i** Because it stopped several times **ii** Ravinder
b Ravinder at 3.58 pm, Sue at 4.20 pm, Michael at 4.35 pm
c **i** 24 km/h **ii** 20.5 km/h **iii** 5

Exercise 13B

1 **a** 39.2°C **b** Day 4 and 5, steepest line
c Day 8 and 9, steepest line
d **i** Day 5 **ii** 4
e 37° C

2 **a** $\frac{15}{2}$ **b** $\frac{25}{8}$ **c** $\frac{15}{16}$ **d** $\frac{2}{25}$
e $\frac{6}{35}$ **f** $\frac{1}{2}$ **g** $-\frac{4}{5}$

3 **a** $2\frac{1}{2}$ km/h **b** 3.75 m/s **c** $2\frac{1}{2}$ km/h

4 **a** AB: 30 km/h, BC: 6 km/h, CD: 0 km/h, DE: 36 km/h (in opposite direction)
b FG: 4 m/s, GH: 16 m/s, HI: 2 m/s (in opposite direction), IJ: 16 m/s (in opposite direction)

5 Rob has misread the scales. The gradient is actually 2. The line is $y = 2x + 2$ when $x = 10$, $y = 22$.

6 **a** 4 m **b** 1 m
c **i** $\frac{4}{3}$ m **ii** 3 m

13.2 Other types of graphs

Exercise 13C

1 **a** Graph C
b Any container with a regular horizontal cross-section

2 **a**
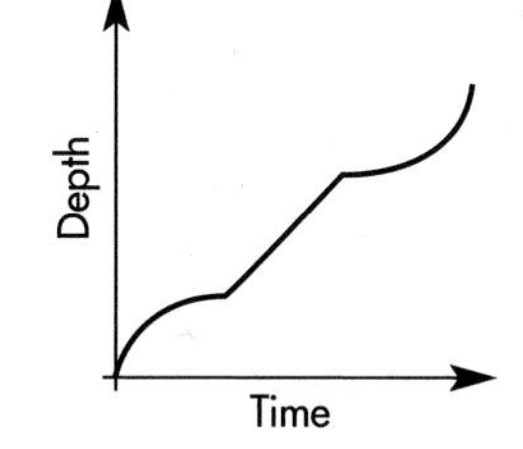

b
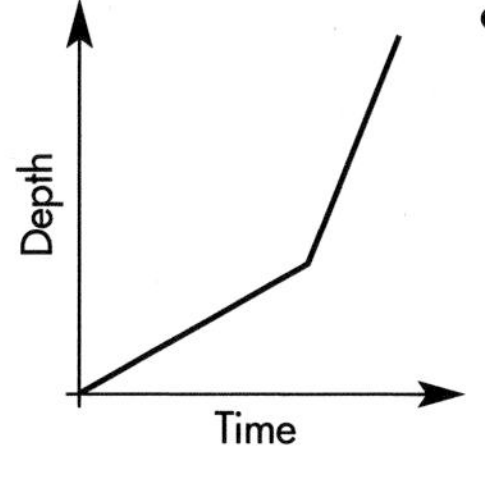

c
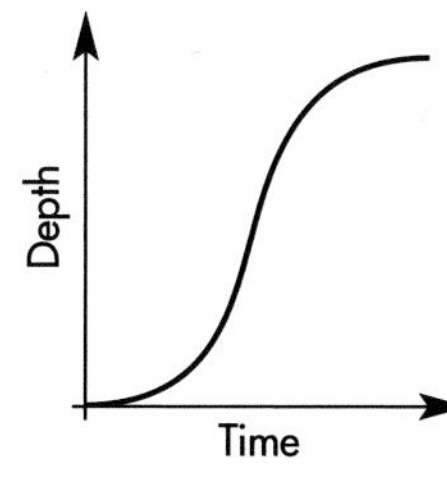

d
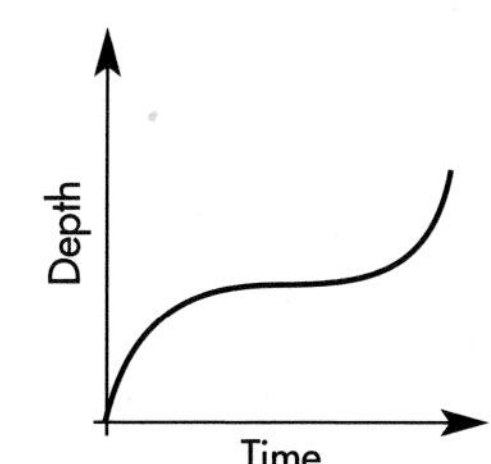

e
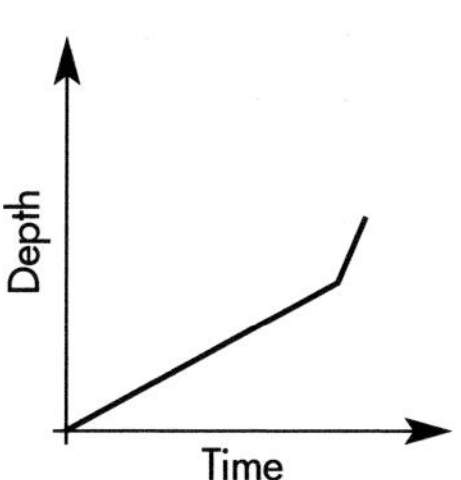

f
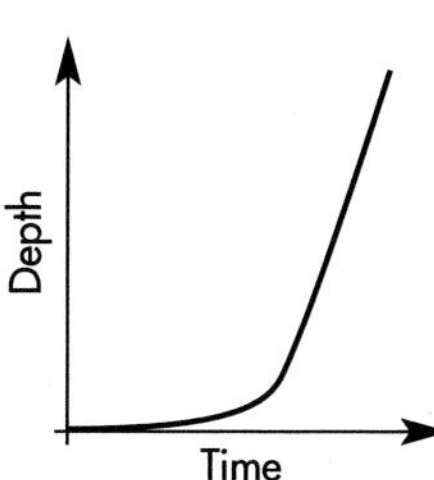

3
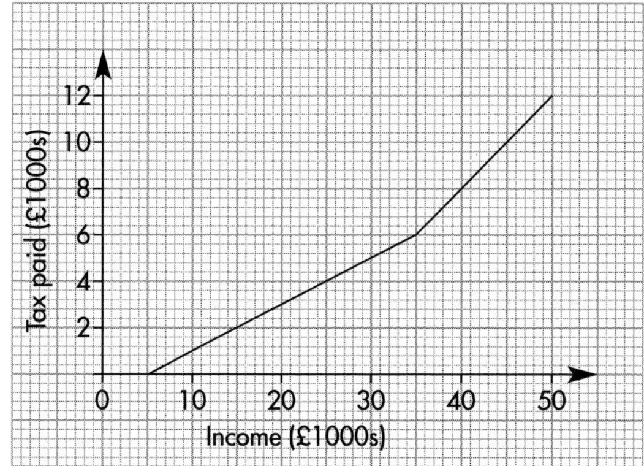

Examination questions

1 a

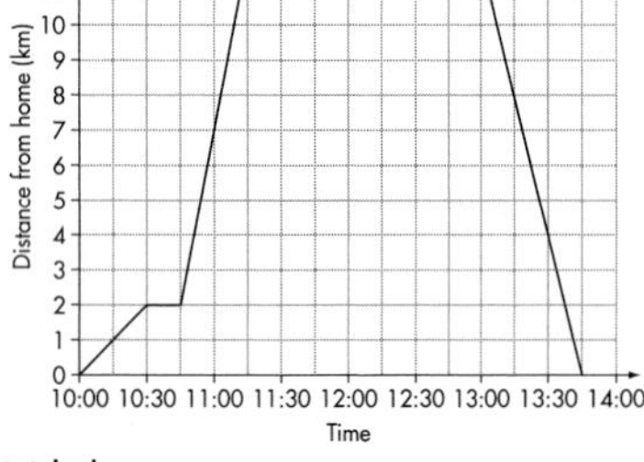

b 16 kph

2 a 15 mins
b 3 km
c D to E , line is steepest.

3 a 150 miles
b 10 minutes
c 50 mph

4 68 mph

5 20 kilometres per litre

6 a Grant
b 93 seconds
c 65m/min
d i Mark
ii Steeper line

7 a

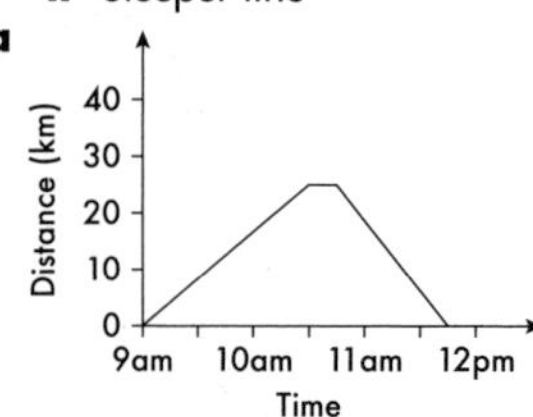

b 25 kph

8 a High
b 30 in
c Line from (18, 18) to (40, 40) and labelled.

Answers to Chapter 14

Quick check

1 29.0

14.1 Line graphs

Exercise 14A

1 a

b About 328 million
c Between 1980 and 1985
d Rising living standards

2 a

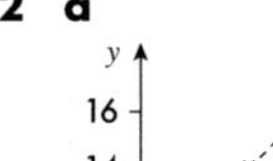
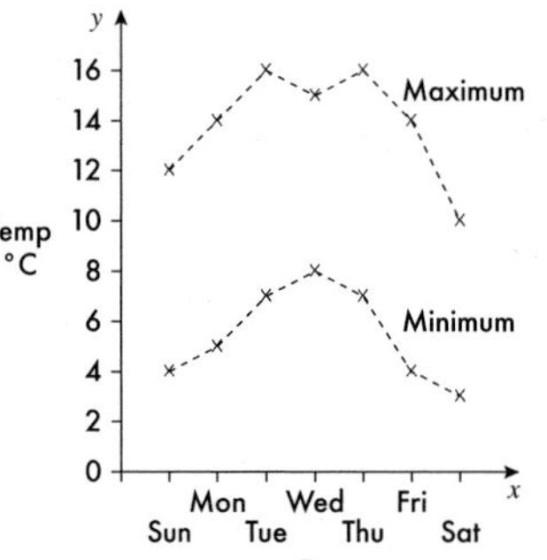

b Smallest difference Wednesday and Saturday (7°), greatest difference Friday (10°)

3 a

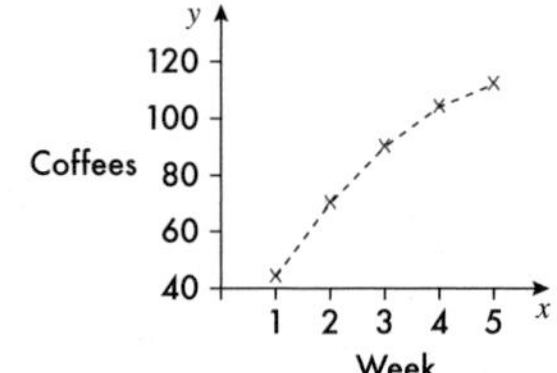

b 119
c The same people keep coming back and tell others, but new customers each week become more difficult to find.

4 Use a graph to estimate about 1040–1050 g.

5 All the temperatures were presumably higher than 20 °C.

14.2 Stem-and-leaf diagrams

Exercise 14B

1 a
2 | 8 9
3 | 4 5 6 8 8 9
4 | 1 1 3 3 3 8 8

b 43 cm **c** 39 cm **d** 20 cm

2 a
0 | 2 8 9 9 9
1 | 2 3 7 7 8
2 | 0 1 2 3

b 9 messages **c** 15 messages

3 a
0 | 7 8 9 9
1 | 0 2 3 4 5 8 8 9 9 key 2 | 3 = 23
2 | 0 3 4 4 6 8
3 | 1

b 18 **c** 24

4 The girls' heights are on the right, 15 | 3 means 153 cm tall. The boys' heights are on the left, 6 | 15 means 156 cm tall.

5 All the data start with a 5 and there are only two digits.

14.3 Scatter diagrams

Exercise 14C

1 **a** Positive correlation, reaction time increases with amount of alcohol drunk
b Negative correlation, you drink less alcohol as you get older
c No correlation, speed of cars on MI is not related to the temperature
d Weak, positive correlation, older people generally have more money saved in the bank

2 **a** and **b**

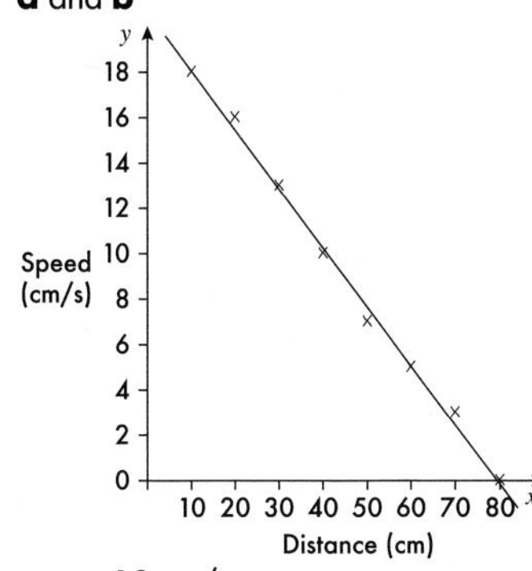

c $c \approx 19$ cm/s
d ≈ 34 cm

3 **a** and **b**

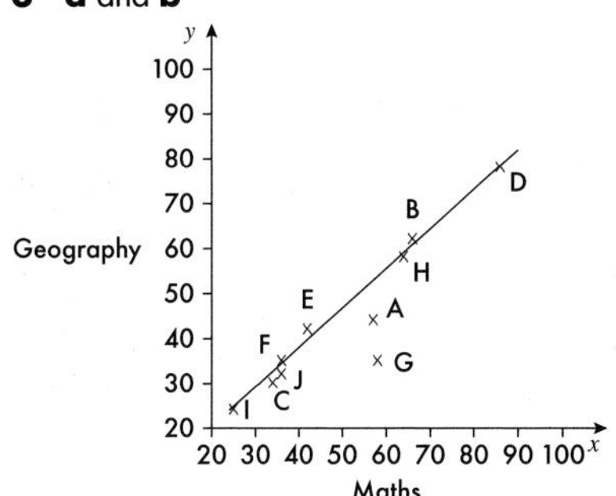

c Greta
d ≈ 70
e ≈ 70

4 **a**

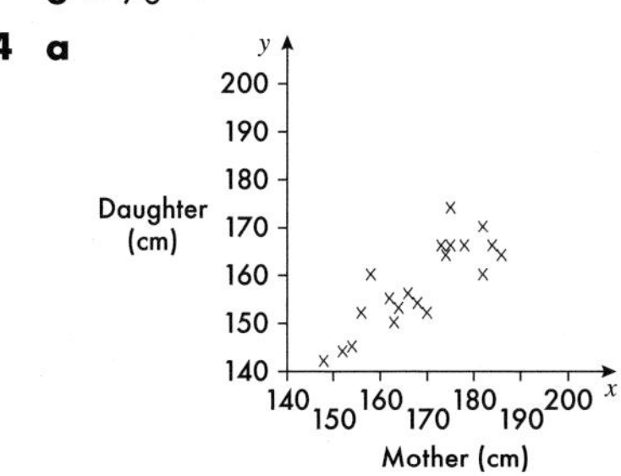

b Yes, usually good correlation

5 **a**

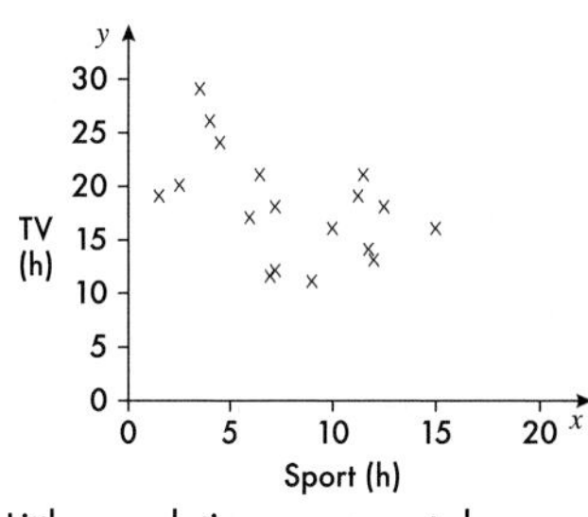

b Little correlation, so cannot draw a line of best fit

6 **a** and **b**

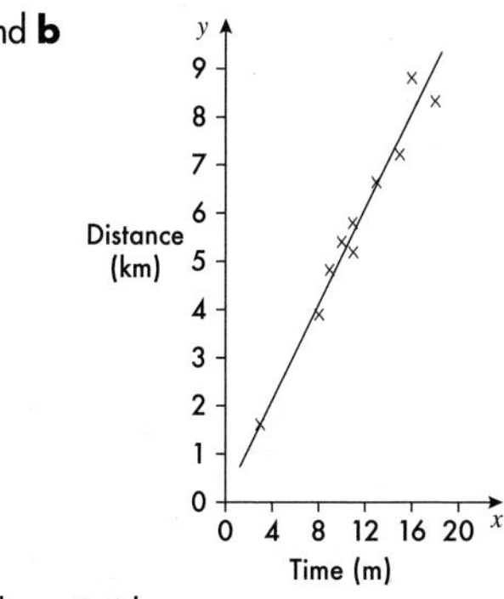

c About 2.4 km
d About 9 minutes

7 About 23 mph

8 Points showing a line of best fit sloping down from top left to bottom right

Examination questions

1

6	5 7
7	0 0 2 6
8	0 2 4 5 7
9	1

6 | 7 represents 67 minutes

2 **a** Negative correlation
b **i** Check students' line of best fit.
ii 7
c He was a good typist.
d No, because the graph shows no further than 10 hours and the correlation could change beyond that point.

3 **a** and **c**

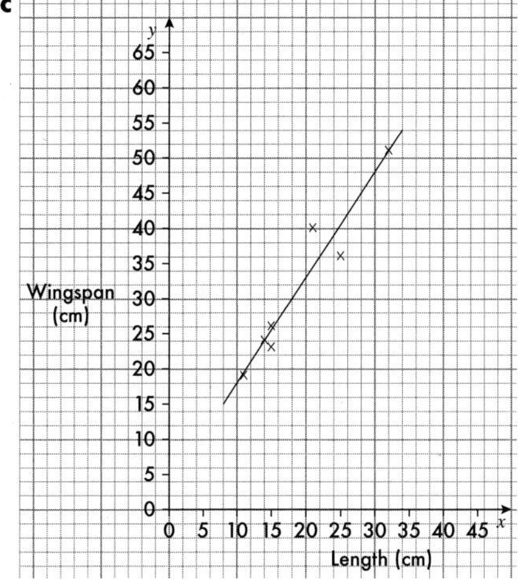

b Strong positive
d 32cm
e No data given is higher than a length of 32 cm and the correlation might not hold at 41 cm.

4 **a** and **b** Check students' scatter diagrams and lines of best fit.
c £7.20 **d** 820 pages

5 **a** 3 **b** 28 **c** 88

6 **a** 17/7 9, 18/1 4 6 6 7 8, 19/0 4
b 186
c There will be 11 members and the middle number will be the 6th in order. This will still be 186.

7 **a** Correct values plotted
b Strong negative
c LOBF drawn
d 75 or whatever value is read from the LOBF
e The data is only valid up to 40 Birth rate

8 £29

Answers to Chapter 15

Quick check

1 **a** Perhaps around 0.6 **b** Very close to 0.1 **c** Very close to 0 **d** 1 **e** 1

15.1 Experimental probability

Exercise 15A

1 **a** $\frac{1}{5}, \frac{2}{25}, \frac{1}{10}, \frac{21}{200}, \frac{37}{250}, \frac{163}{1000}, \frac{329}{2000}$
b 6 **c** 1
d $\frac{1}{6}$ **e** 1000

2 **a** $\frac{19}{200}, \frac{27}{200}, \frac{4}{25}, \frac{53}{200}, \frac{69}{200}$
b 40
c No, it is weighted towards the side with numbers 4 and 5

3 **a** 32 is too high, unlikely that 20 of the 50 throws between 50 and 100 were 5
b Yes, all frequencies fairly close to 100

4 **a** $\frac{1}{5}, \frac{1}{4}, \frac{38}{100}, \frac{21}{50}, \frac{77}{200}, \frac{1987}{5000}$
b 8

5 **a** 0.346, 0.326, 0.294, 0.305, 0.303, 0.306
b 0.231, 0.168, 0.190, 0.16, 0.202, 0.201
c Red 0.5, white 0.3, blue 0.2
d 1
e Red 10, white 6, blue 4

6 **a** Students' answers will vary.
b 20
c Answer depends on students' results
d Answer depends on answer to c

7 **a** 6
b and **c** Answer depends on students' results

8 **a** Caryl, most throws
b 0.43, 0.31, 0.17, 0.14
c Yes, it is more likely to give a 1 or 2

9 **a** Method B **b** B **c** C
d A **e** B **f** A
g B **h** B

10 **a** Not likely **b** Impossible
c Not likely **d** Certain
e Impossible **f** 50–50 chance
g 50–50 chance **h** Certain
i Quite likely

11 Thursday

12 The missing top numbers are 4 and 5; the two bottom numbers are likely to be close to 20.

13 Although you would expect the probability to be close to $\frac{1}{2}$, hence 500 heads, it is more likely that the number of heads is close to 500 rather than actually 500.

14 Roxy is correct, as the expected numbers are: 50, 12.5, 25, 12.5. Sam has not taken into account the fact that there are four red sectors.

15.2 Mutually exclusive and exhaustive events

Exercise 15B

1 **a** Yes **b** Yes **c** No
d Yes **e** Yes **f** Yes

2 Events **a** and **f**

3 $\frac{3}{5}$

4 **a** **i** $\frac{3}{10}$ **ii** $\frac{3}{10}$ **iii** $\frac{3}{10}$
b All except **iii**
c Event **iv**

5 **a** Jane/John, Jane/Jack, Jane/Anne, Jane/Dave, Dave/John, Dave/Jack, Dave/Anne, Anne/John, Anne/Jack, Jack/John
b **i** $\frac{1}{10}$ **ii** $\frac{3}{10}$
iii $\frac{3}{10}$ **iv** $\frac{7}{10}$
c All except **iii** **d** Event **ii**

6 **a** $\frac{3}{8}$ **b** $\frac{1}{8}$
c All except **ii**
d Outcomes overlap

7 $\frac{3}{20}$

8 $\frac{1}{75}$

9 Not mutually exclusive events

10 **a** **i** 0.25 **ii** 0.4 **iii** 0.7
b Events not mutually exclusive
c Man/woman, American man/American woman
d Man/woman

11 **a** **i** 0.95
ii 0.9 (assuming person chooses one or other)
iii 0.3
b Events not mutually exclusive
c Possible answer: pork and vegetarian

12 These are not mutually exclusive events.

15.3 Expectation

Exercise 15C

1 25

2 1000

3 **a** 260 **b** 40 **c** 130 **d** 10

4 5

5 **a** 150 **b** 100 **c** 250 **d** 0

6 **a** 167 **b** 833

7 1050

8 **a** Each score expected 10 times
b 3.5
c Find the average of the scores, which is 21 (1 + 2 + 3 + 4 + 5 + 6) divided by 6

9 **a** 0.111 **b** 40

10 281 days

11 Multiply the number of tomato plants by 0.003

12 400

15.4 Two-way tables

Exercise 15D

1 **a** 23 **b** 20% **c** $\frac{4}{25}$ **d** 480

2 **a** 10 **b** 7 **c** 14% **d** 15%

3 **a**

	1	2	3	4
5	6	7	8	9
6	7	8	9	10
7	8	9	10	11
8	9	10	11	12

b 4

c **i** $\frac{1}{4}$ **ii** $\frac{3}{16}$ **iii** $\frac{1}{4}$

4 **a** 16 **b** 16 **c** 73 **d** $\frac{51}{73}$

5 **a**

1	2	3	4	5	6
1	2	3	4	5	6
2	4	6	8	10	12

b 3 **c** $\frac{1}{4}$

6 **a** The greenhouse sunflowers are bigger on average.

b The garden sunflowers have a more consistent size (smaller range).

7 **a** 40%
b 45%
c No, as you don't know how much the people who get over £350 actually earn

8 Either Reyki because she had bigger tomatoes or Daniel because he had more tomatoes

9

	Men	Women	Children
Left footed	4	3	5
Right footed	21	12	15

10 $\frac{22}{36} = \frac{11}{18}$

11 **a**

Score of second spinner							
	10	10	11	13	15	17	19
	8	8	9	11	13	15	17
	6	6	7	9	11	13	15
	4	4	5	7	9	11	13
	2	2	3	5	7	9	11
	0	0	1	3	5	7	9
		0	1	3	5	7	9
		Score of first spinner					

b 9 or 11 **c** 0
d $\frac{15}{36} = \frac{5}{12}$ **e** $\frac{30}{36} = \frac{5}{6}$

12 **a** 200 **b** $\frac{75}{200} = \frac{3}{8}$
c $\frac{45}{105} = \frac{3}{7}$ **d** $\frac{25}{75} = \frac{1}{3}$

15.5 Addition rules for events

Exercise 15E

1 **a** $\frac{1}{6}$ **b** $\frac{1}{6}$ **c** $\frac{1}{3}$

2 **a** $\frac{1}{4}$ **b** $\frac{1}{4}$ **c** $\frac{1}{2}$

3 **a** $\frac{1}{13}$ **b** $\frac{1}{13}$ **c** $\frac{2}{13}$

4 **a** $\frac{2}{11}$ **b** $\frac{4}{11}$ **c** $\frac{6}{11}$

5 **a** $\frac{1}{3}$ **b** $\frac{2}{5}$ **c** $\frac{11}{15}$
d $\frac{11}{15}$ **e** $\frac{1}{3}$

6 **a** 0.6 **b** 120

7 **a** 0.8 **b** 0.2

8 **a** 0.75 **b** 0.6 **c** 0.5 **d** 0.6
e **i** Cannot add P(red) and P(1) as events are not mutually exclusive
ii 0.75 (= 1 − P(blue))

9 **a** $\frac{17}{20}$ **b** $\frac{2}{5}$ **c** $\frac{3}{4}$

10 Probability cannot exceed 1, and probabilities cannot be summed in this way as events are not mutually exclusive.

11 **a** **i** 0.4 **ii** 0.5 **iii** 0.9
b 0.45
c 2 hours 12 minutes

12 $\frac{5}{52}$ or 0.096 to 3 decimal places

13 **a** $\frac{13}{20}$ as it cannot be square rooted
b $\frac{1}{9}$ as this gives a ratio of red to blue of 1 : 2

15.6 Combined events

Exercise 15F

1 **a** 7
b 2, 12
c P(2) = $\frac{1}{36}$, P(3) = $\frac{1}{18}$, P(4) = $\frac{1}{12}$,
P(5) = $\frac{1}{9}$, P(6) = $\frac{5}{36}$, P(7) = $\frac{1}{6}$,
P(8) = $\frac{5}{36}$, P(9) = $\frac{1}{9}$, P(10) = $\frac{1}{12}$,
P(11) = $\frac{1}{18}$, P(12) = $\frac{1}{36}$
d **i** $\frac{1}{12}$ **ii** $\frac{5}{9}$ **iii** $\frac{1}{2}$
iv $\frac{7}{36}$ **v** $\frac{5}{12}$ **vi** $\frac{5}{18}$

2 **a** $\frac{1}{12}$ **b** $\frac{11}{36}$ **c** $\frac{1}{6}$ **d** $\frac{5}{9}$

3 **a** $\frac{1}{36}$ **b** $\frac{11}{36}$ **c** $\frac{5}{18}$

4 **a** $\frac{5}{18}$ **b** $\frac{1}{6}$ **c** $\frac{1}{9}$
d 0 **e** $\frac{1}{2}$

5 **a** $\frac{1}{4}$ **b** $\frac{1}{2}$ **c** $\frac{3}{4}$ **d** $\frac{1}{4}$

6 **a** 6
b **i** $\frac{4}{25}$ **iii** $\frac{1}{5}$
ii $\frac{13}{25}$ **iv** $\frac{3}{5}$

7 **a** $\frac{1}{8}$ **b** $\frac{3}{8}$ **c** $\frac{7}{8}$ **d** $\frac{1}{8}$

8 **a** 16 **b** 32 **c** 1024 **d** 2^n

9 **a** $\frac{1}{12}$ **b** $\frac{1}{4}$ **c** $\frac{1}{6}$

10 **a**

	1	2	3	4	5	6
1	2	3	4	5	6	7
2	3	4	5	6	7	8
3	4	5	6	7	8	9
4	5	6	7	8	9	10
5	6	7	8	9	10	11
6	7	8	9	10	11	12

b $\frac{1}{18}$ **c** 18 **d** twice

11 0.5

12 You would need a 3D diagram or there would be too many different events to list.

13 $\frac{19}{100}$

14 **a** If he takes Black, striped and spotted with his first three picks he must get one of these with his next pick to make a pair.
b There are only 19 socks left of which only 3 are black
c The events are not independent. The second probability depends on what was taken out the first time
d $\frac{132}{380} = \frac{33}{95}$

Examination questions

1 Veg row, 4, 6 Non-veg row 18, 12
Total row 22, 18

2 He is more likely as there are 13 ways to score 5, 6 or 7. This is a probability of $\frac{13}{25}$ which is just over 0.5.

3 **a** **i** 0.9 **ii** 12
b Yes. She should throw 20 if the dice is fair.

4 The two possibilities, off or not, are not necessarily equally likely chances.

5 **a** $\frac{4}{10} = \frac{2}{5}$ **b** 140

6 **a** 0.126
b All values should be about 250 but 1 and 3 are nowhere near this.
c 100

7 Throw both dice at least 100 times. Record the results in a table. Compare the results with the expected values which for a fair six sided dice would be about 16 or 17 times for each number for 100 throws. The dice with results that are not in line will be the biased dice.

8 **a** 0.15 **b** 0.65

9 once

Answers to Chapter 16

Quick check

1 **a** 17, 20, 23 **b** 49, 64, 81
2 **a** 1 **b** 4 **c** 7

16.1 Number sequences

Exercise 16A

1 **a** 21, 34: add previous 2 terms
b 49, 64: next square number
c 47, 76: add previous 2 terms

2 15, 21, 28, 36

3 61, 91, 127

4 $\frac{1}{2}, \frac{3}{5}, \frac{2}{3}, \frac{5}{7}, \frac{3}{4}$

5 **a** 6, 10, 15, 21, 28
b It is the sums of the natural numbers, or the numbers in Pascal's triangle or the triangular numbers.

6 **a** 2, 6, 24, 720 **b** 69!

7 X. There are 351 (1 + 2 + ... + 25 + 26) letters from A to Z. 3 × 351 = 1053. 1053 − 26 = 1027, 1027 − 25 = 1002, so, as Z and Y are eliminated, the 1000th letter must be X.

8 364: Daily totals are 1, 3, 6, 10, 15, 21, 28, 36, 45, 55, 66, 78 (these are the triangular numbers). Cumulative totals are: 1, 4, 10, 20, 35, 56, 84, 120, 165, 220, 286, 364.

9 29 and 41

10 No, because in the first sequence, the terms are always one less than in the 2nd sequence

11 $4n - 2 = 3n + 7$ rearranges as $4n - 3n = 7 + 2$, so $n = 9$

16.2 Finding the *n*th term of a linear sequence

Exercise 16B

1 **a** 13, 15, $2n + 1$ **b** 25, 29, $4n + 1$ **c** 33, 38, $5n + 3$
d 32, 38, $6n - 4$ **e** 20, 23, $3n + 2$ **f** 37, 44, $7n - 5$
g 21, 25, $4n - 3$ **h** 23, 27, $4n - 1$ **i** 17, 20, $3n - 1$
j 42, 52, $10n - 8$ **k** 24, 28, $4n + 4$ **l** 29, 34, $5n - 1$

2 **a** $3n + 1$, 151 **b** $2n + 5$, 105 **c** $5n - 2$, 248
d $4n - 3$, 197 **e** $8n - 6$, 394 **f** $n + 4$, 54
g $5n + 1$, 251 **h** $8n - 5$, 395 **i** $3n - 2$, 148
j $3n + 18$, 168 **k** $7n + 5$, 355 **l** $8n - 7$, 393

3 **a** 33rd **b** 30th **c** 100th = 499

4
a **i** $4n + 1$ **ii** 401 **iii** 101, 25th
b **i** $2n + 1$ **ii** 201 **iii** 99 or 101, 49th and 50th
c **i** $3n + 1$ **ii** 301 **iii** 100, 33rd
d **i** $2n + 6$ **ii** 206 **iii** 100, 47th
e **i** $4n + 5$ **ii** 405 **iii** 101, 24th
f **i** $5n + 1$ **ii** 501 **iii** 101, 20th
g **i** $3n - 3$ **ii** 297 **iii** 99, 34th
h **i** $6n - 4$ **ii** 596 **iii** 98, 17th
i **i** $8n - 1$ **ii** 799 **iii** 103, 13th
j **i** $2n + 23$ **ii** 223 **iii** 99 or 101, 38th and 39th

5 **a** $\frac{2n+1}{3n+1}$
b Getting closer to $\frac{2}{3}$ $(0.\dot{6})$
c **i** 0.667 774 (6dp) **ii** 0.666 778 (6dp)
d 0.666 678 (6dp), 0.666 667 (6dp)

6 **a** $\frac{4n-1}{5n+1}$
b Getting closer to $\frac{4}{5}$ (0.8)
c **i** 0.796 407 (6dp) **ii** 0.799 640 (6dp)
d 0.799 964 (6dp), 0.799 9996 (7dp)

7 **a** £305 **b** £600 **c** 3 **d** 5

8 **a** $\frac{3}{4}, \frac{5}{7}, \frac{7}{10}$
b **i** 0.666 666 7778 **ii** $\frac{2}{3}$
c for n, $\frac{2n-1}{3n-1} \approx \frac{2n}{3n} = \frac{2}{3}$

9 **a** $8n + 2$ **b** $8n + 1$ **c** $8n$ **d** £8

10 **a** Sequence goes up in 2s; first term is 2 + 29
b $n + 108$
c Because it ends up as $2n \div n$
d 79th

16.3 Special sequences

Exercise 16C

1 **a** 64, 128, 256, 512, 1024
b **i** $2n - 1$ **ii** $2n + 1$ **iii** 3×2^n

2 **a** The number of zeros equals the power.
b 6
c **i** $10^n - 1$ **ii** 2×10^n

3 **a** Even,

+	**Odd**	**Even**
Odd	Even	Odd
Even	Odd	Even

b Odd,

×	**Odd**	**Even**
Odd	Odd	Even
Even	Even	Even

4 **a** $1 + 3 + 5 + 7 = 16 = 4^2$, $1 + 3 + 5 + 7 + 9 = 25 = 5^2$
b **i** 100 **ii** 56

5 **a** 28, 36, 45, 55, 66
b **i** 210 **ii** 5050
c You get the square numbers.

6 **a** Even **b** Odd **c** Odd
d Odd **e** Odd **f** Odd
g Even **h** Odd **i** Odd

7 **a** Odd or even **b** Odd or even **c** Odd or even
d Odd **e** Odd or even **f** Even

8 **a** **i** Odd **ii** Even **iii** Even
b Any valid answer, e.g. $x(y + z)$

9 11th triangular number is 66, 18th triangle number is 171

10 **a** 36, 49, 64, 81, 100
b **i** $n^2 + 1$ **ii** $2n^2$ **iii** $n^2 - 1$

16.4 General rules from given patterns

Exercise 16D

1 **a** 4
b $4n - 3$
c 97
d 50th diagram

2 **a**
b $2n + 1$
c 121
d 49th set

3 **a** 18 **b** $4n + 2$ **c** 12

4 **a** **i** 24 **ii** $5n - 1$ **iii** 224
b 25

5 **a** **i** 20 cm **ii** $(3n + 2)$ cm **iii** 152 cm
b 332

6 **a** **i** 20 **ii** 162
b 79.8 km

7 **a** **i** 14 **ii** $3n + 2$ **iii** 41
b 66

8 **a** **i** 5 **ii** n **iii** 18
b 20 tins

9 **a** 2^n
b **i** $100 \times 2_3^{n-1}$ ml **ii** 1600 ml

10 The nth term is $(\frac{3}{4})^n$, so as n gets very large, the unshaded area gets smaller and smaller and eventually it will be zero; so the shaded area will eventually cover the triangle.

11 Yes, as the number of matches is 12, 21, 30, 39, ... which is $9n + 3$; so he will need $9 \times 20 + 3 = 183$ matches for the 20th step and he has $5 \times 42 = 210$ matches.

12 **a** 20 **b** 120

Examination questions

1 **a** $5n - 1$
b 1.4, 1.5, 1.57 ...

2 **a** Always even
b Could be either odd or even.

3 6, 9, 14

4 **a** **i** 5, 9, 13 **ii** All numbers are odd
b $(1 + 3)^2 - 9 = 4^2 - 9 = 16 - 9 = 7$

5 **a** Even
b Odd

6 **a** **i** 25
b square numbers
c 100

7 The square of any positive fraction less than 1, for example

8 **a** For any value of n, $2n$ is even, so $2n + 1$ is odd
b The square of an odd number will be odd, the square of an even number will be even

9 S, A, N

Answers to Chapter 17

Quick check

1 a 13 **b**

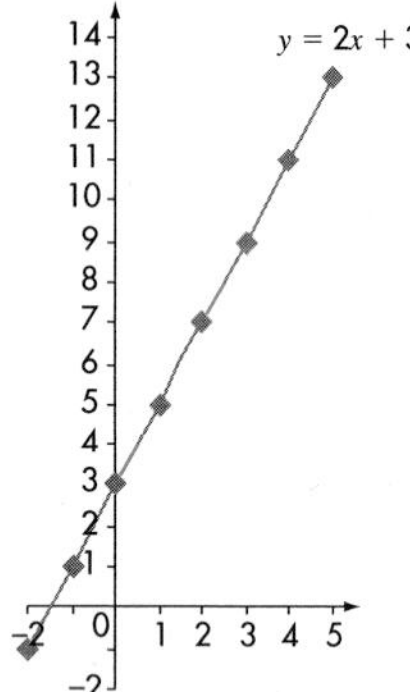

17.1 Linear graphs

Exercise 17A

1

$y = 3x + 4$

2

$y = 2x - 5$

3

$y = \frac{x}{2} - 3$

4

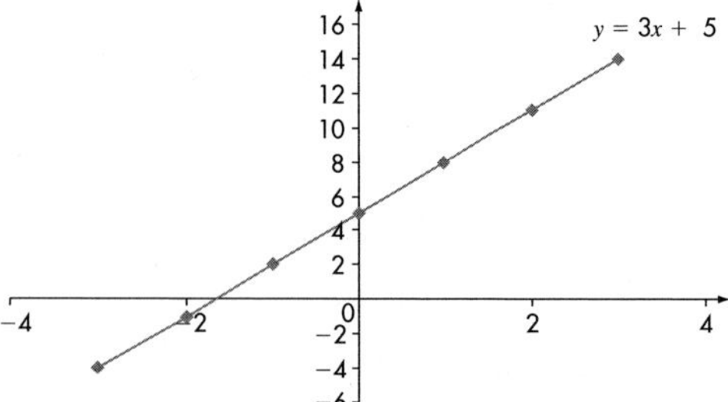

5

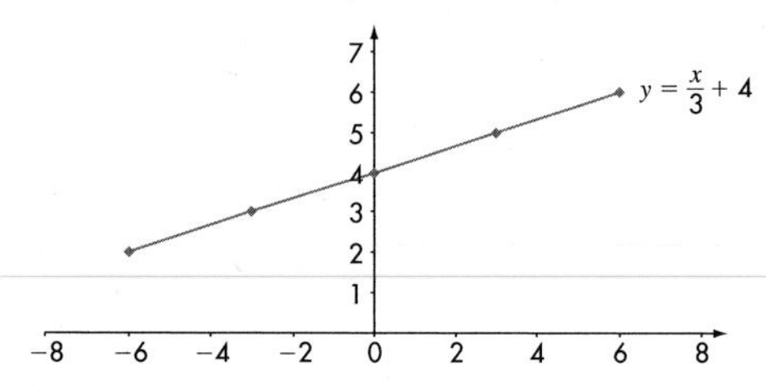

6 a

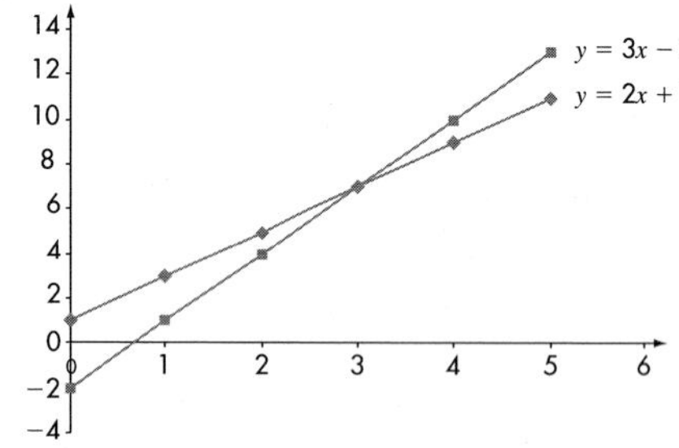

b (3, 7)

7 a

$y = 4x - 5$

$y = 2x + 3$

b (4, 11)

8 a

$y = \frac{x}{2} - 2$

$y = \frac{x}{3} - 1$

b (6, 1)

9 **a**

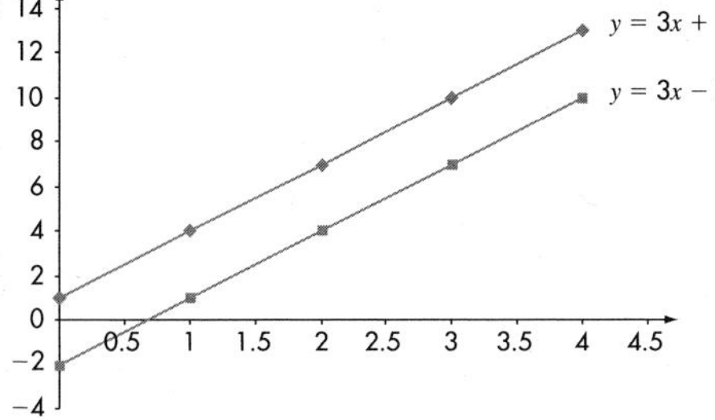

b No, because the lines are parallel

10

x	0	1	2	3	4	5
y	5	4	3	2	1	0

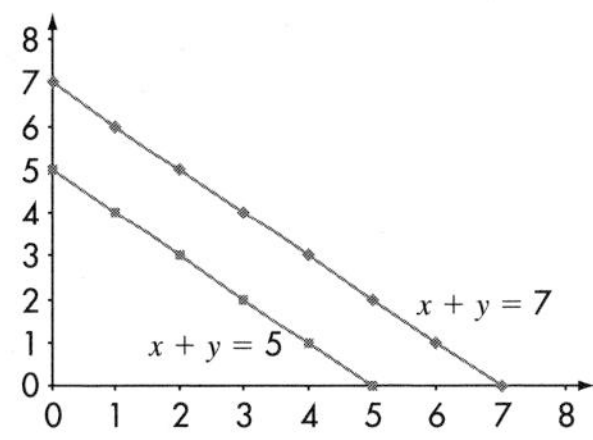

11 **a**

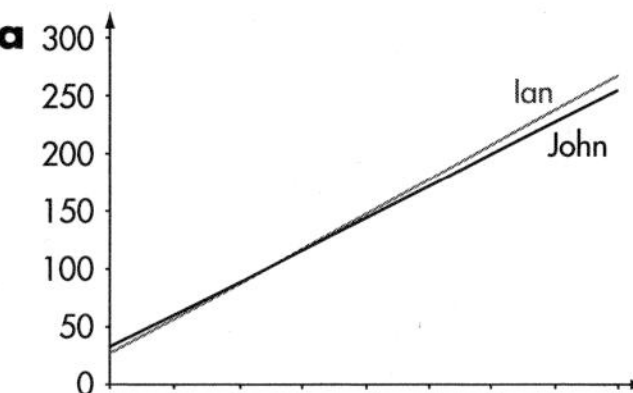

b Ian, as Ian only charges £85, whilst John charges £90 for a 2-house job.

12 **a**

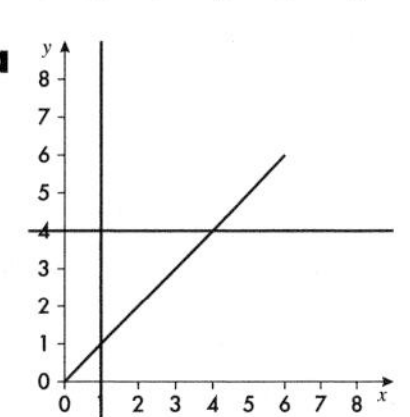

b 4.5 units squared

13

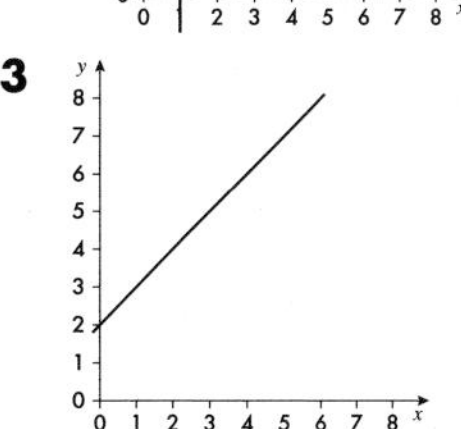

Exercise 17B

1 **a** 2 **b** $\frac{1}{3}$ **c** −3 **d** 1 **e** −2
f $-\frac{1}{3}$ **g** 5 **h** −5 **i** $\frac{1}{5}$ **j** $-\frac{3}{4}$

2

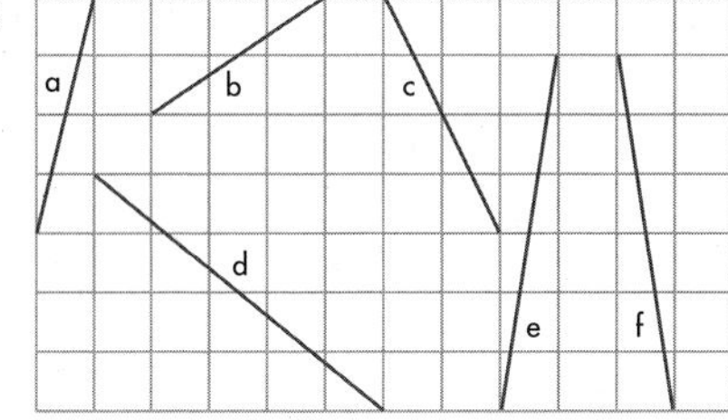

3 **a** 1
b −1 They are perpendicular and symmetrical about the axes.

4 **a** Approx. 320 feet in half a mile (2640 feet), so gradient is about 0.12.
b **i** Because the line on the graph has the steepest gradient.
ii Approximately 550 feet in half a mile so gradient is about −0.21
c BM. Approximately 1200 foot of climbing in 6.3 miles ≈ 190 feet of ascent on average

5 **a** 0.5 **b** 0.4 **c** 0.2 **d** 0.1 **e** 0

6 **a** $1\frac{2}{3}$ **b** 2 **c** $3\frac{1}{3}$ **d** 10 **e** ∞

7 Raisa has misread the scales. The second line has four times the gradient (2.4) of the first (0.6).

8 6 : 4, 8 : 5, 5 : 3, 11 : 6, 2 : 1, 5 : 2

17.2 Drawing graphs by the gradient-intercept method

Exercise 17C

1 **a, b, c, d**

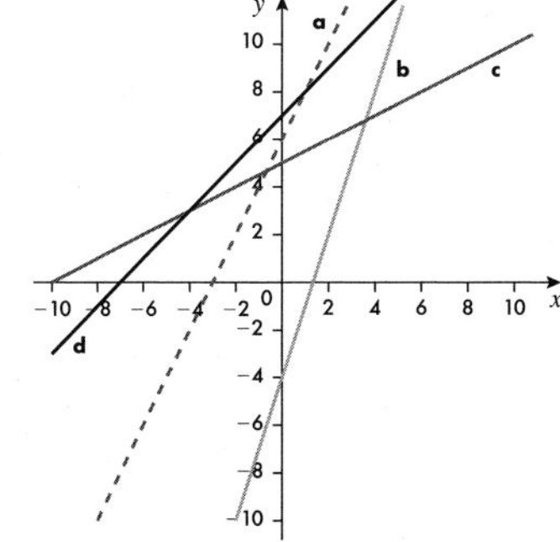

e, f, g, h

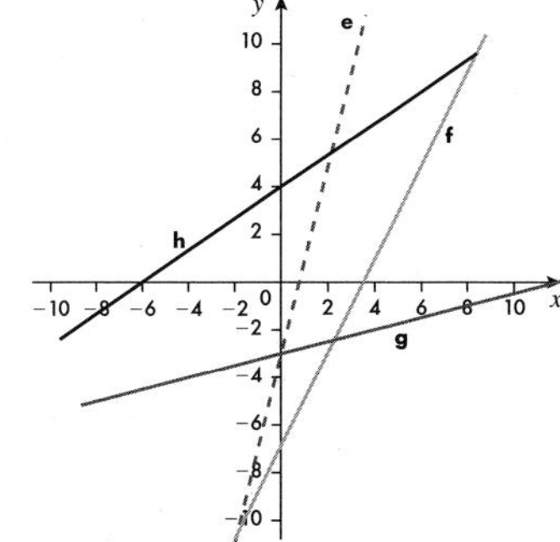

i, j, k, l

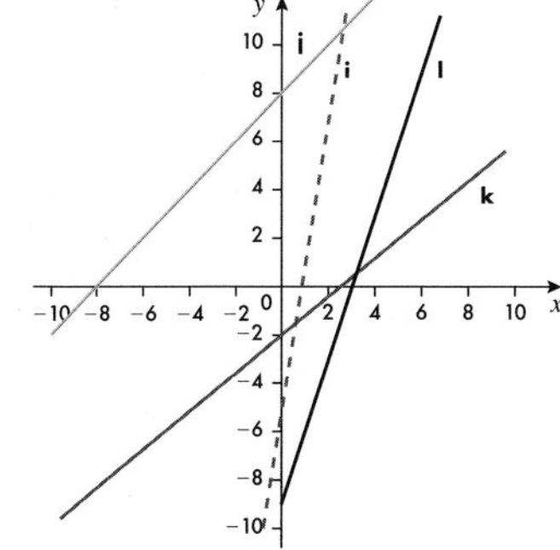

2 a

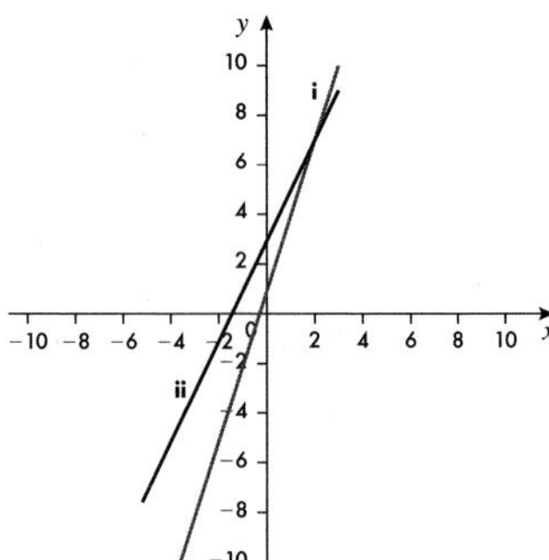

b (2, 7)

3 a

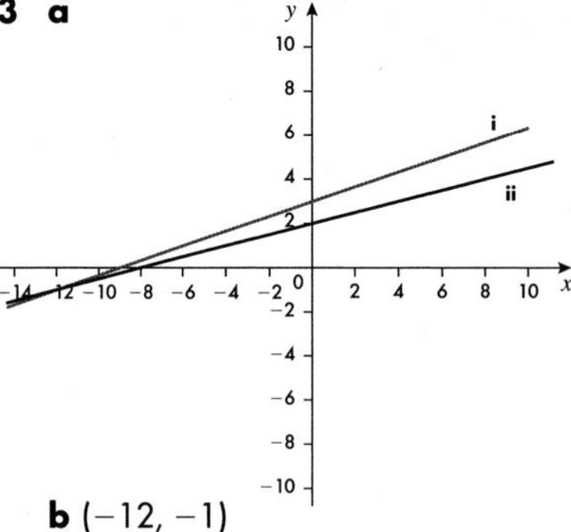

b (−12, −1)

4 a

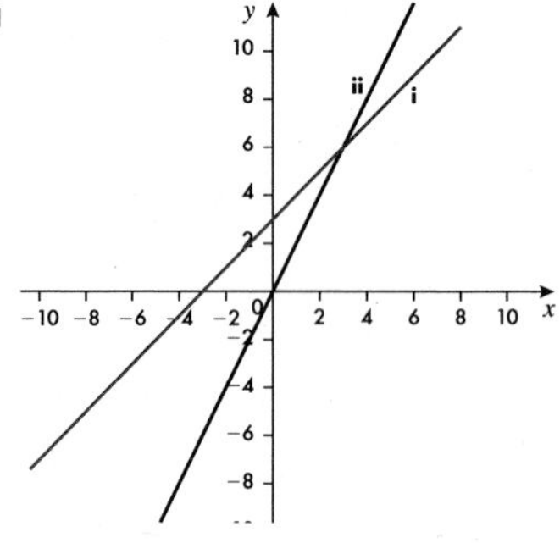

b (3, 6)

5 a They have the same gradient (3).
b They intercept the y-axis at the same point (0, −2).
c (−1, −4)

6 a −2 **b** $\frac{1}{2}$ **c** 90°
d Negative reciprocal **e** $-\frac{1}{3}$

Exercise 17D

1 a, b, c, d

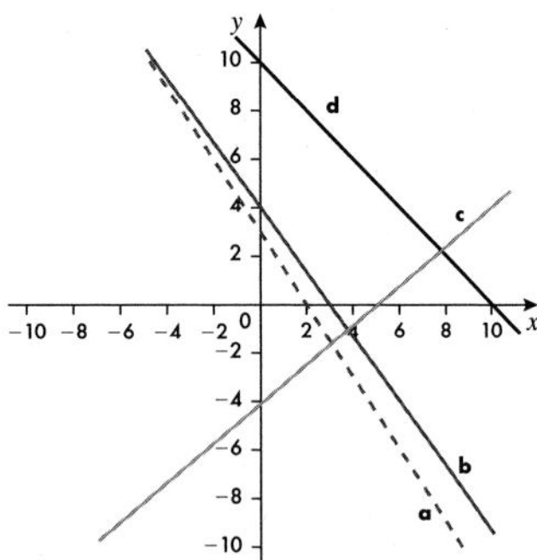

e, f, g, h

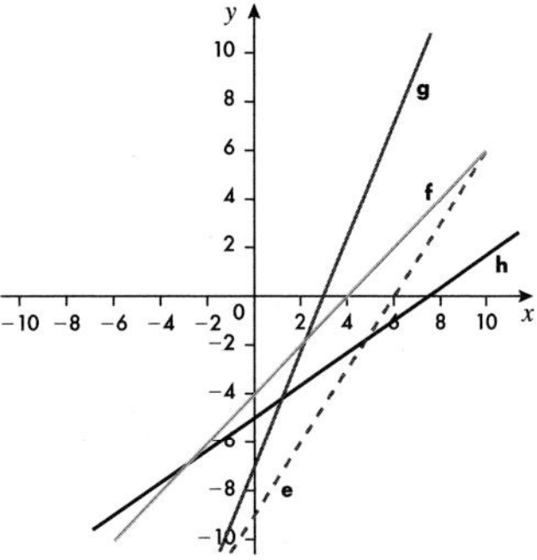

i, j, k, l

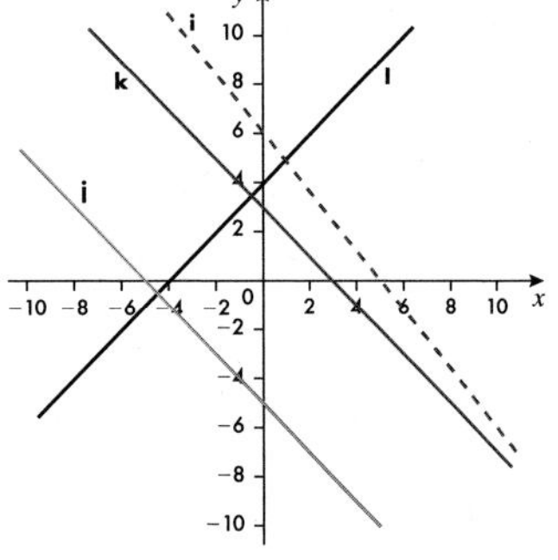

2 a

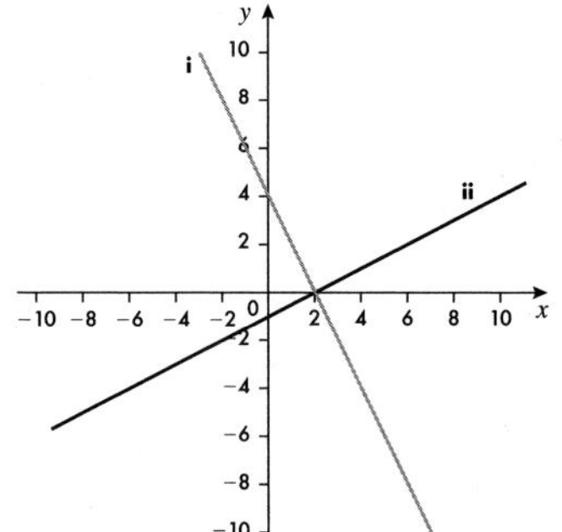

b (2, 0)

3 a

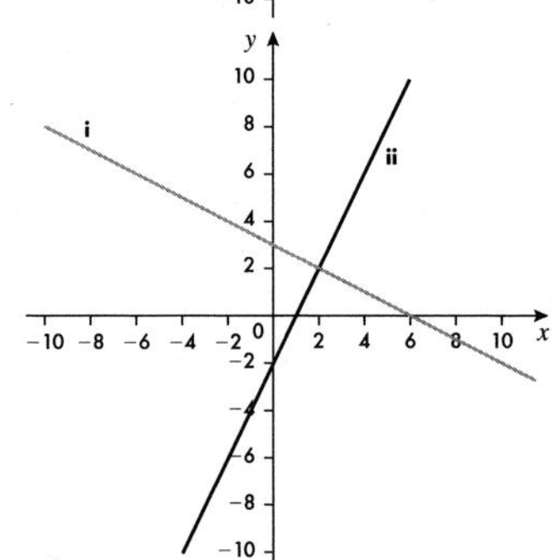

b (2, 2)

4 a

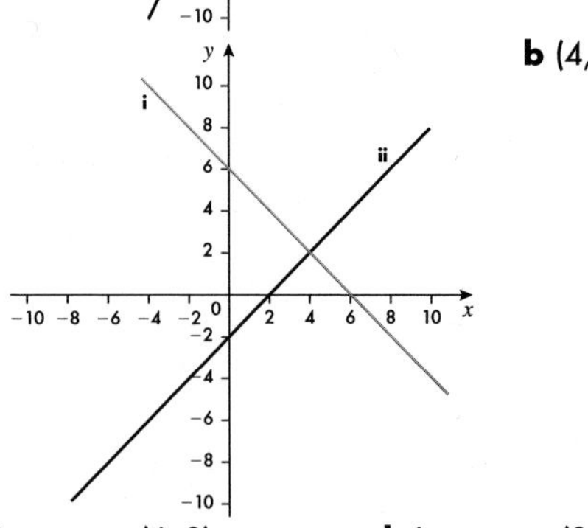

b (4, 2)

5 a Intersect at (6, 0) **b** Intersect at (0, −3)
c Parallel **d** $-2x + 9y = 18$

6 a i $x = 3$ **ii** $x - y = 4$ **iii** $y = -3$
iv $x + y = -4$ **v** $x = -3$ **vi** $y = x + 4$
b i −3 **ii** $\frac{1}{3}$ **iii** $-\frac{1}{3}$

17.3 Finding the equation of a line from its graph

Exercise 17E

1 **a** $y = \frac{4}{3}x - 2$ or $3y = 4x - 6$ **b** $y = x + 1$
c $y = 2x - 3$ **d** $2y = x + 6$
e $y = x$ **f** $y = 2x$

2 **a** **i** $y = 2x + 1, y = -2x + 1$
ii Reflection in y-axis (and $y = 1$)
iii Different sign
b **i** $5y = 2x - 5, 5y = -2x - 5$
ii Reflection in y-axis (and $y = -1$)
iii Different sign
c **i** $y = x + 1, y = -x + 1$
ii Reflection in y-axis (and $y = 1$)
iii Different sign

3 **a** x-coordinates go from $2 \rightarrow 1 \rightarrow 0$ and y-coordinates go from $5 \rightarrow 3 \rightarrow 1$.
b x-step between the points is 1 and y-step is 2.
c $y = 3x + 2$

4 **a** $y = -2x + 1$
b $2y = -x$
c $y = -x + 1$
d $5y = -2x - 5$
e $y = -\frac{3}{2}x - 3$ or $2y = -3x - 6$

5 **a** **i** $2y = -x + 1, y = -2x + 1$
ii Reflection in $x = y$
iii Reciprocal of each other
b **i** $2y = 5x + 5, 5y = 2x - 5$
ii Reflection in $x = y$
iii Reciprocal of each other
c **i** $y = 2, x = 2$
ii Reflection in $x = y$
iii Reciprocal of each other (reciprocal of zero is infinity)

17.4 3-D coordinates

Exercise 17F

1 **a** $A(0, 0, 0)$, $B(0, 3, 0)$, $C(0, 3, 3)$, $D(0, 0, 3)$, $E(3, 0, 0)$, $F(3, 0, 3)$, $G(3, 3, 3)$, $H(3, 3, 0)$
b **i** (0, 1.5, 1.5)
ii (1.5, 3, 1.5)
c (1.5, 1.5, 1.5)
d **i** $\sqrt{18} = 4.24$
ii $\sqrt{27} = 5.2)$

2 **a** **i** (0, 3, 0)
ii (3, 3, 0)
iii (3, 0, 0)
b **i** (0, 3, 2)
ii (–3, 3, 2)
iii –3, 3, 0)
c $\tan^{-1}(2 \div \sqrt{18}) = 28.1$

3 **a** (2, 2, 2)
b (2, 2, 3)
c (0, 2, –1)
d (–4, 3, 3)
e (7, 0, 3)
f (4, –3, 2)

4 **a** (0, 4, 2)
b (2, –3, 0)
c 28 cubic units.

Examination questions

1 **a**

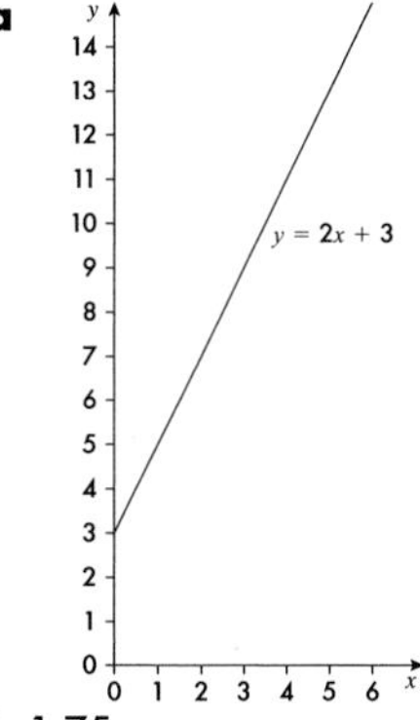

b 1.75

2

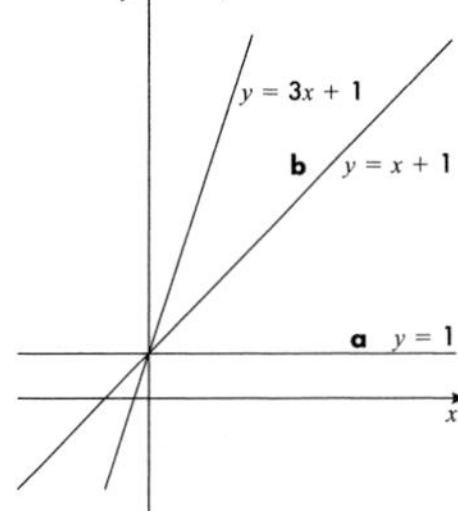

3 **a** $y = -3x + 9$
b -9
c $\frac{1}{3}$

4 (6, 5)

5 Straight line graph from (–4, –13) to (4, 12)

6 12.5 square units

7 **a** Line from (–4, –10) to (4, 14)
b $y = -2x - 1$

8 **a** the line shown is $y = -x$
b $y = x + 1$

Answers to Chapter 18

Quick check

1

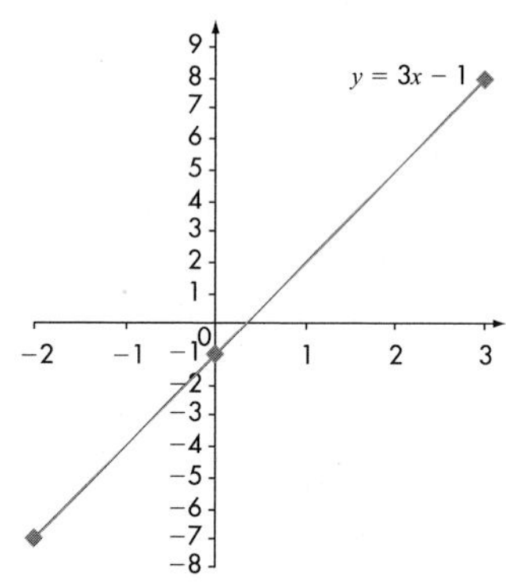

2 $y = 3x + 1$

18.1 Line graphs

Exercise 18A

1 **a** Values of y: 27, 12, 3, 0, 3, 12, 27
b 6.8 **c** 1.8 or −1.8

2 **a** Values of y: 27, 18, 11, 6, 3, 2, 3, 6, 11, 18, 27
b 8.3 **c** 3.5 or −3.5

3 **a** Values of y: 27, 16, 7, 0, −5, −8, −9, −8, −5, 0, 7
b −8.8 **c** 3.4 or −1.4

4 **a** Values of y: 2, −1, −2, −1, 2, 7, 14
b 0.25 **c** 0.7 or −2.7
d

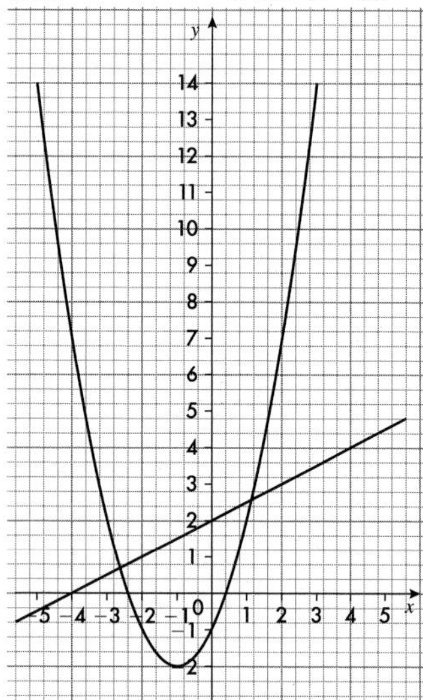

e (1.1, 2.6) and (−2.6, 0.7)

5 **a** Values of y: 18, 12, 8, 6, 6, 8, 12
b 9.75 **c** 2 or −1
d Values of y: 14, 9, 6, 5, 6, 9, 14
e (1, 6)

6 **a** Values of y: 4, 1, 0, 1, 4, 9, 16
b 7.3 **c** 0.4 or −2.4
d

e (1, 4) and (−1, 0)

7 **a** Values of y: 15, 9, 4, 0, −3, −5, −6, −6, −5, −3, 0, 4, 9
b −0.5 and 3

8 Points plotted and joined should give parabolas

9 Points plotted and joined should give a parabola

10 Line A has a constant in front, so is 'thinner' than the rest
Line B has a negative in front, so is 'upside down'
Line C does not pass through the origin

18.2 The significant points of a quadratic graph

Exercise 18B

1 **a** Values of y: 12, 5, 0, −3, −4, −3, 0, 5, 12
b 2 and −2

2 **a** Values of y: 7, 0, −5, −8, −9, −8, −5, 0, 7
b 3 and −3

3 **a** The roots are positive and negative square roots of the constant term
b Check predictions
c Values of y: 15, 8, 3, 0, −1, 0, 3, 8, 15
d Values of y: 11, 4, −1, −4, −5, −4, −1, 4, 11
e 1 and −1, 2.2 and −2.2

4 **a** Values of y: 5, 0, −3, −4, −3, 0, 5, 12
b −4 and 0

5 **a** Values of y: 16, 7, 0, −5, −8, −9, −8, −5, 0, 7, 16
b 0 and 6

6 **a** Values of y: 10, 4, 0, −2, −2, 0, 4, 10, 18
b −3 and 0

7 **a** The roots are 0 and the negative of the coefficient of x
b Check predictions
c Values of y: 10, 4, 0, −2, −2, 0, 4, 10
d Values of y: 6, 0, −4, −6, −6, −4, 0, 6, 14
e 0 and 3, −5 and 0

8 **a** Values of y: 9, 4, 1, 0, 1, 4, 9
b −2
c Only 1 root

9 **a** Values of y: 10, 3, −2, −5, −6, −5, −2, 3, 10
b 0.6 and 5.4

10 **a** Values of y: 19, 6, −3, −8, −9, −6, 1, 12
b 0.9 and −3.4

11 **a** (0, −4), (0, −9), (0, −1), (0, −5), (0, 0), (0, 0), (0, 0), (0, 0), (0, 0)
b (0, −4), (0, −9), (0, −1), (0, −5), (−2, −4), (3, −9), (−1.5, −2.25), (1.5, −2.25), (−2.5, −6.25)
c The y-intercept; the point where the x-value is the mean of the roots

12 **a** $y = (x - 2)^2$ **b** 0

13 **a** $y = (x - 3)^2 - 6$ **b** −6

14 **a** $y = (x - 4)^2 - 14$ **b** −14

15 **a** $y = -(x - 1)^2 - 5$ **b** −5

16 **a** The minimum point is (a, b) **b** (−5, −28)

17 $y = (x - 3)^2 - 7$, $y = x^2 - 6x + 9 - 7$, $y = x^2 - 6x + 2$

18 **a** (−2, −7)
b **i** $(a, 2b - a^2)$
ii $(2a, b - 4a^2)$

Examination questions

1 **a** −3, 7
b Correct graph
c Read from the x-axis

2 **a** −2.2 to −2.3
b −1 and 2

3 **a** 7 and −2
b Correct graph
c 0.6 and 3.4

4 **a** 0.55 **b** −4.25 **c** −6.25

Answers to Chapter 19

Quick check

1 **a** 8 **b** 10

2 **a**

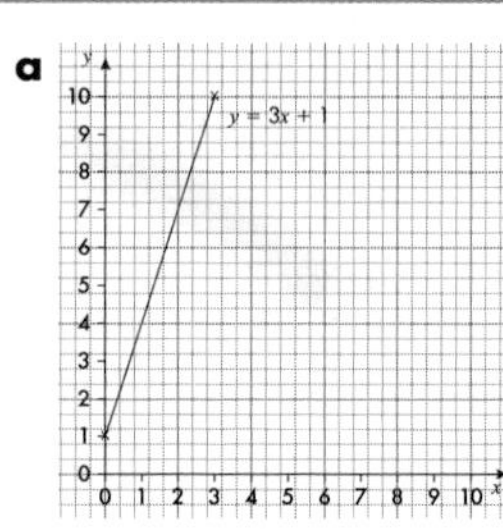

b

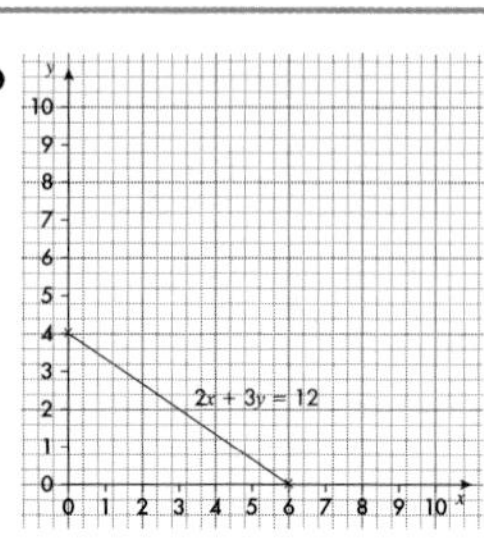

19.1 Solving inequalities

Exercise 19A

1 **a** $x < 3$ **b** $t > 8$ **c** $p \geqslant 10$ **d** $x < 5$
e $y \leqslant 3$ **f** $t > 5$ **g** $x < 6$ **h** $y \leqslant 15$
i $t \geqslant 18$ **j** $x < 7$ **k** $x \leqslant 3$ **l** $t \geqslant 5$

2 **a** 8 **b** 6 **c** 16 **d** 3 **e** 7

3 **a** 11 **b** 16 **c** 16

4 $2x + 3 < 20$, $x < 8.50$, so the most each could cost is £8.49

5 **a** Because $3 + 4 = 7$, which is less than the third side of length 8
b $x + x + 2 > 10$, $2x + 2 > 10$, $2x > 8$, $x > 4$, so smallest value of x is 5

6 **a** $x = 6$ and $x < 3$ scores -1 (nothing in common), $x < 3$ and $x > 0$ scores $+1$ (1 in common for example), $x > 0$ and $x = 2$ scores $+1$ (2 in common), $x = 2$ and $x \geqslant 4$ scores -1 (nothing in common), so we get $-1 + 1 + 1 - 1 = 0$
b $x > 0$ and $x = 6$ scores $+1$ (6 in common), $x = 6$ and $x \geqslant 4$ scores $+1$ (6 in common), $x \geqslant 4$ and $x = 2$ scores -1 (nothing in common), $x = 2$ and $x < 3$ scores $+1$ (2 in common). $+1 + 1 - 1 + 1 = 2$
c Any acceptable combination, e.g. $x = 2$, $x < 3$, $x > 0$, $x \geqslant 4$, $x = 6$

7 **a** $x \geqslant -6$ **b** $t \leqslant \frac{8}{3}$ **c** $y \leqslant 4$
d $x \geqslant -2$ **e** $w \leqslant 5.5$ **f** $x \leqslant \frac{14}{5}$

8 **a** $x \leqslant 2$ **b** $x > 38$ **c** $x < 6\frac{1}{2}$
d $x \geqslant 7$ **e** $t > 15$ **f** $y \leqslant \frac{7}{5}$

9 **a** $3 < x < 6$ **b** $2 < x < 5$ **c** $-1 < x \leqslant 3$
d $1 \leqslant x < 4$ **e** $2 \leqslant x < 4$ **f** $0 \leqslant x \leqslant 5$

Exercise 19B

1 **a** $x > 1$
b $x \leqslant 3$
c $x < 2$
d $x \geqslant -1$
e $x \leqslant -1$
f $x \geqslant 1$

2 **a**
b
c
d
e
f
g
h

3 **a** $x \geqslant 4$
b $x < -2$
c $x \geqslant 3.5$

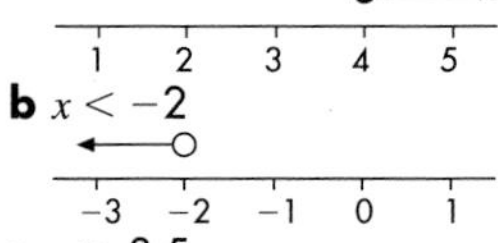

d $x < -1$
e $x < 1.5$
f $x \leqslant -2$
g $x > 50$
h $x \geqslant -6$

4 **a** Because 3 apples plus the chocolate bar cost more that £1.20: $x > 22$
b Because 2 apples plus the chocolate bar left Max with at least 16p change: $x \leqslant 25$
c
d Apples could cost 23p, 24p or 25p.

5 Any two inequalities that overlap only on the integers -1, 0, 1 and 2 – for example, $x \geqslant -1$ and $x < 3$

6 1 and 4

7 **a** $x > 2$
b $x \geqslant 6$

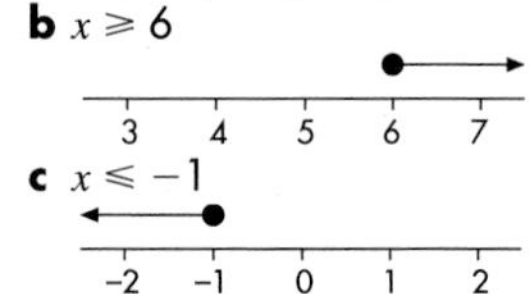

c $x \leqslant -1$
d $x \geqslant -4$

19.2 Graphical inequalities

Exercise 19C

1

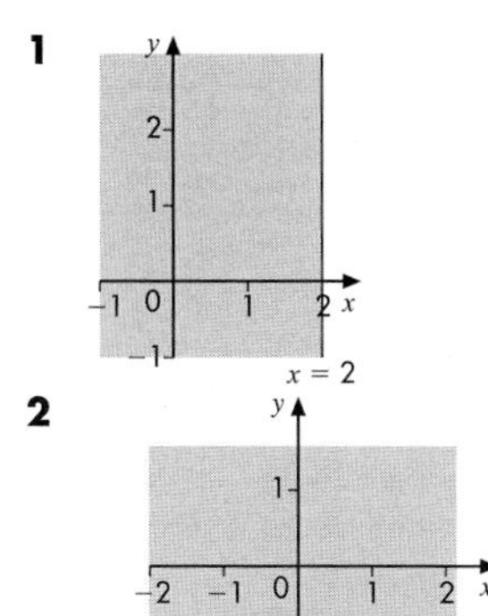

$x = 2$

2

$y = -3$

3

$x = -2$ $x = 1$

4

$y = 4$

$y = -1$

5 a

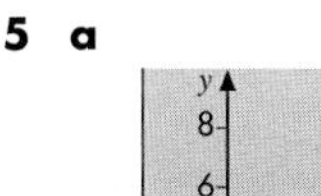

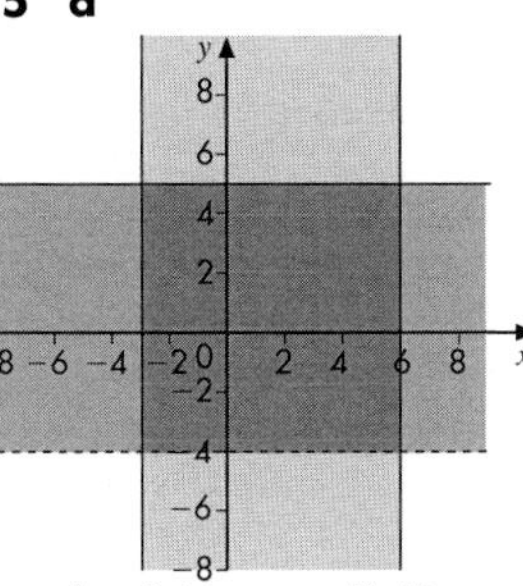

b i Yes **ii** Yes
iii No

6

$y = 2x - 1$

7

8

9

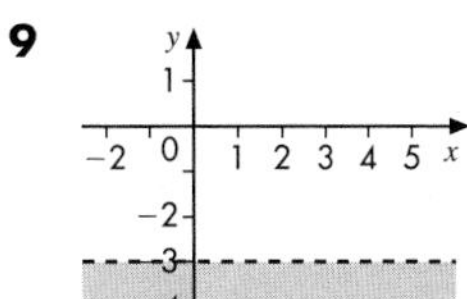

10 a–d

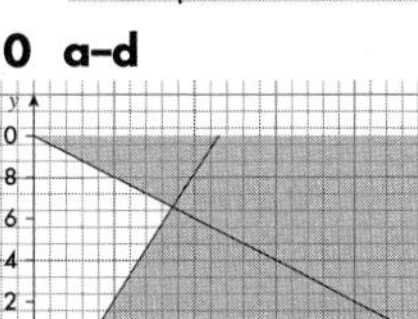

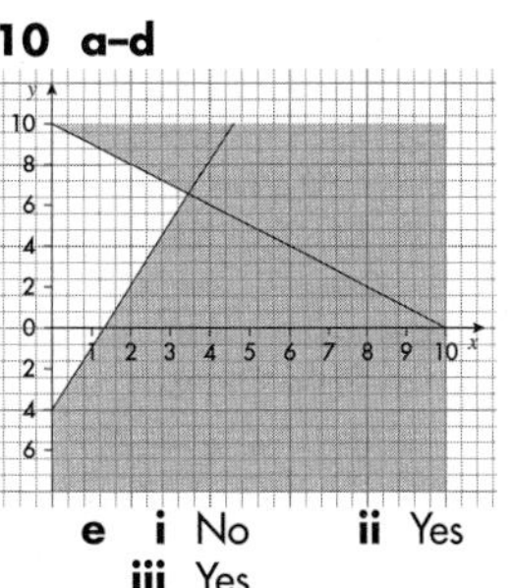

e i No **ii** Yes
iii Yes

11 a–f

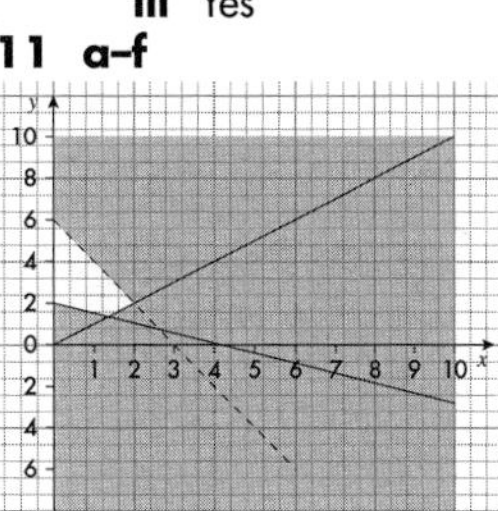

g i No **ii** No
iii Yes

12 a

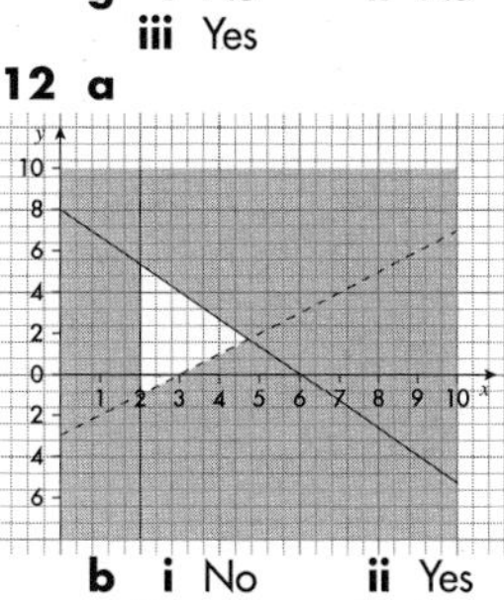

b i No **ii** Yes
iii Yes **iv** No

13 For example, $x \geqslant 1$, $y \leqslant 3$ and $y \geqslant x + 1$. There are many other valid answers.

14 May be true: a, c, d, g
Must be false: b, e
Must be true: f, h

15 Test a point such as the origin (0, 0), so $0 < 0 + 2$, which is true. So the side that includes the origin is the required side.

16 a (3, 0) **b** (4, 5)

Examination questions

1 $x \leqslant 1.4$

2 a $x \leqslant 5$ **b** A line with a solid circle from -2 to 3

3 a $x \leqslant -1$ **b** $y > -3$
c Top side shaded; test a point such as the origin $0 \leqslant 12$

4 a $x > 2$

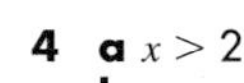

b

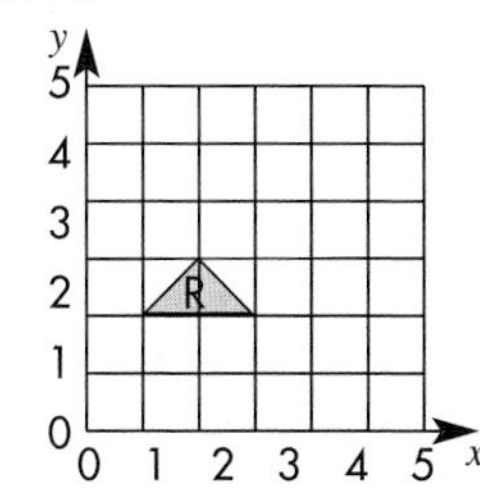

5 Region 1 is graph D, Region 2 is graph C, Region 3 is graph E, region 4 is graph A.

INDEX